15th International Conference "Experimental Fluid Mechanics" (EFM21)

EPJ Web of Conferences Volume 264 (2022)

Liberec, Czech Republic
23-26 November 2021

ISBN: 978-1-7138-5738-9

Printed from e-media with permission by:

Curran Associates, Inc.
57 Morehouse Lane
Red Hook, NY 12571

Some format issues inherent in the e-media version may also appear in this print version.

For additional information, please contact EDP Sciences – Web of Conferences
at the address below.

EDP Sciences – Web of Conferences
17, Avenue du Hoggar
Parc d'Activité de Courtabœuf
BP 112
F-91944 Les Ulis Cedex A
France

Phone: +33 (0) 1 69 18 75 75
Fax: +33 (0) 1 69 28 84 91

contact-edps@webofconferences.org

Additional copies of this publication are available from:

Curran Associates, Inc.
57 Morehouse Lane
Red Hook, NY 12571 USA
Phone: 845-758-0400
Fax: 845-758-2633
Email: curran@proceedings.com
Web: www.proceedings.com

TABLE OF CONTENTS

edp sciences

Web of Conferences

Statement of Peer review

In submitting conference proceedings to *Web of Conferences*, the editors of the proceedings certify to the Publisher that

1. They adhere to its **Policy on Publishing Integrity** in order to safeguard good scientific practice in publishing.
2. All articles have been subjected to peer review administered by the proceedings editors.
3. Reviews have been conducted by expert referees, who have been requested to provide unbiased and constructive comments aimed, whenever possible, at improving the work.
4. Proceedings editors have taken all reasonable steps to ensure the quality of the materials they publish and their decision to accept or reject a paper for publication has been based only on the merits of the work and the relevance to the journal.

Title, date and place of the conference

Experimental fluid mechanics 2021

23.11.2021 - 26.11.2021

North Bohemian Museum Liberec (Czech Republic)

Proceedings editor(s):

Petra Dancova, Jan Novosad, Jaroslav Pulec

TECHNICKÁ UNIVERZITA V LIBERCI
FAKULTA STROJNÍ
Katedra energetických zařízení
Studentská 2, 460 17 Liberec 1

Date and editor's signature

17, avenue du Hoggar - PA de Courtabœuf – BP 112 - 91944 Les Ulis Cedex A (France)
Tél. : 33 (0)1 69 18 75 75 - Fax : 33(0)1 69 07 45 17 – www.edpsciences.org

Experimental Fluid Mechanics 2021

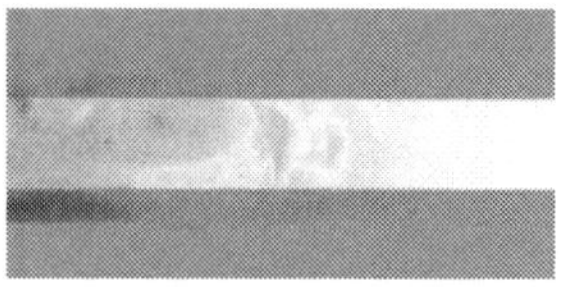

Liberec, Czech Republic, November 23th-26th 2021
efm.kez.tul.cz
efm.conf@gmail.com

15th International Conference "Experimental Fluid Mechanics 2021" was held in the North Bohemian Museum Liberec (Czech Republic), November 23th–26th, 2021.

The EFM conference is focused on obtaining overviews of projects under investigation and experimental methods used in the field of fluid mechanic and thermodynamic. Its aim is to organize a network of research workers who are interested in experimental, and also theoretical, work in the field of fluid mechanic and thermodynamic. The conference should contribute the creation of closer contacts and sharing experience in application of various experimental methods, preparation and implementation of experiments, processing of results and numeric al simulations of experiments.

Keynote lecture for 2021 was:

Dr. Vladimir Kulish from Czech Technical University in Prague: *Viscosity: how much viscous stream buckling can tell us about transitions between flow regimes*

We would like to thank to our sponsors (in alphabetical order)
DANTEC DYNAMICS GmbH (www.dantecdynamics.com),
LAVISION (http://www.lavision.de/en/),
MIT s.r.o. (www.mit-laser.cz),
TSI GmbH (www.tsi.com),
4JTECH (http://4jtech.cz/en/247-2/).

Liberec

North Bohemian Museum

We would like to invite you for the EFM 2022 held in Safari Park in Dvůr Králové nad Labem, November 29 - December 2, 2022. Please visit our web pages efm.kez.tul.cz or for any info contact us via efm.conf@gmail.com or efm@tul.cz.

Safari Park Dvůr Králové nad Labem (conference venue for EFM 2022)

Ultrasonic methods for determining flows and velocity fields

Veronika Barraclough[1*], *Jan* Cizek[1], *Miroslav* Strob[2], *Jan* Novosad[3], *Jaroslav* Pulec[3], *Petra* Dancova[3]

[1]4Jtech, Ltd., Ringhofferova 115/1, 15521 Prague, Czech Republic
[2]Fafulty of Agriculture, Dep. Of Agricultural Machinery and Services, University of South Bohemia in Ceske Budejovice, Branisovska 1645/31a, 370 05 Ceske Budejovice, Czech Republic.
[3]Faculty of Mechanical Enginnering, Dep. Of Power Equipment, Technical University in Liberec, Studentska 1402/2, 46117 Liberec 1, Czech Republic

Abstract. This paper introduces research and development of a modern non-invasive device for determining flows and velocity fields. This device is based on the ultrasonic method. The measured fluid flow is surrounded by appropriate ultrasonic transmitters and receivers, which communicate with each other. A physical principle of this method consists in an interaction of sound waves with a flow, which generates a certain delay in sound waves transmission. The found quantity is thus time delay. The device is being designed as a flowmeter and with advanced and extended postprocessing as a tomograph, which reconstructs a 3D vector field for big volumes. This whole process requires the following: development of an appropriate design of the ultrasonic flowmeter and tomograph, testing the signal transfer and also various postprocessing methods on a measurement accuracy, building of a special verification experimental equipment and building of an electronic device as a control unit and data acquisition system.

1 Introduction

This paper deals with the development of two measuring methods for use in fluid mechanics. Both methods are based on the non-contact measuring principle in which ultrasonic transmitters and receivers are used as sensors, and the sound signal travelling through the flowing medium and the interaction of both of these moving entities is considered. In the case of the first measuring method, it is an ultrasonic flow measurement, in the case of the second method, it is a tomographic mapping of a 3D vector velocity field.

The gauge development involves the following tasks: selecting and testing active gauge elements, i.e. ultrasonic transmitters and receivers. This was the most crucial part of the construction, because several aspects had to be taken into account in the selection: design (especially small dimensions), physical (frequency characteristics, directional characteristics and wave range), microelectronics (control and power supply) and last, but not least, economics. The next step was the construction of meters (to choose the most suitable production technology, especially with regard to the exact placement of ultrasonic transmitters and receivers), prototype design of the control unit and, of course, related activities such as firmware development and control and the measurement software itself. Simultaneously with the construction of the meters themselves, it was necessary to build a verification measuring line and design a set of experiments that verify the sensitivity of measured quantities to different velocity fields (uniform, axially symmetric, deformed), humidity, temperature and temperature field, or turbulence intensity.

Currently, a number of methods are used for flow measurement - from point methods (pressure measurements using multi-hole probes, CTA, etc.) through integral methods (orifice measurements, rotameters, vortex flow meters, etc.) to area / volume methods (PIV). Ultrasonic methods also play an irreplaceable role in this enumeration, but all of them are sensitive to the velocity profile in the measured area and it is therefore necessary in almost all cases to use relatively large installation dimensions of the flow meter.

One of the main innovations of the developed flowmeter is the invariance of the velocity profile of the measured fluid, which significantly reduces its installation dimensions. This can be useful in many technical applications. At the same time, the whole system is designed so that it can be used on pipes of any cross-section (circular, rectangular, multi-walled, general shape).

There are also a number of methods for determining 3D vector fields. The main disadvantages of the methods used today lie either in the disproportionate complexity of the experiment itself (traversing with multi-hole probes), or in very expensive instrumentation in combination with a relatively limited size of the measured area (typically TOMO-PIV, Volumetric PIV). The vector ultrasonic tomograph should be able to eliminate both of these shortcomings. It should be noted that ultrasonic tomography for 3D vector field

* Corresponding author: veronika.barraclough@4jtech.cz

https://doi.org/10.1051/epjconf/202226401001

reconstruction is not an unknown imaging technique and its use, at least from a theoretical point of view, has been described in publications for over 40 years. But despite the undeniable advantages of this experimental method (rapid and non-invasive use) its extension is not significant.

The principle of the ultrasonic vector tomograph, for example, is well described in the theoretical work from Norton [1] or more recent work can be referred to the thesis of Jovanovic [2]. From the point of view of the currently available state of the art technology in the field of ultrasonic flow measurement, resp. velocity fields, it is necessary to mention the publication [3], in which the authors describe a device similar to the proposed ultrasonic tomograph. It is clear from the publication that the proposed measurement principle can be very useful for determining the velocity field.

The authors of this contribution assume the expansion of the HW part of the whole device so that it is possible to determine the impact angle of the ultrasonic beam.

In this area, it is also worth mentioning an older publication in which the authors also deal with the reconstruction of the velocity field using tomographic evaluation of the ultrasonic signal [4], [5] and [6]. The new publication [7] is also worth noting, in which the authors moved the field of application of the ultrasonic tomograph principle to the field of compressible fluid flow.

Fig. 1. Test prototype of ultrasonic flowmeter placed in 4Jtech laboratory, connected to an aerodynamic line.

2 Ultrasonic flowmeter

2.1 Principle

The basic principle of ultrasonic measurement of fluid flow is to determine the flight time of the signal from the transmitter to the receiver, because the sound signal interacts with the medium and this property is successfully used in the calculation of parameters, which are crucial for fluid mechanics. In order to be able to say that a flowmeter is invariant with respect to a velocity field, it is necessary to know the velocity field profile, or to measure and reconstruct it. Such advanced flowmeter, which will measure not only the average flow rate, needs to be designed in a tomo-configuration, such as the one in the figure below. The measurement procedure from the diagram in the figure below is, for example, the transmission of an ultrasonic wave sequentially from all transmitters to all receivers. One measurement contains a matrix of the number of transmitters x the number of receivers of the output signal values.

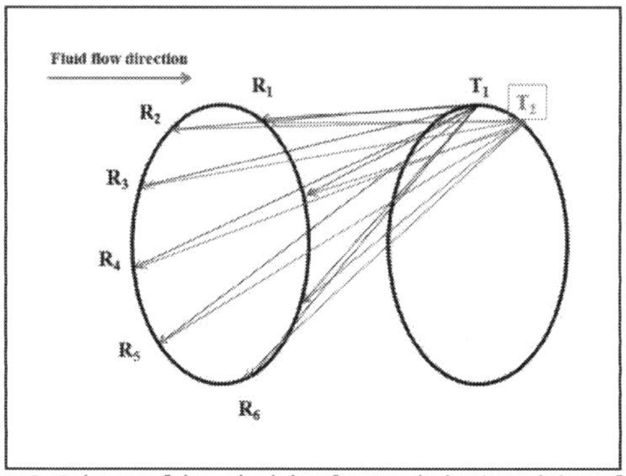

Fig. 2. Schema of the principle of transmission/ receiving of the signal.

2.2 Design

The design consisted of selecting and testing the most suitable active elements, i. e. transmitters and receivers, and selecting the appropriate technology for the production of the flowmeter body. In the case of the selection of transmitters and receivers, the directional characteristics is emphasized most. It is also necessary to take into account the range of the signal and dimensions of the elements. Along with this, it is necessary to think about an easily feasible power supply and control and signal reception. While the individual parameters of course sometimes contradict each other, the result is always a kind of compromise. However, the selection of sensors is the most important part of the entire design process, so special care was taken at the beginning to select and test the appropriate elements. Another part of the design was to choose the technology of the first test piece production. As the work progressed, the original technology was replaced with a new, more accurate technology for production, and with the possibility of a faster change in the basic dimensions of the gauges.

2.2.1 Sensors

The requirements for ultrasonic transmitters/ receivers for use in fluid mechanics are generally conflicting. In order for ultrasonic flowmeters to be suitable for measuring the velocity field, it is necessary to choose transmitters with the highest possible frequencies, ideally in the order of 100-1000 kHz. At such frequencies, the frequencies of the measuring technology no longer collide with the frequencies of possible non-stationarities of the flow field. However, transmitters/ receivers with such high frequencies generally have high power requirements. Another problem with such high frequencies is the weaker power of the transmitters/ receivers or the inappropriate dimensions.

A compromise was therefore chosen for Prototype 1 of the developed ultrasonic flow meter - often using transmitters/ receivers of small dimensions with a frequency of around 40 kHz. Their directional characteristics (specified by the manufacturer) are generally satisfactory and sufficient for the dimensions of Prototype 1 at this time.

Transmitters/ receivers were selected that met the required frequency, size, directional characteristics and price. During the first tests the directional characteristics of the supplied transmitters were verified, because the manufacturers often conceal their procedure for verifying this essential property and the reason is the fact that the directionality is measured in a rough resolution of ten degrees.

Finally, sensors designated MCUST and MCUSR were selected.

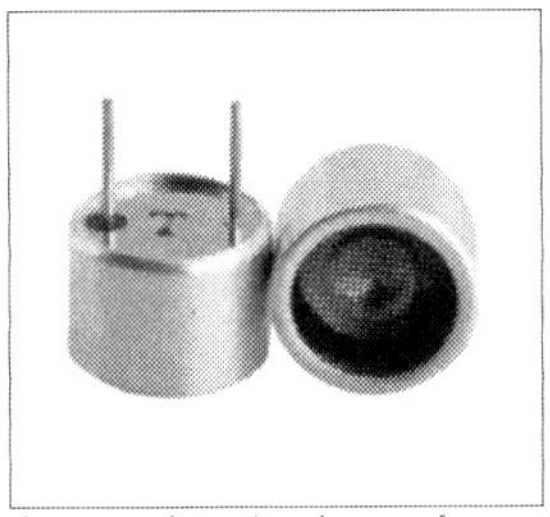

Fig. 3. Ultrasonic transmitters/ recievers chosen for testing.

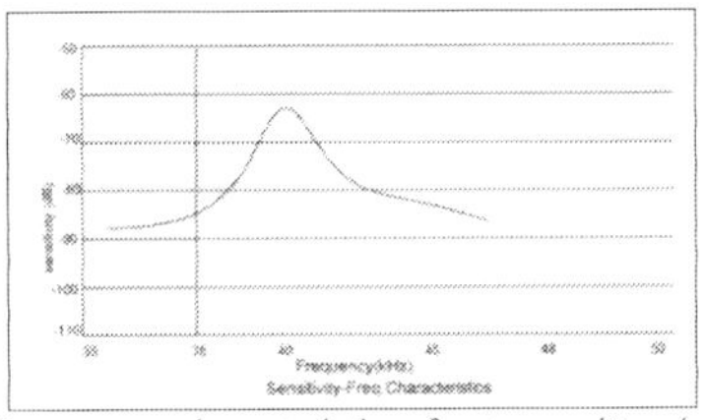

Fig. 4. Frequency characteristics for transmitters/ receivers MCUST/R.

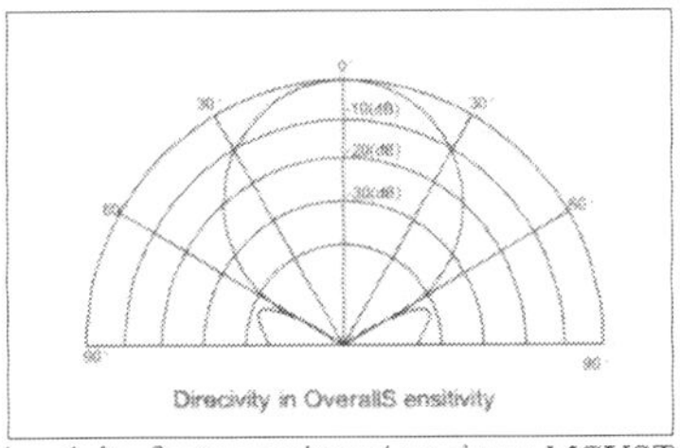

Fig. 5. Directivity for transmitters/ receivers MCUST/R.

2.2.2 Design

Prototype 1 is built for circular piping in a configuration that is least sensitive to the velocity distribution of the fluid flow. To assess the properties of the flowmeters three factors are compared: the hydrodynamic factor, the directional sensitivity factor and the range factor. In general, the more transmitters/ receivers used, the better accuracy is achieved. The highest accuracy is achieved in the arrangement, which the foreign literature refers to as D-Orth, ie double-orthogonal.

Test prototype (see Fig. 1) was made of air ducts and around its perimeter were placed transmitter and receiver holders printed on a 3D printer, while the storage was done in a D-orth configuration. In this arrangement, the first verification experiments were performed, in which the ultrasonic flowmeter was connected to a fan track in the 4Jtech laboratory. Soon after that prototype No. 1 was created, because it was clear that the design of the test gauge is lengthy, demanding to manufacture and also can not achieve the highest possible accuracy geometrically. The sensors holders have an oval surface and it is very laborious to achieve a precise oval, moreover in a precisely determined position, by cutting into sheet metal. The possible production of a custom made ultrasonic gauge would thus be unnecessarily demanding.

The main requirement in the design of prototype No. 1 was the geometric accuracy of the transmitters and receivers location so that for later calculations it is possible to precisely define the distance of the respective pairs (T-R), easy variability of the design (even with regard to later transition to the gauge of non-circular cross-section) and last, but not least, easy to manufacture. In the end, 3D printing seemed to be the most suitable for the prototype.

The considered diameter of the experimental line which the flowmeter of prototype No. 1 was designed for is 200 mm. The only minor obstacle was that the entire 3D flow meter, even with flanges, could not fit on the pad of the usual 3D printer, and it was necessary to divide it into 4 parts and glue them. Although the geometric accuracy in the distribution of ultrasonic flowmeter was slightly lower (the connecting surfaces also have a certain deviation at the height of the printing layer of 0.3 mm), it turned out that despite this small disadvantage, 3D printing is the best possible choice in all possible respects.

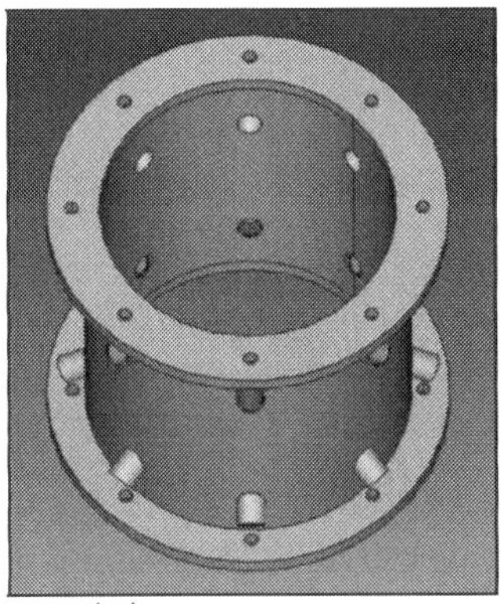

Fig. 6. The gauge design.

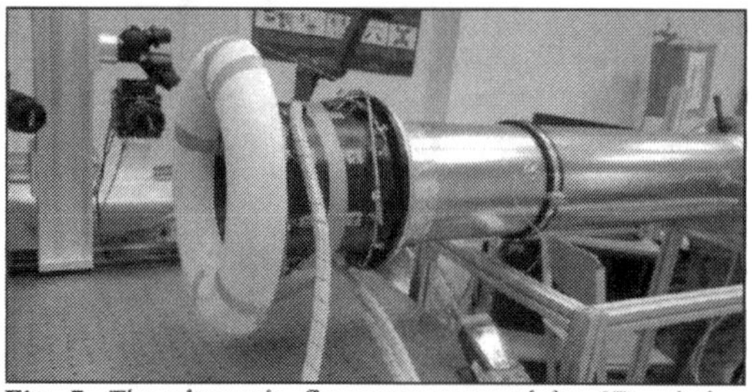

Fig. 7. The ultrasonic flowmeter prepared by 3D printing (together with an experimental line, see chapter 4 bellow).

2.3 Microelectronics

Overall, the control unit has the task of providing the operation of excitation and reception of ultrasonic signals, high-speed sampling of signals, their temporary buffering into the buffer and transmission for further processing.

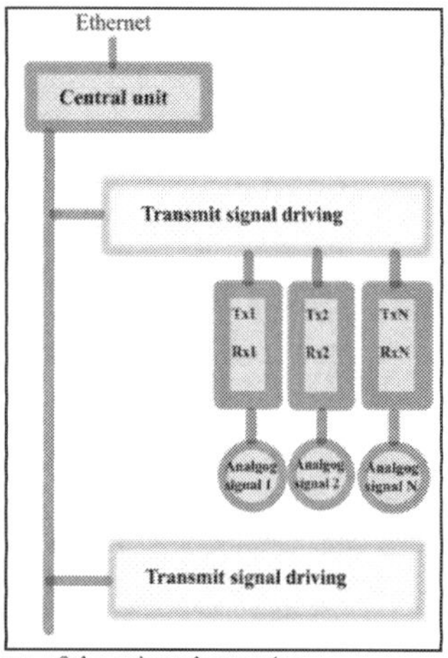

Fig. 8. Schema of the microelectronics.

The above requirements can be summarized, with respect to other parts of the flowmeter, in the following points:

- standard supply level 5 V, 12 V
- compact solution using commonly available components with long-term support
- sampling of analog signals at least 50 MHz
- AD converter resolution min. 8 bits
- at least 4 independent analog channels
- the size of the buffer covering at least the recording of all channels during the flight of the signal through the space of the flow meter
- sufficient computing power to process digital signal processing tasks

From the point of view of electronics, the most problematic part is the signal sampling and subsequent caching. Even with regard to other requirements, it is suitable to use FPGA type circuits, parallel AD converter, DDR type memory.

A standard Ethernet interface was chosen to connect the control unit to the computer. Due to previous experience with FPGA circuits, the Spartan 7 FPGA circuit family was chosen. Circuits of this type are able to process signals at the required sampling levels, they allow the connection of DDR3 type memories. Spartan FPGAs allow the MicroBlaze processor to be synthesized to combine a parallel and sequential structure.

The production of the original prototype was carried out using interconnected modules. The first part consists of a designed module containing the STM32F429 processor, Ethernet connection, IS42S16400J-7TLI memory with 4Mx16 organization, optional LCD display connection and SPI bus.

Power is provided by a standard 12 V adapter. The Ethernet interface operates at a standard speed of 100 Mbit. The board is connected via SPI bus to the FPGA module with a standard connector, data transfer speed is up to 1 Mbit / s. Each analog channel was implemented by one AD converter on a separate printed circuit board. The proposed printed circuit board is shown in the following figure.

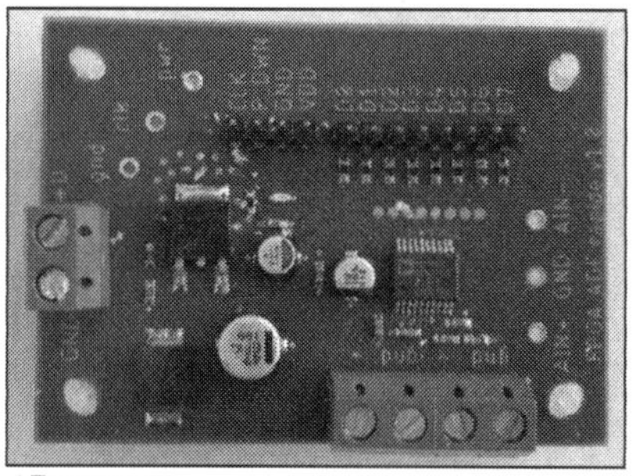

Fig. 9. AD converter.

2.4 First tests

The first verification measurement of a circular flowmeter was performed on a newly assembled experimental line in the laboratory of the Technical University in Liberec (see chapter 4 Experimental line). The evaluation of the obtained data is currently in process, because first it is necessary to obtain statistical data.This requires the evaluation of several hundred files in order to determine the optimal ratio of signal measurements between transmitter-receiver pairs and measurement accuracy.

It is also important to choose a more appropriate way of the signal time measurement approach itself. So far the evaluation is processed in two ways: a more laborious and statistically demanding way, which depends only on the received signal from all receivers and the accuracy of peak correlation, or a simpler simple time determination approach of finding time of flight between the leading edge of the transmitted signal and the beginning of the received signal. The second way can cause entering a significant error in determining the beginning of the received signal (see figure below; in this way, the received signal can be determined with an accuracy of 25 μs, which means that an uncertainty of

4% is introduced into the overall flow of the prototype ultrasonic flowmeter only by this input).

A processing method based on fast Fourier transform, inverse FT, threshold determination and phase shift should help this problem. However, due to the complexity of processing both evaluation methods, this problem has not yet been clearly solved.

Fig. 11 shows the course of signal processing by comparing two received signals (receivers differ in geometric determination, so the signals are correlated).

The essence is to find the maximum value of this correlation and clearly determine the position of the peak of this correlation on the axis of the number of samples (which corresponds to the time axis, at a sampling frequency of 50 Msps, the result is tens of microseconds).

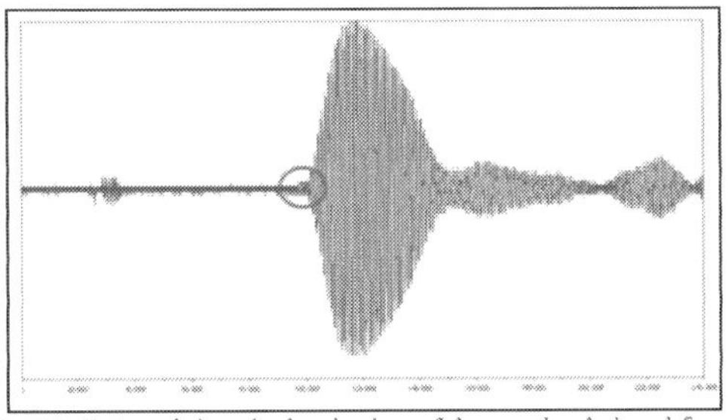

Fig. 10. Determining the beginning of the received signal for accurate flight time determination - a crucial parameter for low measurement uncertainty.

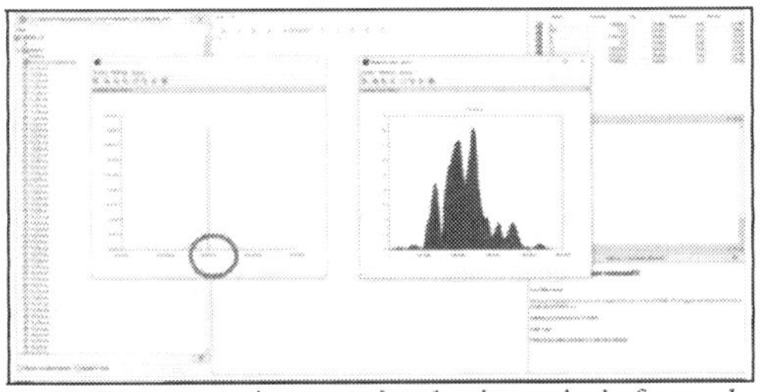

Fig. 11. Data processing procedure by the method of mutual correlation between signals.

3 Ultrasonic tomograph

3.1 Principle

Fig. 2 shows a flowmeter principle in a tomographic arrangement. Each of the transmitters generate a signal for all receivers, which is the principle that a 3D vector tomograph actually works on. However, for vector velocity field reconstruction, it is not sufficient to simply measure the travel time of a signal from one transmitter to one receiver; by evaluating measurements based on time of flight (time-of-flight, abbreviated TOF), one obtains the magnitude of the velocity vector, but not the direction.

The time-of-flight of the signal between the transmitter and the receiver is expressed by a curve integral containing elements with the speed of sound and the flow rate, which after linearization has the shape

$$(\tau_0 - \tau) \cdot c_0^2 = \int_\Gamma (\Delta cn + v).ds \qquad (1)$$

where τ is time-of flight, τ_0 is reference time-of-flight, c is speed of sound and c_0 reference speed of sound, v is the flow velocity, s is vector tangent to the Γ trajectory, n is vector normal to the wave, Γ is sound ray trajectory.

However, the right side does not lead to a clear solution.

If a substitution is done with the Helmholtz' decomposition of the velocity field into a solenoidal and a curl-free part, and this equation is solved on a vector field, information about the solenoidal component of the flow cannot be obtained. The right part of the equation only provides information about the longitudinal interaction of the sound wave with the velocity field. For a complete reconstruction of the velocity field, it is also necessary to obtain information about the transverse interaction.

Put into practice, it is actually a measurement of the incident angle of the speed of sound vector. There are sophisticated precision optical methods that can be used to obtain this information, for example Braun and Hauck suggest the use of a Schlieren method [6], but this method can only be used with a tomograph with optical access, which would limit its use.

In her work, Jovanović [2] proposes a more easily applicable method. It is based on the TOF method, but this time between one transmitter and three receivers on the other side. The principle of the method can be seen in following figure.

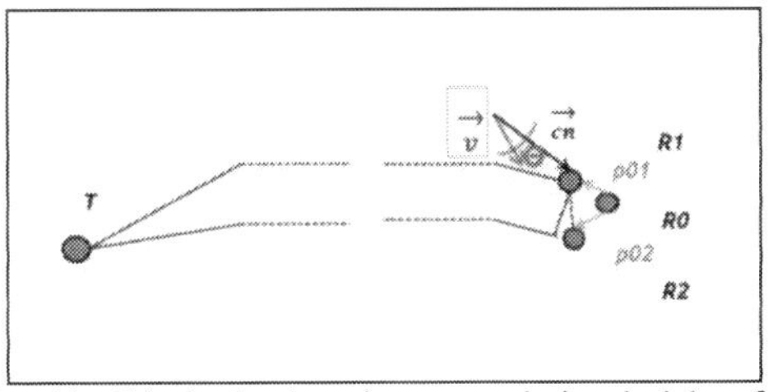

Fig. 12. Tripol showing the geometrical principle of ultrasonic tomograf measurement.

The travel times of the audio signal that interacted with the detected velocity field are obtained from the three receivers; from the difference of the detected times on the receivers R1-R0 and R2-R0, while in the case of the receiver R0 the magnitude of the velocity vector is evaluated, then distances are found and from them the angle of the velocity vector of the flow field by simple trigonometry.

3.2 Design and tests

The prototype, on which selected transmitters and receivers were tested, eventually consisted of one transmitter and five receivers in the configuration, see the photo below.

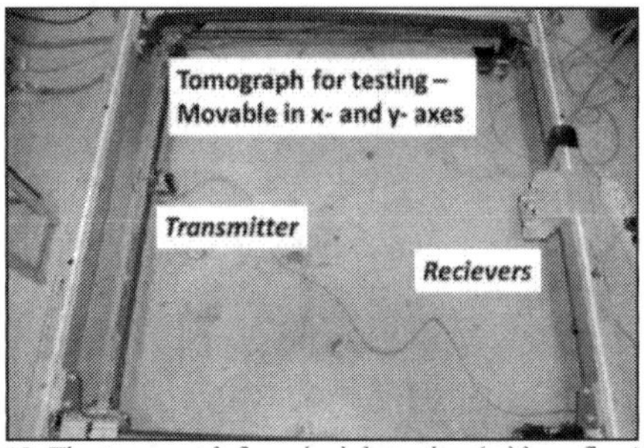

Fig. 13. The tomograph for principle testing (with no flow).

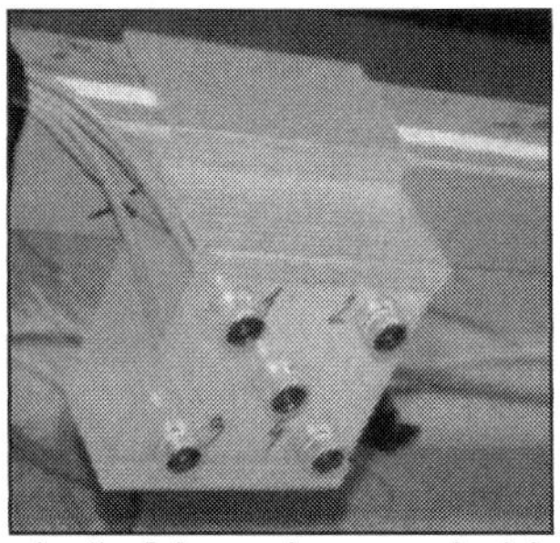
Fig. 14. The detail of the receivers organized in a rosette.

In addition to verifying the tomograph's operating principle, this test equipment was designed to quantify the effect of configuration on measurement uncertainty; the number of transmitters and receivers is one of the criteria, as well as, for example, determining their position relative to others as accurately as possible. Directionality was tested here (and thus the absolute distance and possible absolute size limit of the tomograph; for tomographic reconstruction of the velocity field and determination of the highest possible accuracy of determining the velocity vectors, it is important that one transmitter emits to as many receivers as possible).

The selection of sensors was similar to that one of the ultrasonic flowmeter. The main conditions for selection were high directionality and the expected range of min. 0.5m. However, the originally selected great transmitters and receivers from the Taiwanese manufacturer Pro-wave proved to be unsuitable due to the extremely high supply voltage requirements even for a relatively small tomograph; the power requirements for a tomograph which is intended to serve, for example, wind tunnel measurements for the automotive industry, are unimaginable. Thus, the already proven MCUST and MCUSR transmitters and receivers have been chosen for the smaller tomograph, which have already had several tests on the ultrasonic flowmeter. Although their directional characteristics are not as magnificent as the Pro-wave, tests on a test tomograph have shown solid results.

The tomograph for real operation on the aerodynamic line will have 24 transmitters and receivers. At the turn of the spring and summer of 2022, the first tests will take place, and the measurement of the wake behind the bluff body will follow immediately.

4 Experimental line

In addition to the flowmeter and the tomograph, it was necessary to design and build an experimental line, with the purpose to verify the functionality of the developed ultrasonic flowmeter. This will also be used to measure velocity fields using an ultrasonic 3D tomograph.

The line will serve as three different methods of measuring the flow and velocity fields of fluids: PIV, the ultrasonic flowmeter and tomograph, and was built as follows. The flowmeter and tomograph as developed meters will be measured for sensitivity to different velocity fields, humidity, and temperature fields. Comparative measurement of flow fields by the Particle Image Velocimetry method will be used to validate the set flow regime.

The line consists of the following components:

At the beginning of the measuring line (section A-B) elements will be placed that will ensure the setting of the required flow regime, i.e. even and uneven, or deformed velocity profile and elements to ensure turbulent flow of the required turbulence intensity. Turbulent screens for determining the effect of turbulence intensity on the measured flow will be made as gratings (with partitions 25 mm wide and 50 mm apart, 2 vertical, 2 horizontal) - the grating intensity designed in this way can achieve turbulence intensity Tu approx. 10-15% (at a distance of about 50 mm behind the screen - along the length of the ultrasonic flowmeter part, the intensity of turbulence will decrease exponentially with a coefficient of -5/7; it is desirable that this value be as high as possible at the location of ultrasonic transmitters and receivers).

In section B-C there is a glass pipe part for PIV measurement, a tested ultrasonic flowmeter (see photo below) or an ultrasonic tomograph will be installed here.

Section A-B is now omitted for verification measurements of the ultrasonic flowmeter and the line itself, and the glass part and the ultrasonic part are immediately fitted with a lead made of hardened polystyrene.

The IRIS 200 orifice, indicated in the figure below, was used to measure the flow. The flow is evaluated by measuring the pressure difference at the orifice using a TESTO 512 differential pressure gauge.

A KU200 valve for flow control, and a probe for measuring the temperature of the fluid (thermocouple type K) are installed in the section behind the measuring orifice. The line was powered by a radial fan CRMT / 4-225 / 90.

The glass part for measuring by the PIV method and the ultrasonic part do not have the same length. Thus, a control calculation and comparison of pressure losses for given lengths and material of these parts was performed.

The pressure losses of the glass part are 30% higher, but in absolute numbers it is only units of pascals, i.e. practically negligible. In this context, it was decided to produce a total number of three ultrasonic flowmeters of different lengths (the goal is to achieve the shortest possible part for the specified diameter of the measured pipe and determine the limit dimensions to achieve the required accuracy, or as a bonus to find a functional relationship between part lengths and accuracy).

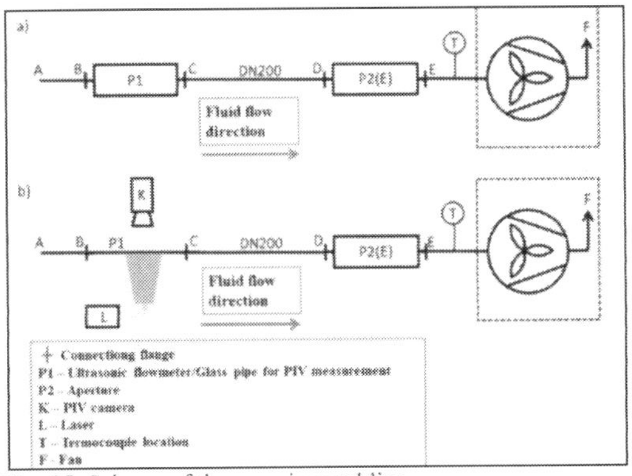

Fig. 15. Schema of the experimental line.

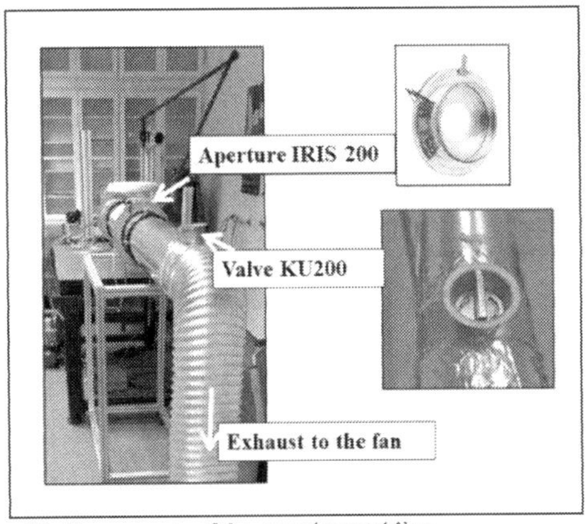

Fig. 16. Components of the experimental line.

Fig. 17. The experimental line prepared for PIV measurement.

After the completation of the experimental line, the first verification measurements were performed. The flow was regulated by the KU200 aperture, which allows the aperture position to be set in the range of 0 ° (fully open) to 90 ° (closed). The flow results and the calculated velocity of the line are shown in the following graphs. The highest speed achieved on the experimental line when using the installed fan was 7.66 m / s at a flow rate of 0.24 m3 / s. Due to the fact that the full range of speeds required for testing ultrasonic devices (up to 20 m / s) has not been reached, an additional fan will be added to increase the transport pressure before the next measurement.

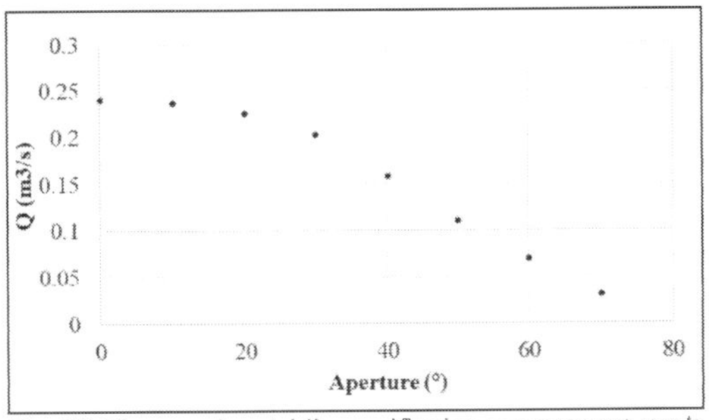

Fig. 18. The experimental line verification measurement result – the flow.

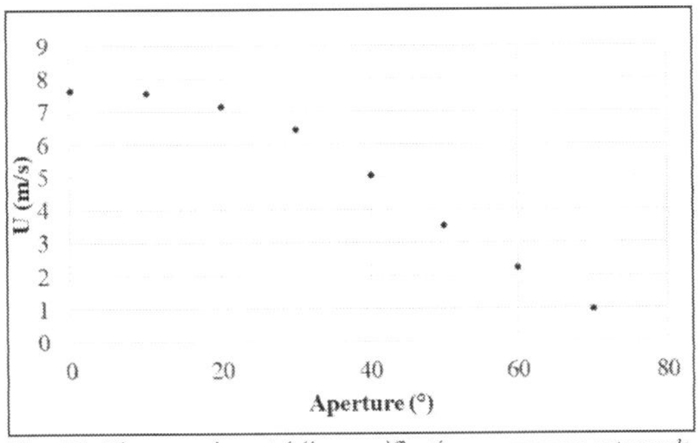

Fig. 19. The experimental line verification measurement result – the calculated velocity.

4.1 The error analysis

The default equation consists of speed of sound, time delay and geometry. It is necessary to determine partial uncertanties.

$$v_a = \frac{C - \frac{L}{\Delta t + \frac{L}{C}}}{\cos\alpha} \quad (2)$$

V_a is an axial velocity, L a sound travel distance, c speed of sound, α an angle between the gauge axis and the soudn wave direction and Δ tis the time difference.

Sound velocity uncertainty with respect to temperature measurement velocity is:

$$u_B(c) = \pm 0{,}18\ m.s^{-1} \quad (3)$$

The uncertainties of measuring the geometry of the flowmeter were estimated as (resulting from the meters and their capabilities):

$$u_B(L) = \pm 0{,}14\ mm$$
$$u_B(\alpha) = \pm 0{,}29\,° \quad (4)$$

The uncertainty of the determined measurement uncertainty Δt was determined as:

a) the uncertainty of type A_t of repeated measurements, where:

$$u_A(\Delta t) = \sqrt{\frac{\Sigma(xi - \bar{x})^2}{N.(N-1)}} = \pm 45\ ns \quad (5)$$

b) The type B uncertainty of Δt in the autocorrelator, we assume an error uniformly distributed over an interval of ± 10 samples, then:

$$u_B(\Delta t) = \frac{\Delta}{\sqrt{12}} = \frac{20}{100.10^6.\sqrt{12}} = \pm 57\, ns \quad (6)$$

c) The combined uncertainty is as follows:

$$u_c(\Delta t) = \sqrt{u_A(\Delta t)^2 + u_B{}^2(\Delta t)} = \pm 72,6ns \quad (7)$$

Based on all partial uncertainties, the velocity uncertainty for highest experimental line velocity was determined to be:

$$v = (7,69 \pm 0,23)\, m.s^{-1} \quad (8)$$

5 Conclusion

The paper was devoted to the description of the development of flow and velocity field meters in fluid mechanics and all activities related to this development.

First of all, it was the preparation of an experimental line in the laboratory of the Technical University in Liberec, where the first verification tests of both the line itself and the ultrasonic flowmeter were performed. The line was designed as variable because the measurement will be performed by several different methods for many configurations. It was necessary to consider generators of various velocity field uniformity disorders, and to further consider the influence of different temperatures or humidity on the flow measurement, thus ensuring that it is possible to change these quantities for measurement purposes. The measurement plan also includes flowmeters of various cross-sections and extensive tomograph measurements, including 3D velocity field measurements behind the model - a bluff body body.

Another task was to prepare the ultrasonic flowmeter itself in terms of construction. The first test piece was made for use in measurements in the 4Jtech laboratory, but the production technology was gradually reworked into the first functional prototype for verification measurements at the Technical University. The production technology had to be modified with emphasis on higher geometric accuracy of transmitters and receivers location.

Simultaneously with the advancing construction, it was necessary to prepare the meters electronically. PCBs for signal excitation were made and the correct excitation signal was designed. A far greater challenge for the specialist was the design and construction of a control unit ensuring all the flow of data and energy between the transmitters, receivers and the evaluation unit (computer). Of course, it is possible to purchase a commercial DAQ system and save several months of research work, but a significant disadvantage of these data collections is their rapid obsolescence, when once acquired, a not cheap unit will eventually cease to be operated by the supplier, which will result in stopping firmware updates at the moment when the equipment ceases to be manufactured. The prototype of the control unit was therefore assembled on a programmable FPGA circuit and put into operation.

This part also intersects with the construction of the ultrasonic tomograph control unit, which is built on the same basis, but must serve a significantly higher number of channels (while the ultrasonic flowmeter currently has 8 transmitters and 8 receivers and accurate measurement requires 8x8 combinations, an ultrasonic tomograph to be tested on the experimental line, is designed for an array of 24 transmitters and 24 receivers which means a minimum of 24x30 triple signals that will need to be received and processed). The tomograph was tested in a miniature and electronically simplified version and without flow to determine the approximate size of the test tomograph for the experimental line, and the effect of geometry on the rough estimate of the uncertainty of the angle measurement, which is one of the two basic parameters for the ultrasonic tomograph principle.

Acknowledgement to Technology Agency of the Czech Republic for supporting FW02020070 project.

References

1. S.J. Norton: Tomographic reconstruction of 2-D vector fields: application for flow imaging, In: Geophys. Journal, **97**, pp. 161-168 (1988).
2. I. Jovanovic: Inverse Problems in Acoustic Tomography: Theory and Applications, doctoral thesis, Ecole Polytechnique de Lausanne, Switzerland (2008).
3. J.-N. Liu, B.-X. Wang, Y.-Y. Cui, H.-Y. Wang : Ultrasonic tomographic velocimeter for visualization of axial flow fields in pipes, In : Flow Meas. and Instr., Volume **41**, pp. 57-66 (2015).
4. S. Teerawatanachai, K. Komiya, and H. Sasamoto : Estimating the velocity profile of air flow by means of the ultrasonic translation time computed tomography method, *Proceedings IECON '91: 1991 International Conference on Industrial Electronics, Control and Instrumentation*, pp. 2379-2384, vol.3, doi: 10.1109/IECON.1991.238972 (1991).
5. K. Komiya, S. Teerawatanachai, Ultrasonic tomography for visualizing the velocity profile of air flow, In : Flow Meas. and Instr., Vol. **4**, Issue 2, pp. 61-65 (1993).
6. Braun, H., Hauck, A.: Tomographic reconstruction of vector fields, *IEEE Transactions on Signal Proceeding*, vol. **39**, no. 2, (1991).
7. R Otero Jr et al : Coupled velocity and temperature acoustic tomography in heated high subsonic Mach number flows, In : Meas. Sci. Technol. **30** 105901 (2019).

An experimental analysis of droplet charging in a wind tunnel

Ondřej Bartoš[1]* and *Adam* Huněk[1]

[1]CTU in Prague, Energy Engineering Department, Technická 4, Prague 6, 166 07, Czechia

Abstract. The aim of this paper is to introduce an experimental analysis of droplet charging that occurs due to liquid film atomization on two different surfaces. The atomization of the droplets from the liquid film at high speed is still not satisfactory understood. The phenomena can be found in steam turbines, wind turbines, aeroplanes and other technical applications. The liquid film atomization is associated with the electrostatic charge of the formed droplets. The paper describes the experiment when the coated and non-coated surfaces were tested for an estimation of the droplet size distribution function and the droplet charge. A new wind tunnel was developed for the study of liquid film atomization in similar conditions in steam turbines. The coarse droplets are formed in the steam turbine from the liquid films on the blades and inner casings. The coarse droplets formed on the non-moving blades don´t exactly follow the bulk flow and they collide with the moving blades. These collisions cause erosion and corrosion processes which have an unfavourable effect on the reliability and the efficiency of steam turbines. The tunnel is equipped with the standard instrumentation for the measurement of the flow properties and for the analysis of the size distribution function of the droplets. Two measurements methods were used to attain the size of the droplets, along with their photogrammetry and light scattering. The first tested surface is polished stainless steel and the second is the electrical isolating and hydrophobic coating. The results from the measurement suggest the possibility of the better understanding the erosion processes in steam turbines.

1 Introduction

The presence of the liquid phase in the flowing wet steam has a very unfavourable effect on the operating characteristics and reliability of steam turbines [1]. Electrostatic charge could be involved in the formation of a liquid phase in the flow section of steam turbines on one side, and on the other side, of the erosion process. Knowledge of the electrostatic charge phenomena in steam turbines is therefore essential, especially these days when such understanding has the potential to enhance steam turbine performance. The presence of an electrostatic charge in the flowing steam has been proven by the measurement of steam turbines [2,4], which showed that the charge distribution in the flow was not homogeneous and was probably influenced by the operation and chemical mode of the thermal cycle.

The wind tunnel, located at the Institute of Energy Engineering, Faculty of Mechanical Engineering of the Czech Technical University in Prague, was developed to study liquid film disruption and to research the electrostatic charge in the flowing air or steam.

The paper describes the measurement of the electrical current on two surface properties. The experiment simulates a situation when manufacturers of steam turbines use the coating to modify the coarse droplet distribution that allows for an extended blade life. The existence of the electrical charge in the steam turbines is well documented since steam turbines expand to the wet stem. A similar situation occurs during aeroplane charging in the flow through rain.

2 Droplet charge

Several known methods of droplet charging have been investigated (diffuse bipolar and unipolar, electrostatic field charging, water droplet charge dropping on the probe surface). The measured reality best described the model based on the disintegration of the water film that forms on the surface of the blade and the probe. This proposal is based on work on an electric double-layer [3,4]. The water that collects on the electrostatic probe is in turbines created directly during spontaneous condensation and therefore we can assume higher concentrations of chemical impurities [4] than in the bulk condensate and in the wind tunnel the liquid used is mostly standard fresh water. According to the derivation at work [2], the charge magnitude of the resulting drop is more dependent on geometric characteristics than on electrochemical properties and can be estimated at approximately $\mathbf{q} = 3{,}3 \cdot 10^{-15}$C. Since it is obvious that the radius of entrained droplets is crucial, it was determined according to the Weber number [5]. Geometric characteristics are further responsible for the erosion processes in the turbines. Erosion of the steam turbine blades can be defined as a progressive loss of material from the solid surface bodies caused by collisions with the water droplets which possess high kinetic energy.

* Corresponding author: ondrej.bartos@fs.cvut.cz

The blades are most often subject to this type of erosion due to the high relative velocity between the coarse droplets and themselves. The swept droplets don't follow the bulk flow and they are insufficiently accelerated before colliding with the blades. Erosion reduces efficiency, turbine power and decreases time between overhauls. The problem of the formation of water droplets due to condensation of water vapour in the last stages of the turbine is the subject research for several decades, but nowadays is even more urgent due to the rise of the use of the renewable energy sources. The serious impact on reducing the life of turbines is the reason for the constant interest and efforts to develop new measures to reduce blade wear.

3 Measurement

The experiment was carried out in the wind tunnel with the two blades placed in the air flow. The first blade was coated with electrical insulating and hydrophobic paint and second was polished stainless steel. The overall view of the tunnel is in Fig. 1.

Fig. 1. The wind tunnel.

3.1. The wind tunnel

The motivation for the wind tunnel assembly arose from a previous measurement [5] of the coarse droplets in the steam turbine under operational conditions. The measurement was performed in three steam turbines in the Czech Republic. The amount of the acquired coarse droplets was very low due to their low number density level. Due to difficulties connected with measurements in an operating steam turbine, a wind tunnel was designed and manufactured for the analysis of the coarse droplet formation from the liquid films. The aim of the design was to create, as closely as possible, the conditions in a steam turbine while utilising the advantages of working in the laboratory.

The wind tunnel is designed as a classical CD nozzle, but the planned operational regime is mainly subsonic or transonic. An aerofoil NACA0008 is placed 50 mm behind the nozzle throat. The aerofoil simulates the blade in the turbine and it is possible to remove or replace it. There is a groove on the aerofoil which supplies liquid to the surface. The liquid is pumped to the aerofoil through the dosing pump with a flow from 1 ml/min to 500 ml/min. The tunnel is equipped with four large optical windows that provide good optical access for the measurements. It is possible to operate the tunnel with steam in continuous mode or with compressed air in periodic mode.

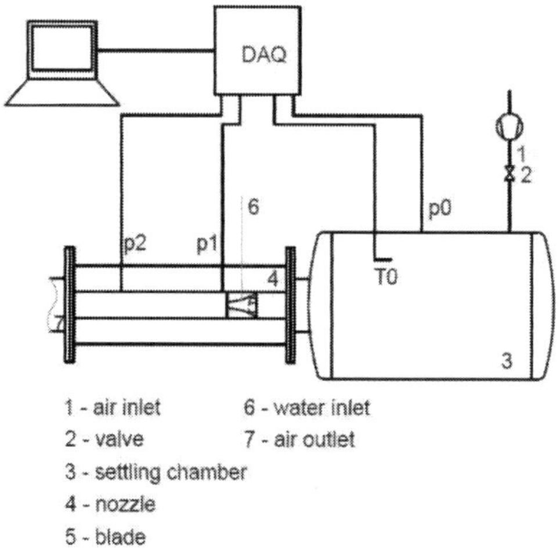

Fig. 2. The wind tunnel measurement schema.

3.2 The size distribution measurement

The droplet distribution function measurement was performed on a Spraytec instrument (Malvern, Inc.). Both blades were tested under the same conditions in the wind tunnel. The measuring position of the laser beam was about 20 cm behind the trailing edge of the blade. Due to the principle of measurement (scattering of light on the droplets), the conditions for evaluation were not always met, therefore the results are calculated only for the states when the conditions for evaluation were acceptable. In Fig. 3, the volumetric or mass fraction distribution is shown. It is possible to see the bi-modal distribution.

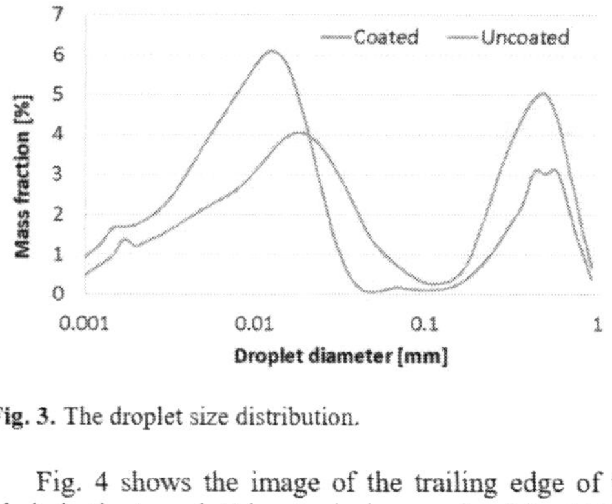

Fig. 3. The droplet size distribution.

Fig. 4 shows the image of the trailing edge of the blade in the tunnel. It is a typical example of liquid film break up. The image was taken by the developed photogrammetric probe [6]. The situation on the image indicates the formation of small particles and randomly formed large droplets.

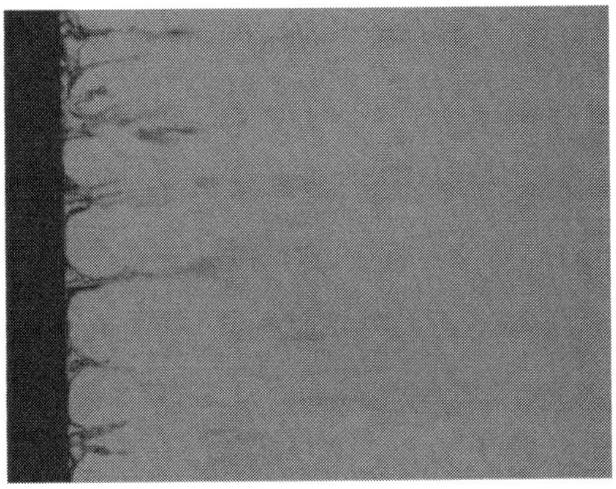

Fig. 4. An example of liquid break-up on the trailing edge of the blade.

3.3 Charge measurement

The electrostatic probe is schematically shown in Fig. 5. The impacting droplets carrying the el. charge produce the el. current. The recorded electric current is the result of the interaction between the droplets and the active part of the probe. A similar process also occurs as the droplets leave the probe. The projection area of the active part of the probe is 20 x 2 mm. This probe can measure across the nozzle channel. The flow field influence for different positions of the active part can be considered constant for all positions of the active surface of the probe.

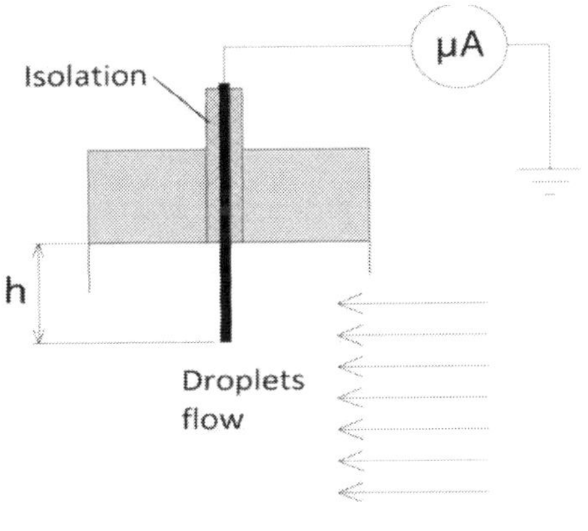

Fig. 5. Electrostatic probe measurement schema.

3.4 Measurement accuracy

The experiments presented in this paper are more qualitative analyses of the phenomena, whereas classical error analysis is not possible to suggest. The coefficient of variation for the droplet optical measurement is smaller than 1% for a monodisperse sample. The error of the micro-amperemeter is smaller than 1% of the span.

4 Data processing

The electric current I generated by the water droplets carried in the steam flow upon impact on the probe can be expressed by the following formula (1)

$$I = AcN_v \int_{rk,min}^{rk,max} \eta_z(r_k)\alpha(r_k)q(r_k)\varphi(r_k)dr_k \quad (1)$$

Where $\mathbf{r_k}$ is the radius of the droplet, **A** is the projection of the probe surface perpendicular to the flow axis, **c** is the steam velocity, $\mathbf{N_V}$ is the number density of droplets in 1m^3, $\eta_z(r_k)$ is the drop impact efficiency on the probe, $\alpha(r_k)$ is the probability of the charge transfer from the droplet to the probe, $\varphi(r_k)$ distributing the droplets in the stream by the radius.

Due to existence bi-modal distribution, the data of the size distribution function of the droplets were divided into two groups and the Sauter mean diameter was computed for both groups. The edge diameter between small and large droplets was estimated at 60 μm (approximate saddle between modes).

$$D_{32} = \frac{\sum d^3}{\sum d^2} \quad (2)$$

For the simplified analysis of the measured el. current the eq. 1 may be applied. The dependence on the diameter can be reduced by two Sauter diameters of the two groups from the droplets' distribution measurement. The charge transfer efficiency and impacting efficiency could be in this simplified situation expected to be a unity. The density level of the droplets was determined according to the volumetric concentration and the measured mass flow of the liquid on the blade surface. In all cases 1ml/min. The cross-section of the electrostatic probe is 6% of the nozzle channel. The final simplified equation can be written as,

$$I \approx e(\dot{R}_s q_s + \dot{R}_b q_b) \quad (3)$$

where **e** is the elementary charge $-1{,}602\text{e}10^{-19}$ C, **R** is the collection rate [s^{-1}] of the el. stat. probe of small and large droplets. **q** is the mean droplet charge expressed in the elementary charge. The ratio between mean charge of the small and large droplets $\mathbf{q_b/q_s}$ is an important parameter which can be useful for practical application. Large droplets are mostly responsible for the erosion of the turbine blades.

5 Results

Three measurements of the coated and uncoated blade were chosen for the data processing. In Table 1, the result of the Sauter mean diameters for both droplets group are presented alongside as their ratio. In Fig. 6, the time dependent profiles of the current measurement are presented. The el. current seems similar for all measurements with a change of polarity. Direct data processing is not possible because the el. probe proves

Table 1. Droplet size distribution results.

	Diameter D_{32} [μm]		Mass Ratio [%]	
Group	Small	Large	Small	Large
Coated 1	5.3	386.2	72.4	27.6
Coated 2	5.0	458.8	88.3	11.7
Coated 3	4.8	353.3	75.6	24.4
Uncoated 1	7.9	332.2	59.8	40.2
Uncoated 2	5.8	136.8	75.9	24.1
Uncoated 3	5.2	209.4	84.2	15.8

impossible to distinguish a charge which is coming with the droplets from the blade and one which is leaving the probe with the droplets. From this reason, we are only able to make similarity comparisons. The best fit between eq. 3 and measured data was determined for the ratio $(\mathbf{q_b}/\mathbf{q_s}) \approx 10^5$.

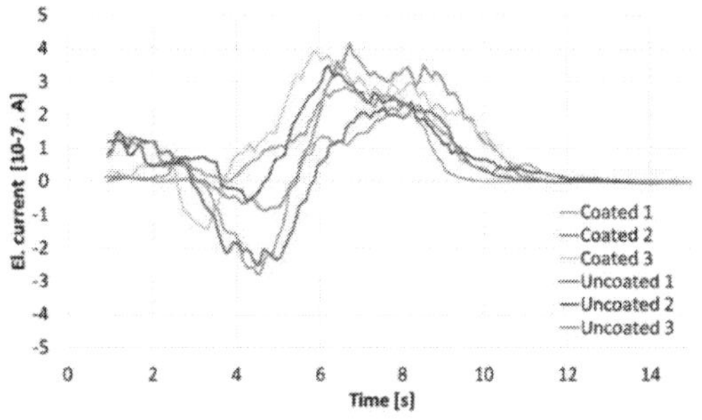

Fig. 6. Electrostatic probe measurement.

6 Discussion

The explanation of the result is not direct and it is necessary to realise the source of the charging which is the expected double layer disruption of the liquid film. The presence of the el. charge on the droplets is already proven. But the polarity and charge distribution in respect to droplet diameter is still not clear. In Fig. 7, the mean droplet charge profile for a coated and uncoated blade is presented. Both dependencies have similar profiles, but we expect the charge ratio is $(\mathbf{q_b}/\mathbf{q_s}) \approx 10^5$, but the mass ratio of both droplets groups $\mathbf{m_b}/\mathbf{m_s}$ (D_{32}) is 3.10^5. From this perspective, the coated surface produces more charged particles and the larger particles have a lower charge mass ratio. Moreover, the large droplets mean charge $\mathbf{q_b}=10^{10}$ of elementary charge is very close to the Rayleigh limit of the droplet charge which is, in our case for large droplets, in order 10^{12} units of elementary charge. In Fig. 4, it is possible to see the excess of the charge at the end of the liquid columns and the explosive production of the small particles at their tips.

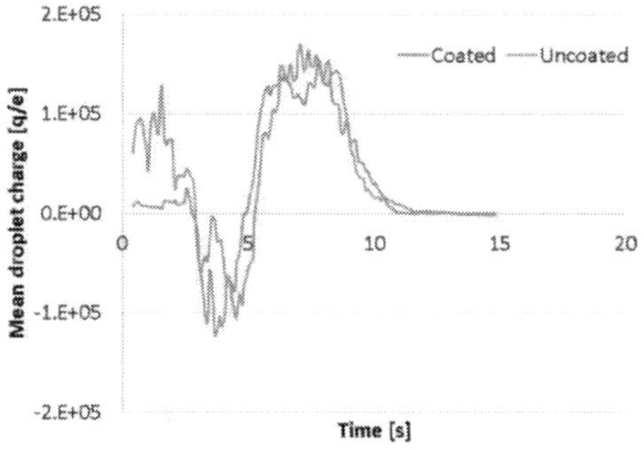

Fig. 7. Small droplet mean elementary charge.

7 Summary

The results presented in this paper are selected examples and they are presented as temporarily results. The complexity of the phenomena is extremely extensive. The formation of the droplet charge involves aerodynamic, chemical, surface and electrostatic forces. For this reason, this paper only presents the current progress in the data processing and the example of the possible result.

We gratefully acknowledge the support of the Grant Agency of the Czech Technical University in Prague, Grant No. SGS 21/154 and National Centre for Energy TN01000007.

References

1. Moore, M.J., Sieverding, C.H.: Two-phase flow in turbines and separators, Theory, Instrumentation, Engineering, Hemisphere Publishing corporation, 1976.
2. Lindsay, W.T., Behavior of impurities in steam turbines, Power engeneering, Vol. 5, pp. 68-72, 1975.
3. Cross, J.A., Electrostatics, principles, problems and applications, Adam Hilger, Bristol,1987.
4. Iribarne, J.V., Mason, B.J., Electrification accompanying the bursting of bubbles in water and dilute aqueous solution, Transaction of Faraday society, Vol. 63, pp.2234-2245, 1967.
5. O. Bartoš, X. Cai, M. Kolovratník, EPJ Web of Conferences. 2014, vol. 67, ISSN 2100-014X
6. Měšťanová and O. Bartoš. An analysis of the liquid films break-up using the computer vision technique. Acta Polytechnica 61(1):253–269, 2021, DOI:10.14311/AP.2021.61.0253

Validated numerical simulation of airflow in child respiratory airways

Miloslav *Belka*[1*], *Milan* Maly[1], *Ondrej* Cejpek[1], *Jakub* Elcner[1], *Frantisek* Lizal[1], *Jan* Jedelsky[1] and *Miroslav* Jicha[1]

[1]Brno University of Technology, Department of thermodynamics and environmental sciences, 616 69 Brno, Czech Republic

Abstract. Both pulmonary airway geometry and breathing pattern evolve from birth to adulthood. These gradual changes significantly influence airflow dynamics and subsequently particle transport and deposition. Regarding the airway structure, the differences between infants and adults are most profound till 5 years of age. Since that age the child airway geometry can be downsized from adult lungs. The objective of the present work was to investigate airflow patterns in child airways. The digital model of male respiratory airways was downsized to dimensionally correspond to a 5–year old child. Airflow simulations with properly validated RANS k–ω turbulence model were performed under steady inhalation boundary conditions. The inspiratory flow rate was 12.5 L/min. The velocity profiles just upstream of the first bifurcation were validated by Laser–Doppler anemometry (LDA) experiments. A replica of the same geometry as the one used in the simulation was 3D printed and part of the trachea was replaced by a glass tube. LDA measurements in several points upstream of the first bifurcation were carried out. The validated flow patterns were compared to the ones obtained in adult lungs.

1. Introduction

According to [1], about 300 million children worldwide breath polluted air that exceeds international guidelines sixfold. The effects of indoor and outdoor pollution cause nearly 600 000 deaths of children under 5 years old every year [1]. Moreover, an early life respiratory illnesses and exposure to pollutants were correlated with increased risk of airway obstructions [2] and asthma [3] in adulthood. Children are more vulnerable to polluted air compared to adults due to their weaker immune system and higher respiratory rates taking in more air relative to their body weight. Also, the narrow airway calibre is more susceptible to obstructions from any inflammatory disease [4].

Therefore, the research of particulate matter transport and deposition in juvenile population is crucial. Although a lot of attention was paid to healthy adults, children received less interest. One of the reasons is the limited access of researchers to pediatric medical images [5]. It is also not entirely possible to extrapolate results from adults to children as there are differences in lung morphology and breathing regimes affecting the subsequent airflow dynamics [6].

Respiratory airways continue to develop after birth and the changes are rapid until the age of roughly 5 years. During this period the airways change not just in dimensions but also in shape, e.g. 4–month–old subject has shorter nasal turbine region and narrower nasopharynx compared to adults [5]. Since age of 5 years the respiratory airways can be down-scaled from adult lungs with reasonable accuracy [7]. The breathing patterns differ significantly from adults as well. Up to the age of 20 months the nasal breathing is preferential. Also, tidal volume is lower and respiratory rate is higher compared to adults [4].

Because of ethical and practical reasons, it is not feasible to study particulate matter transport and deposition in children using *in* vivo methods. However, there has been a significant progress in various *in silico* models, such as the use of computational fluid dynamics to solve equations of fluid flow and particle movement. The geometry used in these simulations can be both idealized or realistic. Tsega [8] analysed numerically airflow differences between infant, child, and adult in idealized symmetric geometries of tracheobronchial airways, in particular G3–G6 and G9–G12. It was found that velocity, pressure and wall shear stress decreases with increasing age indicating that infants are more susceptible to airway damages. Deng et al. [9] investigated deposition in the same parts of the tracheobronchial airways in infant, child and adult. The deposition efficiency decreased with increasing age showing that infants and children are more vulnerable to harmful effects of pollutants. Apart from idealized models, realistic geometries can be based on CT scans. Oakes et al. [6] created subject–specific models of infant, child and adult. They carried out simulations of entire respiratory cycle and highlighted dissimilarities between age–dependent peak pressure drops and velocities. Xi et al. [5] developed several upper respiratory airway models spanning the period from infancy (4–month–old) to childhood (5–year–old). The models were compared and employed in CFD simulations. The influence of age and airway

* Corresponding author: belka@fme.vutbr.cz

abnormalities on pressure drop and regional aerosol deposition were described.

The aforementioned studies focused on either upper airways or tracheobronchial tree. However, it is essential to include both regions to accurately describe the flow features as the flow through upper respiratory and glottis is quite complex and its influence propagates downstream to at least first bifurcation [10]. Moreover, the models need experimental validation. This validation should be subject–specific as inter–subject variability can be significant and such a validation is not optimal [11]. The objective of this work is to compare flow dynamics in realistic models of 5–year–old child and healthy male adult. The replicas of both models were used in experiments to validate obtained flow fields.

2. Material and methods

2.1. Models

A realistic replica of healthy human adult has been developed in the past [12]. The model was employed in properly validated numerical simulations [10]. To study the effect of age on flow dynamics the model was down–scaled to correspond to respiratory airways of 5–years–old child (Fig. 1.).

The down–scaling was executed in two steps. This is reasonable approach as there are differences in growth of different airway regions, e.g. mouth–throat region and conducting airways. Also, there are no studies that would include dimensions of both upper and lower pediatric airways. Considering inter–subject variability, it is not optimal to group data from various studies together. Moreover, different teams use different methologies and publish dimensions of different regions. Therefore, the tracheobronchial tree was down–sized with a scaling factor of 0.54 based on the trachea length [13]. After that, the upper respiratory airways were scaled down based on the length of the nasal cavity from nostrils to the end of septum [14]. The scaling factor of 0.8 was used which was slightly higher than average. The reason was that the head of the adult model was somewhat smaller than average and down–scaling of it based on the average dimensions would result in unrealistically small head for a 5–year–old child. The number of bronchial airways was decreased because of its small dimensions and 3D printing restrictions.

The child model was modified for the sake of experiments. A 2–mm–thick envelope was added around the model to permit a 3D printing of the model. A part of the trachea just upstream of the first bifurcation was substituted by a 1–cm–long glass tube (Fig. 2.). The cross section of this trachea part was approximately circular and therefore, the substitution with glass tube had minor effect on geometry. The inner diameter of the tube was carefully chosen to match the inner diameter of the trachea.

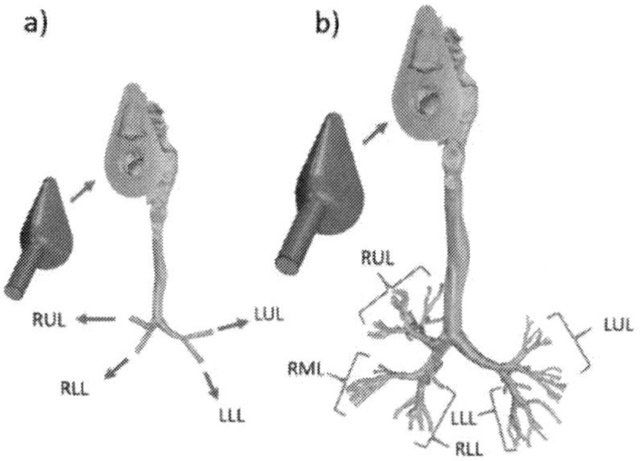

Fig. 1. Child (a) and adult (b) digital model of respiratory airways

A mask was created around the head to facilitate prescription of boundary conditions in the simulations and to permit hose connection during the experiments.

2.2. Numerical simulations

The digital geometries were imported into a commercial solver Star CCM+ version 2019.2 (Siemens PLM, Texas, USA). A computational grid consisting of polyhedral cells together with prism cells at the walls was generated. The first cell thickness at the wall was determined to satisfy the $y^+ < 1$ condition. Meshes containing $7.32\cdot10^6$ and $5.31\cdot10^6$ cells were selected for the child and adult geometry, respectively, based on a grid independency study.

A RANS k–ω SST turbulence model with low–Re corrections was used as a good trade–off between accuracy and computational costs. It proved to be able to capture mean–flow dynamics in the respiratory airway models reasonably well [15]. Moreover, this low–Reynolds–number model accurately predicted flows from low–level to high–level breathing spanning the range of Reynolds number from 500 to 10^4 [16]. The pressure–velocity coupling was done using the SIMPLE algorithm. Linearized equations describing the air flow were solved according to the Gauss–Seidel iteration method.

The simulations were performed at steady state inhalation representing sedentary conditions. The total flow rate through the child model was 12.5 L/min based on the information in [17]. The flow division between different lung lobes was based on [18] and can be found in Table 1. The stationary inhalation of 15,1 L/min was chosen for the adult case and flow division into different outlets were selected according to [12]. It is worth noting that there is a constriction in left bronchi resulting in lower flow rate into the left lung compared to the average right/left lung flow ratio.

The pressure boundary condition was prescribed at the mask inlet and outward pointing velocities were set at the outlets in conformity with the experiments. The no–slip condition was assumed at the walls. The values of turbulent intensities at the inlets were set to 1 % and the value of ratio of turbulent viscosity to laminar viscosity was 10 %.

Table 1. Flow rates through respiratory airway models

	child airways	adult airways
	flow rate (L/min)	
LUL	3,16	1,7
LLL	2,58	2,8
RUL	2,61	3,4
RLL	4,15	3,4
RML		3,8
total	12,5	15,1

2.3. Experimental measurements

LDA measurements (Fig. 2.) were carried out to validate CFD simulations. Velocities of a particle–laden flow were measured in several points in the trachea. A part of trachea was therefore substituted by a glass tube to provide optical access. Aerosol particles were generated in a condensational monodisperse aerosol generator TSI 3475 (TSI Inc., Shoreview,MN, USA) utilizing heterogeneous condensation of Di–Ethyl–Hexyl–Sebacate (DEHS) vapour on NaCL particles. The resulting particle size was measured using process aerosol monitor TSI 3375 (TSI Inc., Shoreview,MN, USA). The particle size was in a range of 2,5–3 μm to satisfy a Stokes number (Stk) < 0.1 condition to assure that particles closely follow the streamlines.

The flow through the model was driven by a vacuum pump. The flow division was set up by a flowmeter with regulating valve downstream of each outlet. A filter was inserted just upstream of each flowmeter to collect all the DEHS particles that penetrate the model.

2D LDA FlowExplorer (Dantec Dynamics A/S, Skovlunde, DK) was employed to measure point–wise velocities. The measurements were performed in several points in a coronal plane 10 mm above the first bifurcation. The LDA system contained two built-in diode pumped solid state lasers with wavelengths of 660 nm and 785 nm. The beam from each laser was split into two parallel beams with a separation distance of 60 mm and power of 30 mW. The beam crossing formed a measuring volume with dimensions of 0.1 × 0.1 × 0.8 mm. The wavelengths of 660 and 785 nm were utilized to measure the axial and radial velocity, respectively. The LDA system was used in a backscatter mode and had a lens focal length of 150 mm. The LDA signals were processed using a Dantec P80 processor and the data acquisition was controlled using BSA Flow software v5.2.

3. Results

3.1. CFD and experiments comparison

The obtained axial velocities by both experiments and simulations were compared (Fig. 3.). Both the methods agreed reasonably well. The experimental velocity towards the left wall (positive distance from the axis) decreased more rapidly and are less symmetrical compared to simulations. The size of the measurement volume also prohibited to measure velocities closer than 1 mm to the wall.

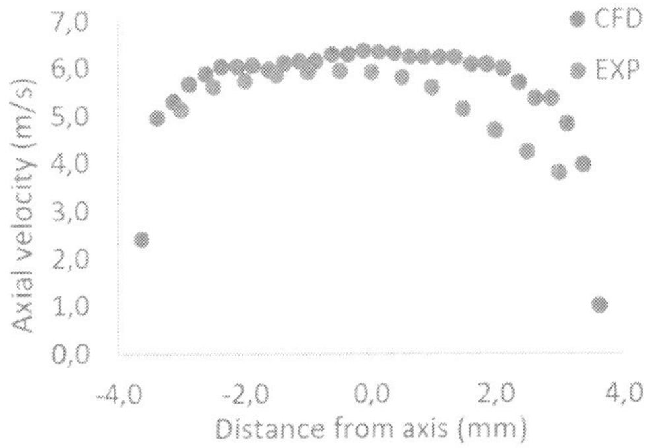

Fig. 3. Comparison of axial velocities obtained by CFD and LDA measurements. The velocities studied in coronal plane of the child airway model. Positive distance from axis corresponds to left side of the model.

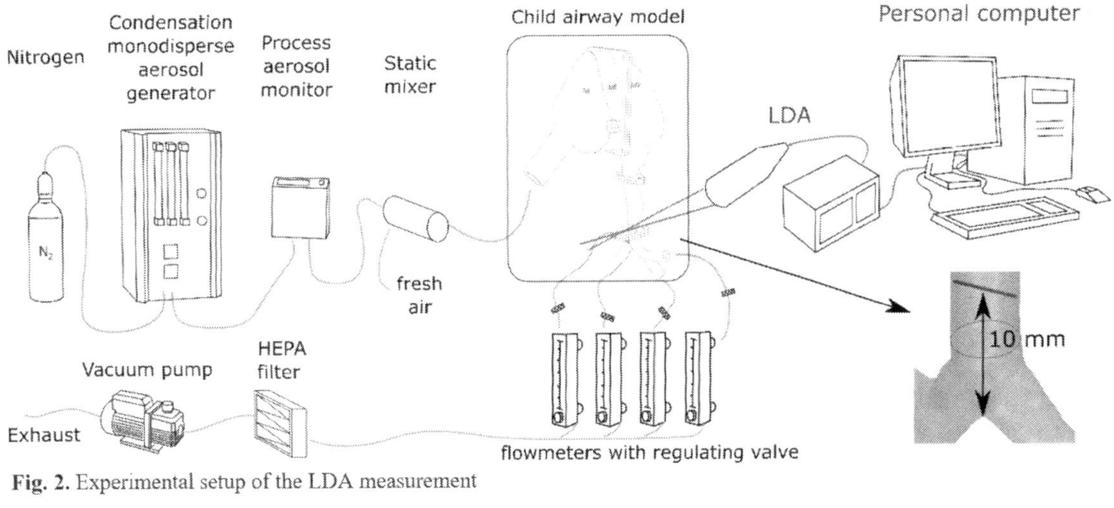

Fig. 2. Experimental setup of the LDA measurement

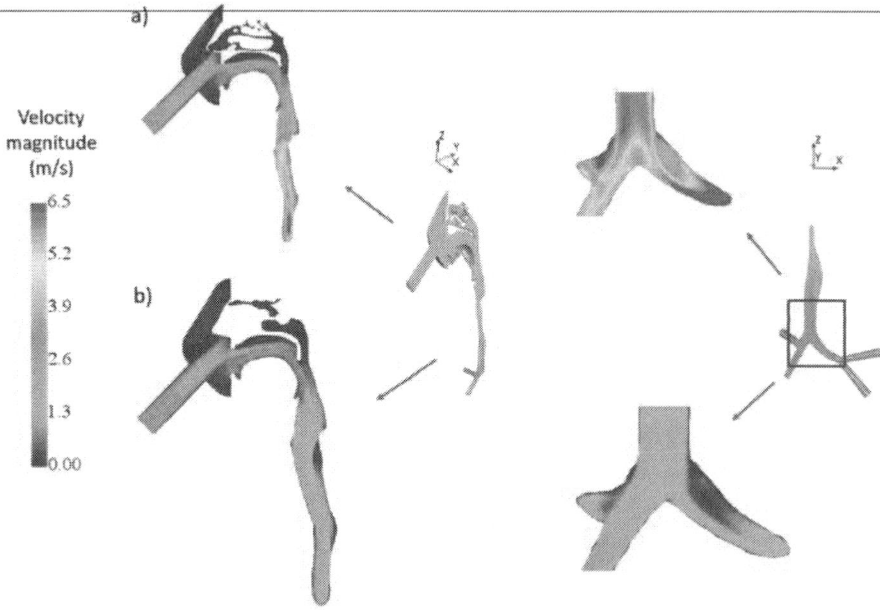

Fig. 4. Velocity magnitudes in sagittal and coronal plane of a) child and b) adult model

3.2. Child and adult airway flow dynamics comparison

Results of numerical simulations, in particular velocity magnitudes and wall shear stresses, are discussed below. Comparison of velocity magnitudes in sagittal and coronal plane are depicted in Fig. 4. The planes were created at the centreline of the first bifurcation. The Reynolds numbers ($\rho UD/\mu$) at the trachea inlet were 1913 and 1210 for the child and adult model, respectively, indicating the flow is in laminar regime. The velocity trends were similar in both the models. Majority of the air flowed through the mouth due to high pressure loss of the nasal cavity. The air in the trachea was skewed towards the front trachea wall because of the glottis geometry. Zones of flow reversal can be seen in daughter branches of the first bifurcation (right side of Fig. 4.)

The velocity magnitudes differed in the two models. The velocities were higher in the case of the child model compared to the adult case. The maximum velocities were approximately three times higher in the child model. This was caused by a significant increase in airway calibers and a small increase of inhalation flow rate comparing the child case to the adult one. The inhalation flow rate is affected by both respiratory rate and tidal volume.

The wall shear stress expresses a force exerted on the airway wall by a fluid motion. It is proportional to the velocity gradient next to the wall [19]. The wall shear stress distribution at the model walls is depicted in Fig. 5. It can be seen that the wall shear stress was high at the bifurcations and glottis area. The highest wall shear stress was in the left bronchi bifurcation. It is worth noting that the model encompasses a constriction in the left bronchi that generates high velocities. Comparing the models, the wall shear stress was higher in the child model. The

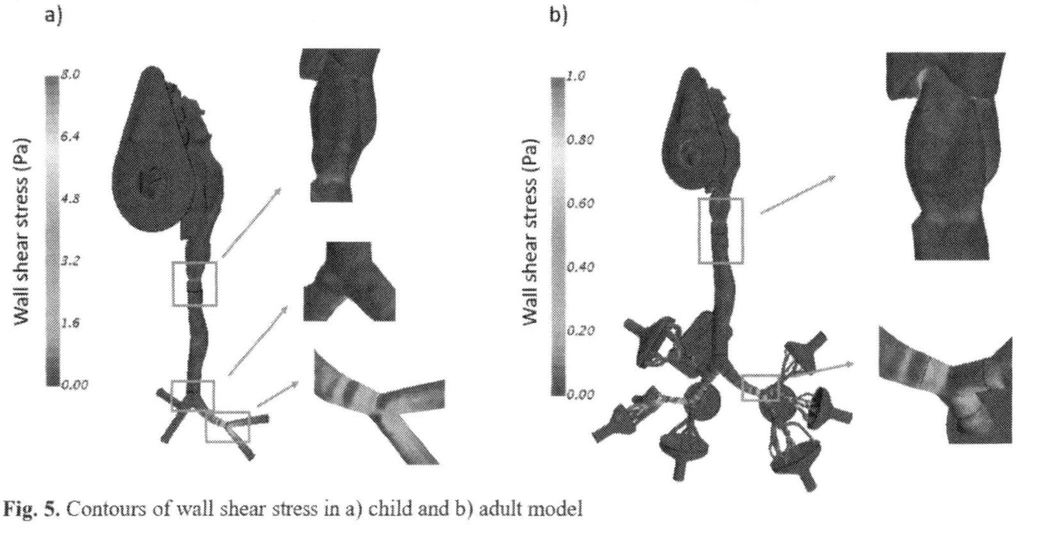

Fig. 5. Contours of wall shear stress in a) child and b) adult model

maximum values were 7.4 and 0.96 in the child and adult model, respectively. The strong forces exerted on the child airway walls by the fluid flow indicate their higher susceptibility to wall damages compared to adults.

4. Conclusions

The model of child airways was created by downscaling an adult model. The child model included upper respiratory airways and conducting airways down to 2nd generation. The model was employed in properly validated CFD simulations.

The LDA point–wise velocity measurements in the lower part of the trachea were carried out. The measured axial velocities were compared to the ones obtained from simulations. Reasonably good agreement was found.

The results of numerical simulations in the child model were compared to the ones from adult model. The velocity trends were similar, but the velocity magnitudes were higher in the child model. This was caused by a significant increase of airway calibres compared to small increase of inhalation flow rate from childhood to adulthood. Regions of increased velocity could result in enhanced particle deposition due to inertia and thus, higher risk for children inhaling polluted air.

The forces exerted on the airway walls by the flowing fluid, i.e. wall shear stresses, were studied in both the models. The zones of high wall shear stress corresponded to the zones of high velocities, e.g. glottis area or bifurcations. The maximum values of wall shear stress were 7.4 and 0.96 in the child and adult airways, respectively. The roughly eight-fold increase in the child airways compared to the adult ones indicate higher child vulnerability to wall damages.

Acknowledgement. This work was supported by the Czech Science Foundation under the grant GA20-27653S.

References

1 UNICEF, *Clear the air for children.* (United Nations Children´s Fund, New York, USA, 2016).
2 D. A. Stern, W. J. Morgan, A. L. Wright, S. Guerra, and F. D. Martinez, Lancet **370** (9589), 758 (2007).
3 N. A. Clark, P. A. Demers, C. J. Karr, M. Koehoorn, C. Lencar, L. Tamburic, and M. Brauer, Environmental health perspectives **118** (2), 284 (2010).
4 I. Amirav and M. T. Newhouse, Expert Rev Respir Med **2** (5), 597 (2008).
5 J. X. Xi, X. H. Si, Y. Zhou, J. Kim, and A. Berlinski, Respiratory care **59** (2), 263 (2014).
6 J. M. Oakes, S. C. Roth, and S. C. Shadden, Annals of Biomedical Engineering **46** (3), 498 (2018).
7 P. Das, E. Nof, I. Amirav, S. C. Kassinos, and J. Sznitman, Plos One **13** (11) (2018).
8 E. G. Tsega, Comput Math Method M **2018** (2018).
9 Q. H. Deng, C. Y. Ou, J. Chen, and Y. G. Xiang, Sci Total Environ **612**, 339 (2018).
10 J. Elcner, F. Lizal, J. Jedelsky, M. Jicha, and M. Chovancova, Biomech Model Mechan **15** (2), 447 (2016).
11 C. Darquenne, W. J. Lamm, J. M. Fine, R. A. Corley, and R. W. Glenny, Journal of Aerosol Science **99**, 27 (2016).
12 F. Lizal, J. Elcner, P. K. Hopke, J. Jedelsky, and M. Jicha, Proceedings of the Institution of Mechanical Engineers Part H-Journal of Engineering in Medicine **226** (H3), 197 (2012).
13 M. H. Dave, M. Kemper, A. R. Schmidt, C. P. Both, and M. Weiss, Pediatr Anesth **29** (8), 782 (2019).
14 L. Golshahi, M. L. Noga, R. B. Thompson, and W. H. Finlay, Journal of Aerosol Science **42** (7), 474 (2011).
15 P. Koullapis, S. C. Kassinos, J. Muela, C. Perez-Segarra, J. Rigola, O. Lehmkuhl, Y. Cui, M. Sommerfeld, J. Elcner, M. Jicha, I. Saveljic, N. Filipovic, F. Lizal, and L. Nicolaou, European Journal of Pharmaceutical Sciences (2017).
16 C. Kleinstreuer and Z. Zhang, International Journal of Multiphase Flow **29** (2), 271 (2003).
17 M. Roy and C. Courtay, Radiation Protection Dosimetry **35** (3), 179 (1991).
18 N. Jahani, S. Choi, J. Choi, K. Iyer, E. A. Hoffman, and C. L. Lin, J Appl Physiol (1985) **119** (10), 1064 (2015).
19 B. Sul, A. Wallqvist, M. J. Morris, J. Reifman, and V. Rakesh, Comput Biol Med **52**, 130 (2014).

Dual-capillary dilatometer for density measurements of supercooled water

Aleš Blahut[1*], *Jiří* Hykl[1], *Pavel* Peukert[1], and *Jan* Hrubý[1]

[1]Institute of Thermomechanics of the Czech Academy of Sciences, Dolejškova 1402, 182 00 Prague 8, Czech Republic

Abstract. An apparatus tailored to accurate density measurements of supercooled water, i.e. liquid water in a metastable state below the freezing point temperature, was recently developed at the Institute of Thermomechanics of the Czech Academy of Sciences. The apparatus, dual-capillary dilatometer, is described, together with the measurement procedure and the evaluation methodology. The primary result of the dual-capillary method are relative densities with respect to the density at a reference temperature and given pressure. In order to calculate absolute densities, densities at the reference temperature as a function of pressure are needed. For calculation of such pressure dependence of density, so called thermodynamic integration involving literature thermodynamic data and our experimental results is used. The dual-capillary dilatometer was successfully employed in density measurements of ordinary water, heavy water and seawater. The data in the temperature range from 238.15 to 303.15 K at pressures from atmospheric up to 200 MPa are presented and compared with respective IAPWS formulations of thermodynamic properties. The data for ordinary water are also compared with an accurate equation of state for supercooled water of Holten et al. (2014) revealing good mutual agreement. The expanded uncertainty of relative densities acquired by means of the dual-capillary method is estimated to be lower than 50 ppm.

1 Introduction

Recently, an apparatus for accurate density measurements of supercooled water, i.e. liquid water in a metastable state below the freezing point temperature, was developed at the Institute of Thermomechanics of the Czech Academy of Sciences [1, 2]. The primary aim of the apparatus was to provide reliable density data with low uncertainty for supercooled water not only at atmospheric pressure but also at high pressures and contribute thus to the ongoing research of metastable water.

Usually, the measurement of thermophysical properties of water in the supercooled state is a compromise between accuracy and a level of supercooling. The persistence of liquid state below the freezing point temperature has a probabilistic character. In order to trigger freezing, formation of a nucleation center is needed. The purer is a liquid, the lower is its volume, and the smaller is the departure from the freezing point temperature, the creation of such nucleation center is less probable. In other words, more time is available for measurement before the liquid freezes.

The small volume of liquid for various measurement techniques is usually achieved in capillaries with small inner diameter [3, 4] or in emulsions, e.g., where water is dispersed in the form of small droplets in a hydrophobic solvent [5, 6]. However, decreasing the volume pronounces surface effects which distort the recognition of bulk properties. Some studies were discussing a possible influence of capillary inner diameter on results of density measurements with supercooled water [3, 7]. It seems that water enclosed in a capillary with diameter of around 300 μm represents well bulk water, while keeping considerable supercooling capabilities at the same time [3]. Measurements carried out in emulsions approach the homogeneous nucleation limit but they are less accurate than measurements in capillaries because of uncertainty connected with subtracting emulsion matrix properties.

Highlights of the developed apparatus, dual-capillary dilatometer, are the high accuracy exceeding previous measurements in supercooled region and the possibility of measurements at high pressures. The estimated relative expanded uncertainty of measurements is lower than $5\cdot10^{-5}$. On the other hand, compared to literature data, e.g. [3, 5], only moderate level of superccoling was achieved (approximately 20 K below freezing point temperature).

Even though the dual-capillary dilatometer was specifically designed for ordinary water, it can be used for wider variety of substances. Straightforward is the application for water isotopologues or aqueous solutions. Up to this date the apparatus was successfully employed within three measurement campaigns including ordinary water (H_2O), heavy water (D_2O) [8] and a sample of IAPSO seawater standard. The results for D_2O together with preliminary results for H_2O and seawater standard are presented in this paper.

* Corresponding author: blahut@it.cas.cz

2 Measurement principle

Measurement method is based on measuring height changes of a liquid column situated inside a capillary. The change of the column height is induced by switching temperature of liquid from a reference temperature T_{ref} to a temperature of interest T. Due to practical reasons, the change of temperature is realized only for a part of the capillaries situated in a heat exchanger, as schematically depicted on the right hand side of Fig. 1. The left hand side of the figure illustrates a zone held at a constant reference temperature. In reality, the zone corresponds to an optical cell which enables observation of the column heights through high-pressure sapphire windows. In order to eliminate an effect of the transition zone between the optical cell and the heat exchanger, two capillaries are used: a long one (L) and a short one (S). Mercury is used in both capillaries as a movable plug separating the investigated liquid from a pressurizing gas. At the same time, mercury serves also as a contrast substance for better identification of the position of liquid menisci. Two measurement states with different menisci positions are distinguished in Fig. 1 by black and gray colors. Δa represents the length difference of two capillaries, Δz_S and Δz_L stand for menisci position differences in respective capillaries.

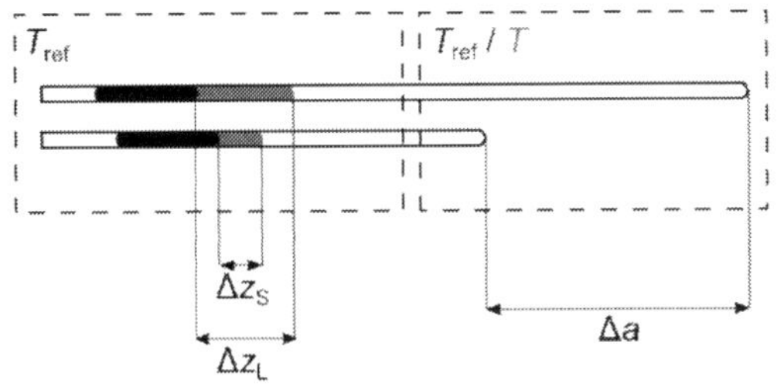

Fig. 1. Illustration of the dual-capillary method principle (rotated 90 ° counterclockwise compared to the position of capillaries in the apparatus).

The primary result of the measurement using dual-capillary method is the relative density η, i.e. the ratio of density at the temperature of interest T and density at the reference temperature T_{ref} at a given pressure p:

$$\eta(T,p) = \frac{\rho(T,p)}{\rho(T_{ref},p)}. \quad (1)$$

Based on the measurement of menisci position differences, Δz_S, Δz_L, and fixed length difference of capillaries, Δa, the relative density can be calculated in first approximation with the idealized equation

$$\eta = 1 - \frac{\Delta z_L - \Delta z_S}{\Delta a}. \quad (2)$$

It should be noted that the differences Δz_S and Δz_L are negative in case of liquid contraction.

In order to improve accuracy of the measurement description, several correction factors are considered [8]. The corrective terms stand for dilation of capillary material (fused silica), dilation of a reference glass scale, variation of cross-section area of capillaries along their length [2], or hydrostatic pressure correction. Variation of temperature and pressure from nominal values is also considered.

3 Description of apparatus and measurement procedure

Since a design of apparatus was introduced in detail before [1, 8], here only a brief description is given.

Long and short capillaries are manufactured of fused silica, which offers a very low coefficient of thermal expansion, compared, e.g., to the borosilicate glass or stainless steel commonly used in experiments with supercooled water. Prior to the measurements both capillaries are sealed tightly at the one end and filled with liquid using a special device [1]. During the filling procedure a sample is degassed by controlled stepwise evacuation while being stirred with a magnetic bar. Consequently, the degassed sample is sucked into evacuated capillary. Finally, a small mercury column is introduced in the capillary from the open end in order to prevent contact of liquid with a pressurization gas during the measurement.

For the measurement, capillaries are fixed vertically inside a high-pressure containment which consists of thermostated high-pressure optical cell and two tube heat exchangers. The pressurization is carried out with a help of inert gas (nitrogen) whereas the capillaries are pressurized both from inner and outer side thanks to the movable mercury plug. The actual state of apparatus enables density measurements up to 200 MPa.

Locations of menisci in capillaries are observed through sapphire windows using high-resolution CCD camera (piA2400-12gm, Basler) equipped with telecentric lens (TECHSPEC®, Edmund Optics). The resolution of camera is calibrated by means of in-situ installed contact reticle (Edmund Optics) with declared line to line accuracy 2 µm. Acquired images are semi-automatically evaluated with an image processing code yielding positions of menisci in capillaries z_L and z_S.

The measurements are carried out at isobaric conditions. In order to obtain relative density for a certain temperature, two measurement states have to be realized. In reference state the entire vertical profile of capillaries is maintained at the reference temperature T_{ref}, while at measuring state only the bottom heat exchanger is switched to the temperature of interest T. The main reason of switching procedure is to avoid freezing when temperature of interest is within the supercooled region. After a certain time needed for stabilization, camera images are acquired and temperature switched quickly back to T_{ref}. Usually three images are taken per each measurement state.

Temperatures at different locations of apparatus are monitored with Pt100 sensors connected to thermometry bridge (CTR6500, ASL). Pressure is measured using accurate transducers Digiquiartz® 415K-101 or 430K-101 (Paroscientific, Inc.) depending on pressure range.

4 Thermodynamic integration

As mentioned before, the dual capillary dilatometer provides relative densities. In order to convert relative densities into absolute ones, density at the reference temperature as a function of temperature is needed. E.g., available reference equations of states can be used for this purpose. However, within our studies with water, heavy water and seawater, an alternate approach called thermodynamic integration was used. The approach benefits from availability of highly accurate sound speed data for selected liquids [9-12]. In this case the thermodynamic integration may yield more accurate densities at high pressures compared to contemporary reference equations of state.

For the determination of reference densities, the following pair of differential equations needs to be solved.

$$\left(\frac{\partial \rho}{\partial p}\right)_T = \frac{1}{w^2} + T\frac{\alpha_p^2}{c_p}, \tag{3}$$

$$\left(\frac{\partial c_p}{\partial p}\right)_T = -\frac{T}{\rho}\beta_p\,. \tag{4}$$

In these equations, w is speed of sound, c_p is specific isobaric heat capacity, α_p denotes isobaric expansivity, and β_p is the second order thermal expansion coefficient. In favorable cases, accurate values of w are available in the literature at the conditions of interest, i.e. at the reference temperature and simultaneously as a function of pressure. Heat capacity and density at the reference temperature and atmospheric pressure are needed as initial conditions for the integration. Concerning liquids involved in this work, respective reference equations recommended by IAPWS [13-15] can be sufficiently used for calculation of these single point data. Particular values of isobaric expansivity α_p or the second order thermal expansion coefficient β_p can be calculated from experimental relative densities as follows

$$\alpha_p = \frac{1}{v}\left(\frac{\partial v}{\partial T}\right)_p = -\frac{1}{\eta}\left(\frac{\partial \eta}{\partial T}\right)_p, \tag{5}$$

$$\beta_p = \frac{1}{v}\left(\frac{\partial^2 v}{\partial T^2}\right)_p = 2\alpha_p^2 - \frac{1}{\eta}\left(\frac{\partial^2 \eta}{\partial T^2}\right)_p. \tag{6}$$

There are many options how to solve the set of differential equations Eq. 3 and Eq. 4. In our studies numerical integration using fourth order Runge-Kutta method was applied.

5 Results and discussion

Dual capillary dilatometer was used for density measurements with ordinary water, heavy water and IAPSO standard seawater sample. Relative expanded uncertainties ($k = 2$) of primary experimental results, i.e. relative densities, are estimated to be lower than $5\cdot10^{-5}$.

5.1 Specification of liquid samples

Ordinary water was prepared using two step purification system consisting of reverse osmosis (Rowapur) and purification unit (Neptune Analytical). According to the unit specification, outlet water reaches resistivity 18.2 MΩ·cm, is free of particles larger than 0.2 μm and has an organic carbon content below $5\cdot10^{-9}$.

Heavy water (D_2O) sample was purchased from Sigma-Aldrich. According to the certificate of analysis provided by the supplier for particular batch, deuterium fraction among hydrogen isotopes was declared to be higher than 0.9999. Heavy water sample was used without any purification, except for degassing during the capillary filling procedure.

IAPSO Standard Seawater (OSIL, batch 161) with certified practical salinity 34.995 was used for density measurements. However, during the controlled degassing procedure the salinity gently increased due to evaporation of pure water. Based on a VLE calculation, the practical salinity of seawater involved in measurements was estimated to be 35.016 which corresponds to the absolute salinity 35.181 g kg^{-1} [16].

5.2 Absolute density results

Density results for all investigated liquids are shown in graphical form in figures Fig. 2 to Fig. 4. Resulting densities were obtained by combining experimental relative densities and reference density values from thermodynamic integration as follows

$$\rho(T,p) = \rho_{\mathrm{TD}}(p)\,\eta(T,p)\,, \tag{7}$$

where ρ_{TD} denotes density values obtained by thermodynamic integration at a reference temperature T_{ref}. For ordinary water and seawater T_{ref} = 293.15 K, whereas heavy water experiments were carried out with different reference temperature 298.15 K.

In all these figures open circles mark nominal experimental temperatures and gray dashed lines represent melting curves of associated liquids.

As can be seen, the measurements carried out with water covers the broadest pressure range, from atmospheric pressure up to 200 MPa. Also for ordinary water most efforts were made to achieve the deepest supercooling possible. Because these efforts usually lead to destruction of laboriously prepared capillaries, decreasing the temperature to a limit was done only for several isobars. The deepest supercooling, 22 K below the freezing point, was achieved at 100 MPa. In case of usual measurements, the range of supercooling varied roughly between 10 and 15 K.

The general feature for water, heavy water and seawater is an existence of density maximum and the fact that this maximum vanishes with increasing pressure, particularly temperature of maximum density moves to lower temperatures as discussed in section 5.4.

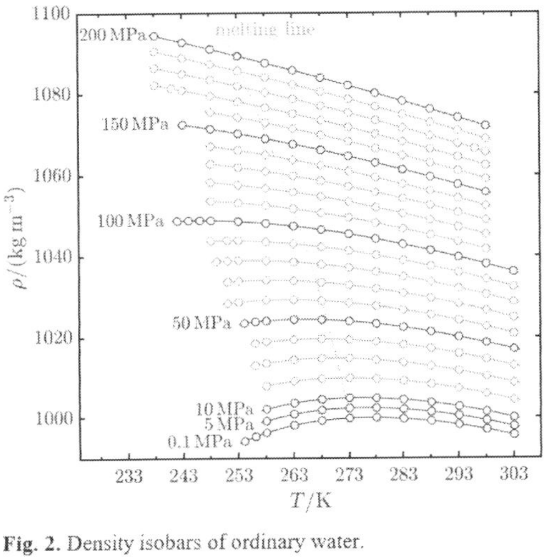

Fig. 2. Density isobars of ordinary water.

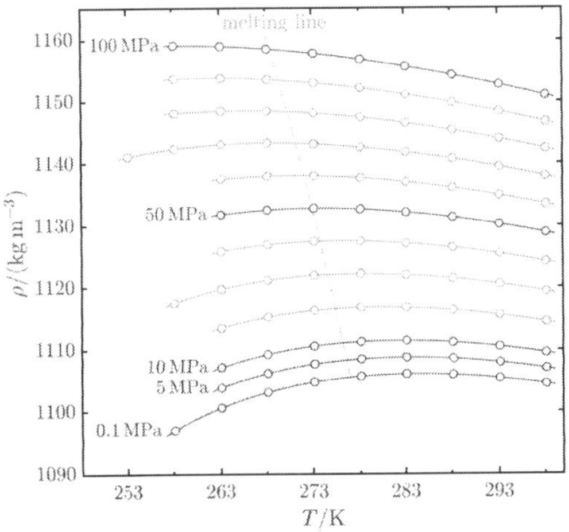

Fig. 3. Density isobars of heavy water.

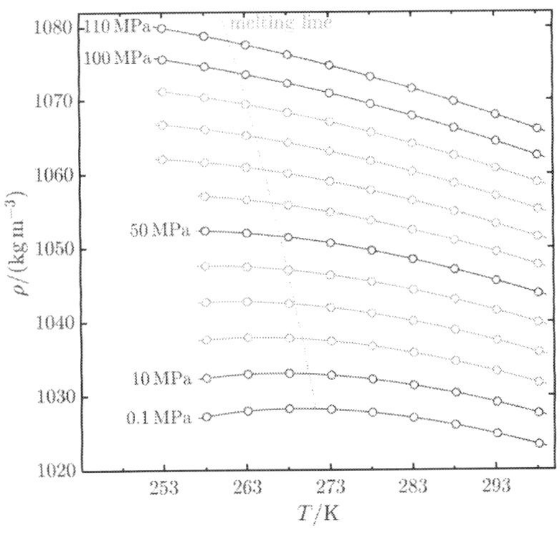

Fig. 4. Density isobars of standard seawater (absolute salinity 35.16504 g·kg^{-1}).

5.3 Comparison with reference equations of state

In this section the new experimental densities are compared with relevant reference equations of state (EoS). For pure water, results are compared with two equations, namely IAPWS-95 [13] and EoS of Holten et al. [17] from 2014, which is specifically designed to describe supercooled region. Results of heavy water are compared to recent EoS of Herrig et al. [14] and seawater densities are confronted with TEOS-10 equation [18], which is principally based on EoS developed by Feistel [15].

In Fig. 5 and Fig. 6 experimental densities are compared to IAPWS-95 formulation and Holten EoS, respectively. The agreement with Holten equation is very good, up to 120 MPa Holten equation even agrees with our densities within relative expanded uncertainty of our data (~$5\cdot 10^{-5}$). The mutual agreement with IAPWS-95 in supercooled region is somehow poorer, the deviations are more pronounced with increasing pressure.

What is not visible due to limited resolution is the fact, that Holten EoS also agree better than IAPWS-95 with our data at 273.15 K and above at high pressures. In particular, differences from Holten EoS for all experimental pressures are within $\pm 20\cdot 10^{-6}$, whereas in case of IAPWS-95 the differences range within $\pm 60\cdot 10^{-6}$.

It should be noted that at atmospheric pressure above melting point the Holten equation provides the same densities as IAPWS-95, but differences between these equations are increasing with pressure. At high pressures in the limited temperature range of our studies Holten equation may be considered more reliable than IAPWS-95, e.g., because more accurate speed of sound data was used within equation development and quality of speed of sound data influences the quality of density derivative with respect to pressure (as can be inferred from Eq. 3 in section 4).

The excellent agreement with Holten equation above melting point curve proves to certain extent reliability of our results.

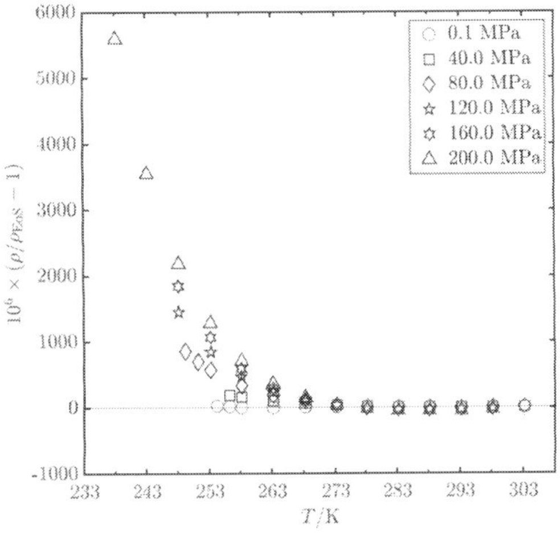

Fig. 5. Relative deviations of H_2O densities from this work from densities calculated with IAPWS-95 formulation [13].

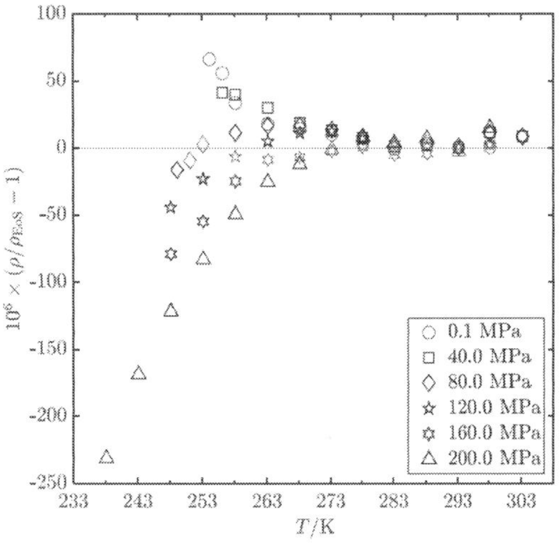

Fig. 6. Relative deviations of H_2O densities from this work from densities calculated wtih Holten EoS [17] for several isobars.

Recently developed equation for heavy water [14] agrees quite well with our data as can be seen in Fig. 7. However, it should be noted that our preliminary results were used for development of this equation. Comparing our density data with the equation in detail, there is an obvious difference in slope of density vs temperature from 273.15 K to 298.15 K. This deviation seems to be systematic for all experimental pressures. As discussed previously [8], the deviation is probably an artifact of the equation. The slope of density vs temperature of our data at the atmospheric pressure agree considerably better with relevant literature datasets of the highest metrological quality rather than with the EoS.

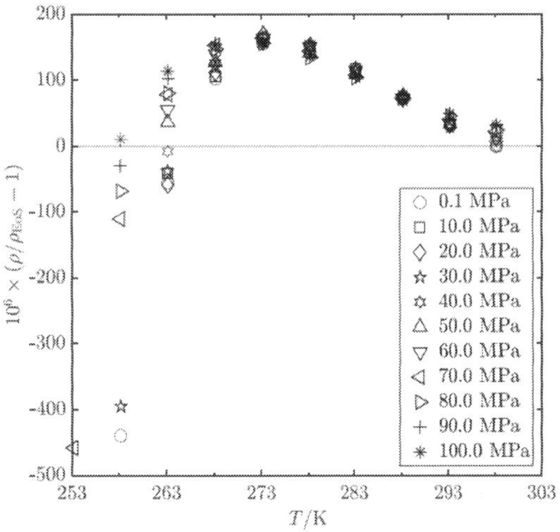

Fig. 7. Relative deviations of D_2O densities from this work from densities calculated wtih equation of Herrig et al.[14].

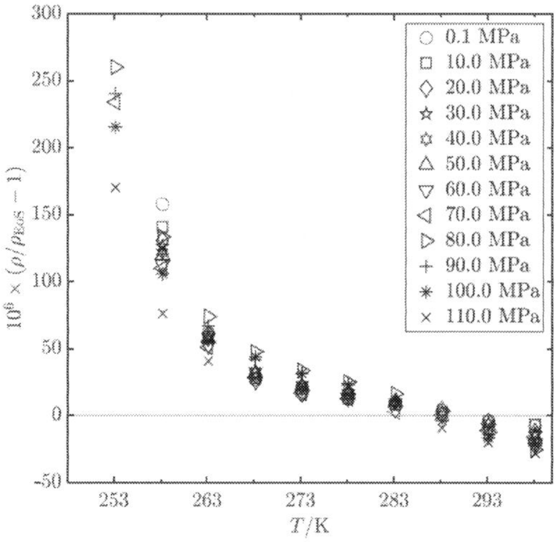

Fig. 8. Relative deviations of seawater densities from this work from densities calculated with TEOS-10 equation [15, 18].

Fig. 8 reveals systematic deviation of our data from reference equation for seawater, TEOS-10. Above melting point temperature, the slope of density vs temperature is slightly different, which is reflected also in the deviation of isobaric expansivities α_p. The deviation in the range of (1 to 2)$\cdot 10^{-6}$ K^{-1} is comparable to very small volume thermal expansion coefficient of fused silica. In the stable region the mutual agreement is within experimental expanded uncertainty of our data. In the supercooled region deviations are increasing with decreasing temperature showing thus limited extrapolation capability of TEOS-10.

5.4 Temperatures of maximum density

Water at atmospheric pressure exhibits a density maximum which is located close to 4 °C. The existence of such a density maximum is considered anomalous among all various liquids. However, the same anomalous behavior could be expected for liquids similar to pure water, e.g., dilute aqueous solutions or water isotopologues (water molecules that differ only in their isotopic composition).

As can be seen in figures Fig. 2 to Fig. 4, the density maxima are moving to lower temperatures (and further below the melting point temperature) with increasing pressure. Therefore, measurements in the supercooled region broads the observation range of density maximum phenomenon. E.g., for pure water the density maximum at 30 MPa is already located below the freezing point. Level of supercooling attained in the experiments for pure water enabled to identify density maxima for isobars up to 100 MPa.

In Fig 9. temperatures of maximum density as a function of pressure are compared for water, heavy water and seawater are shown. Temperatures correspond to density maxima in figures Fig. 2 to Fig. 4. However, for evaluating the temperature of maximum density, only relative density data are sufficient. Temperature of maximum density is then calculated from condition of zero first derivative of relative density. Practically, the particular temperature dependence of relative density may be given by a curve smoothing (or interpolating) experimental data.

Fig. 9 shows interesting feature of three investigated aqueous liquids, i.e. similar evolution of temperatures of density maximum with pressure. The individual pressure dependencies are roughly shifted by just a constant.

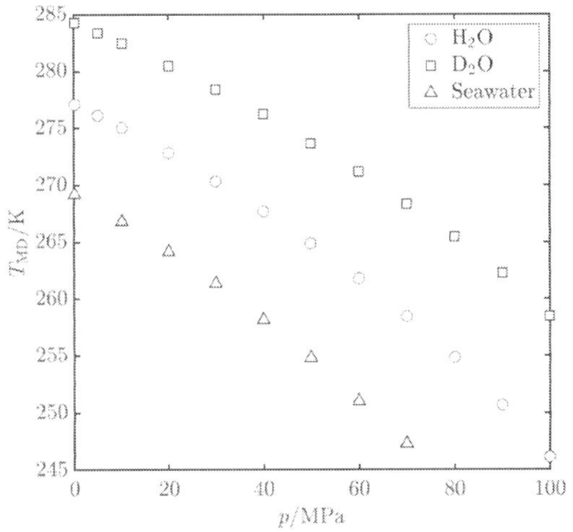

Fig. 9. Comparison of temperatures of maximum density as a function of pressure for all three investigated liquids. Points corresponding to seawater above 40 MPa were calculated from relative densities extrapolated to lower temperatures.

6 Conclusions

As seen the in comparison section 5.3, our results agree well with current reference equations in stable state, but agreement is somehow poorer in the supercooled region. This reflects the fact that validity range of these reference equations (except for Holten equation) is usually from one side limited by the melting curve. The deviation in supercooled region could be thus explained by limited extrapolation capabilities of these equations. At the same time considerable agreement above melting point line validates our experimental method and shows the potential of combining relative density measurement using dual-capillary dilatometer and tool of thermodynamic integration.

The data presented within this contribution may be involved in development of new reference equations of state in order to improve their extrapolation capabilities or improve a description of thermodynamic properties along melting line.

The presented similar course of density maxima for water, heavy water and standard seawater in section 5.4 may be a basis for some future efforts in sense of theorem of corresponding states for aqueous solutions. The reference state would not be a critical point but a curve of maximum density.

Acknowledgement: The study was supported by the Czech Science Foundation grant no. GA19-05696S and the institutional support RVO: 61388998. Dr. Aleš Blahut gratefully acknowledges funding within "Support for International Mobility of Researchers of the Institute of Thermomechanics, Czech Academy of Sciences, part II", no. CZ.02.2.69/0.0/0.0/18_053/0017555 of the Ministry of Education, Youth and Sports of the Czech Republic funded from the European Structure and Investment Funds (ESIF).

References

1. J. Hrubý, J. Hykl, P. Peukert, B. Šmíd, EPJ Web of Conferences 25, 01026 (2012)
2. P. Peukert , M. Duška , J. Hykl, P. Sladký , Z. Nikl, J. Hrubý, EPJ Web of Conferences 92, 02067 (2015)
3. D. E. Hare, C. M. Sorensen, J. Chem. Phys. **87**, 4840 (1987)
4. H. Kanno, C. A. Angell, J. Chem . Phys. **70**, 4008 (1979)
5. O. Mishima, J. Chem. Phys. **133**, 144503 (2010)
6. D. H. Rasmussen, A. P. MacKenzie, J. Chem. Phys. **59**, 5003 (1973)
7. J. V. Leyendekkers, R. J. Hunter, J. Chem. Phys. **82**, 1447 (1985)
8. A. Blahut, J. Hykl, P. Peukert, V. Vinš, J. Hrubý, J. Chem. Phys. **151**, 034505 (2019)
9. K. Fujii, *12th Symposium on Thermophysical Properties (Boulder, Colorado, USA, 1994)*
10. C.-W. Lin, J. P. M.Trusler, J. Chem. Phys. 136, 094511 (2012)
11. K. Meier, A. ElHawary, 17th International Conference on the Properties of Water and Steam (Prague, Czech Republic, 2018)
12. F. Fehres, S. Rudtsch, IAPWSWG-TPWS Meeting (online presentation, 2021)
13. IAPWSR6-95 (2018), Revised Release on the IAPWS Formulation 1995 for the Thermodynamic Pro perties of Ordinary Water Substance for General and Scientific Use (2018), www.iapws.org
14. S. Herrig, M. Thol, A. H. Harvey, E. W. Lemmon, J. Phys. Chem. Ref. Data 47, 043102 (2018)
15. R. Feistel, Deep-Sea Res. I55, 1639 (2008)
16. F. J. Millero, R. Feistel, D. G. Wright, T. J. McDougall, Deep-Sea Res. I55, 50 (2008)
17. V. Holten, J. V. Sengers, M. A. Anisimov, J. Phys. Chem. Ref. Data 43, 043101 (2014)
18. IOC, SCOR and IAPSO, 2010: *The internationalt hermodynamic equation of seawater --- 2010: Calculation and use of thermodynamic properties. Intergovernmental Oceanographic Commission, Manuals and Guides No. 56, UNESCO (English), 196pp.*

Polymeric hollow fibers: a modular heat exchanger for thermal management systems of battery modules in electric vehicles

Jan Bohacek[1*], *Jiri* Hvozda[1], *Krystof* Mraz[1], *Alexander* Vakhrushev[2], and *Ebrahim* Karimi-Sibaki[2]

[1]Heat Transfer and Fluid Flow Laboratory, Faculty of Mechanical Engineering, Brno University of Technology, Technicka 2896/2, 616 69 Brno, Czech Republic
[2]Christian-Doppler Laboratory for Metallurgical Applications of Magnetohydrodynamics, Dept. of Metallurgy, Montanuniversitaet Leoben, Franz-Josef Strasse 18, 8700 Leoben, Austria

Abstract. The liquid-cooling system based on polymeric hollow fibers (ø1 mm) embedded inside durable polydicyclopentadiene is proposed for thermal management of li-ion cylindrical batteries from their surface. In the present work, a dramatically improved design is presented, eliminating the following drawbacks: I) the thermal resistance of the heat exchanger is minimized by bringing fibers into direct contact with the negative terminal/cylindrical shell, II) the fabrication process (combination of extrusion and rapid injection molding) is significantly simplified and accelerated as the essential component of the heat exchanger is nearly planar, and III) inlet/outlet manifolds were refined to enhance modularization of the cooling system of the whole battery pack. The pressure loss of the heat exchanger is favorably low, about 100 Pa, corresponding to 10 l/min of the coolant circulating in the entire battery pack of a virtual electrical vehicle. The heat exchanger was stacked with 18650-type lithium-ion cells, which were repeatedly charged/discharged. With the coolant inlet temperature of 20 °C and the C-rate of 1 C, the maximum temperature of the cells during cycling was between 26 °C and 22 °C in the given range of flow rates (5–45 ml/min). Temperature spreads were 10 °C and 4 °C.

1 Introduction

Well-designed battery thermal management systems (BTMS) ensure optimal thermal conditions for batteries [1]. Temperature variations are caused by heat generation in the battery due to Ohmic losses, concentration gradients (mixing of species), and activation (interfacial energy). Undesired thermal effects, such as capacity/power fade [2], self-discharge [3], thermal runaway [4], electrical imbalance, and low-temperature performance [5], can affect batteries' performance, lifetime, and safety if not maintained at around 25 °C [6]. At elevated and low temperatures, different phenomena accelerate ageing of the battery. As a result, an active BTMS is needed, which involves a cooling system. Many different approaches are known, from which conventional liquid systems, systems with vapor compression cycles (VCC) [7] and air-cooled systems [8] are commonly involved in existing electric vehicles. The air-cooled systems are seen with low/medium power battery packs, those designated for home charging. On the contrary, battery packs of the newest powerful electric vehicles (EV), particularly those demanding super-charging, necessitate discharging a considerable amount of heat; thus, involve the liquid-cooled systems or systems with VCC. In addition, it is more an objective of research than an engineering application to explore the potential of phase-change materials (PCM) [9], heat pipes (HP) [10], thermoelectric coolers [11], and thermo-acoustic refrigerators [12]. Apparently, future developments seem to stay with the liquid-cooled systems, systems with VCC or a combination of both.

A sound design of a heat exchanger, an essential component of the liquid-cooled system, necessarily requires knowledge of the heat generated in the battery. Battery pack designers rely on data from experiments [13] as well as simulations [14].

Mini-channel cold plates, also known as serpentine channels [15] or cold plates [16], are the most used heat exchangers for temperature control of prismatic and pouch cells. Often, the word "serpentine channels" is used when referring to the well-known cooling strategy applied in TESLA cars [17]. Aluminium and even copper are frequently entering ingredients, which are raw materials burdened with a very high carbon footprint. Furthermore, the weight-based standards bestowed on car makers make them literally fight for every gram of a vehicle to meet CO_2 targets. Therefore, plastics may be the way to go due to their favorably low density. Concerns linked with an undesirably increased thermal resistance due to the low thermal conductivity can be overcome, e.g. by considering thinner walls of the heat exchanger. Note that metal heat exchangers often must be lagged with electrical insulators to avoid short circuits, which is unnecessary with plastic counterparts. Further note that the thickness of such thermal insulators is often the same size as the wall

* Corresponding author: jan.bohacek@vut.cz

thickness of the polymeric hollow fiber. Therefore, the advantage of highly conductive materials such as aluminium goes to waste. Then, the polymeric heat exchanger offers the same or better thermal performance and more compact dimensions.

In the present work, a plastic heat exchanger is presented for surface cooling of cylindrical 18650-type Li-ion cells (Figure 1). It is a dramatically improved version of the heat exchanger published in [18, 19]. Coolant channels are made of polypropylene hollow fibers [20, 21] with an inner diameter of 0.84 mm, frozen permanently in a polydicyclopentadiene housing. Unlike the preliminary design, the fibers protrude halfway where the cells are in direct contact with the heat exchanger. Redesigned inlet/outlet manifolds promote flexibility of the cooling system. An assembly of several heat exchangers fills the space between the cells completely. The mass-weighted ratio of the heat exchanger-to-battery is just about 1:5. Each cell is firmly clamped within the flexible fibers; moreover, little protrusions in the heat exchanger prevent the cells from slipping out.

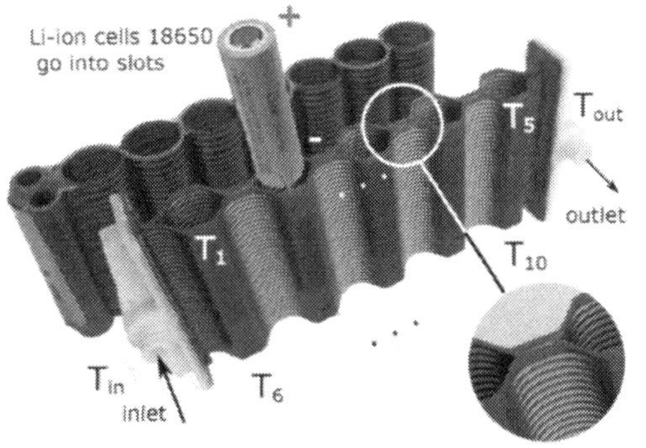

(a)

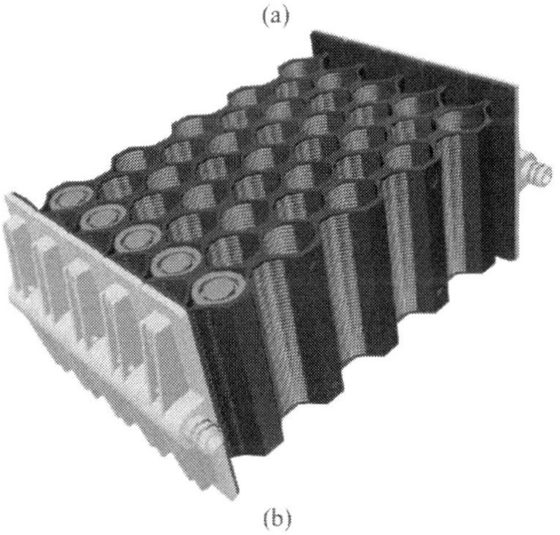

(b)

Fig. 1. A polymer heat exchanger with hollow fibers as coolant channels (a) – the preliminary (back) [18] and the progressed (front) prototype; a compact assembly for a battery module (b) with the first row of Li-ion cells inserted.

2 Experiments, results and discussion

2.1 Pressure drop

As displayed in Figure 2, the heat exchanger reveals favorably low pressure drops in the wide range of flow rates. For example, the flow rate of 15 ml/min would roughly correspond to 1.0 l/min per battery module of TESLA Model S [22] if the battery module was virtually assembled according to Figure 1b. The flow characteristics were examined in the experiment, which is shown in Figure 2. The pressure drop measurements were realized with the BD SENSORS pressure measurements device (max. 1000 Pa of a pressure difference). The pressure drop is extremely low due to many heat exchangers connected in parallel assuming negligible losses in manifolds. The flow distribution into individual heat exchangers is currently a subject of CFD simulations and optimization to improve the manifold design. The goal is to achieve a uniform distribution of the flow and a small pressure drop on manifolds compared to the pressure drop in hollow fibers.

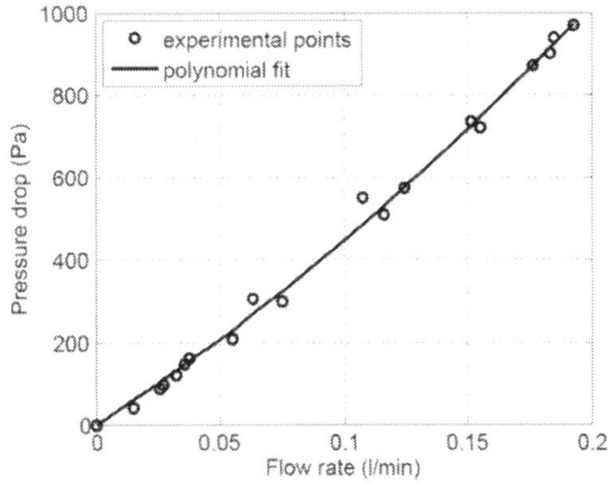

Fig. 2. A pressure drop of the heat exchanger shown in the foreground of Figure 1a.

2.2 Thermal performance

The Li-ion cells (18650 cell type with the capacity of 2200 mAh) connected in series were subjected to a computer-controlled charge/discharge cycling in the entire range of state-of-charge (SOC). The West Mountain Radio Computer Battery Analyzer CBA IV with the 500W amplifier and the charge controller were used for that purpose. A constant current (CC) of 2.2 A, corresponding to 1 C-rate, was considered. 1 C means that the battery is charged to the full capacity exactly in one hour. Charging was stopped when the maximum voltage of 4.2 V of the Li-ion cell was reached. Likewise, discharging was stopped when the voltage dropped to 2.8 V. The voltages are following the manufacturer's specification. The inlet temperature of the coolant (a distilled water) was 20 °C. The flow rates were 5, 10, 20, and 45 ml/min. Such small flow rates were adjusted by the precise flow-regulating needle

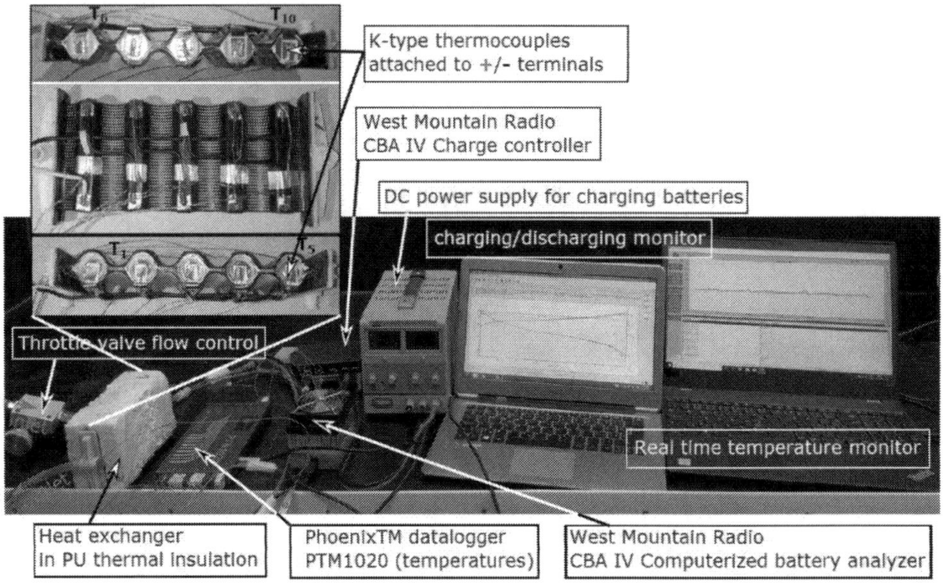

Fig. 3. A photo of the experimental set-up with a detail of wiring Li-ion cells 18650 and the location of K-type thermocouples.

valve (HAM-LET Group H-1300 series), a filtration component, and the KROHNE flow meter VA40V/R. The experimental setup is shown in Figure 3.

K-type thermocouples were used to record temperatures of positive and negative tabs of each 18650, as shown at the right top of Figure 3. The thermocouples were electrically insulated from the tabs to avoid the false reading of the temperatures. As the highest temperatures can be expected to occur at the tabs, the temperature is measured therein. Note that wiring has to have a low enough electric resistance to avoid parasitic Joule heating.

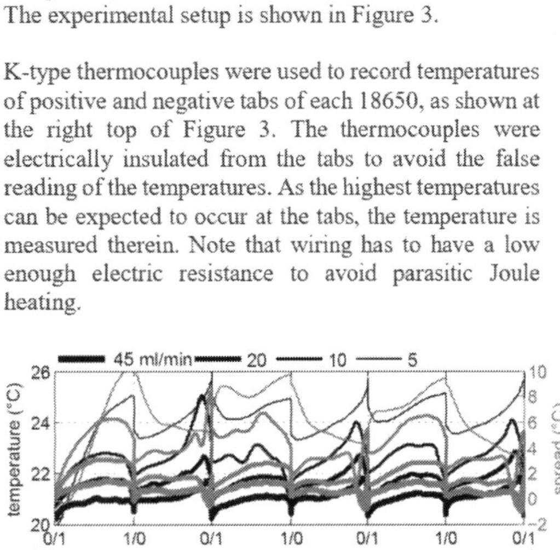

Fig. 4. The thermal performance (at the top) of the heat exchanger assessed by means of the mean temperature (black) and the temperature spread (magenta) of the cells in the range of flow rates (5–45) ml/min. Corresponding voltage and current during charging/discharging (at the bottom).

As shown in Figure 4, the cells' mean temperature and temperature spread during cycling were not exceeding 26 °C and 10 °C, respectively, for the lowest flow rate of 5 ml/min. Concerning the highest flow rate of 45 ml/min, the numbers significantly improved, namely 22 °C and 4 °C. In addition to temperature curves, the voltage and the current are shown at the bottom of Figure 4.

3 Further comments and future outlook

To obtain a reliable heat exchanger, particularly ensuring a perfect bonding between the polymeric fibers and the polydicyclopentadiene (PDCPD) thermoset, appropriate parameters of the rapid injection molding (RIM) had to be adjusted. The final quality of the heat exchanger was inspected with the help of computed tomography. A 3D CT scan with a planar section provides a closer look at the crucial fiber-PDCPD interface in Figure 5.

Cylindrical Li-ion cells have by one order of magnitude higher effective thermal conductivity in the axial direction than in the radial direction. Moreover, nowadays battery manufacturers and customers from the EV automotive seem to shift from the type 18650 over the 2170 to the "tabless" 4680. Hence, the increasing tab surface area and the higher conductivity along the axis of the can suggest switching to the tab cooling in the near future.

In the present work, a plastic heat exchanger was introduced for a cylindrical type of Li-ion cells. However, our lab also has an ongoing research on heat exchangers for prismatic or pouch cells. In Figure 6, the first photo shows organized fibers just before the RIM, and the second photo shows the final planar heat exchanger with the precast manifold.

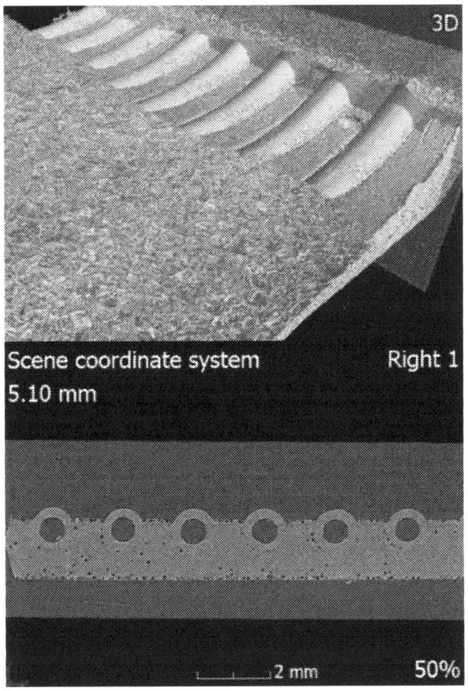

Fig. 5. A detailed view of the 3D CT scan of the heat exchanger, in which the fibers protrude outside the PDCPD matrix (top) and the planar section showing an intact wall thickness of the fibers (bottom).

Fig. 6. Parallel fibers before the RIM process at the top and the final planar heat exchanger at the bottom.

4 Conclusions

A plastic heat exchanger was introduced to be used in battery modules of electric vehicles. Since the aluminium counterparts must be electrically insulated using special foils (~0.1 mm thick), the proposed plastic alternative comprising many polymeric hollow fibers (~0.1 mm wall thickness) provides the same or even more favorable thermal resistance. Hence, the plastic heat exchanger shows a competitive thermal performance and the corresponding dimensions are advantageously smaller.

This work was supported by the Ministry of Education, Youth and Sports of the Czech Republic under OP RDE grant number CZ.02.1.01/0.0/0.0/16_019/0000753" ""Research centre for low-carbon energy technologies""". Computational resources were supplied by the project " ""e-Infrastruktura CZ""" (e-INFRA LM2018140) provided within the program Projects of Large Research, Development and Innovations Infrastructures.

References

1. S. Arora, Design of a modular battery pack for electric vehicles (*Doctoral thesis*) Swinburne University, Hawthorn, Australia (2017).
2. D. Aurbach, B. Markovsky, G. Salitra, E. Markevich, Y. Talyossef, M. Koltypin, L., Nazar, B. Ellis, and D. Kovacheva, J. Power Sources, 165, 491 (2007).
3. D. Aurbach, J. Power Sources, 146, 71 (2005).
4. G.-H. Kim, A. Pesaran, and R. Spotnitz, J. Power Sources, 170, 476 (2007).
5. A. Vlahinos and A. Pesaran, in 2002 SAE Future Car Congress, Computer Optical Disc, Crystal City, Arlington, VA, p. 1 (2002).
6. T.M. Bandhauer, S. Garimella and T.F. Fuller, *J. Electrochem. Soc.* **158**, R1-R25 (2011).
7. N. Javani, I. Dincer, G.F. Naterer and B.S.Yilbas, *Appl. Therm. Eng.* **64,** 471-82 (2014).
8. R. Sabbah, R. Kizilel, J.R. Selman and S. Al-Hallaj, *J. Power Sources*, **182**(2), 630-8 (2008).
9. R. Ataur, M.N.A. Hawlader and H. Khalid, *Int. J. Automot. Technol.* **18**(5), 875–82 (2017).
10. J. Smith, R. Singh, M. Hinterberger and M. Mochizuki, *Int. J. of Therm. Sci.* **134**, 517–29 (2018).
11. C. Bolsinger and K. P. Birke, *J Energy Storage* **21**, 222-30 (2019).
12. M. Senga and S. Hasegawa, *J. of Appl. Phys.* **119**, 204906 (2016).
13. J.S. Hong, H. Maleki, S.A. Hallaj, L. Redey and J.R. Selman, *J. Electrochem. Soc.* **145**, 1489-501 (1998).
14. S. Panchal, M. Mathew, R. Fraser and M. Fowler, *Appl. Therm. Eng.* **135**, 123–32 (2018).

15. J. E, S. Xu, Y. Deng, H. Zhu, W. Zuo, H. Wang, J. Chen, Q. Peng and Z. Zhang, *Appl. Therm. Eng.* **145**, 552-68 (2018).
16. J. Wang, Y. Gan, Y. Liang, J. Liang, M. Tan and Y. Li, *Appl. Therm. Eng.* **151**, 475-85 (2019).
17. W. Prochazka, Beyond Teardown - AVL Series Benchmarking Tests: Getting from low level parameter to target orientation, *Proc. of Battery Engineering 2017* (Mainz) 162-73 (2017).
18. J. Bohacek, M. Raudensky and E. Karimi-Sibaki, *Appl. Therm. Eng.* **159** 113940 (2019).
19. J. Bohacek, M. Raudensky, T. Kroulikova and E. Karimi-Sibaki, *Int. J. of Therm. Sci.* **146**, 106060 (2019).
20. M. Raudensky, I. Astrouski and M. Dohnal, *Appl. Therm. Eng.* **113,** 632-8 (2017).
21. J. Bohacek, M. Raudensky, I. Astrouski, and E. Karimi-Sibaki, *Energy* **229**, 120571 (2021).
22. V. Nilsson, Development of a Mathematical Model for the Cooling of an Electric Car Battery (Master's thesis) Chalmers University of Technology, Sweden (2017).

High-Speed Temperature Measurement

Patrik Bouchal[1*], *Jiří* Hejčík[1], and *Miroslav* Jícha[1]

[1]Brno University of Technology, Department of Thermodynamics and Environmental Engineering, 616 69 Brno, Czechia

Abstract. In certain applications where temperature changes rapidly, e.g., compressor cylinder or explosion chamber, it is necessary to obtain temperature data with a sample rate high enough to register the change, i.e., the temperature probe response time must be very low. Type E coaxial thermocouples with response time in the range of μs can be an example of such a temperature probe. This paper describes several approaches for high-frequency temperature data obtained using coaxial thermocouples. Individual approaches are described together, as well as their shortcomings. Furthermore, the possible method of data evaluation is discussed.

1 Introduction

Fast response thermocouples are used in wind tunnels, explosion chambers, and other cases where the temperature change is sudden [1,2]. They are robust and offer better longevity in these types of environments compared to thin-film thermometers that are also sometimes used [3]. Regardless of the application, they must always be connected to hardware capable of acquiring signals at a high sample rate to be able to register the change. In day-to-day instances, the thermocouples would be connected to modules designed specifically for them, that is, with built-in cold junction compensation (CJC). However, their low sample rate (up to 100 S/s) makes them unusable in the applications described in this paper. For this reason, oscilloscopes are often used [3]. They offer a very high sample rate (in the range of GHz) and an adjustable range, meaning that with a low bit rate analog-to-digital converter (ADC) they do not produce very noisy data even when measuring low voltage coming from a thermocouple. On the other hand, they seem to be more prone to picking up noise from their surroundings. Newer versions of voltage modules seem to produce promising results; nevertheless, in research they are barely used. In this research, several NI voltage modules were considered. Module 9222 appeared to be the best option due to its high sample rate (up to 500 kS/s), but 16 bits with ±10 V range made the ADC noise higher than the signal itself. The best results were achieved with the 24-bit ±5 V module NI 9234 with IEPE (Integrated Electronics PiezoElectric Signal Conditioning) turned off paired with a signal amplifier.

2 Materials and Methods

2.1 Experimental setup

For this paper, an experimental setup (Fig. 1.) has been built. It consists of a water supply, two water boilers, a three-way valve, and a temperature measurement set-up. Two tanks, one with hot water and the other with cold water, provide 2 L/s of water. A three-way valve periodically switches between the two tanks, serving as a source of temperature that varies over time. The valve is controlled by a computer with LabVIEW software installed to make the switch every 5 seconds.

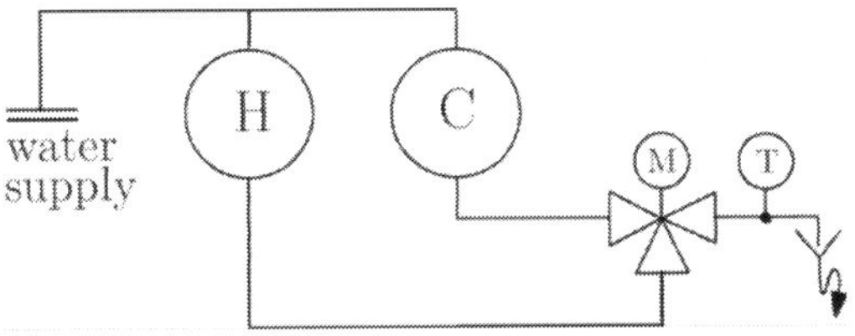

Fig. 1. Experimental setup

The temperature measurement setup consists of two Müller MT 19 type E coaxial thermocouples (Chromel - Constantan) from Dr. Müller Instruments. The probe is 26 mm long and is equipped with M2 thread on its 1.9 mm outer diameter. The thermocouple offers a response time of 3 μs and a sensitivity of 60 μV/K. Because the two metals are insulated one from another by a very thin layer of nonconductive material, the thermocouple joint is formed only at the tip of the probe by lightly sanding it with sand paper.

The first thermocouple is attached to the pipe surface and the other one is placed inside a zero point thermostat with a constant temperature of 0 ° C. The two are connected differentially. The voltage generated by the thermocouples is amplified and collected by the data acquisition system (DAS) (Fig. 2). This type of differential connection compensates for the lack of cold junction compensation, which is usually necessary when using thermocouples [1].

* Corresponding author: patrik.bouchal@vutbr.cz

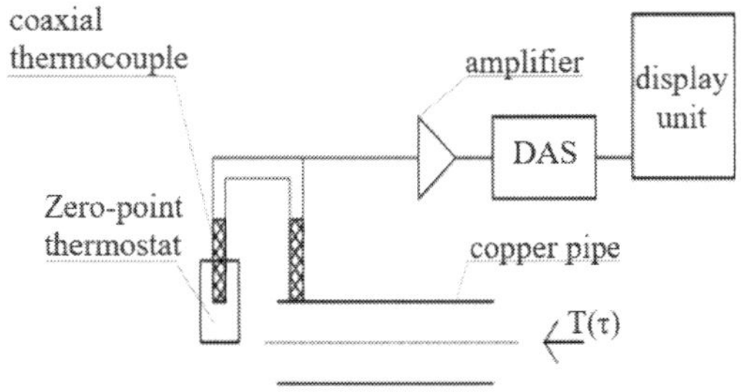

Fig. 2. Temperature measuring setup

2.2 Data acquisition

In this paper, two types of DAS were used. Both were voltage modules made by National Instruments. The ±10 V module NI 9222 with 16-bit ADC and a maximum sample rate of 500 Ks/s was compared with a NI 9234 module that offers a sampling rate of 50 kS/s, a 24-bit ADC, and a measuring range of ±5 V. Because it is made primarily for sound and vibration, it has built-in IEPE that had to be turned off. For data acquisition, a virtual instrument (VI) was made using LabVIEW 2020 software because it works well together with the NI module line. For both modules, data were collected at a sample rate of 11.8 kS/s.

Because thermocouples produce such a low voltage (hundreds of μVs), it can be helpful to amplify the signal [1,5].

The amplifier amplifies only the core signal, thus increasing the signal-to-noise ratio, making the signal more readable. Although in theory it is true, practical use shows that the amplifier itself can be a source of noise that can sometimes be even higher than the noise from the ADC (Fig. 6.) [1], making the amplifier redundant or even harmful. This noise can be present, especially when powering the amplifier with an alternating current (AC) system. Powering it with a direct current (DC) system in the form of a battery pack or a laboratory power supply can solve the problem.

It is also necessary to properly shield the amplifier to ensure it does not pick up any noise from its surroundings.

In this work, the Müller Voltage Amplifier MVA 10 was used. Its advantage is the ability to switch between ×200 and ×2000 gain, however, in this study, only the ×200 option was used. The VI was altered when reading the amplified data so that it divides them by gain before writing them into a file.

2.3. Data Processing

2.3.1 Smoothing

As a tool to make noisy data more readable, simple exponential smoothing was used. It was chosen over a simple moving average that weights the past data equally, whereas exponential smoothing exponentially decreases weight over time. A simple MATLAB script was used to get the smoothed value of the voltage U_i and was applied 5 times to achieve the maximum possible smoothing. The script takes the original voltage reading in the current time-step $U_{0,i}$, multiplies it by a smoothing constant α and adds a voltage reading from the previous time-step U_{i-1} multiplied by the difference of the smoothing constant from 1 (1). The value of a smoothing constant α was set at 0.9. Because the function only works from the second time-step, the smoothed voltage in the first time-step U_1 value was set to the original value in the first time-step $U_{0,1}$ (2).

$$U_i = \alpha \cdot U_{0,i} + (1-\alpha) \cdot U_{i-1} \quad (1)$$

$$U_1 = U_{0,1} \quad (2)$$

2.3.2 Calculating the temperature

The electromotive force (EMF) or a voltage (U) produced by the thermocouple must be converted to temperature. For this robust datasheet, it is best done by a polynomial equation (3) for which a MATLAB script was written. The coefficients (v_0; p_i; q_i) are slightly different for different temperature ranges. But because in this application the temperature never exceeds one temperature range, the coefficients are constant. The calculated temperature is the temperature difference from the cold junction of the thermocouple. Due to the zero-point thermostat, in this case, the value is the total temperature. The only variable in the equation is the voltage U.

$$T_i = \frac{(U_i - v_0)\cdot(p_1 + (U_i - v_0)\cdot(p_2 + (U_i - v_0)\cdot(p_3 + p_4\cdot(U_i - v_0))))}{1 + (U_i - v_0)\cdot(q_1 + (U_i - v_0)\cdot(q_2 + q_3\cdot(U_i - v_0)))} \quad (3)$$

3 Results

All the measurements were performed with the sample rate of 11.8 kS/s for 50 seconds. The pipe was completely cooled down by cold water at the beginning of every measurement to have comparable results and then the three-way valve started to switch the water source every 5 seconds.

The first measurement was performed with module NI 9222 without any amplification. The signal oscilated around constant value of 0.6 mV with an amplitude of 0.6 mV. The core signal from the thermocouple was so low that the module wasn't able to register it and showed only the white noise of the ADC.

As the module couldn't register the core signal, it needed to be amplified. For that reason the voltage amplifier was connected to the circuit as shown if Fig. 2 to achieve better results. It was powered by the included power supply that plugs into an electrical outlet. The gain was set to ×200 but the module could not read such a signal and has gotten out of the measurement range even though the signal was not supposed to exceed it. To ensure that the system worked properly, the amplifier was connected to a DC power supply in the form of two differentially connected laboratory power supplies. This approached solved the issue and the signal was properly registered but the noise was still prominent with the same amplitude of 0.6 mV but now there was noticeable tendency of the main signal.

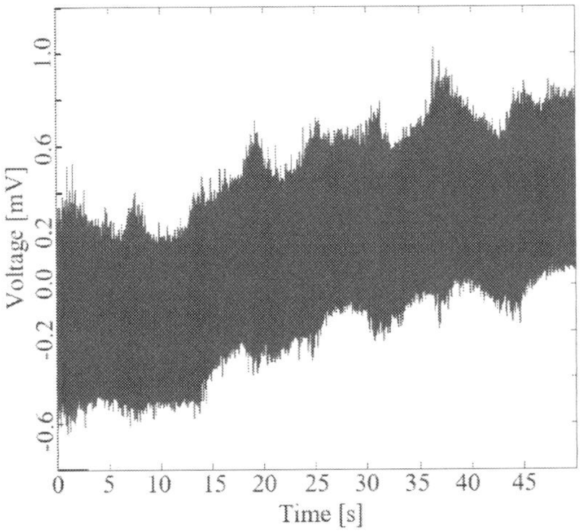

Fig. 3. Signal acquired by NI 9234 with the thermocouple directly touching the copper pipe

The next module NI 9234 was evaluated. Although the amplifier was used, the signal was still very noisy (Fig. 3). Because the registered noise was no longer white, the thermocouple might have been picking up noise coming from the 3-way valve power system by directly touching the copper pipe. To test this theory, a strip of clear adhesive tape was adhered to the surface of the pipe where the thermocouple was mounted so that it does not touch the pipe directly and does not pick up this noise.

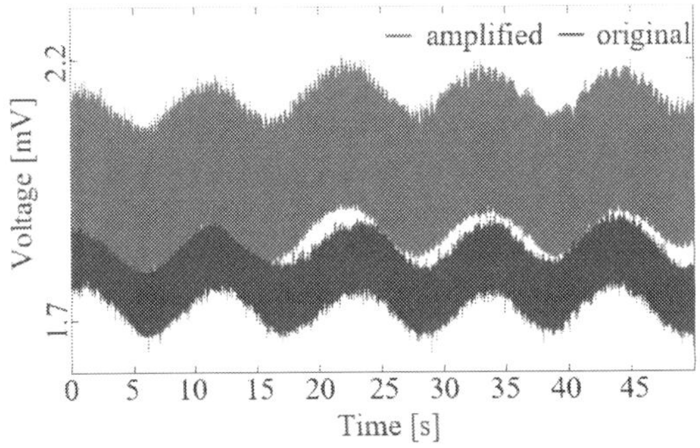

Fig. 4. Signal with amplifier noise and the original signal

The NI 9234 module was then used with and without the amplifier. The sampling rate remained the same. The amplifier was powered by the included power supply to see if there was any difference in behaviour. The amplified signal and the original signal from that module are shown in Figure 4. The noise associated with the amplified signal is higher than in the original data and does not appear to be white, meaning that the included power supply could be a source of noise.

As a confirmation that the differential connection of the second thermocouple mounted on the zero-point thermostat, acting as compensation for cold junction, is needed, a measurement with only a single thermocouple was done. The signal from the thermocouple is set a bit higher, which was to be expected, and the local peaks are more pronounced (Fig. 5.).

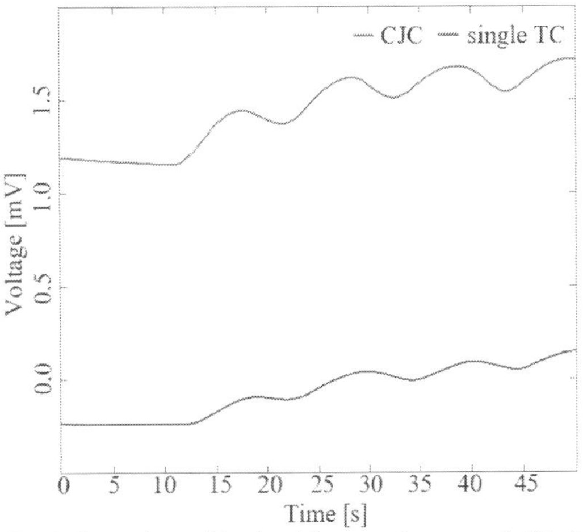

Fig. 5. Comparison of the signal from one thermocouple (blue) and two differentially connected thermostats where one is on the zero-point thermostat (red).

Finally, the signal acquired by module 9234 without an amplifier was processed by the exponential smoothing described above. It was compared to the signal acquired by the same module with an amplifier that was not smoothed. The surface temperature was then calculated from both sets of data using the polynomial equation (3). The smoothed signal is set slightly below the amplified signal and has a more noticeable noise (Fig. 6). The amplified signal appears to be cleaner, which is caused by the resolution. The local peaks are so close together that they form a line.

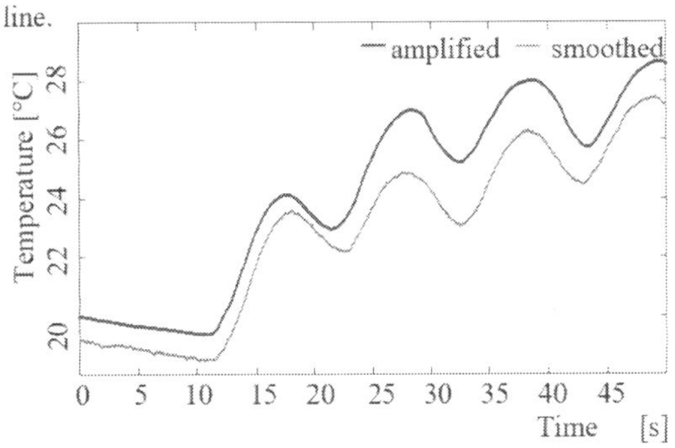

Fig. 6. Temperature deducted from amplified and smoothed data

4 Discussion and Conclusions

Coaxial thermocouples are widely used in situations where the temperature suddenly changes because they offer a response time close to 1 μs [1,2,3]. Appropriate hardware is needed to acquire the data fast enough to register a sudden temperature change.

In this study, the National Instruments modules NI 9222 and 9234 were used with and without signal amplification as data acquisition systems for type E coaxial thermocouples. The thermocouple measured the change in surface temperature of a copper pipe through which variable temperature water flowed.

The main challenge when measuring temperature at a high sample rate turned out to be noise. The measurement of high sample rate temperature in different applications is the subject of several studies, but few mention this issue or how they overcome it [3]. There is a different type of noise to consider when working with this setup. One is the so-called white noise that has a constant power spectral density. This noise in this application comes from the resolution of the data acquisition system. This is affected by the measurement range of the DAS and the bit rate of the analog-to-digital converter. As can be seen in Figure 3., although it has a high sampling rate, it is not very fitting for this kind of measurement because the voltage from the thermocouple is very low, the module has a large measuring range coupled with 16 bit ADC. For that reason, even an amplified signal is not strong enough to be readable by the module.

Another source of noise can be any type of AC electricity supply system. In this case, when the amplifier was powered by the included AC/DC adapter, it introduced a lot of noise, as can be seen in Figure 6. Although that could be filtered out by a low-pass or a band-pass filter, the sensitivity of the module could be negatively affected.

The last source of noise observed in this paper came from the 3-way mixing valve. Although it is powered by a 24 V DC system, it is connected to the grid, introducing another AC noise that the thermocouple was able to pick through the copper pipe. This issue was solved by adhering a strip of clear adhesive tape to the surface of the pipe.

To reduce the noise, the exponential smoothing method was proposed. Although it can smooth the data, it is quite demanding for computational power and the results do not appear to be very accurate, as can be seen in Figure 8. The best results were achieved with the amplifier powered by a laboratory DC power supply because it did not introduce any noise into the system. Another possible solution is to power it with the battery pack.

Lastly, the need for cold junction compensation was tested. For instance, where only the temperature difference is important, it might seem like there is no need to have cold junction compensation. As in Figure 7, the signal is offset, which makes sense, but the signal peaks are much less pronounced. This corresponds to previous research work [6]. The signal from thermocouple differentially connected with one in the zero-point thermostat is stronger and therefore easily readable by the DAS.

Acknowledgement

This article was supported by the project "Computer Simulations for Effective Low-EmissionEnergy" funded as project No. CZ.02.1.01/0.0/0.0/16_026/0008392 by the Operational Program Research, Development, and Education, Priority axis 1: Strengthening capacity for high-quality research and by FSI-S-20-6295.

Nomenclature

U	[mV]	Thermoelectric voltage
α	[-]	Smoothing constant
T	[°C]	Temperature
$v_0; p_i; q_i$	[-]	Polynomial constant

References

1. J.D. Nash, D.R. Cardwell, M.J. Zelman, J.N. Moum, J Atmos Ocean Technol, **16**, 1474 (1998)
2. S. K. Manjhi, R. Kumar, Exp. Heat Transf, **33**, 282 (2019)
3. H. Olivier, H. Gronig, B. Schluze, ICIASF, 1995
4. L. Reichelt, U. Meingast, U. Renz, Int. J. Heat Mass Transf, **45**, 579 (2002)
5. R. L. Denton, Bull Am Meteorol Soc, **32**, 214 (1951)
6. K. J. Irimpan, N. Mannhil, H. Arya, V. Menezes, J. Int. Meas. Confed., **61**, 291 (2015)

https://doi.org/10.1051/epjconf/202226401007

Spray in cross–flow: comparison of experimental and numerical approach

Ondrej Cejpek[1*], *Milan* Maly[1], *Miloslav* Bělka[1], *Jaroslav* Slama[2], *Ondrej* Hajek[1], *Frantisek* Prinz[1], and *Jan* Jedelský[1]

[1]Energy Institute, Faculty of Mechanical Engineering, Brno University of Technology, Technická 2896/2, 616 69 Brno, Czech Republic
[2]Provyko s.r.o., Vinařská 558/3a, Pisárky, 603 00 Brno, Czech Republic

Abstract. The spray behaviour and droplet trajectories in realistic conditions are of crucial importance in many industrial, agricultural and chemical applications. Droplet characteristics and spray trajectory in chemical applications (e. g. flue gas scrubbing, CO_2 capture in spray column) determine the amount of mass involved in the gas scrubbing process, mass trapped by the flow or attached to the walls. Knowledge of the droplet behaviour can improve a nozzle design and scaling, increase the process efficiency, minimize the process liquid and blow away the fraction. In this study, experiments with pressure swirl nozzle in cross–flow of air were performed at one nozzle injection pressure (0.5 MPa) and several cross–flow velocities (8, 16, 32 m/s). The results on droplet trajectories are compared with numerical results obtained by ANSYS Fluent. Two Lagrange approaches for spray modelling were used. Injection of droplet groups and Linearized Instability Sheet Atomization (LISA) model incorporated within ANSYS Fluent were used to represent the spray. The CFD results of spray penetration and droplet trajectories are compared with experimental data. A simple analytical model is able to well predict trajectories of large droplets, but fails to predict trajectories of small droplets. The LISA model yields a better accuracy for spray in cross-flow prediction.

1 Introduction

Pressure swirl atomizers are used in many industrial, chemical and agricultural applications. Cross-flow air simulates the conditions in combustion chambers, chemical processes dealing with gas scrubbing, agricultural spraying and many other realistic applications. Pressure-swirl atomizers are designed to produce a rapid break-down of liquid in a two-phase droplet flow. They are used to increase the specific surface area of the fuel and thereby achieving a high rate of mixing and evaporation. In most combustion systems, a reduction in mean drop size leads to higher volumetric heat release rate, easier ignition, a wide burning range and lower exhaust concentrations of the pollutant emissions [1]. In the gas scrubbing application, the increase in the interfacial area is also beneficial. The knowledge of spray behaviour and droplet trajectory is also crucial in order to determine and reduce the blown-away fraction.

This study combined the experimental and numerical approach to investigate the pressure swirl atomizer in the cross-flow condition. However, the experiments are usually time consuming and, in some cases, unrealizable. These drawbacks can be off-set by numerical modelling; this can be deployed beneficially to determine the spray characteristics in realistic conditions. Different numerical simulations are compared in this study.

Amedorme et al. [2] modelled the pressure-swirl atomizer in still ambient conditions. The Eulerian modelling of two-phase flow was employed in STAR-CD. Both internal and external flows were solved. Deshpande et al. [3] performed a computational analysis of hollow cone liquid injection into cross-flow. Lagrangian-Eulerian point parcel treatment of the spray was used in OpenFOAM. The authors classified the spray into the near and far field spray region, characterized by the drag force. The penetration of spray cone by the cross-flow was observed for high cross-flow velocities. Zhang et al. [4] numerically investigated the flow field structure of hollow cone spray injected into cross-flow. The Scale-Adaptive Simulation was employed to predict the turbulent flow. The atomization was modelled by Linear instability analysis and Taylor analogy breakup theory. Many different scaled vortices were identified around the hollow cone. Models for heat and mass transfer were also added to the sprays in cross-flow. Sun et al. [5] compared different models for heat and mass transfer for a spray injected to the cross-flow. Different corelations yield very different results on droplet evaporation. Dikshit et al. [6] simulated the pressure atomizer for small scale gas turbines applications. The study was mainly devoted to understanding of secondary droplet breakup.

Many different simulations of pressure-swirl atomizers and sprays with various physical models were performed. However, a simple and reliable model of

* Corresponding author: Ondrej.Cejpek@vutbr.cz

spray behaviour for the rapid prediction of droplet trajectories has not been developed and validated yet.

2 Experimental setup

Tests were performed at Spray research laboratory, Brno University of Technology.

2.1 Wind tunnel

A specially designed wind tunnel in blow down arrangement was used to simulate the flow condition. The flow parameters were probed using Laser Doppler anemometry prior to the measurement. The cross-flow velocity (v) and turbulent intensity (Tu) are outlined in Figure 1. Profiles in horizontal plane are presented. The vertical v and Tu profiles are symmetrical.

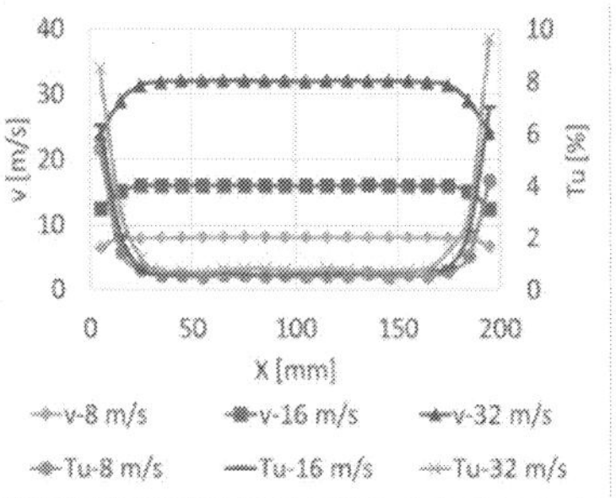

Fig. 1. Profiles of velocity and turbulent intensity in the wind tunnel test section

2.2 PDA measurement setup

Fiber based PDA system (Dantec Dynamics A/S Skovlunde, Denmark) was used to determine the droplet velocity and diameter. Three planes at axial distance Z = 10 mm, 15 mm and 20 mm from the nozzle tip were measured. At least 30 000 samples were acquired in each measurement position or for 10 s measurement duration. The Dantec BSA software 5.2 was used to control the measurement. PDA measurement setup is outlined in Figure 2. Two velocity components along with droplet size were measured in a coincidence mode. The laser output was set to 300 mW. Lenses with 500 mm focal length were used on both transmitting and receiving optics. A scattering angle of receiving optic was set to 70° degrees with slit of 100 µm.

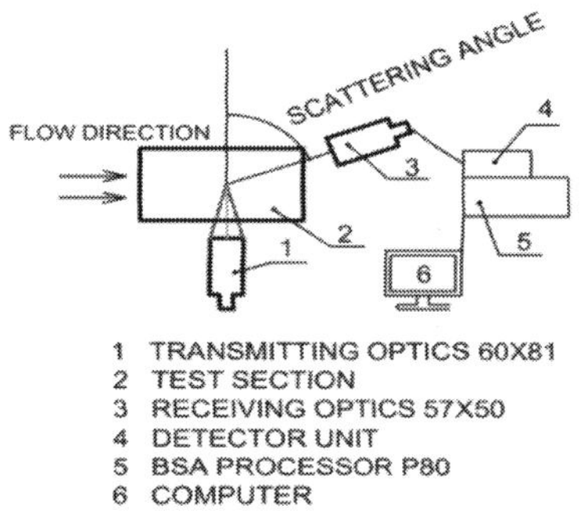

Fig. 2. PDA measurement setup

2.3 High speed visualization

A high-speed camera (FASTCAM SA-Z, Photron, Japan) was used with back illumination by pulsed LED light with 400 ns pulse duration (HPL3-36DD18B, Lightspeed Technologies, USA) to capture the liquid sheet behaviour. 4000 spray images were captured by the camera with a frame rate of 60000 and shutter speed of 1 µm.

2.4 Tested atomizer

A small pressure-swirl atomizer was tested in this study; for dimensions see [7]. The droplets in the range from 0 to 140 µm were created by the breakup of a thin conical liquid sheet.

3 Numerical setup

Description of numerical approaches, boundary condition and computational setup is provided in this chapter.

3.1 Boundary conditions

Boundary conditions were derived from experimental data. Droplet characteristics (velocity, size) were obtained from the PDA measurement. These boundary conditions are outlined in Table 1. Droplets were separated into 4 groups according to their Stokes number. For each droplet group, the velocity magnitude and swirl fraction were determined. The swirl fraction is defined as the ratio of swirl velocity component to the overall liquid velocity inside the spray cone. Since this value is very low, as shown in Table 1, the swirl velocity component can be neglected.

Table 1. Experimental data

Droplet group number [–]	Average group diameter [µm]	Velocity magnitude [m/s]	Swirl fraction [m/s]
1	12.5	18.6	0.006
2	30	20	0.006
3	60	23	0.007
4	110	26	0.008

From instantaneous images (Figure 3), the liquid sheet breakup length was determined to be 7.5 mm with a liquid sheet radius at the breakup length of 4.2 mm. The spray cone angle (SCA) is 60°. All these conditions are determined for the inlet pressure of 0.5 MPa and steady conditions / air flow velocity of 16 m/s.

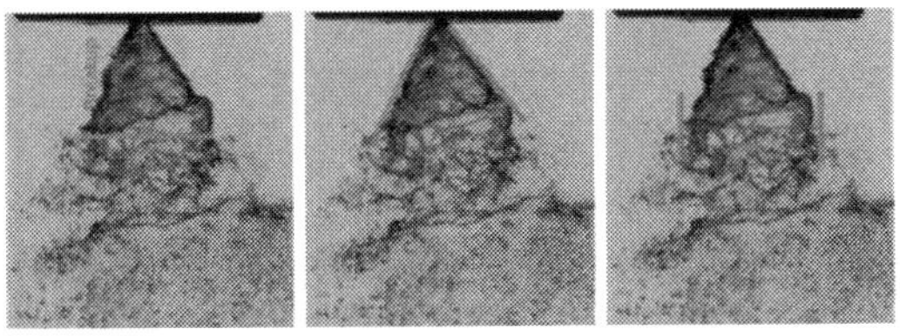

Fig. 3. High speed visualization for air flow of 16 m/s.

3.2 Analytical model

A simple analytical model was developed in MATLAB® software. This model can be used to determine the individual droplet trajectories; this can help in determining the amount of mass involved in the gas scrubbing process, mass trapped by the flow or attached to the walls. The model solves motion equations (Equation 1 and 2) by calculating the drag force and acceleration of droplet, see Equation 3 and 4. and allows for the setting of the initial droplet velocity angle, diameter and cross-flow velocity magnitude and direction. Drag coefficients, C_d, for different ranges of the Reynolds number, *Re*, incorporated within the model, are described by Equations 5 and 6 [8]. The Runge-Kutta method for solving the differential equation is used within the analytical model.

$$d^2x/dt^2=-(\pi d^2 C_d \rho_g/8m)dx^2/dt \quad (1)$$

$$d^2y/dt^2=g-(\pi d^2 C_d \rho_g/8m)dy^2/dt \quad (2)$$

$$F_d=18\mu C_d Re/[(\rho_l-\rho_g)/\rho_l] \quad (3)$$

$$a=F_d(v-u_i)+g[(\rho_l-\rho_g)/\rho_l] \quad (4)$$

$$Re<1$$

$$C_d=(24/Re) \quad (5)$$

$$1<Re<10^5$$

$$C_d=(24/Re)(1+0.173Re^{0.657})+0.413/(1+16300Re^{-1.09}) \quad (6)$$

3.3 Ansys Setup

3.3.1 Geometry

To simulate the droplet behaviour, the 600 mm long duct with a rectangular cross-section was designed. The geometry is the same as that for the wind tunnel. The injector body was placed 250 mm from the exhaust and inserted 35 mm to the cross-flowing air. The axis origin was set to the nozzle tip. The geometry is illustrated in Figure 4.

Fig. 4. Geometry of computational domain

3.3.2 Grid independence test

Four differently sized meshes were tested. The node size was 10, 8, 5 and 2 mm with 204268, 304112, 706689 and 7342555 nodes respectively. For the final computation, the mesh with the node size of 5 mm was selected. There was no difference for the nodes of 5 mm and smaller. The droplet trajectories for different mesh sizes and 16 m/s cross-flow velocity is outlined in Figure 5.

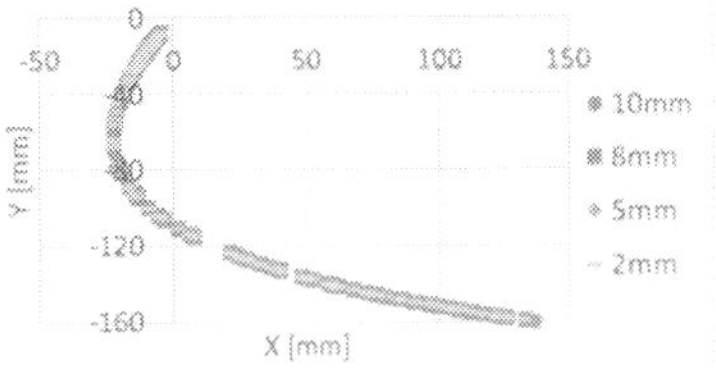

Fig. 5. Comparison of droplet trajectories for differently sized mesh

3.3.3 Mesh

The Ansys mesher was used to generate the structured/unstructured mesh. From the grid independence test, a 5 mm base size was selected. From this setup, the cross-section contains 40×40 nodes. The area of interest for droplet trajectory is in the middle of the model. A standard treatment of wall function was used. The experimentally determined velocity profile is uniform, 12 mm from the walls. The droplet injection was carried out almost 40 mm from the wall.

3.3.4 Physical models

Steady simulation with coupled solver was performed in Ansys fluent. For the turbulence, two-equation *k-ε* (Realizable) model was selected with scalable wall function. The second order upwind solution for momentum, turbulent kinetic energy, momentum and turbulent dissipation rate was used with presto pressure solution. All models were selected to achieve the simplest case possible.

3.3.5 Boundary conditions

The inlet air velocity was set constant to 8, 16, 32 m/s with turbulent intensity 0.8% at the blue frame in Figure 6. The pressure outlet was set to the opposite side of the model. Other faces were set as walls with no slip boundary and a standard roughness model with roughness constant of 0.5. The droplets that will hit the wall will escape from the simulation domain.

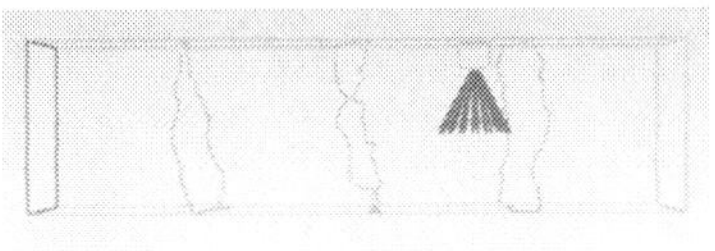

Fig. 6. Boundary conditions

3.3.6 Solution methods

At least 1000 iterations were computed for each case. The continuity, ε and k residuals were lower than 10^{-2}, x, y, and z velocity residuals are lower than 10^{-4} after 1000 iterations.

3.4 Droplet injection

In this section, two methods of droplet injection in ANSYS are described. The Linearized Instability Sheet Atomization (LISA) model and the hollow-cone droplet injection model.

3.4.1 LISA model

The Ansys includes a model for the pressure swirl atomizer simulation (LISA) developed by Schmidt et al. [9]. The LISA model can be divided into two stages, film formation and sheet breakup and atomization (Figure 7)

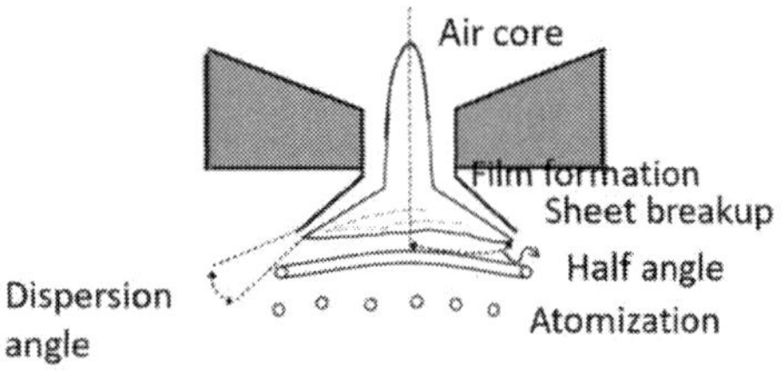

Fig. 7. Schematic process of atomization by LISA model

Sheet constant (S_C) can be derived by equation 5, where L_b is the breakup length, U is the liquid sheet velocity and β is the maximum growth rate. L_b and U can be determined from experimental data. The β can be found in [9] for liquid sheets. It is computed by equation 6, where ρ_L represents the liquid density, γ is the liquid surface tension and ρ is the ratio of air to liquid density.

$$L_b = U{\cdot}S_C/\beta \tag{5}$$

$$\beta = 0.385\rho^{1.5}(\rho_L U^3/\gamma) \tag{6}$$

The ligament constant (C_L) can be derived from equation 7, where d_L is the ligament diameter at the breakup length and K_S is the wave number corresponding to the maximum growth rate. K_S can be determined by equation 8, where h is the liquid film thickness at the breakup length. h is derived from equation 9, where r_0 is the radius of exit orifice, h_0 is the liquid film thickness at the exit orifice and SCA is the spray cone angle.

$$d_L=2\pi C_L/K_s \tag{7}$$

$$d_L=\sqrt{8h/K_s} \tag{8}$$

$$h=r_0h_0/r_0+L_b\sin\,(SCA/2) \tag{9}$$

The default values in the fluent are $S_C = 12$, $C_L = 0.5$. From the experimental data the values are $S_C=12.7$ and $C_L = 1.27$. There is also a dispersion angle, which defines the amplitude of liquid sheet oscillation. Its value is initially set to 6°.

3.4.2 Droplet hollow-cone injection

For the droplet injection, a cone type injector was used with the boundary conditions outlined in Table 1. Droplets were separated to the groups according to Table 1. The droplet injection was set to the breakup length position on the circle (cone periphery) with outer radius of 4.2 mm. For each droplet group, 720 streams of droplets uniformly distributed on the cone periphery were injected. This corresponds to two droplets streams for every 1° of the cone periphery; the cone angle is 60°.

The spherical drag law is used for droplets injected to the cross-flow. The spherical drag law is expressed by equations 8 and 9. Other ranges of Re can be found in [10].

$$1<Re<10$$

$$C_d=1.22+29.1/Re-3.8/Re^2 \tag{10}$$

$$1000<Re<5000$$

$$C_d=0.52-1662.5/Re+5.41/Re^2 \tag{11}$$

4 Results and Discussion

Discrete droplet tracks with hollow cone type injection are illustrated in Figure 8.

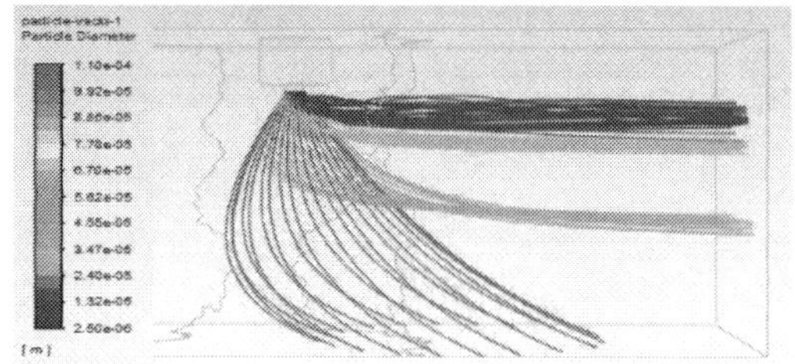

Fig. 8. Discrete droplet tracks for hollow-cone type injector for 8 m/s

The LISA model for the pressure swirl nozzle was employed to compute the track of particles (Figure 9). The droplet size range is from 20 μm to 150 μm. However, the experimental data showed a large fraction of small droplets below 20 μm as these droplets represent as much as 40% of all measured droplets. However, their volume is responsible for only 1.2% of liquid mass. The droplet distribution in LISA model is determined by Rosin-Rammler distribution, which is in agreement with experimental data. Figure 10 shows particle tracks created by analytical model for particle diameter of 110 μm in 16 m/s cross-flow.

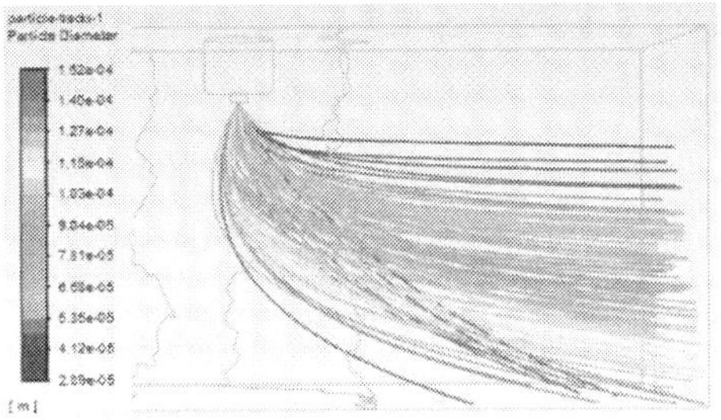

Fig. 9. Particle tracks for LISA for 16 m/s

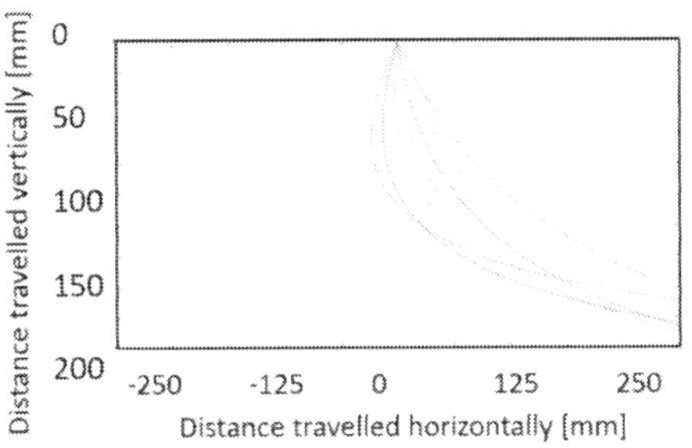

Fig. 10. Particle tracks by analytical code for 16 m/s

4.1 Spray penetration

Spray penetration (penetration length) was determined as the maximum vertical distance that droplets reach after travelling 150 mm downwind from the nozzle. Penetration length for all models and the cross-flow velocity of 16 m/s is illustrated in Figure 11.

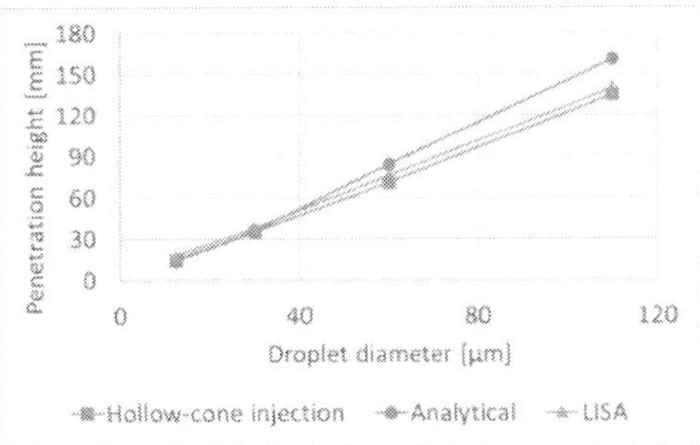

Fig. 11 Penetration length determined by different models for cross-flow velocity of 16 m/s

The penetration height predicted by the hollow-cone injection is almost the same for LISA and analytical models. Some differences occur for large droplets; this can be caused by the oversimplicity of the analytical model.

Figure 12 outlines the penetration of droplets (60 μm) for different v and the used numerical approach.

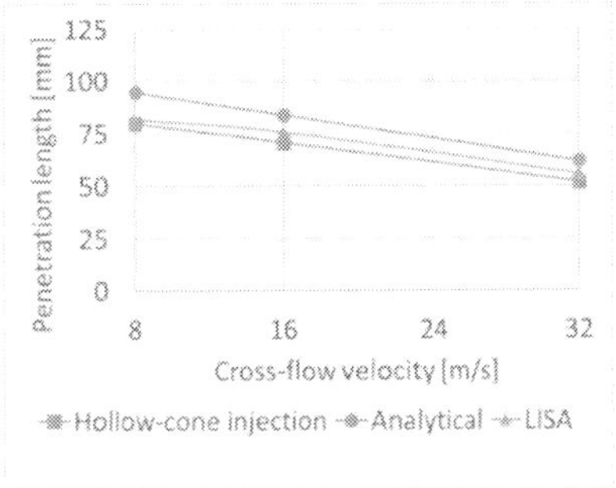

Fig. 12 Comparison of penetration length for different cross-flow velocities and different numerical approaches for 60 μm droplet

4.2 Spray trajectory

The spray trajectory for hollow-cone injection, analytical model and LISA are compared in Figure 13. The trajectory predicted by LISA agrees well with the hollow-cone injection while the trajectory predicted by the analytical model is slightly different; however, with maximum difference below 5%. This small variance can be attributed to only one-way coupling of the analytical model, while both CFD models used a two-way coupling approach.

4.3 Comparison of experimental and numerical results

In Figure 14, the particle trajectories derived by LISA and the analytical model are compared with streamlines derived from experimental PDA data. Streamlines from windward and leeward sides are compared for droplet range of 40 – 80 μm; the LISA model predicts the trajectories really well. analytical models yield higher deviations, which can be caused by oversimplicity of the model with the high-speed image of the spray. It is apparent from the Figure 14 that numerical results and droplet trajectories lie within the high-speed image of the spray. The LISA model is accurate for prediction of droplet trajectories.

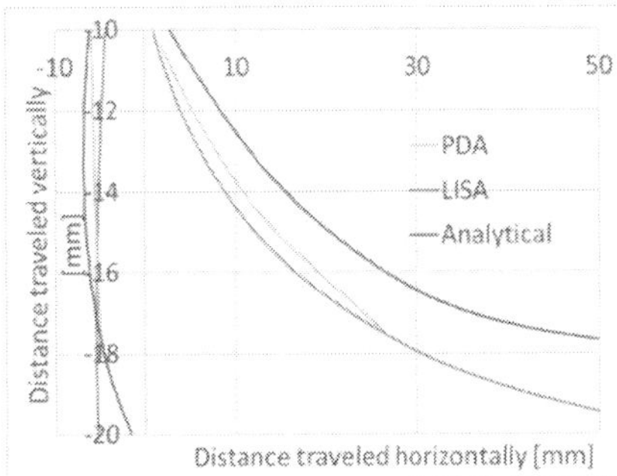

Fig. 13 Comparison of droplet trajectories (40 – 80μm) from leeward and windward sides in cross-flow velocity of 16 m/s for the analytical model, LISA and experimental streamlines created from PDA data

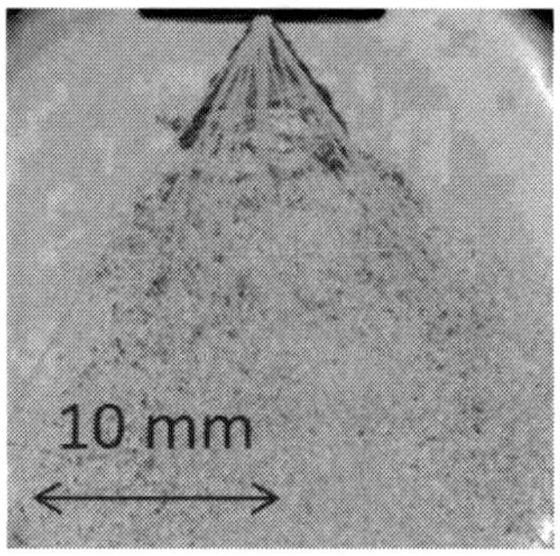

Fig. 14 Comparison of droplet trajectories predicted by LISA and high-speed visualization for 16 m/s

In Figure 15, the droplet diameter predicted by LISA is compared with experimentally obtained PDA data. The results are compared for two axial distances (Z = 15 and 20 mm) from the nozzle tip for cross-flow velocity of 16 m/s. As it can be seen from Figure 15, the deviation is slightly larger for Z = 15 mm. This can be caused by inaccurate prediction of small droplets by the LISA model. However, for Z = 20 mm, the small droplets are usually blown away by the cross-flow and the deviation is much smaller. This is even more apparent for 8 m/s cross-flow velocity as the deviations between the PDA and LISA data were more significant, while for 32 m/s, the results show a good match even for Z = 15 mm.

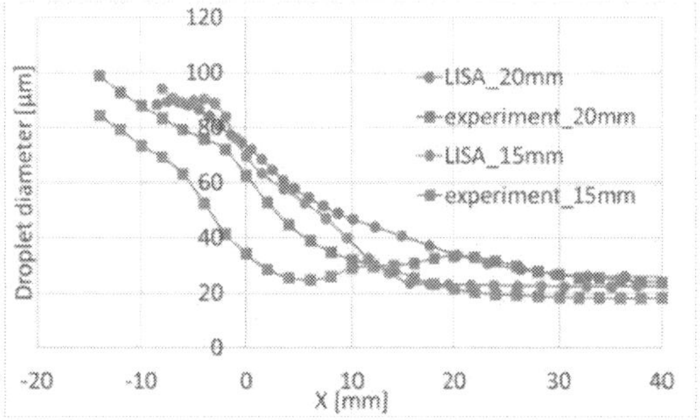

Fig. 15 Comparison of droplet diameter predicted by LISA and obtained by PDA measurement for two axial locations (15 and 20 mm) from the nozzle tip for 16 m/s

5 CONCLUSIONS

The spray behaviour in cross-flow conditions was investigated in this study mainly devoted to the droplet trajectories and behaviour in cross-flow using different approaches. The injection of droplets by the LISA model, a hollow cone injector in Ansys Fluent and a simple analytical model were used to compute the spray/droplet trajectories. Boundary conditions were derived from the experimental results obtained by the PDA and high-speed visualization of the spray in cross-flow. Predicted spray penetration agrees well for all numerical models. Small differences occur in the prediction of penetration of large droplets affected by other physical phenomena, which are not incorporated in the analytical model. With decreasing droplet diameter and increasing cross-flow velocity, the penetration height also decreases. The LISA model predicted the droplet trajectories well.

Droplet trajectories were compared with the experimental data. The LISA method agrees well with the experimental data while the analytical model yields a larger deviation. This can be caused by oversimplicity of this model. A droplet diameter predicted by LISA was compared with experimental results for two planes of 15 and 20 mm from the nozzle tip. The LISA method is not able to predict the behaviour of small droplets (<20 µm) and is more accurate further away from the nozzle where small droplets are drained away.

Nomenclature

a	Acceleration [m·s^{-2}]
C_d	Drag coefficient [–]
d	Droplet diameter [m]
d_L	Ligament diameter at the breakup length [m]
Fd	Drag force [N]
g	Gravitational acceleration [m·s^{-2}]
h	Liquid film thickness at the breakup length [m]
h_0	Liquid film thickness at the exit orifice [m]
K_S	Wave number corresponding to the maximum growth rate [–]
L_b	Breakup length [m]
m	Droplet mass [kg]
r_0	Radius of exit orifice [m]
Re	Reynolds number [–]
S_C	Sheet constant [–]
SCA	Spray cone angle [°]
Tu	Turbulent intensity [%]
U	Liquid sheet velocity [m·s^{-1}]
u_l	Droplet velocity [m·s^{-1}]
V	Cross-flow velocity [m·s^{-1}]
x	Distance in x direction [mm]
y	Distance in y direction [mm]
Z	Axial location from the nozzle [mm]
Greek characters	
β	Maximum growth rate [–]
γ	Liquid surface tension [N·m^{-1}]
ρ	Ratio of air to liquid density [kg·m^{-3}]
ρ_g	Density of gas [kg·m^{-3}]
ρ_l	Density of liquid [kg·m^{-3}]

Acknowledgment

The authors acknowledge the financial support from the project no. LTAIN19044 program INTER–EXCELLENCE (INTER–ACTION), project Quality Internal Grants of BUT (KInG BUT), Reg. No. CZ.02.2.69 / 0.0 / 0.0 / 19_073 / 0016948, which is financed from the OP RDE and from No. FSI-S-20-6295 founded by Faculty of Mechanical Engineering, Brno University of Technology.

References

1 J. H. Kim, R. D. Reitz, and S. W. Park, Proceedings of the Institution of Mechanical Engineers, Part D: Journal of Automobile Engineering **224** (8), 1113 (2010).

2 S. K. Amedorme and J. Apodi, AIP Conference Proceedings **2016** (1), 020026 (2018).

3 S. Deshpande, Atomization and Sprays **21** (2011).

4 H. Zhang, H. Zhang, and B. Bai, Heat Transfer Engineering, 1 (2021).

5 H. Sun, B. Bai, and H. Zhang, Heat Transfer Engineering **35** (6-8), 664 (2014).

6 S. Dikshit, D. Kulshreshtha, and S. Channiwala, *Numerical simulation of Pressure Swirl Atomizer for Small Scale Gas Turbine Combustion Chamber*. (2017).

7 M. Maly, J. Jedelsky, J. Slama, L. Janackova, M. Sapik, G. Wigley, and M. Jicha, International Journal of Heat and Mass Transfer **123**, 805 (2018).

8 R. L. C. Flemmer and C. L. Banks, Powder Technology **48** (3), 217 (1986); R. Turton and O. Levenspiel, Powder Technology **47** (1), 83 (1986).

9 D. Schmidt, I. Nouar, P. Senecal, J. Hoffman, C. Rutland, J. Martin, and R. Reitz, *Pressure-Swirl Atomization in the Near Field*. (1999).

10 S. A. Morsi and A. J. Alexander, Journal of Fluid Mechanics **55** (2), 193 (1972).

New measurement method for in-field measurements of liquid phase drift from cooling towers

Jan Cizek[1], *Veronika* Barraclough[1*], *Jiri* Houzvicka[1], *Martin* Dostal[1], *Pavel* Prochazka[2], and *Vaclav* Uruba[2]

[1]4Jtech s.r.o, Ringhofferova 115/1, 15521 Prague, Czech Republic
[2]Institute of Thermomechanics, Czech Academy of Sciences, Dolejskova 1402/5, 18200 Prague, Czech Republic

Abstract. This contribution introduces a newly proposed thermal based method for measurements of drift for in-field applications on real cooling towers up to an eliminator efficiency of approximately 0.0001% of circulating water flow. This method should thus become the main alternative to the methods used nowadays. The main advantage of this newly proposed measurement procedure should be its easy preparation and implementation and quick analysis of measurement results. The article summarizes the development of an innovative probe, and why this new probe and the entire method is needed for the industry. The design of this new probe consists of two main directions of development pathways: the electronic part and the aerodynamic part. The first one lies in the development of a sensor (and the accompanying electronics) and sums up the theoretical principle of the method (the calculation of droplet sizes scatter based on statistics, and the calculation of power needed for the sensor). The aerodynamics part derives from the desired efficiency and accuracy of the measurement and is based on precedent modelling and calculations of flow, containing droplets of various sizes through various uniquely shaped channels. The contribution also demonstrates an experiment made thus far, showing quantity measured and the drift evaluation process.

1 Introduction

Drift is a cluster of water droplets released from cooling tower to surroundings. The drift poses a problem for infrastructure in vicinity of its source and tower structure itself, as it could be damaged, but it also represents important problem for environment. Water droplets released from cooling tower contain the same chemicals and the same biology as the circulating water. This is a potential risk for environment and also health.

The drift also causes considerable water loses; at least, the water loss can be 0.0007-0.0014 15 $m^3/h/m^2$, which, in the case of cooling towers in Temelín, represents a lower hundreds of cubic meters per one cooling tower per day.

Determining a drift rate from a real cooling towers in operation becomes very important part of the overall cooling tower operation. In line with these facts multiple states worldwide work or already accepted legislation, which sets up required minimal drift eliminator performance rate – e.g. for France 0.01 %, Spain 0.001 %.

The drift eliminator efficiency is one of the two crucial characteristic parameters. The drift emission from a cooling tower is usually expressed as a percentage of the total circulating water flow rate. It can also be expressed as a mass per unit time related to water or air flow (as defined in [1]).

There are several methods used for eliminators drift measurement. The bellow diagram (Fig. 1) shows their overall classification. More of these methods could be find in [2],[3] and [4].

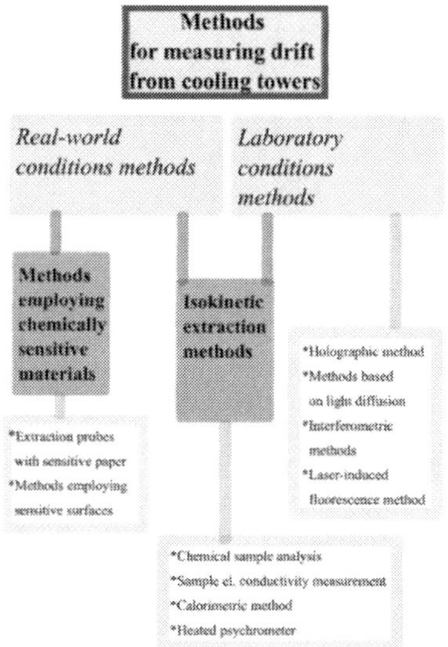

Fig. 1. The state of the art methods for measuring drift from cooling towers.

* Corresponding author: veronika.barraclough@4jtech.cz

All these methods, recently used, for determining drift from cooling tower or drift eliminators certification are long used but, slow, quite complex and costly. The entire industry would accept any new method, which could be used in both lab and field environment at lower cost, if it provides reliable results.

1.1 Calorimetric based principle for the new method

The method is to be based on calorimetric principle. The calculations for this principle are as follows:

First, the total amount of water captured in the new probe in the dependency on sampling time must be determined.

$$m_{DRIFT} = \dot{m}_{DRIFT} \cdot t_{SAMP} = \mu \cdot \dot{V}_{CIRC} \cdot \rho_{WATER} \cdot \frac{A_{PROBE}}{A_{UNIT}} \cdot t_{SAMP} \qquad (1)$$

Assumed and intended values are: cross-section area A_{PROBE} = 10 x 10 mm, typical value of water flow rate V_{CIRC} = 15 $m^3/h/m^2$, density of water ρ_{WATER} = 997 kg/m^3, drift eliminator efficiency μ = 0,001% and a sampling time of 20 minutes (t_{SAMP} = 1200s). The total heat needed for evaporation simply comes out from this value and specific heat of water on boiling:

$$Q_{DRIFT} = m_{DRIFT} . L_{water} \qquad (2)$$

2 Theoretical analysis of the drift

The above reasons led to an idea of new measurement methodology to be developed to determine the amount of the drift via the net of simple sampling probes working on the calorimetric principle. The method will be able to determine the drift with sensitivity at least of 0.001 % of the circulating water and will require max 8 hours in a common cooling tower/cell. This sensitivity was chosen because it is a threshold required by some of the legislations in Europe and also worldwide (Green Building CTI Japan, LEED – 0.002 % drift rate for counterflow towers and 0.005 % for cross-flow towers). This new proposed method started with exact theoretical analysis of the drift.

2.1 The efficiency of the new designed probe

The efficiency of eliminators used in the cooling towers to capture the cooling liquid is one of their two main characteristic parameters. It is defined as the ratio of the drift (i.e the volume of water that passes through the eliminator unit area per unit time and eventually leaves the cooling tower in the form of droplets) to the total water flow through the cooling tower unit area.

Although this parameter is illustrative and very well understood by the general public, it is not a correct parameter. The total volume of water that is not captured by the eliminator, in addition to its geometry and the magnitude of the velocity in the eliminator space, also depends on the size of the droplets that the eliminator is to separate from the flowing air. This fact follows from the principle on which the eliminator separates droplets from the flowing air. It is basically only a curved channel, when due to centrifugal accelerations in each individual bend of the eliminator the droplets cease to be able to follow the flowing air and hit the walls of the duct, where they gradually merge up to drops which, by their weight, exceed the tangential stress and more or less in the form of a water film flow back into the cooling circuit. Correctly defined efficiency should therefore be based on the spectrum of droplets that the eliminator is unable to capture at a given speed.

Although the above-mentioned approach to the definition of eliminator efficiency is inaccurate, it is not possible to use another principle for the initial design, mainly because it is not possible to obtain the necessary experimental data for commonly available eliminators.

Therefore, the efficiency defined by the world's renowned manufacturers was used in the design of the thermal probe. The second mentioned approach (based on the distribution of the spectrum of drops below and above the eliminator) was then used to get at least a basic idea of how many and which droplets stick to the walls of the eliminator.

According to the valid certification standards used in Europe, currently manufactured eliminators must meet the condition that their efficiency, at the air velocity of 3.5 $m.s^{-1}$, reaches at least 0.007% of the total water flow through the cooling tower; the value 0.007% is the limit value for which the eliminator is still declared compliant. The new designed probes for measuring its efficiency have to be designed for lower value of this. The value 0.001% was chosen as the limit value that the method should still be able to detect.

2.3 Considerations of a measurement success rate with respect to droplet size

With regard to the above information and the calculations, which will be carried out later in the chapter 3 describing the probe design, the power for the heated element was set at 100 mW to start with; this information, even maybe posted here in this chapter is a bit premature, is needed for another theoretical consideration of great importance: *measurement success rate with respect to droplet size.*

With water captured in the probe at 0.001% of drift and cooling tower water load of 15 $m^3/h/m^2$ the evaporation of a 1200 second sample collection will occur in approximately 113 s.

The water content is delivered to the probe in the form of droplets of various sizes of the order of about 5 μm to 100 μm. A droplet with a diameter of 100 μm has a total volume three orders of magnitude higher than a droplet with a size of 5 μm, but the probability of its occurrence is on the other hand several orders of magnitude lower than with droplets of the order of 5 - 20 μm. By right, there was a concern that the random presence of large droplet could significantly affect the measurement, respectively, debase it. It is obvious, that although the big droplets which will pass through the eliminator are not statistically relevant by their number, they might have significant impact on the total water volume of the captured water.

So, an estimation (with use of some measured data from drift eliminators, calculations of water collected in a probe of considered effectiveness and with help of Rosin-Rammler particle-size distribution statistics [5]) was processed.

First of all, for some idea, a model with one large droplet with considered diameter 100 μm brought into a group of small droplets was tested. The result was surprising, even this one single droplet represents 11 % of the total water volume. This phenomenon might have significant impact on the method preciseness, especially when total droplet count is low.

After that, a calculation of droplets quantity, which will be captured by the probes in 1200 s time period, was then processed. It was based on data measured in the drift eliminator by IPI method.

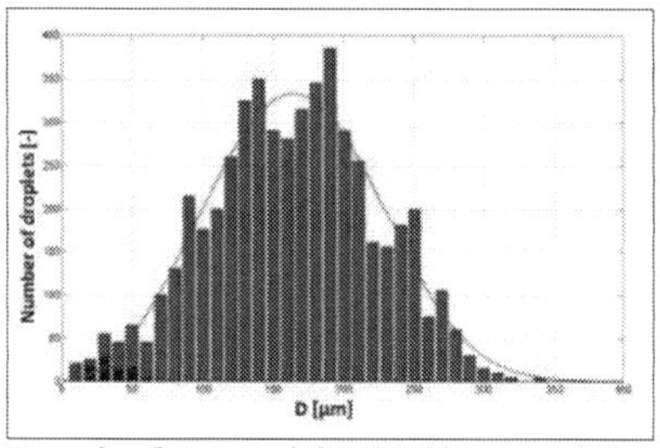

Fig. 2. Example of measured droplets histogram, below (blue) and above (black) eliminator.

The histogram shown above (Fig. 2) is based on data obtained by IPI (Interferometric Particle Imaging) optical measurements. It is clear from it that the number of recorded droplets in the space above the eliminator in the area of the order of 0.1x0.1x0.005 m in the total sum reaches the value of 250.

Normalized distribution of the spectrum of drops above the eliminator based on the IPI measurement is shown in Fig. 3.

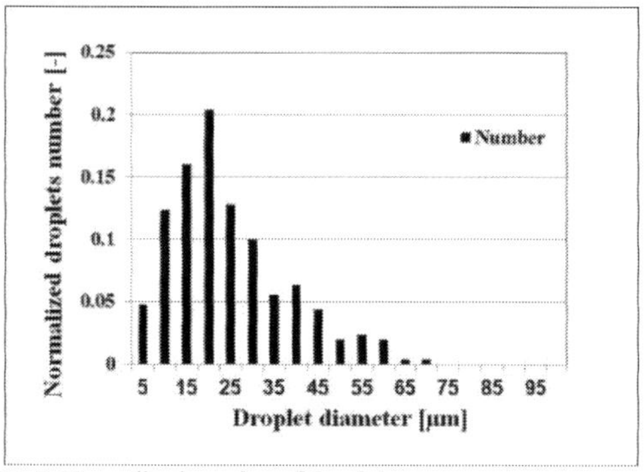

Fig. 3. Normalized number of droplets.

Then, when considering one probe and the calculated volume of water collected in probe for specified time plus Rosin-Rammler distribution, the time, for which a drop will definitely appear in the probe, was determined (Fig. 4).

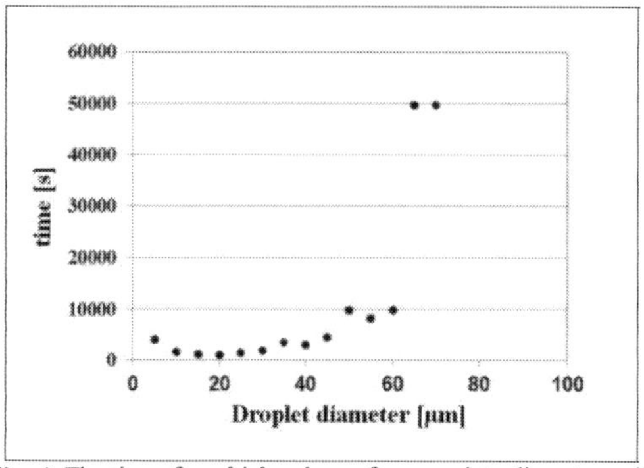

Fig. 4. The time, for which a drop of appropriate diameter, will definitely appear in the probe.

It showed that the impact of big droplets on the method preciseness will be negligible, because their appearance in the probe is very unlikely in comparison to smaller ones (the most common occurrence is recorded in droplets with the diameter of 20 μm).

3 Probe design

3.1 The design of microelectronics part of the new probe

The probe itself consists of two parts, which have to be designed: microelectronics component and aerodynamics proposal of the channel. Schematic illustration of such probe see in Fig. 5. From the microelectronics point of view, an active element together with whole control unit was designed. The main parameter for the active element, which has to be prepared from the theoretical analysis, was required power.

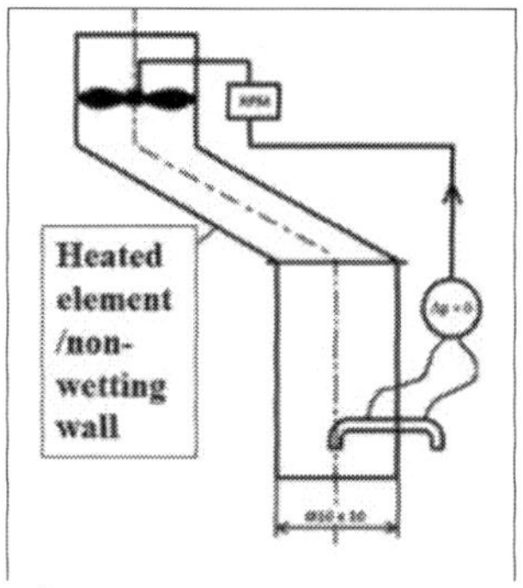

Fig. 5. Probe scheme.

There were several preconditions laid down for sample collection. On one hand the element can evaporate water in a suitable time (short enough so that the value is not affected by other droplets introduced into the probe space) and on the other hand that its temperature does not exceed a reasonable limit.

The cross section of the probe was chosen to be the size 10 x 10 mm, the flow of uncaptured water through the probe therefore corresponds to a value of 0.01 - 0.02 g/hour, resp. 0.17 - 0.33 mg/min.

For the design value of probe cross-section area (A_{PROBE} = 10 x 10 mm), typical value of water load (V_{CIRC} = 15 m^3/m^2/h), density of water (ρ_{WATER} = 997 kg/m^3) the drift eliminator efficiency (μ = 0.001%) and a sampling time of for instance 20 minutes is the total amount of water in the probe m_{drift} = 5mg. The total heat needed for evaporation is then Q_{drift} = 11.32 J.

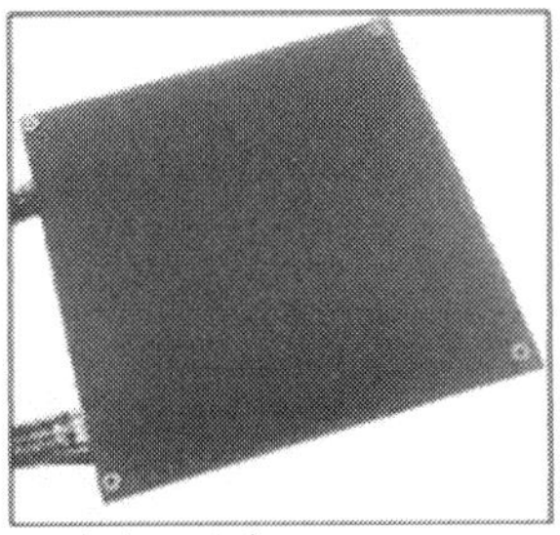

Fig. 6. Custom made heated element.

This heated element (see Fig. 6) is driven by control unit. PCB of this control unit is shown below (Fig. 7).

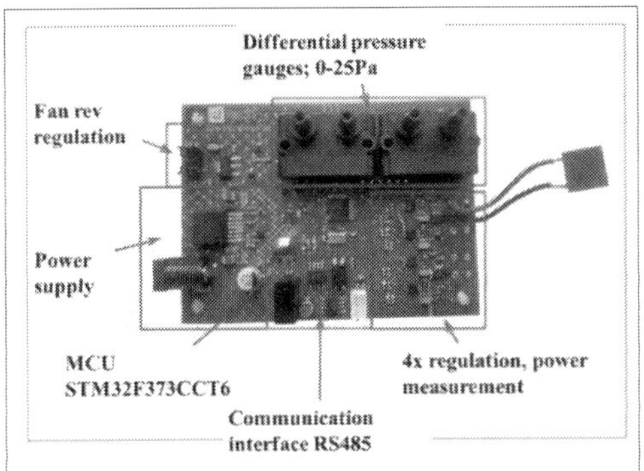

Fig. 7. PCB of the probe control unit. Apart from the heated element, it also controls the pressure gauges and the fan. The latter two components are required to comply with certain measurement conditions.

3.2 Probe shape design

The design of the hardware part of the probe started with the simple scheme (see Fig. 2). In principle, it should be a curved channel, the shape of which is based on the shapes of highly efficient eliminators with known efficiency values. On the other hand, there is another aspect that must be taken into account: at the same time, the probe must be an easily manufacturable channel, not only for the purposes of a prototype of the probe, but also for an experiment using optical methods. So far, the last design deviated from the originally proposed scheme and settled on a probe of uncomplicated shape with incandescent elements located on the side opposite the inlet opening. One of the tested designs see in Fig. 8.

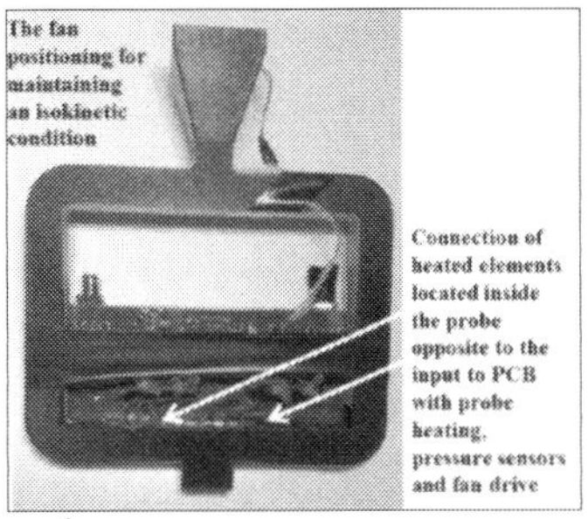

Fig. 8. The probe.

3.3 Verification tools

Given that this is an innovative approach to measuring the drift, it was also necessary to design or possibly expand existing experimental or theoretical methods that will serve as tools to verify the theoretically designed principle of the probe. Chapter 3.3 Verification tools deals with their design, development and testing.

Two options were chosen from the available tools for verifying the probe principle. The first of these was the Interferometric Particle Imaging (IPI) method, the second was another optical method using similar experimental equipment Particle Image Velocimetry (PIV).

In the case of IPI, it is a non-invasive optical method providing mainly information on the size of individual droplets below and above the probe channel. The results obtained may generally include the integral value of the cooling tower drift as well as the efficiency of the eliminator (or in our case the probe channel) according to particle size, correlation between particle size and even speed, etc. The great advantage of this method is that the minimum measurable droplet sizes are almost order of magnitude smaller compared to other methods.

Similarly, the other optical PIV method is a non-invasive method using the same laboratory laser and the same high-speed camera. In principle, this is nothing more than taking two images of the same space with a very short time interval (in the order of hundreds of microseconds depending on the expected velocity), the setting of which is the basic input parameter for obtaining relevant data. From these duplicates, the flow rate in the probe can then be determined using complex algorithms used to process the image. The probe efficiency can be obtained from the subsequent purely mathematical processing after calculating the behavior of the two-phase flow.

3.3.1 Measuring stand for IPI and PIV measurement

It is an open wind tunnel - a discharge line with a drive. The fan is driven by a motor with speed control. The output cross section is 250mm x 250mm, the speed of the air flow is continuously adjustable in the range of 4-25m/s. Two-phase flow is provided by a droplet generator. The track was designed as variable for easy exchange of the measured channel so that it is possible to test different variants of the channel. Experimental setup see in Fig. 9.

Fig. 9. The detail of an experiment setup for IPI and PIV veryfication measurements.

3.3.1.1 Experiment results from IPI and consequent evaluation method development

Measured data from IPI first served as a underlying material for a program, which is being developed for later droplets interferograms automatic detection. This program is called *IPIDET*. Its development will be described later in this subchapter. Thus, the data from IPI experiment have not yet been acquired primarily for the purpose of evaluating the drift, but rather setting the image detector. This means that the images do not contain interferograms of droplets taken before entering the canal and after leaving it, which would be the right approach to measure the number of droplets before and after the probe (from which the capture rate is evaluated), but directly in it.

The settings of the measuring apparatus were as follows: the lens has a focus of 105mm, the aperture value is set to F2.8 and the distance laser plane - front lens is approximately 200 - 210mm (depending on the configuration) and this distance is measured with an accuracy of about 5 percent. The picture (Fig. 10) shows interferograms in channel No. 1., manually labeled.

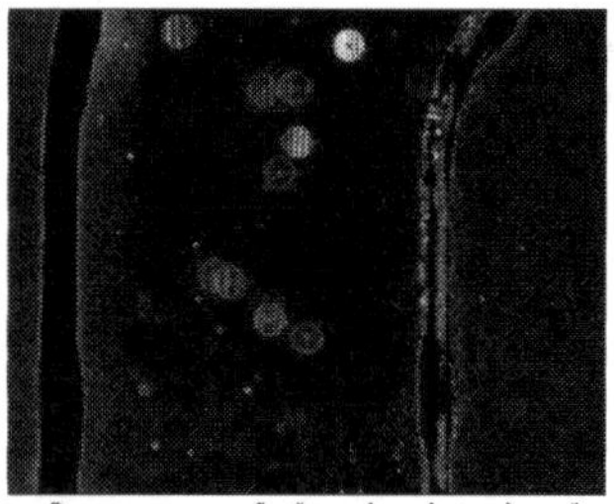

Fig. 10. Interferograms of the droplets in the channel No. 1.

The goal of the developed software called IPIDET (acronym composed of IPI Detection), which is created in the python programming language, is to automatically detect droplets interference images from measurements obtained by the IPI method and then calculate the number of their minimum / maximum.

The whole image processing algorithm consists of two main parts:

1. Detection of interference images (droplets).
2. Edge counting.

For example, an image obtained using IPI might look like this (Fig. 11):

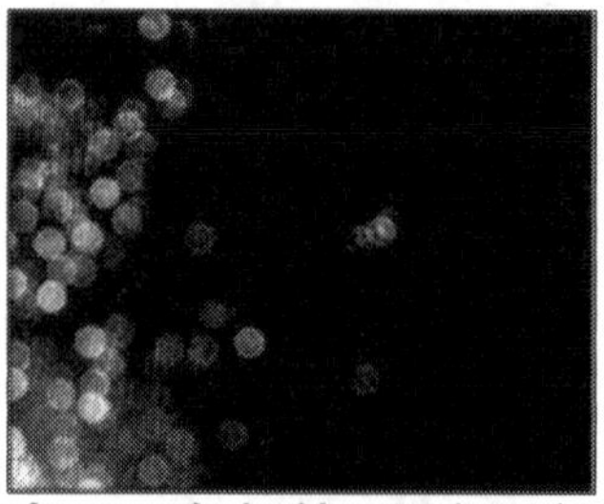

Fig. 11. Interferogram obtained by experimental IPI method.

It can be seen that there is a large overlap of images, some of which are overexposed, others underexposed. Of course, it depends on the skills and especially the experience of the experimenter to be able to prevent similar-looking images by precisely setting the measurement technique, however, this could not be expected from the first data sets available for programming and it was necessary to obtain data for program creation. The advantage may have been that it was necessary to deal with worse input data at the outset and tune the detector to various picture difficulties. Therefore, when most images have one overexposed area containing a lot of droplets, while the rest is underexposed, the detector has been programmed from the beginning, for example, in the way that it does not ignore these dark areas that also contain droplets. Therefore, before the detection itself, the program performs image preprocessing. Pretreatment is performed using CLAHE (Contrast Limited Adaptive Histogram Equalization).

Further, in order to detect the droplet image, the images must first be resampled to reduce the computation time. The image is then correlated with a circular image of approximately the same size as the droplets. The result of this correlation looks like this: the image is still quite noisy and the so-called peaks, which then need to be searched, are not so well defined (Fig. 12 - left). Therefore, the second correlation is performed using the Gaussian kernel (Fig. 12 - right).

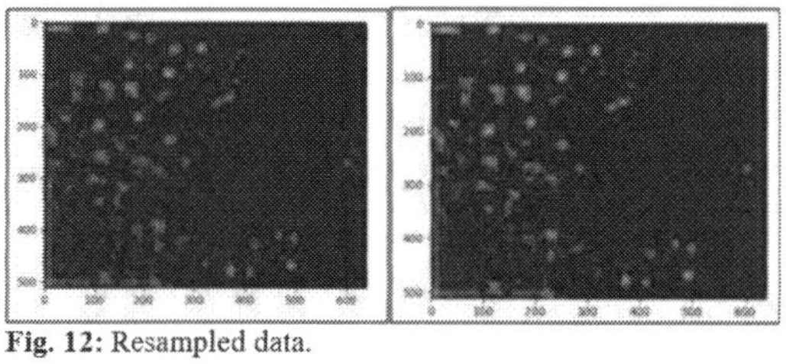

Fig. 12: Resampled data.

In such image, it is finally possible to detect individual peaks for further processing.

Droplet slices are extracted from the image using the coordinates obtained in the previous section. To calculate the number of droplet interference image fringes, the following procedure is done with the image:

1. | 2D FFT |.
2. Standardization.
3. Finding adjacent peaks (except for the DC component in the middle).
4. Calculation.
5. The droplet size calculation follows.

3.3.1.2 Experiment results from PIV and consequent evaluation method development

The velocity field was measured using the time-resolved PIV method. The experiment was carried out for the first proposed channel geometries, which were still based on the original idea (as was shown in Fig. 5). The figures below show the channels that were prepared for optical measurements using plexiglass (models of the channels see in Fig. 13).

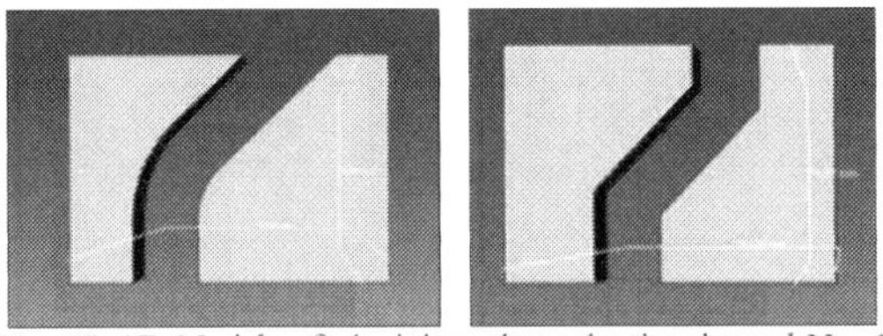

Fig. 13. 3D Models of plexiglass channels: the channel No. 1 (left) and the channel No. 2 (right).

The experiment was performed with the following equipment in the configuration described below.

The measuring device consists of a New Wave Pegasus laser and a CMOS camera. The Phantom V611 camera with a resolution 1,280 x 800 pixels is capable of capturing double images with a frequency of up to 3 kHz (full resolution) and uses an internal memory of 8 GB. The data were obtained and subsequently processed in Dynamic Studio software. The evaluated velocity field consisting of 77 x 47 vectors was taken at a frequency of 100 Hz, one record contained 1,000 double frames representing 10 s in real time. The time between pulses was set at 50 μs. SAFEX tracer particles, oil droplets with a diameter of 1 mm, were used.

The experiment results prepared for the later numerical evaluation of the particle trajectory are shown in the following figures (Fig. 14 and 15). Q 45% (or 70%) means 45% (or 70%) of the maximum fan speed.

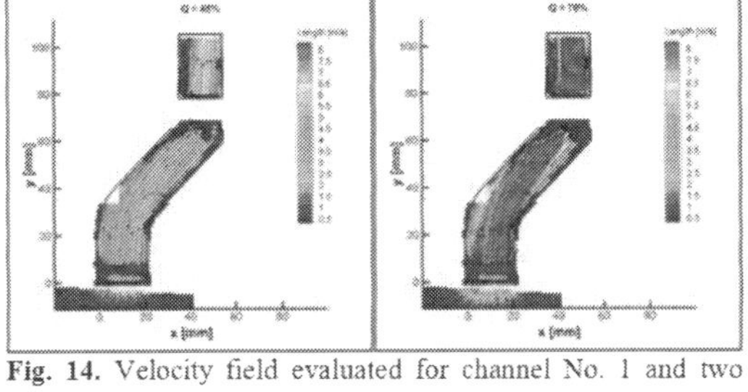

Fig. 14. Velocity field evaluated for channel No. 1 and two flow values.

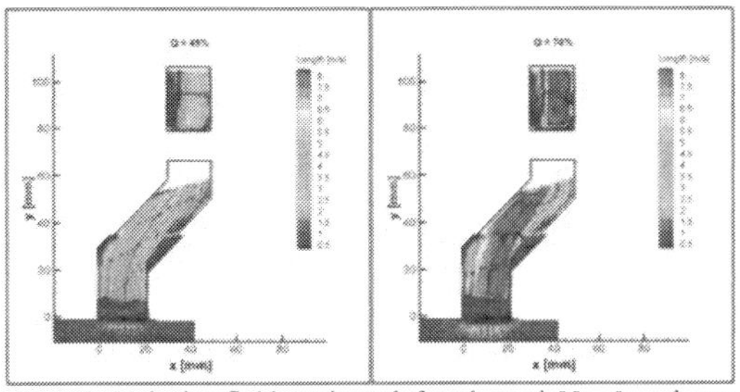

Fig. 15. Velocity field evaluated for channel No. 2 and two flow values.

Such results are useful for the real probe design itself (as they help with the aerodynamics), but as they are, they need another theoretical calculation for the probe efficiency determination.

The search for this problem was performed with the help of L. Novakova dissertation thesis [6]. The efficiency of capturing the liquid phase by the probe can be determined by calculating the trajectories of the droplets carried by the air stream. The trajectory calculation is being performed under certain necessary assumptions, for example: the velocity field is not affected by the presence and movement of droplets, mutual interactions of droplets are not considered, the probability of droplets with a given radius is constant along the entire entrance or the effect of water film on walls is not considered. The trajectory of the particle motion was calculated based on the solution of the equation of motion. As the input data, detailed information of the flow in the channel, using the velocity field measured by the PIV method are used. The algorithm is still in progress.

4 First experiments and experimental verification method

4.1 First experiments

First tests with heated elements were performed in a lab. Three configurations were measured in simple configuration to prove the principle of the designed probe: a) dry heated element, b) heated element with water droplet attached and c) heated element covered by water film. Also different power values for evaporation were set – 50 mW, 100 mW and 150 mW. Results are shown in the charts bellow (Fig. 16).

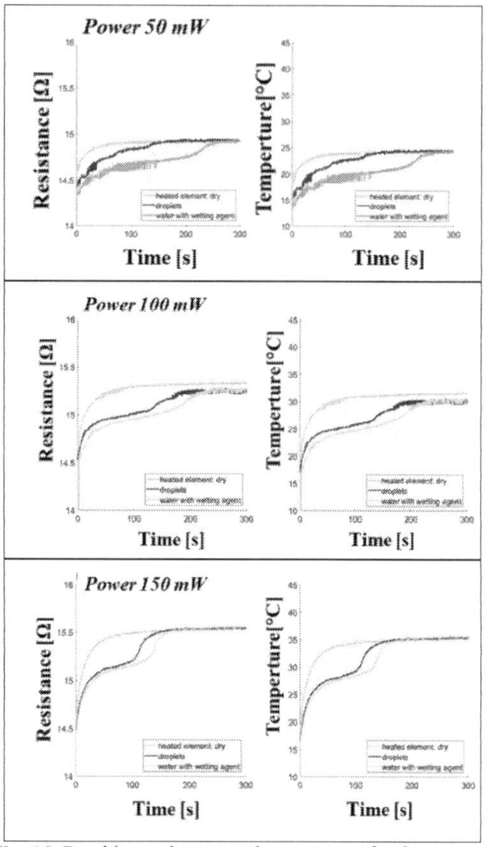

Fig. 16. Resulting resistance and temperature for three sets of power on heated element in configuration: dry, with droplet attached to the surface and with the water film covering the element.

The repeatability experiments are planned in the spring of 2022. So far, measurements have been made in a small laboratory tower equipped with an ultrasonic fog generator (Fig. 17).

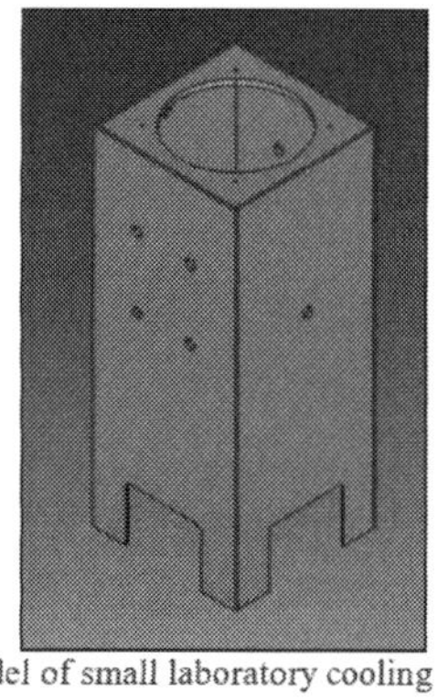

Fig. 17. 3D model of small laboratory cooling tower (height 700 mm).

The testing was performed for a flow velocity in the tower of 2 m.s^{-1}, which was measured at the point of probe entrance with a Testo 425 anemometer. The probe was equipped with a fan necessary to maintain the isokinetic conditions at the outlet and four heated elements located opposite the entrance to the probe; it is possible that in the ongoing tests this configuration will be extended by other positions of the heated elements. The output of the measurement monitored so far was the change in resistance of the heated element over time. A large number of experiments have been performed, where the droplets collection takes place for different times and each of these collections is completed by evaporation of one set value of glow power and one set value of evaporation time. Data processing is in progress at the time of submission and further experiments are planned, so no specific results will be given here.

The probe was further tested for mechanical endurance under cooling tower conditions. The real conditions in which the probe will operate represent a burden for the control and power electronics. The most important were the water resistance tests, both for the microelectronics itself and for the fan providing the isokinetic condition. For real operation, the electronics must withstand at least two days of such a load, which has been proven by above described testing.

4.2 HGBIK probe

The effectiveness of the probe in the experiment will be compared with the method that is certified by the highest authorities in the discipline for the measuring of the drift - the HGBIK method (Heated Glass Beads Isokinetic). Its production in accordance with the CTI-140 standard [7] also belongs to the range of activities that needed to be carried out during the development of the probe.

Method is based on the calorimetric principle with subsequent chemical analysis of the captured sample. It is characterized by isokinetic extraction of air samples containing water droplets with so-called tracers. The measured quantity is the total mass flow. The HGBIK method is tested and introduced in the certification practice of measuring the efficiency of the drift eliminators, therefore this was chosen for the verification measurement of the new developed probe. It is one of the most widely used methods in the industry. Schematic illustration see in Fig. 18, finished probe in Fig. 19.

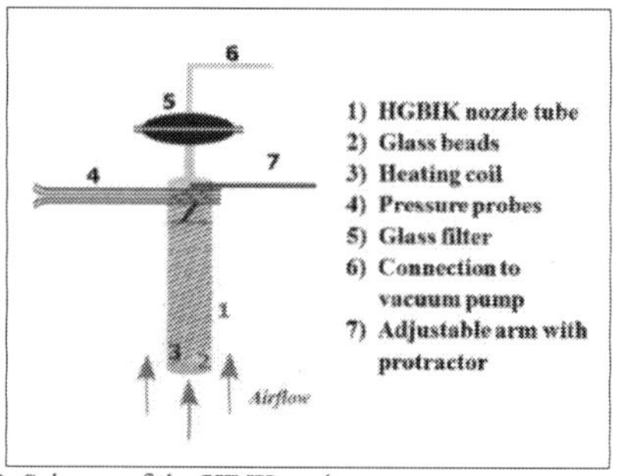

Fig. 18. Schema of the HBIK probe.

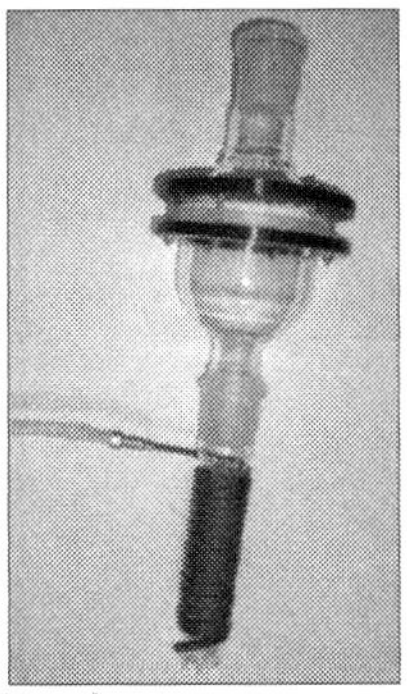

Fig. 19. The HGBIK probe.

5 Conclusion

The purpose of this paper was to present the development of a new probe for measuring drift in real and laboratory conditions. This is a development from the initial theoretical analysis of the drift (its probable capture in the probe of the selected cross-section, size and frequency of droplets) through practical tasks both in terms of microelectronics and selection of a suitable aerodynamic channel shape for sampling air flow, to the development of the methods, which will verify the effectiveness of the meter.

These are advanced experimental methods of PIV and IPI and subsequent processing of the achieved results. In the case of the PIV method, it is a calculation of the efficiency of the probe based on PIV experimental data. On the basis of PIV measured data, the geometry of the channels was modified, which was a bonus output of this experimental part. In the case of the IPI method the programming of a droplets detector and the calculation of their size isbeing developer. The IPI method uses the same measuring device as the PIV method, but its output is a little bit different and the results will have to wait a while until the program for detecting droplets and calculating their size IPIDET is finalized. This is an integral part of the whole development, because the entire development is more or less focused on the size of the drift droplets.

By completing the calculation of droplet trajectories from the PIV data and finalizing the IPIDET program, two methods for determining the efficiency of the probe will be obtained. After its full deployment in a real experiment in the cooling tower, another method will be added, which will verify the probe. It is the method of certified isokinetic sampling by the HGBIK probe.

All three methods are either so important or so interesting from the point of view of scientific contribution that they are all gradually used for verification and experimentation. By measuring using the IPI method it is possible to obtain an accurate distribution of drift droplet size and thus reduce the measurement uncertainty of this quantity. Using of PIV, an overview of the velocities of the droplets and especially the direction of their movement within the two-phase flow, which clearly determines the ideal place of capture and thus the location of the heated elements , will be obtined. And using the HGBIK certified isokinetic sampling method, the new probe, compared to the results from the measurement with HGBIK prope, will gain maximum confidence.

References

1. CTI Bulletin BUL-109: Nomenclature for Industrial Water-Colling Towers (version 2015).
2. J. Cizek, L. Novakova, J. Nozicka: *Reducing Liquid Phase Drift from Cooling Towers*, Gradient, Prague (2009).
3. J. Cizek, L. Novakova: New methods for drift eliminators performance evaluation, CTI TP11-11, (2011).
4. A. Roffman, L.D. Van Vleck: The State-of-the-Art of Measuring and Predicting Cooling Tower Drift and Its Deposition, Journal of the Air Poll. Contr. Assoc., 24:9, 855-859, DOI: 10.1080/00022470.1974.10469982 (1974).
5. P. Rosin, E. Rammler : The law governing the fineness of powdered coal, Journal Inst. Fuel, Vol. 7, pp. 29-36, and discussion, pp. 109-122 (1933).
6. L. Nováková: Methodology for determining the effectiveness of liquid phase eliminators (published in czech), Dissertation thesis, Prague (2010).
7. CTI ATC-140 Isokinetic drift test code (2011).

Two-wavelength digital holographic interferometry for unambiguous range extended measurements in fluid mechanics

Gramoz Çubreli[1,2*], *Pavel* Psota[2], *Ahmad* Kouta[1], and *Petra* Dančová[1]

[1]Faculty of Mechanical Engineering, Technical University of Liberec, Studentská 2, 461 17 Liberec 1, Czech Republic
[2]Faculty of Mechatronics, Informatics and Interdisciplinary Studies, Technical University of Liberec, Studentská 2, 461 17 Liberec 1, Czech Republic

Abstract. Non-contact optical methods such as digital holographic interferometry are highly suitable in measurements where the phenomena is fast, performed in transparent or semi-transparent environment and mustn't be obstructed as when applying local contact techniques. Such specific application can be studying dynamic events during transonic and supersonic blade flutter. Fast, sensitive and rather easy access to the phase information make these techniques very attractive in the study of phase objects/phenomena. However, since light's phase is bounded to a repetitive cycle of 2π radians, the range of measurement is limited to one cycle of the phase, limiting applications to small gradient phenomena. This paper presents a new interesting way of by-passing this limitation, while still keeping noise values low, by introducing a second laser with a close value wavelength, giving rise to a new interferometric pattern with an extended unambiguous range of measurement. Image acquisition is done simultaneously for both wavelengths and all reconstructions are digitally performed. The principle and preliminary results are included in this paper.

1 Introduction

For fast occurring phenomena, where the nature and properties of the fluid must be studied without disturbances from the observer, non-contact optical methods such as digital holographic interferometry present themselves as reliable, fast and accurate techniques. Such technique require nonetheless that the liquid under study to be transparent or semi-transparent, in order for the optical rays to pass through and be recorded. Phase imaging holographic interferometry provides high flexibility and means for highly sensitive measurement of phenomena with optical path variations.

Interferometric methods offer very high precision measurements which lie in the range of the nanometric scale [1]. One clear advantage is the fast numerical reconstruction.

Digital holographic interferometry is being used more and more due to its clear advantages over the conventional holographic technique. Probably the most important advantage lays in its power to acquire and process images/information in real time. It offers the possibility of easily acquiring images through CCD cameras, complex and fast numerical processing, storage of large data, numerical image reconstruction and real time presentation of the phenomena. Digital holographic interferometry finds a very wide range of applications, notably in medical imaging where the study of biological specimens has been made easier and better. The range of application is quite wide, in industrial and scientific applications, notably in quality control in micro-electro-mechanical systems (MEMS) fabrication, ultraprecise manufacture, research in ophthalmology etc. [2]

In this paper, an off-axis two-wavelength digital holographic interferometry technique for resolving the issue of phase unwrap ambiguity for the case of a resistor is presented.

When the object we measure has sudden profile changes that go beyond the range limit allowed for a specific wavelength, a phase ambiguity arises. If this ambiguity isn't solved in the phase unwrapping process, it will give faulty result, not mapping the true profile of the phenomenon we are looking at. The phase is proportional to the surface profile (when trying to image surface profiles from reflection), to the integral of the refractive index for transparent media, otherwise it is linked to the wavelength via respective formula. [3] Two-wavelength interferometry offers an alternative solution to solve the phase ambiguity problem. [3]

This method enables a fast and efficient way of imaging the phase profile for a multitude of applications. It makes it possible to measure in real time phenomena that cause large phase changes without ambiguity. Thus, by using two lasers or more, the range of unambiguity can be increased compared to using a single laser [1]. The disadvantage of this technique is that any noise present in the phase profile of any of the wavelengths will get amplified by a magnification factor equal to one of the synthetic wavelength [4]. Off-axis configuration spatially separates the holographic images away from the undiffracted zero order. [5,6]

* Corresponding author: Gramoz Çubreli gramoz.cubreli@tul.cz

Two-wavelength interferometry can be used in many areas of optical metrology, such as for the measurement of the thickness and surface profile/inspections of objects [3]. It can be used in research dealing with samples having large topographical changes and biological samples [7].

Due to its extended unambiguity range, high-speed acquiring and large field of view, the method will be applied for investigation of high-speed flow in planar blade cascades, namely for studying dynamic events during transonic and supersonic blade flutter.

In interferometry, specifically phase-shifting interferometry, at least three intensity maps are needed to resolve the phase map, while in two-wavelength interferometry it is six but can be decreased to two. [8]

2 Principle of two-wavelength interferometry

In single-wavelength technique, when the object of investigation is thicker than the laser wavelength (in our case the temperature range that can be mapped surpasses the range of measurement) used for research, the phase gets wrapped and suffers from 2π ambiguities. The range of measurement is directly connected to the wavelength. For smaller wavelengths, the range decreases, while for longer wavelengths it increases [4]. Longer wavelengths produce fewer fringes, reducing the number of 2π ambiguities [9]. Phase has a periodicity of 2π radians and the phase of light can be expressed as a multiple of this 2π periodicity. The repetitive nature of light propagation as wave in the form of a sinusoidal wave makes the range of clear unambiguous measurement to be limited to one cycle of repetition, i.e. one wavelength. Due to Nyquist criteria, the real range of unambiguity is limited to a half-wavelength [10]. In other words, the wrapping is a direct consequence of the 2π periodicity of the $arctan$ function used to extract the phase profile from the light that is being measured [4,11]. During the phase unwrapping process, the absolute phase difference between points of measurement in time and space must be less than the value of π radians for a better unwrapping [12]. There are cases when the difference is much higher, causing multiple wrapping simultaneously. Care must be taken when dealing with such situations.

The sole idea of using a two-wavelength technique comes from the limitations mentioned above. In two-wavelength or multiple-wavelength interferometry, the unambiguous range of measurements gets extended by what is called the synthetic wavelength [1,11]. The general graphical concept representation is shown in the following fig.1.

A requirement that must be fulfilled for such optical technique is that the optical path from all beams must be the same. This condition can create some difficulties in the building of the optical setup, and sometimes it can turn out to be expensive [2]. Hence, appropriate geometrical setup must be found in order to accommodate this condition.

A disadvantage of this technique is that, despite ambiguity being solved (reduced), the range of unambiguous measurement can be extended at the cost of increasing noise by the factor of the ratio between the synthetic/equivalent wavelength and the shortest wavelength Λ_{eq}/λ_1, thus decreasing the sensitivity of the technique for the same factor. In length measurement, this is reflected as increase of uncertainty measurement. [1,4]

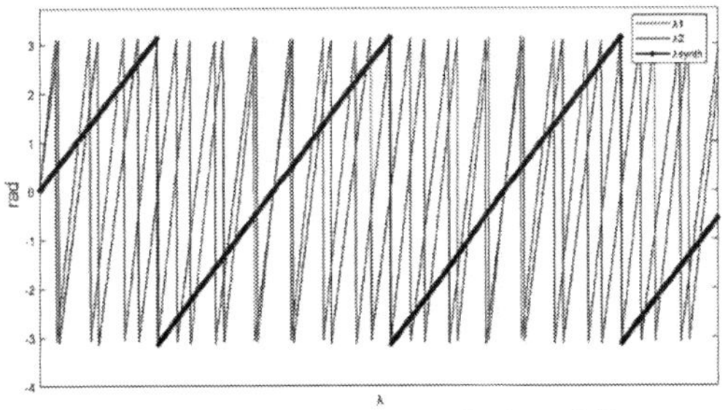

Fig.1. Graphical representation of the difference of the measurement range covered by smaller (λ_1, λ_2) and larger (λ_{synth}) wavelength. [9]

Let the change in refractive index that the wave experiences while passing through a heat field be expressed as $\varphi(x, y)$, then the interference pattern on the CCD camera is proportional to:

$$I(x,y) \sim \cos\left[2\pi\varphi(x,y)\left(\frac{1}{\lambda_1}-\frac{1}{\lambda_2}\right)\right] \quad (1)$$

, which leads to what is called the synthetic wavelength (or equivalent/beat wavelength) [13,14]:

$$\frac{1}{\Lambda_{eq}} = \frac{1}{\lambda_1} - \frac{1}{\lambda_2} = \frac{\lambda_2 - \lambda_1}{\lambda_1\lambda_2} \Rightarrow \Lambda_{eq} = \Lambda_{12} = \frac{\lambda_1\lambda_2}{|\lambda_2 - \lambda_1|} \quad (2)$$

From (2), it can be seen that the smaller the difference in the dominator, the larger the synthetic wavelength will be. This implies that, if the two lasers have a smaller difference in wavelengths, then the new range of unambiguous measurement will increase. [5,9]

For the case of multiple-wavelength interferometry, the synthetic wavelength can be defined as [15,6]:

$$\Lambda_{eq} = \frac{\lambda_1\lambda_n}{|\lambda_n - \lambda_1|} \quad (3)$$

, where subscript n represents the number of lasers used.

3 Unwrapping

Phase unwrapping has been extensively studied and adapted for many different applications. Different techniques that require phase unwrapping process are magnetic resonance [16], x-ray crystallography [17], synthetic aperture radar (SAR) [18] etc.

There exist many different techniques to extract the phase distribution from the fringe pattern, such as phase-shifting profilometry, Fourier transform profilometry, windowed Fourier transform profilometry etc. Highest resolution and accuracy are provided through the first technique, phase-shifting profilometry [19]. Phase-shifting interferometry is not suitable for measuring dynamic phenomena because it requires at least three holograms to be recorded in order to solve the equations to retrieve the coefficients [11].

Phase unwrapping process has some requirements when it comes to the noise and the discontinuity the phase might have. The easiest way is usually to use one of the 2π phase unwrapping algorithms, but in certain cases not the same algorithms are valid for different kinds of wrappings. Unwrapping algorithms usually are heavy computational wise and can be subject of failure where the object or phenomena under study has large irregularities.

In the cases when the phenomena is restricted to a small confinement in the image, the edge of the image where no change has occurred can be chosen as the starting point for phase unwrapping, since interferometric measurement require at least one point where the phase change is known. [11]

The basic idea of phase unwrapping consists on dividing the phase image into horizontal line, which are then unwrapped pixel by pixel by adding or subtracting the offset (2π radians) when necessary. This is done separately for each of the lines. Once this is done, the same process happens, but vertically. At the end the phase image will be unwrapped in 2D [12].

By having the difference of the two wrapped phase profiles created from both wavelengths, a new wrapped phase profile arises. This new phase profile ''belongs'' to a much greater wavelength, which we called previously as the synthetic wavelength. The synthetic wavelength is used to resolve the ambiguity by detecting the average phase shift (by calculating the fringe order) while using the lower laser wavelength to measure the temperature.

Based on the dependence of the path, there are path dependent and path independent phase unwrapping techniques. In path dependent techniques, image processing is used to detect the edges and the phase ambiguities, and from that to calculate the offset that must be added or subtracted, while in path independent techniques the regions where error could be caused are eliminated before the phase unwrap process starts. [9]

Spatial phase unwrapping, from its name, uses the relationship between the phase information of the spatial neighboring pixels [19,20]. There exist different kinds of spatial unwrapping algorithms, notably the Goldstein's method, Flynns' method etc.

Determining the fringe order from spatial neighboring pixels is not possible in the spatial phase unwrapping since it is based on the information that neighboring pixels have.

The necessity of temporal phase unwrapping plays in when we need to unwrap a general phase map with large discontinuities. The fact that each pixel is unwrapped independently from its neighbor makes that any present noise in one of the pixels not to spread to other pixels during the process of unwrapping. There are also different kinds of temporal phase unwrapping algorithms, and the advantage of using temporal unwrapping stands from the fact that phenomena with large phase changes (in the case of profile measurements coming from high discontinuities) can be studied. The simplest one is the Gray-code temporal phase unwrapping. Other techniques, for example that use additional wrapped phase maps that differ from their fringe order to unwrap temporally the phase map, can be categorized into three groups: multiwavelength (heterodyne), multifrequency (hierarchical) and number-theoretical approach. A detailed review over these three techniques has been published, according to which, this group outperforms the Gray-code algorithm in accuracy, pattern efficiency and unambiguous range. [8]

The interference pattern on the CCD camera can also be expressed as [11,10]:

$$\begin{aligned} I(x,y) &= A(x,y) + \\ &+ \sum_{l=1}^{2} B_l(x,y)\cos\{\varphi_l(x,y) + 2\pi f_l x\} = \\ &= A(x,y) + \\ &\quad +B_1(x,y)\cos\{\varphi_1(x,y) + 2\pi f_1 x + 2\pi f_1 y\} + \\ &\quad +B_2(x,y)\cos\{\varphi_2(x,y) + 2\pi f_2 x + 2\pi f_2 y\} \end{aligned} \tag{4}$$

, where $A(x,y)$ is the background intensity and pattern brightness, $B(x,y)$ is the intensity modulation and f_1, f_2 are the carrier frequencies. This is known as spatial carrier frequency interferometry. The relation between the interference phase and the refractive index variation is given in (5). If we assume no variation of the refractive index n in the direction of propagation, then the integral simplifies in multiplication only:

$$\varphi(x,y) = \frac{2\pi}{\lambda}\int_L \Delta n(x,y,z)ds = \frac{2\pi}{\lambda}\Delta n(x,y)L \tag{5}$$

, where ds denotes the differential distance along the line of integration L. The wrapped phase is extracted by:

$$\varphi(x,y) = \arg\{I(x,y)\} = \arctan\left(\frac{Im\{I(x,y)\}}{Re\{I(x,y)\}}\right) \tag{6}$$

The basic idea of unwrapping consists of shifting back to its place each phase jump by adding or subtracting integer multiples of 2π radians, in order to draw the phase as a continuous function (without the jumps, causing this characteristic sawtooth/triangle form of the phase), which can be expressed as:

$$\theta(x,y) = \varphi(x,y) + 2\pi N(x,y) \tag{7}$$

The main idea is to retrieve as fast and accurately possible the fringe order term $N(x,y)$ for each of the pixels of the camera. We can make of use the extended range of unambiguous measurement from the synthetic wavelength and the accurate measurement of single wavelength phase that doesn't suffer from noise amplification to unwrap the original phase profile. There are several ways of how to do it and different technique corresponding to different applications. The most straightforward way of unwrapping is to make use of the synthetic phase profile in order to extract the average phase shift (fringe order). By making use of the average operator ⟨ ⟩ the procedure can be mathematically represented as:

$$N = round\left(\frac{\langle \varphi_{synth}\frac{\Lambda_{eq}}{\lambda_2}\rangle - \langle \varphi_2\rangle}{2\pi}\right) \tag{8}$$

Other unwrapping procedures are for example the hierarchical temporal phase unwrapping which uses one of the wavelength's phase map, usually the one with the shortest wavelength. It unwraps the phase map with the help of additional wrapped phase maps which have different fringe orders. Temporal processing is for relatively slowly developing phenomena.

It is worth saying that, for any kind of phase unwrapping algorithm, if the process is successful, the final accuracy will be identical. The difference between hierarchical and multiwavelength unwrapping lays in that that the first uses one of the wavelengths for the process of unwrap, while the later uses the synthetic wavelength to start the process of unwrapping. [14]

The information that was measured from both wavelengths is retrieved in the reconstructed phase field. Both phase information have different dimensions in the Fourier spectrum, thus they must be cropped and resized in order to be used for the obtention of the synthetic phase field. This means that the size of pixels in both fields must be the same. This is achieved by resizing one of the phase fields, using the equation:

$$\Delta\xi_2' = \frac{\lambda_1}{\lambda_2}\Delta\xi_2 \tag{9}$$

, where $\Delta\xi_2$ denotes the sampling interval before compensation. Linear interpolation between neighbouring pixel values is adopted to determine the value of the compensated pixel.

4 Optical setup

Fig.2. represents the experimental lensless Fourier optical setup used to conduct this experiment. Two fiber coupled DFB (distributed feedback) laser diodes having different wavelengths are used. Light is split into the reference (RB1, RB2) and object beams (OB1, OB2) by means of fibersplitters. The object beams are combined and are collimated by a lens to then fall onto the diffuser. Scattered light from the diffuser passes through the measured area (MA) to combine with reference beams and be recorded by the camera (CAM). Both reference beams form angles with respect to the axis of propagation. It is these angles that introduce the spatial carrier frequencies that make it possible to separate phase information.

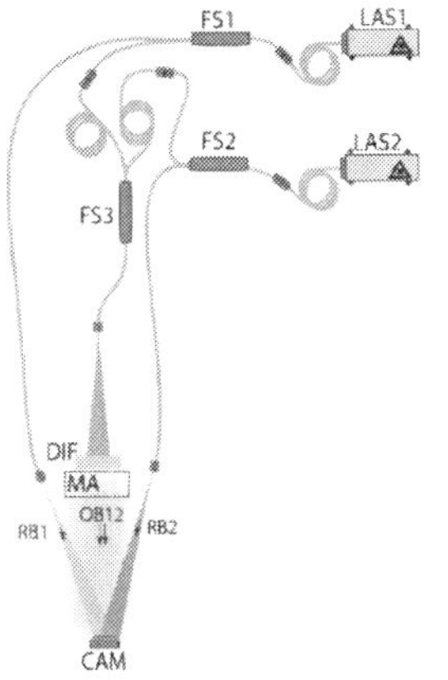

Fig.2. The two-wavelength lensless measuring system: LAS-laser, FS-fiber splitter, DIF-diffuser, MA-measured area, OB-object of investigation, RB-reference beam, CAM-camera .

Two lasers of wavelengths $\lambda_1 = 773nm$ and $\lambda_2 = 780nm$ were employed for the measurement.

According to (2), this gives rise to a synthetic wavelength of $\Lambda_{eq} = 86134.28nm$

5 Results and discussion

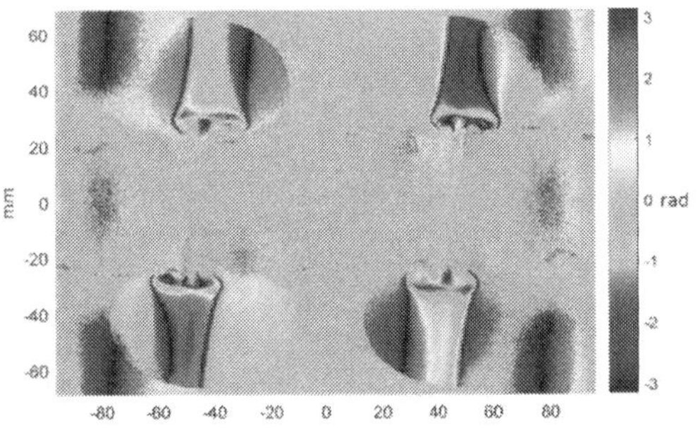

Fig.3. Reconstructed phase field map at some time instant.

Fig.3. represents the reconstructed phase field map recorded by the CCD camera at some time instant. Each recorded phase field is represented in a conjugated pair. The introduction of spatial carriers through the introduction of the angle between each of the reference wave and the axis of propagation has introduced enough separation so that the set of conjugated phase pairs don't overlap and can be filtered out confidently (i.e. off-axis arrangement). Any of the identical but conjugated pair carries the same valuable phase information used for further analysis. Any sudden change of the color red-blue indicates a 2π jump.

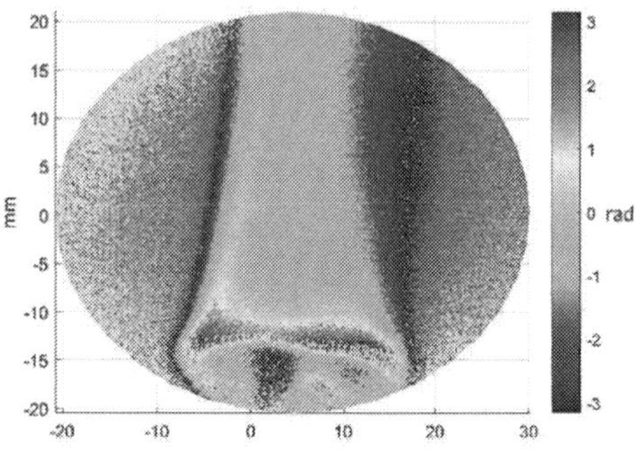

a)

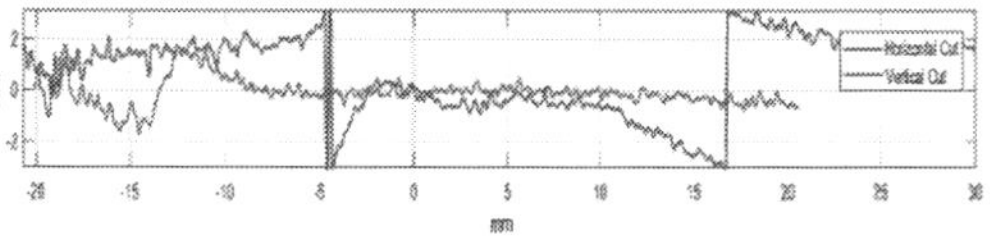

b)

Fig.4. a) Reconstructed phase field map at some time instant from λ_1 and b) plot of the phase profile run across the phase field horizontally and vertically through the origin axis.

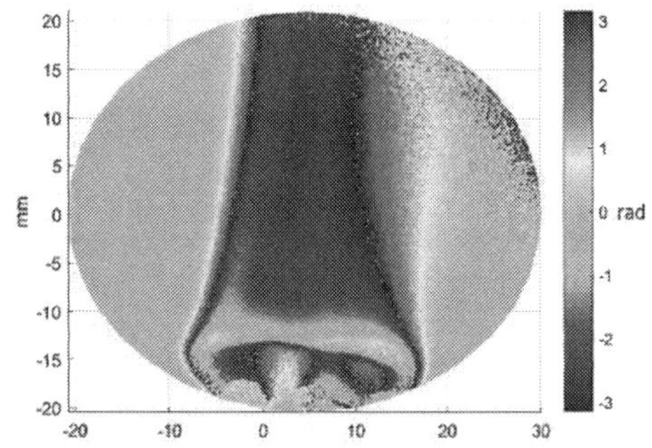

a)

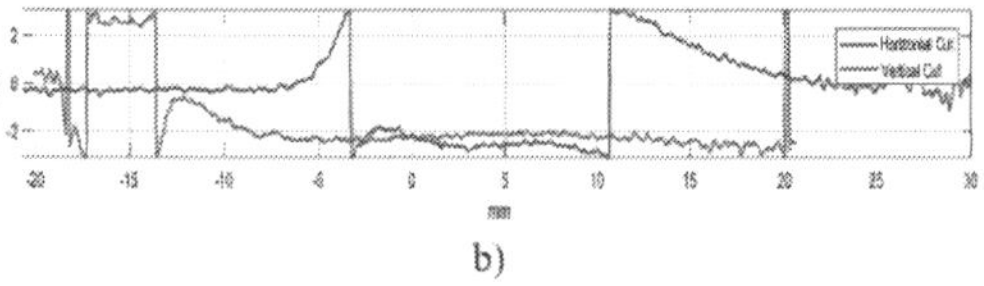

b)

Fig.5. a) Reconstructed phase field map at some time instant from λ_2 and b) plot of the phase profile run across the phase field horizontally and vertically through the origin axis.

Fig.4. and fig.5. represent the reconstructed phase field map for each of the wavelengths. The background is masked out and only the physical field of view is shown. For different wavelengths, the camera "sees" different phase values, thus making the profile look different as what is seen from the other wavelength. The phase profile was plotted for both phase field maps, passing by the origin axis for both horizontal and vertical profiles. Clear phase jumps can be seen in the horizontal cut of fig.4. as indicated in fig.4.b) while the phase profile across the vertical cut of the phase field is smoother and no noticeable jump can be observed.

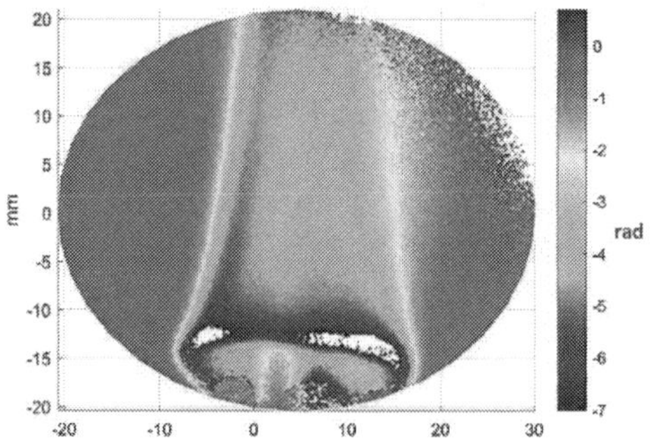

a)

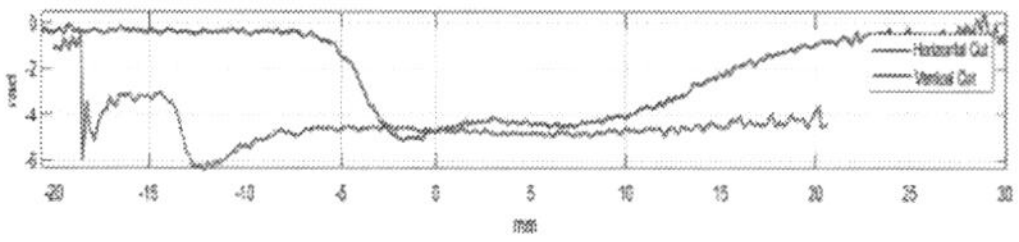

b)

Fig.6. a) Corrected phase field map at some time instant from λ_2 and b) plot of the phase profile run across the phase field horizontally and vertically through the origin axis

Fig.6. represents the corrected phase map coming from λ_2 based on (7). The phase field that had 2π jumps is now corrected and represents visually a continuous phase map.

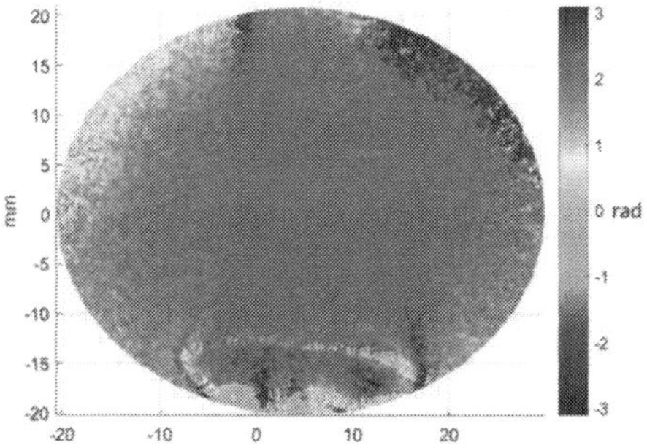

a)

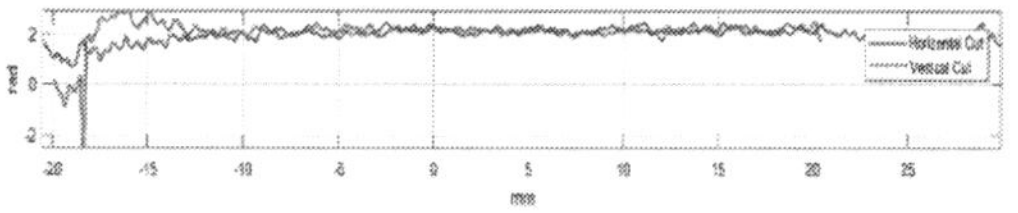

b)

Fig.7. a) Synthetic phase field map at some time instant and b) phase profile run across the phase field.

Fig.7. represents the fictional synthetic phase map which the CCD camera would have "seen" if a laser with wavelength Λ_{eq} would have been used. In our case, since the heat gradient wasn't high, the synthetic phase map manifests itself with relatively smooth flat profile. It is worth noting that even the synthetic phase map can present phase jumps if the range of measurement is exceeded. The synthetic phase map however obscures phenomena that cause relatively small phase changes.

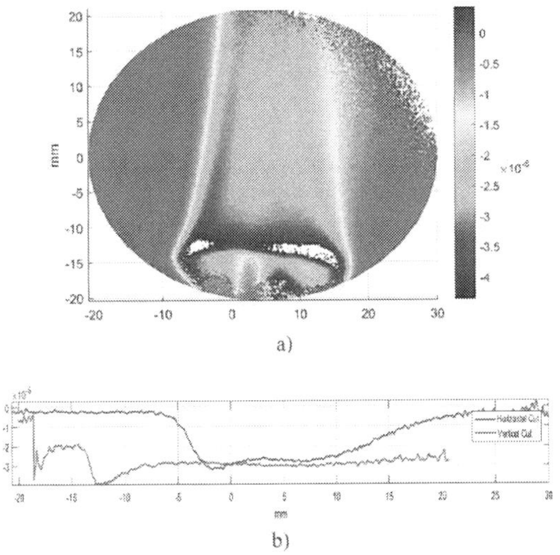

Fig.8. a) Refractive index field map at some time instant and b) refractive index profile run across the field.

Through the phase field map of either λ_1 or λ_2, the refrative index distribution can be obtained via (5) assuming a 2D distribution field of the refractive index. A distribution of the refractive index at some instant is shown in fig.8. Notice that digital holographic interferometry is able to detect very small changes in the refractive index cause by external influence (in this case heat gradient).

Through the determination of the refractive index, many other important parameters such as the temperature, density, fluid velocity etc. can be determined. This allows for the study of fluid and gases and phenomena related to. An interesting application of such technique could prove useful for the study of shockwaves around blade cascade in wind tunnels. Flow in wind tunnels are supersonic and demonstrate high levels of pressure and velocities. Such high gradients cause many fold wrapping of the phase. The use of a two-wavelength technique could therefore help in studying e.g. dynamic events during transonic and supersonic blade flutter.

Fig.9. represents the temperature change field calculated by making use of λ_2. It was calculated using the following mathematical relation :

$$T(x,y) = \frac{\lambda_2 \cdot \varphi(x,y)}{2\pi L \frac{dn}{dT}} \tag{10}$$

, where $L = 20mm$ is the estimated object length and $\frac{dn}{dT} = -0.9617 \cdot 10^{-6}\,°\mathrm{C}^{-1}$ [15] is the change of refractive index per unit temperature.

But, this temperature was measured as a reference to the starting temperature $T_0 = 20°\mathrm{C}$ in our case. Thus, the absolute temperature can be calculated as:

$$T_{abs}(x,y) = T_0 + T(x,y) \tag{11}$$

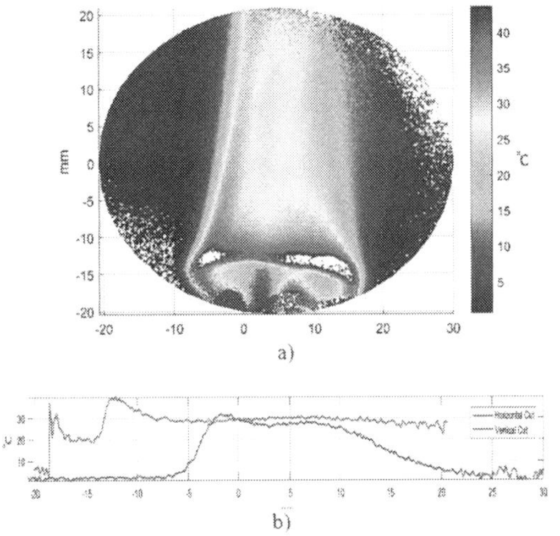

Fig.9. a) Reconstructed temperature field map $T(x,y)$ at some time instant and b) temperature profile run across the field.

6 Conclusions

This paper introduces a new approach to measure dynamic processes in fluid mechanics using lensless Fourier digital holographic interferometry with extended dynamic range. The investigation specifically deals with measuring the temperature field created from a resistor. The temperature is estimated through the optical phase, which was captured by a CCD camera. The captured phase is wrapped due to 2π limitations coming from the bounded nature of light and the two-wavelength technique deals in solving the phase ambiguity and determining the absolute phase values.

A key factor in this investigation is the use of two wavelengths and the recording of digital holograms from both wavelengths in one shot. We showed that the spectral separation of the phase information from both wavelengths can be achieved when the hardware of the experimental arrangement (angles of the reference waves) is properly adjusted [11]. The difference between the measured phase fields yields in a synthetic phase that has significantly larger dynamic range of measurement [21]. Such range covers large changes of the measured quantity as it was demonstrated in our investigation. However, the synthetic phase is more influenced by noise. By combining the synthetic phase and the phase obtained from the single wavelength, we can achieve the same accuracy as single wavelength technique but with a significantly higher range of measurement. This method is applicable to dynamic processes in fluid mechanics and could be applied for further studies of dynamic events during transonic and supersonic blade flutter in wind tunnels.

The advantages of such technique are its simplicity, fast acquisition, processing and displaying results, even in real time and in combination with the two-wavelength technique the unambiguous range of measurement is considerably increased without amplifying the noise.

Acknowledgment

This research was supported by the Ministry of Education, Youth and Sports of the Czech Republic – program Inter-Excellence, project No. LTAUSA19036.

References

[1] Meiners-Hagen, K., Schödel, R., Pollinger, F. & Abou-Zeid, A. Multi-wavelength interferometry for length measurements using diode lasers. *Meas. Sci. Rev.* **9**, 16–26 (2009).

[2] Guo, T., Li, F., Chen, J., Fu, X. & Hu, X. Multi-wavelength phase-shifting interferometry for micro-structures measurement based on color image processing in white light interference. *Opt. Lasers Eng.* **82**, 41–47 (2016).

[3] Turko, N. A., Eravuchira, P. J., Barnea, I. & Shaked, N. T. Simultaneous three-wavelength unwrapping using external digital holographic multiplexing module. *Opt. Lett. Vol. 43, Issue 9, pp. 1943-1946* **43**, 1943–1946 (2018).

[4] Turko, N. A. & Shaked, N. T. Simultaneous two-wavelength phase unwrapping using an external module for multiplexing off-axis holography. *Opt. Lett.* **42**, 73–76 (2017).

[5] Parshall, D. & Kim, M. K. Digital holographic microscopy with dual-wavelength phase unwrapping. *Appl. Opt. Vol. 45, Issue 3, pp. 451-459* **45**, 451–459 (2006).

[6] Jaedicke, V. *et al.* Multiwavelength phase unwrapping and aberration correction using depth filtered digital holography. *Opt. Lett.* **39**, 4160–4163 (2014).

[7] Khmaladze, A., Kim, M. & Lo, C.-M. Phase imaging of cells by simultaneous dual-wavelength reflection digital holography. *Opt. Express* **16**, 10900 (2008).

[8] Zuo, C., Huang, L., Zhang, M., Chen, Q. & Asundi, A. Temporal phase unwrapping algorithms for fringe projection profilometry: A comparative review. *Opt. Lasers Eng.* **85**, 84–103 (2016).

[9] Warnasooriya, N. & Kim, M. K. Quantitative Phase Imaging Using Multi-Wavelength Optical Phase Unwrapping. *Adv. Lasers Electro Opt.* (2010) doi:10.5772/8668.

[10] Onodera, R. & Ishii, Y. Two-wavelength interferometry that uses a Fourier-transform method. *Appl. Opt.* **37**, 7988–7994 (1998).

[11] Psota, P., Doleček, R., Lédl, V. & Vít, T. Dynamic interferometric measurement with extended unambiguity range in flow measurement. *EPJ Web Conf.* **180**, 02087 (2018).

[12] Cheng, Y.-Y. & Wyant, J. C. Two-wavelength phase shifting interferometry. *Appl. Opt.* **23**, 4539–4543 (1984).

[13] Psota, P., Cubreli, G., Kredba, J., Stasik, M. & Ledl, V. Two wavelength digital holographic interferometry for investigation of dynamic processes in fluid mechanics. in *5-6th Thermal and Fluids Engineering Conference (TFEC) - American Society of Thermal and Fluids Engineers* (Begell House Publishers, Inc., 2021).

[14] Wagner, C., Osten, W. & Seebacher, S. Direct shape measurement by digital wavefront reconstruction and multi-wavelength contouring. *Opt. Eng.* **39**, 79–85 (2000).

[15] Kreis, T. *Handbook of Holographic Interferometry: Optical and Digital Methods. Handbook of Holographic Interferometry: Optical and Digital Methods* (Wiley, 2004). doi:10.1002/3527604154.

[16] Chavez, S., Xiang, Q. S. & An, L. Understanding phase maps in MRI: a new cutline phase unwrapping method. *IEEE Trans. Med. Imaging* **21**, 966–977 (2002).

[17] Momose, A. Recent Advances in X-ray Phase Imaging. *Jpn. J. Appl. Phys.* **44**, 6355–6367 (2005).

[18] Pritt, M. D. Phase unwrapping by means of multigrid techniques for interferometric SAR. *IEEE Trans. Geosci. Remote Sens.* **34**, 728738 (1996).

[19] Yin, W. *et al.* High-speed three-dimensional shape measurement using geometry-constraint-based number-theoretical phase unwrapping. *Opt. Lasers Eng.* **115**, 21–31 (2019).

[20] Lu, L., Jia, Z., Luan, Y. & Xi, J. Reconstruction of isolated moving objects with high 3D frame rate based on phase shifting profilometry. *Opt. Commun.* **438**, 61–66 (2019).

[21] K, H. & F, C. Two-wavelength interferometry: extended range and accurate optical path difference analytical estimator. *J. Opt. Soc. Am. A. Opt. Image Sci. Vis.* **26**, 2503 (2009).

Pool boiling heat transfer on minichannels with porous structure

Norbert Dadas[1], *Robert* Pastuszko[1*], and *Robert* Kaniowski[1]

[1]Kielce University of Technology, Faculty of Mechatronics and Mechanical Engineering, al. 1000-lecia Państwa Polskiego 7, PL-25-314 Kielce, Poland

Abstract. The work focused on the investigations and comparison of pool boiling heat transfer on the specimens with in 5 mm deep and 1 mm wide parallel mini-channels to the specimens with additional porous material (copper foam or mesh) between the fins. The experiments were carried out with boiling FC-72 at atmospheric pressure. Many works show that enhanced surfaces with open minichannels improve heat transfer performance substantially. The use of additional porous material led to the highest heat transfer coefficients in the range of 100 to 230 kW/m^2. The coefficients obtained were 15% higher and the superheat was 15% lower compared to the plain smooth minichannel without porous material.

1 Introduction

The widespread miniaturization of components, mainly electronic ones, and increasing their efficiency lead to the intensification of the heat they generate. The heat from the microprocessor can reach 2000 – 3000 kW/m^2, and in some cases maximum heat fluxes up to 15000 kW/m^2 have been reported [1]. Cooling by phase change during boiling ensures efficient passive heat dissipation and fluid circulation [2]. Extension of the heat transfer surface in the form of micro- and minichannels is very common, but the boiling mechanism is not fully described. The use of boiling on surfaces with minichannels allows for obtaining much higher heat transfer coefficients compared to forced convection. The liquid enters the channels at random locations [3]. Porous structures are added to increase the surface of heat transfer and to initiate a capillary pressure that directs the fluid flow. The pores in the material increase the area of bubble nucleation [4]. The decrease in temperature on the heating surface is due to the formation, growth, and detachment of the vapour bubble [5]. Partial filling of channels with porous surfaces in the form of meshes or copper foam should direct the flow in the minichannels. The added filling will allow the liquid to be sucked in, and the formation, growth, and detachment of bubbles will occur in the gaps between the porous material.

2 Experimental setup

The drawing of the experimental setup for testing the pool boiling of working fluids is shown in Fig. 1.

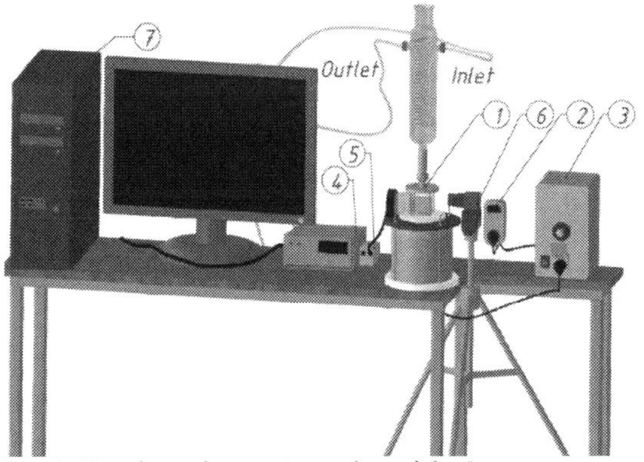

Fig. 1. Experimental setup, 1 – main module, 2 – wattmeter, 3 – autotransformer, 4 – data logger, 5 – light, 6 – high speed camera, 7 – PC.

The autotransformer with a wattmeter allows determining the power supplied to the cartridge heater placed in the copper cylinder of the main module (Fig. 2.) Two sheathed K-type thermocouples (NiCr-NiAl) are placed in boiling liquid above the specimen (T1, T2). They allowed the saturation temperature of the fluid to be checked. The stability of temperature during the experiment indicated that the saturation temperature was maintained. The condenser allowed the fluid to return to the glass vessel, which made the main module work as a thermosiphon. Four K-type thermocouples (T5 – T8) were placed at specified distances in the heating cylinder and two in the grooves under the specimen (T3, T4).

* Corresponding author: tmprp@tu.kielce.pl

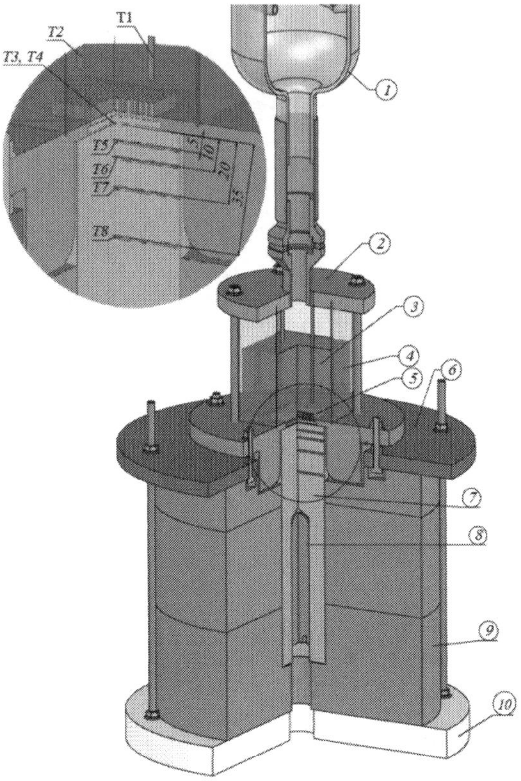

Fig. 2. Visualization of the main module of the test stand; 1 – condenser, 2 – top flange, 3 – boiling liquid, 4 – glass vessel, 5 – specimen, 6 – annular flange, 7 – copper cylinder, 8 – cartage heater, 9 – insulation, 10 – bottom flange. Dimensions in millimeters.

They allowed the determination of a one-dimensional heat flow in the copper cylinder. The heat flux was calculated according to the following formula:

$$q = \frac{\lambda_{Cu} \cdot \pi d^2}{4A_{bs}} \cdot \frac{T_8 - T_5}{\delta_8 - \delta_5} \tag{1}$$

The extrapolated superheat, defined as the temperature difference between the reference surface and the saturated liquid, was as follows:

$$\Delta T = \frac{T_3 + T_4}{2} - q\left(\frac{\delta_{bs}}{\lambda_{Cu}}\right) - \frac{T_1 + T_2}{2} \tag{2}$$

According to Newton's law of cooling, for one- dimensional heat conduction, the heat transfer coefficient was defined as:

$$\alpha = \frac{q}{\Delta T} \tag{3}$$

The MC sample contained pieces of copper mesh or foam in various configurations. The dimensions of specimen are shown in Fig. 3, and an example of the arrangement of porous structures is shown in Fig. 4.

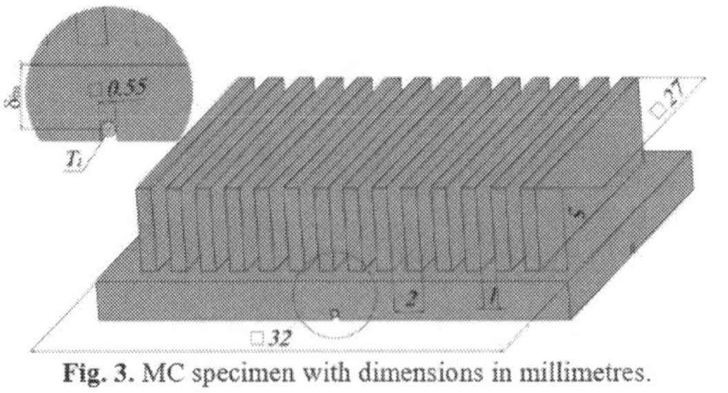

Fig. 3. MC specimen with dimensions in millimetres.

Fig. 4. Example of the wire mesh arrangement for the MCC-M-7.6-0.5-2 specimen.

The foam and mesh fillings had the following variable parameters: pitch (p_p), thickness of a single layer (δ_p), number of meshes placed side by side in the channel (n_p), length of the filling (l_p) and mesh aperture (a_p). Fig. 5 shows the surface MCC-M-3.5-0.5-2 with markings described in Table 1.

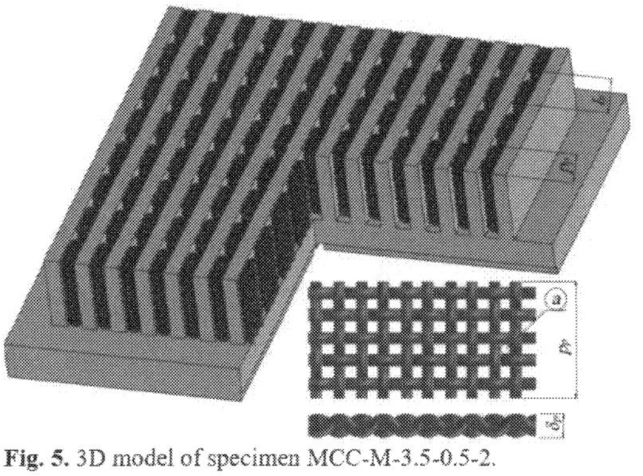

Fig. 5. 3D model of specimen MCC-M-3.5-0.5-2.

Specimens having a porous structure inside the minichannels are marked with M or F (mesh or foam). For specimens with porous material pieces in the minichannels, the code consists of three variables from Table 1: pitch, thickness of a single layer, number of meshes placed side by side in the channel.

Table 1. Specimen code and specification.

Sample code	p_p, mm	δ_p, mm	n_p	l_p, mm	a_p, mm
MC	-	-	-	-	-
MCC-M	-	1	1	27	0.7
MCC-F	-	1	1	27	0.25
MCC-M-3-0.5-2	3	0.5	2	2	0.3
MCC-M-3.5-0.5-2	3.5	0.5	2	2.5	0.3
MCC-M-7.6-0.5-2	7.6	0.5	2	3.8	0.5
MCC-F-3.6-0.3-3	3.6	0.3	3	1.5	0.25
MCC-F-3.5-0.5-2	3.5	0.5	2	2	0.25
MCC-F-4.2-0.5-2	4.2	0.5	2	2	0.25

During the tests, the linearity of the temperature values was checked in all specimen. The coefficient $R^2 \approx 0.99$ indicated one-dimensional thermal conductivity (Fig. 6).

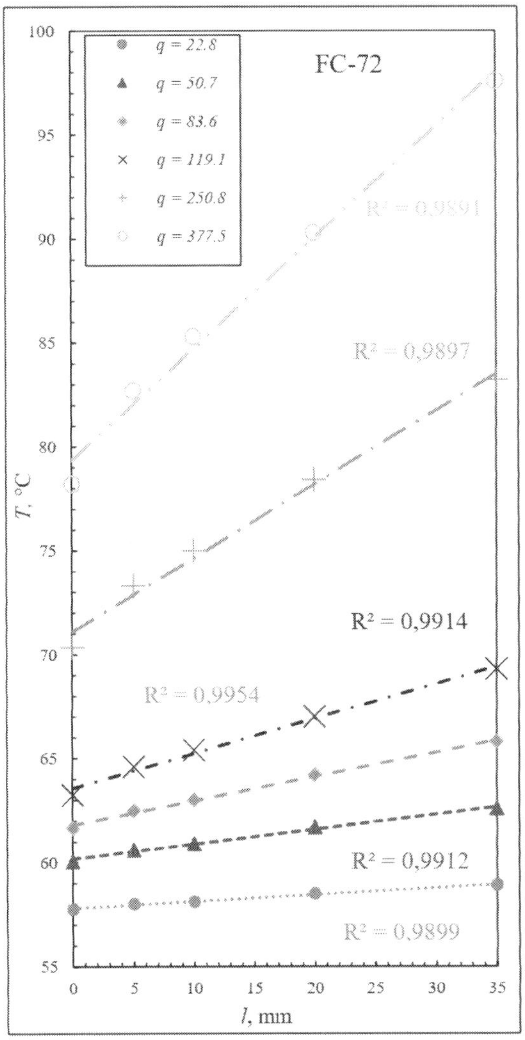

Fig. 6. Temperature vs. thermocouple location in the heating cylinder, MCC-F-4.2-0.5-2, boiling FC-72.

3 Results

The surface with open minichannels (MC) significantly improved the heat transfer coefficient compared to the plain surface (Figs. 7 and 8). The highest heat transfer coefficient for MC was about 25 kW/(m²K) – it was five times better than for the plain surface. Additional porous fillings improved HTC at a heat flux less than 220 kW/m²; the best results were obtained at 170 kW/m² for specimen MC: HTC was about 20 kW/(m²K) for MCC-M-3.5-0.5-2 and about 23 kW/(m²K) for MCC-M-3-0.5-2. For this range of heat flux, the additional mesh structure improved HTC by 15% compared to open minichannels. Minichannels with foam fillings had better performance than empty minichannels (about 15% for heat flux near 110 kW/m²). The superheat obtained was two to three times lower than for a plain smooth surface for heat fluxes smaller than 50 kW/m². Comparing the specimens with the additional filling to MC, no significant differences in superheat were observed for heat fluxes less than 120 kW/m². From the lowest values of heat flux to about 110 kW/m², the superheat values were comparable. When this value was exceeded, the MC specimen had lower superheat, but specimens with additional filling were able to withstand higher heat flux. For MCC-F-3.5-0.5-2, the heat flux obtained was about 400 kW/m² before the boiling crisis occurred. It was 25% higher than for the MC surface.

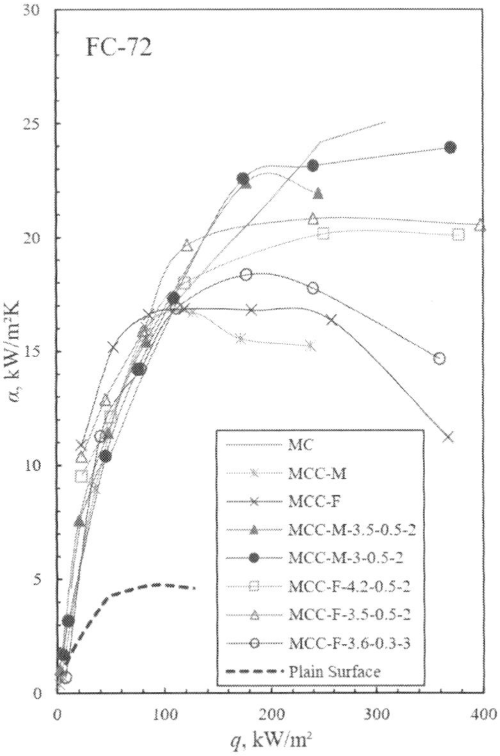

Fig. 7. Boiling curves for FC-72, heat transfer coefficient vs. heat flux.

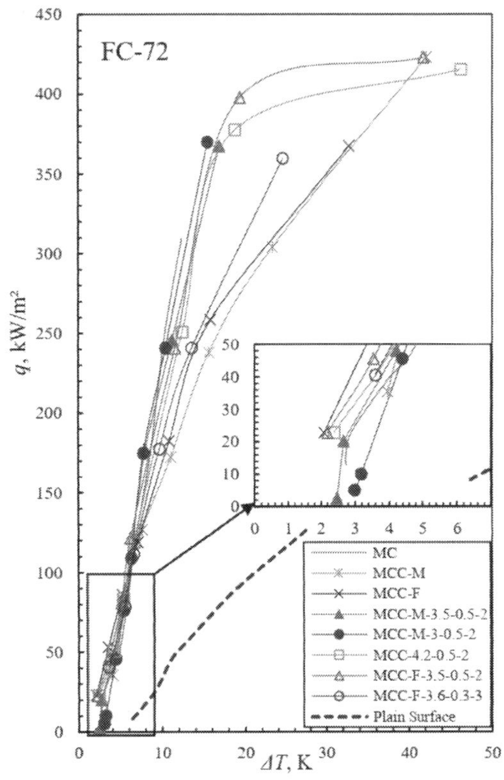

Fig. 8. Boiling curves for FC-72, heat flux vs superheat.

4 Conclusions

Based on the research carried out, the following conclusions can be drawn:

- The use of the surface with open minichannels significantly increased HTC and critical heat flux during pool boiling, compared to the plain surface. This occurs because surfaces with minichannels enable an increase in the number of nucleation centres and effective vapour removal. Most bubbles form between fins, but such formations are random.
- Additional porous structures in the spaces between the fins allow systematizing both bubble formation and liquid inflow. Most bubbles nucleated, grew, and detached in the channels created by porous filling. The porous filling acted as a capillary structure that improves suction of the working fluid into the minichannel space.
- Additional porous structures delayed the boiling crisis even at higher superheats. Stabilizing the flow of the working fluid and a larger area of extended surface increased the HTC.
- Minichannels fully filled with porous material demonstrated worse performance than open minichannels. This is probably due to the smaller empty spaces between the additional fillings. The flow of the liquid and the formation and growth of the bubbles occur in this case in a similar manner as on surfaces with capillary porous structures.

The tests were carried out with only one working fluid and one size of the specimens with minichannels. Sometimes, even small changes can significantly improve the heat transfer coefficient or superheat. Further tests including other working fluids and channel widths will be performed in the future.

Nomenclature

A – area, mm²
a – aperture diameter, mm
d – copper cylinder diameter, mm
l – length, mm,
MC – minichannel,
n – number of mesh layers,
p – pitch, mm,
T – temperature, K,
q – heat flux, kW/m²,

Greek symbols

ΔT – difference of temperature, K,
λ – thermal conductivity, W/(mK)
δ – thickness, mm,
α – heat transfer coefficient, W/(m²K),

Subscripts

p – porous
Cu – copper
bs – base of sample
1–8 – numbers of thermocouples

Reference

1. K.C. Leong, J.Y. Ho, K.K. Wong, Appl. Therm. Eng. **112** (2017),
2. L. L. Manetti, A. S. O. Henriques Moita, R. R. de Souza, E. M. Cardoso, Int. J. Heat Mass Transf. **152** (2020),
3. A.K. Das, P.K. Das, P. Saha, Appl. Therm. Eng. **29** (2009),
4. C. M. Patil, S. G. Kandlikar, Heat Transf. Eng. **35** (2014),
5. D. Eok Kim, D. In Yu, D. W. Jerng, M. Hwan Kim, H. Seon Ahn, Exp. Therm. Fluid Sci. **66** (2015)

https://doi.org/10.1051/epjconf/202226401011

Simcenter STAR-CCM+ software for CFD and heat transfer calculations in minichannels

Norbert Dadas[1], *Artur* Piasecki[2], and *Magdalena* Piasecka[1*]

[1] Kielce University of Technology, Faculty of Mechatronics and Mechanical Engineering, al. 1000-lecia Panstwa Polskiego 7, PL-25-314 Kielce, Poland
[2] Echo Investment S.A., Al. Solidarnosci 36, 25-323 Kielce, Poland

Abstract. The paper focuses on CFD modelling and numerical simulation of heat transfer during fluid flow in minichannels. The data from the experiment were adopted into Simcenter Star-CCM+ software. The test section comprised seven or twenty-one parallel minichannels of rectangular cross sections. Each minichannel was 1 mm height. During the experimental series, the temperature measurements of the heated wall were recorded due to an infrared camera. Measurement data was compared with the results obtained from numerical calculations. Fluorinert FC-72 was used as a working fluid in experiments and simulations. An increase in the values of the heat transfer coefficient was observed for the test section with 21 minichannels compared to that with seven minichannels. Numerical simulations helped to identify how the change in the number of minichannels affects the intensification of heat transfer during flow without the time- and cost-consuming experimental tests

1 Introduction

The miniaturization of devices that generate large heat fluxes in various industries requires the search for high-performance heat exchangers, including phase change and minigeometries [1, 2]. A review of the relevant literature and selected publications covering heat transfer during cooling liquid flow in rectangular minichannels was presented in [3, 4]. The results of experiments conducted in the test section with an annular minigap are described in [5]. The review of the literature leads to the conclusion that, although much has been written recently on heat transfer during flow in minichannels, the literature does not offer any generalized conclusions concerning heat transfer during liquid flow in such miniature systems, while experiments required complicated apparatus and efforts.

CFD simulations can be used to minimize the number of experimental studies by selecting only those with a significant improvement in parameters. The model for the lifetime of the EGR coolers was proposed in [6]. CFD calculations helped simulate the temperature of the metal during the heating and cooling cycles. In [7] the 3D model of shell side flow in rod-baffle heat exchangers with spirally corrugated tubes was compared to rod-baffle heat exchangers with plain tubes. It was concluded that 10% higher heat transfer values were obtained when spirally corrugated tubes were applied in rod-baffle heat exchangers compared to plain tubes.

Preparing a correct numerical simulation can lead to many problems. Large differences between elements require appropriate mesh selection and optimization [8]. To verify the model, the results of the numerical calculations should be compared with the known values resulting from the correlation or directly obtained from the experiment. With deviations not exceeding 10%, convergence can be considered good, ensuring the credibility of the simulation [9].

2 Experimental setup

The data from experiments conducted on a research set-up with the test section comprising a group of minichannels were adopted for CFD numerical simulations. The main objective of this paper was to determine how the number of minichannels influences heat transfer during flow by analyzing the values of the heat transfer coefficient. The aim was achieved on the basis of the results from the numerical computations, obtained using the data from the experimental investigations. In the calculations, two variants of the test section that differed in the number of minichannels were applied. In one construction variant, the test section consisted of seven minichannels, while in the other 21 minichannels were tested. The side walls of the minichannels were formed within a Teflon spacer. The main elements of the test section are shown in Fig. 1. The most important information about the dimensions of the main elements of the test section is listed in Table 1.

* Corresponding author: tmpmj@tu.kielce.pl

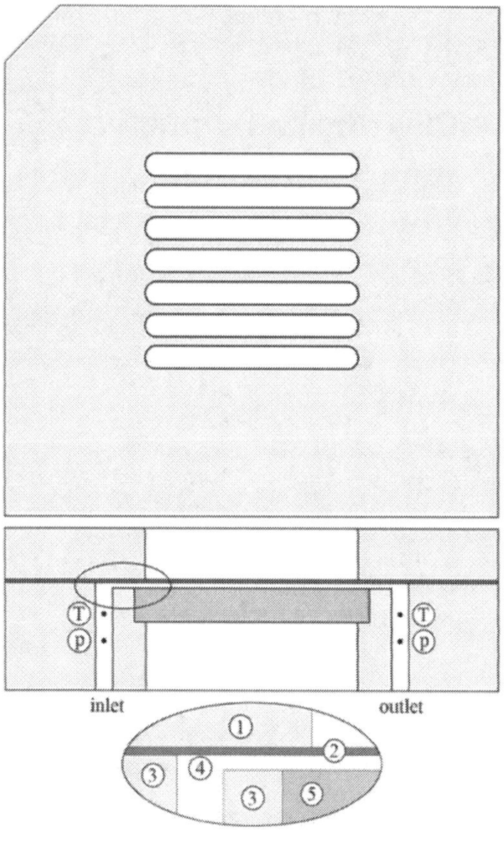

Fig. 1. Main elements of the test section, 1 – top cover, 2 – heater (heated foil), 3 – channel body, 4 – minichannel, 5 – glass panel.

Table 1. Dimensions of the main elements of the test section.

Element of the test section	Dimensions (mm) (height x length x width)
Top cover	9 x 90 x 90
Heated foil	0.1 x 90 x 40
Channel body	18 x 90 x 90
Area of minichannels	1 x 57 x 38
Single minichannel (7 channels in a group)	1 x 57 x 4
Single minichannel (21 channels in a group)	1 x 57 x 0.86
Glass panel	6 x 43 x 38

The materials of the main elements of the test section are as follows (see Fig. 1):

- 1, 3 – aluminum alloy PA6,
- 2 – Haynes-230 alloy,
- 5 – glass panel.

To perform the CFD simulation physical properties of the materials were entered into the program (Table 2). These data were required for the equations and the appropriate procedure used in the numerical computations.

Table 2. Physical properties of the materials used in the CFD calculations.

Material	Density ($kg \cdot m^{-3}$)	Dynamic viscosity ($Pa \cdot s$)	Specific heat ($J \cdot kg^{-1} \cdot K^{-1}$)	Thermal conductivity ($W \cdot m^{-1} \cdot K^{-1}$)
FC-72	1680	$6.4 \cdot 10^{-4}$	1100	0.057
Aluminum	2702	-	903	237
Haynes -230	8970	-	400	8.9÷14.4 for temperature range from 293.1 (K) to 673.1 (K)
Glass	2500	-	840	1.4
Teflon	2160	-	702	0.21

During the experimental sets, Fluorinert FC-72 was circulating as the working fluid in the main loop. One wall of the minichannels was a thin foil resistively heated. Recording the temperature on the outer surface of the heated foil was realized by infrared thermography. The reverse surface of the foil with visible two-phase fluid flow was observed through a glass panel. The temperature and pressure of the cooling fluid were measured at the inlet and outlet of the channel. There were no significant changes in the ambient air temperature; it was within the range of +/- 0.2 K throughout the measurement. The experimental parameters were captured as signals from K-type thermocouples, pressure meters, flow meter, ammeter and voltmeter. Data collected for a selected setting of the supplied heat flux, taken for the calculations, are shown in Table 3. Additionally, captured images from an infrared camera were converted to obtain a temperature distribution on the outer surface of the heater.

Table 3. The main experimental parameters.

Specified variables	
Temperature of FC-72 fluid at the minichannel inlet (K)	286.05
Temperature of ambient air (K)	294.45
Atmospheric pressure (kPa)	95.47
Overpressure at the minichannel inlet (kPa)	93.01
Overpressure at the minichannel outlet (kPa)	94.17
Mass flow rate (kg/h)	21.99
Heat flux supplied to the heater (W)	94.79

3 Numerical calculations

Numerical calculations were provided in Simcenter Star-CCM+ software, version 2020.2.1 Build 15.04.010. The elements of the 3D model were meshed using a polyhedral grid. The number of cells, faces, and vertices of the individual elements is presented in Table 4.

Table 4. Data of meshed regions.

Region	Cells	Faces	Vertices
Top cover	1 998 550	9 565 776	6 498 512
Heater	6 127 213	26 059 982	16 858 042
Channel body	597 633	2 871 269	1 957 548
Top spacer	4 826 177	20 313 609	12 328 173
Bottom spacer	3 864 620	17 093 891	10 888 766
Glass panel	143 079	661 636	438 453
Fluid FC-72	1 587 691	7 007 002	4 457 513

Due to the large differences in the thickness of the elements, the mesh had additional prism layers implemented. This option allowed for significantly densification of the mesh at the points of contact between elements, increasing the quality of the calculations. Several versions of the meshes were tested. It was assumed that a precise mesh selection significantly affects the quality of the results. The residuals were flattened out after 6000 iterations, but to be sure of the correctness of the numerical calculations, they were proceeded to 7500. The absolute errors in the solutions of particular variables ranged from 10^{-6} (energy) to 10^{-13} (turbulent dissipation rate) and the continuity was about 10^{-8}.

Mesh dependency studies based on GCI were carried out to estimate the numerical accuracy resulting from mesh resolution. To define the accuracy discretization error, the model was meshed with three grids with different sizes to obtain a range of 3 to 15 million cells. The results were widely discussed in [10].

Table 5 presents the general assumptions adopted for the two continua: the fluid, Fluorinert FC-72, and the solid, a heater (a thin heated foil) made of Haynes-230 alloy. The rest of the solid continua was similar in assumptions to the heater.

Table 5. The general assumptions adopted for the fluid and the solid in the calculations by Simcenter Star-CCM+ software.

Continua	General assumptions
Fluid: Fluorinert FC-72	Circumferential heat flux averaging
	Constant density
	Gradients
	Gravity
	Implicit unsteady
	K-epsilon turbulence
	Liquid
	Realizable k-epsilon two-layer
	Reynolds-averaged Navier-Stokes
	Segregated flow
	Segregated fluid temperature
	Solution interpolation
	Three dimensional
	Turbulent
	Two-layer all y+ wall treatment
	Wall distance
Solid: Haynes-230 alloy	Constant density
	Gradients
	Implicit unsteady
	Segregated solid energy
	Solid
	Solution interpolation
	Three dimensional

4 Results

The data from the experimental set (see Table 2) were entered into the Simcenter Star-CCM+ numerical program and the general assumptions adopted for the fluid Fluorinert FC-72 and the heater made of Haynes-230 (see Table 5) were made. After computations, the temperature of the outer heated foil surface was compared with that recorded by an infrared camera during the experiment. The temperature on the outer heated surface versus distance from the minichannel inlet, obtained from the experiment with seven minichannels on the basis of IR measurement and from numerical computations due to Simcenter Star-CCM +, are given in Fig. 2.

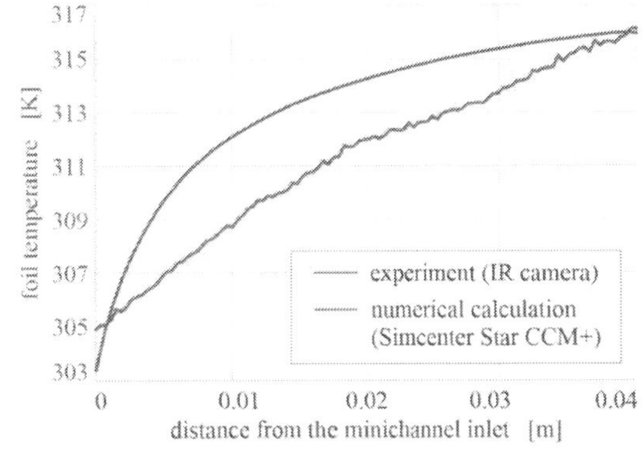

Fig. 2. Temperature on the outer heated surface versus distance from the minichannel inlet, obtained from the experiment with seven minichannels on the basis of IR measurement and from numerical computations due to Simcenter Star-CCM+.

When analyzing the results shown in Fig. 2 it is observed that the highest temperature difference between the results of the experiment and the numerical calculation reached 3 K. It can be underlined that the accuracy of the infrared camera is about 2 K. Furthermore, the values of the temperatures at the ends of the minichannel are very similar. The assumptions introduced into the numerical program regarding the heat exchange between the test section and the environment might result in even better compliance of both temperature dependences.

After numerical calculations and validation of the results gained for the test section with seven minichannels (carried out as a comparison between the foil temperature values obtained from the experiment and the numerical calculation), the number of minichannels was greatly increased.

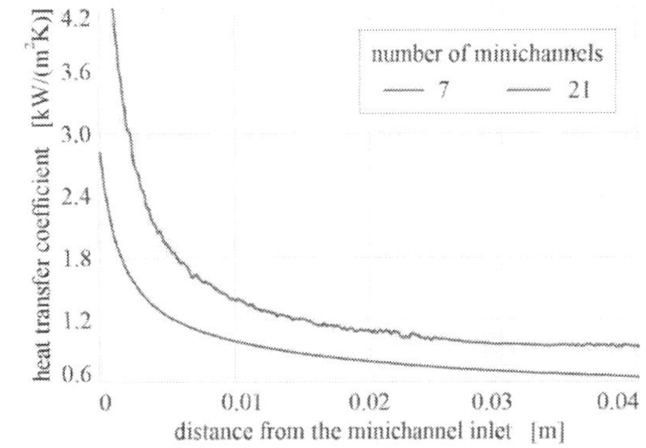

Fig. 3. Heat transfer coefficient versus distance from the minichannel inlet obtained from numerical computations due to Simcenter Star-CCM+ software; the results for seven and 21 minichannels.

The experimental data listed in Table 3 were used in the numerical calculations for the test section comprising 21 channels. The heat transfer coefficient (HTC) versus distance from the minichannel inlet is illustrated in Fig. 3. The HTC values determined for the test section with seven minichannels are also presented in this figure.

When analyzing the dependences shown in Fig. 3, it can be noticed that the increase in the number of minichannels in the test section results in a 30% higher HTC. Changing the number of minichannels in the test section also affects the fluid pressure and velocity in each of the channels. Furthermore, these parameters also affect the values of the temperature and the heat transfer coefficient.

Figure 4 shows the distributions of temperature (Fig. 4a), absolute pressure (Fig. 4b), and velocity (Fig. 4c) obtained from numerical computations. The data refer to the longitudinal cross section of the central minichannel, whereas the test section with seven channels was assumed in the calculations.

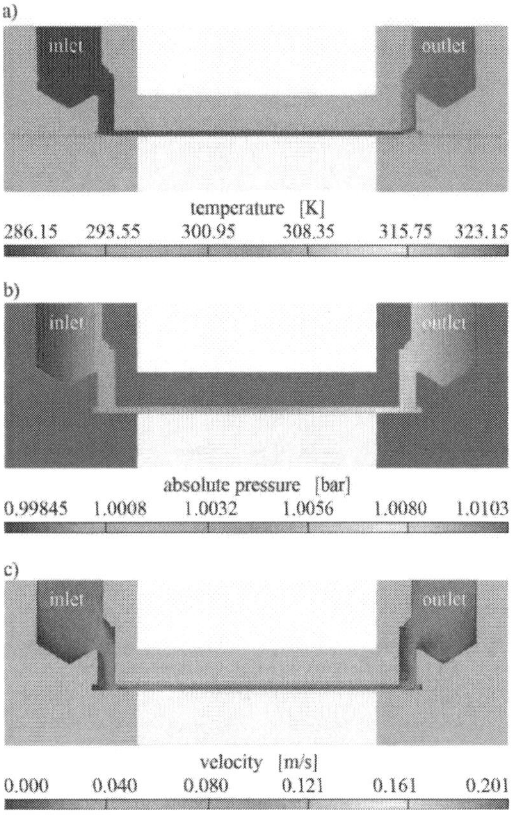

Fig. 4. Results from the numerical calculations obtained in the longitudinal cross section of the central minichannel showing the distributions of: a) temperature, b) fluid absolute pressure and c) fluid velocity; screenshots from Simcenter STAR-CCM + software.

5 Conclusions

The main aim of this paper was to present CFD modelling and numerical computations concerning boiling heat transfer during fluid flow in minichannels. For this purpose, the selected experimental data were adopted for numerical calculations. The calculations were performed using Simcenter Star-CCM+ software. The experimental data were used to validate the numerical results. Data from the numerical computations collected for the test section with seven minichannels were compared with the heater temperature experimentally measured. In further numerical simulations, the number of minichannels was increased to 21. Such calculations helped determine how the change in the number of minichannels affects the intensification of heat transfer during flow.

The authors noted the following comments and observations:

- Increasing the number of minichannels in the test section causes increasing obtained heat transfer coefficient values (up to 30%);
- Validation of the experiment with numerical simulations ensures that physical models and grid convergence enable acquired the reliable results;
- CFD simulations helped to identify how the change in the number of minichannels affects the intensification of heat transfer during flow without the time- and cost-consuming experimental tests.

Acknowledgments

The research reported herein was supported by a grant from the National Science Centre, Poland, No. UMO-2018/31/B/ST8/01199.

The calculations were made using the Simcenter Star-CCM+ software developed by Siemens PLM Software Inc., provided by GMSystem which is a supplier of Siemens Digital Industries Software in Poland.

References

1. M. Klugmann, P. Dąbrowski, D. Mikielewicz, Appl. Therm. Eng. **181** (2020)
2. R. Kaniowski, R. Pastuszko, EPJ Web Conf. **180**, No. 02041 (2018)
3. B. Maciejewska, M. Piasecka, A. Piasecki, Heat Transf. Eng. **43** (2021)
4. M. Piasecka, B. Maciejewska, P. Łabędzki, Energies **13** No. 6647 (2020)
5. M. Piasecka, T. Musiał, A. Piasecki, EPJ Web Conf. **213** No. 02066 (2019)
6. B.G. Holland, T.L. McKinley, B.R. Storkman, SAE Int. J. Engines **8** (2015)
7. J.J. Liu, Z.C. Liu, W. Liu, Int. J. Therm. Sci. **89** (2014)
8. J. Wajs, T. Kura, D. Mikielewicz, E. Fornalik-Wajs, J. Mikielewicz, Energy **238** (2022)
9. P. Dąbrowski, R. Kumar, Int. Commun. Heat Mass Transf. **129** (2021)
10. M. Piasecka, A. Piasecki, N. Dadas, Energies **15** No. 536 (2022)

2D numerical investigation of liquid dispersion in rotating packed bed and its comparison with experimental measurements using high-speed camera

Jakub Elcner[1*], *Ondrej* Hajek[1], and *Miroslav* Jicha[1]

[1]Brno University of Technology, Faculty of Mechanical Engineering, Energy Institute, Technicka 2896/2, Brno 616 69, Czech Republic

Abstract. Study deals with liquid dispersion of water in unstructured wire mesh of rotating packed bed, experimentally investigated using high speed camera. Dispersed water in outer region of rotating packed bed were captured by high-speed camera and droplet size were investigated using image processing. Numerical simulations were performed on two – dimensional plane with similar geometry characteristics (diameter and distance between each wire). Volume of fluid method was used to describe a multiphase behaviour of phenomena and turbulence was modelled using k-epsilon model. Results of both were compared on basis of frequency distribution of individual size fraction of dispersed water in outer region of rotating packed bed and its velocities. Discussion was focused on possibility of 2D numerical investigation and possibility of its comparison with experimental measurement.

1 Introduction

Carbon dioxide contributes to global warming with approximately 60% and Carbon Capture and Storage technology (CCS) or Carbon Capture, Utilization and Storage (CCUS) becomes a key strategy for its reduction if we leave aside sustainable sources. It is today merely accepted that the CCS or CCUS technology for coal power plants is too expensive and it's cost of investment and operation is too high to be widely implemented unless the price of emission permits increases to about 60 to 80 EUR/tCO2. Therefore, the CCS (CCUS) technology is aimed at several intense industrial branches like cement, refineries, steel works and many small and medium industries. In the complete system of CCS technology, the most expensive is the CO2 capture that represents about 70% of the total cost for required 90% CO2 reduction. It´s quite logic that research and development focus on intensification of the capture process. According to many studies the most advanced and mature technology when comparing with other methods is chemical absorption, which is on the Technology Readiness Level 9 (TRL9), i.e. close to commercialization although it has its weaknesses on the absorption side itself (the need for intensification), and also on the solvent regeneration side. Ammonia NaOH or different amines (most often Monoethanolamine MEA) are the most used solvents. Both have some positives and negatives, but MEA became a reference liquid. Static columns or any spray systems like scrubbers are limited by gravitation and the process cannot be further intensified. Therefore, the aim is to intensify the absorption process. This can be reached by using a rotary absorber called Rotating Packed Bed (RPB). Depending on its diameter and speed, and considering dynamic forces, the rotary absorber can develop up to twenty times the gravitational acceleration and subsequently ensure up to one order of magnitude higher transfer of CO2 to the solvent, with dimensions ten times smaller than static columns or scrubbers.

Main task in the development of Rotating Packed Bed is to clarify the phenomena that occur during their operation e.g. influence of geometry, rotational speed, inlet behaviour, liquid dispersion, primary solvent distribution, and type of packing (wire mesh, metal foam or others) on liquid distribution, and liquid disintegration and film formation, i.e. phase interface inside rotating packing. The most suitable way to assess the influence of various parameters and to propose an optimal solution for RPB is to use computational fluid dynamics (CFD) to simulate the process inside the packing and validate it by experiments on identical geometry. Our previous research [1], based on experiments performed by Yang et al. [2] focused on RPB has dealt with effect of multiple liquid inlets and its affect on mass transfer inside the packing represented by liquid holdup. Thanks to own experimental model of RPB with wired packing, we are now able to perform experiments and have direct comparison/validation of our numerical research.

* Corresponding author: elcner@fme.vutbr.cz

2 Methods

2.1 Rotating packed bed

The main object of the RPB is a wired mesh consisting of 34 m long grid, twisted in 48 layers. The inner diameter of the mesh is 150 mm, and the outer diameter is 300 mm. The grid is 30 mm wide. The wire mesh is attached to the housing by six spikes of 8 mm in diameter. The diameter of single wire of the grid is 0.8 mm and the distance between the axis of the individual wires forming the grid is 3.95 mm. The housing is then a plastic enclosure with external dimensions of 400 x 405 x 100 mm, which houses a wire mesh attached to the rotor shaft, thus separating the space into three parts. Inner region - Wired mesh - Outer region. The inner region contains the fluid inlet in the form of a pipe of 0.54 mm diameter, oriented at an angle of 45° from horizontal plane of RPB. The outer region contains the fluid outlet with a height of 35 mm. All dimensions of the RPB are shown in Figure 1.

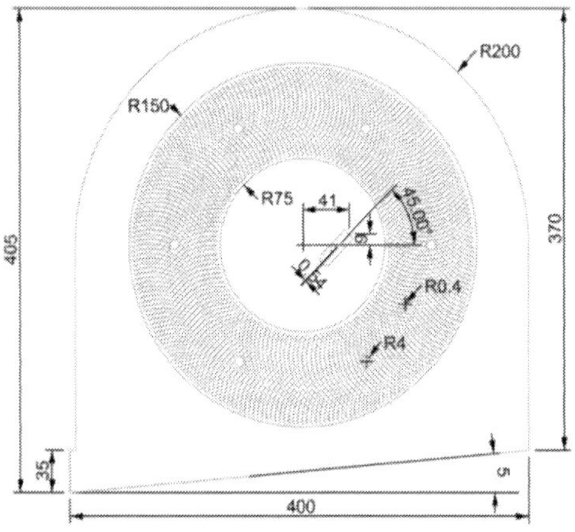

Fig. 1. Dimensions of rotating packed bed

2.2 Experimental setup

The experiments were performed on a test rig, which scheme can be found in Figure 2. The wired mesh was made of 1.4301 stainless steel. Its porosity was 0.82 and specific surface area was 874 m^2/m^3. The packing was connected to a motor of rotary speed of 10Hz. This value corresponds to the real speed of the wired mesh of 409.5 rpm. The flow rate of the water injected by the inlet was 3.8 kg/h and it flowed through a nozzle located in half of thickness of wired mesh. At a distance of 115 mm (horizontal) and 90 mm (vertical) from the axis of rotation of the wired mesh, the lower left corner of the measured area was located, and an image was captured using a Photron FASTCAM SA-Z type 2100K-M-16GB high-speed camera. The area of 23.4 x 23.4 mm (1024 x 1024 px) was recorded at a frame rate of 20,000 fps and a shutter speed of 1.25 μs, for the time of 0.05 s. This produced 1000 contiguous frames.

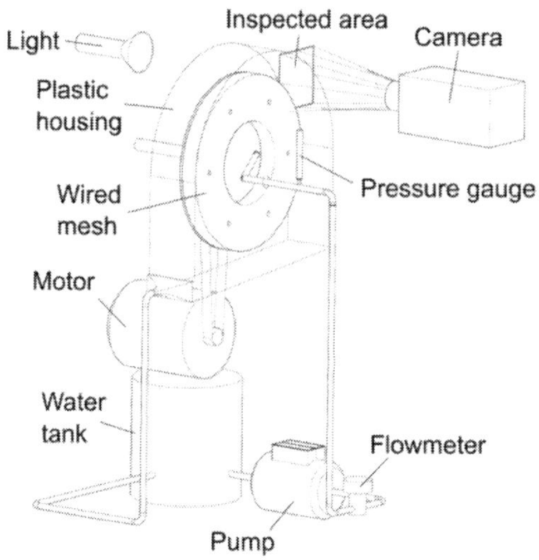

Fig. 2. Test rig

2.3 Numerical setup

2.3.1 Geometry and mesh

In order to reduce the hardware requirements, a two-dimensional representation of the RPB region was created containing an inner region with an inlet boundary condition of dimension corresponding to diameter of the nozzle. The wired mesh region is represented by circular obstacles of 0.8 mm in diameter, at a distance of 3.95 mm, helically arranged in 48 rows of geometry with inner and outer diameters ranging from 150 to 300 mm, respectively. The area of the inner region was constrained (see Figure 3 green line) in order to reduce the final number of control volumes and thus reduce the computational requirements.

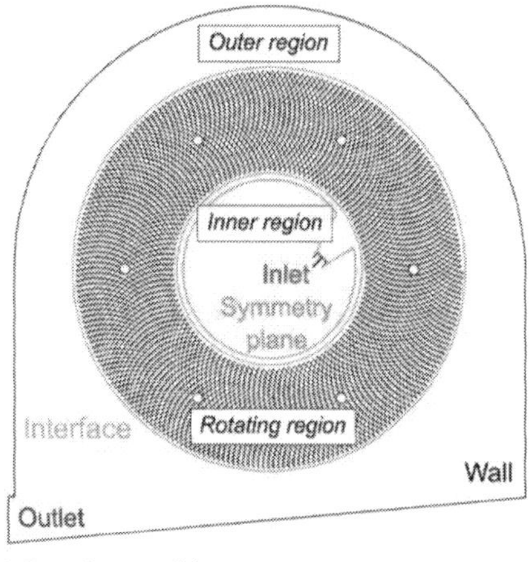

Fig. 3. Boundary conditions

A computational mesh of 5.5 million cells (0.2 inner region; 1.4 mil wired mesh; 3.9 mil outer region) was then created on this geometry. Since the Volume of Fluid method (VOF) has specific requirements for describing the interfaces between the phases and the aim of the research was to capture the smallest possible droplets formed due to dispersion, it was necessary to choose the smallest possible element size of the computational mesh. The base size was chosen to be 0.25 mm with the inner and outer region being further refined to 0.125 mm. Due to the large number of circular objects, the surface curvature value was reduced from the standard 36 points per circle to 18 pts/cycle to maintain the circular shape of the wires and reducing the number of cells in computational mesh. The resulting computational mesh is shown in Figure 4.

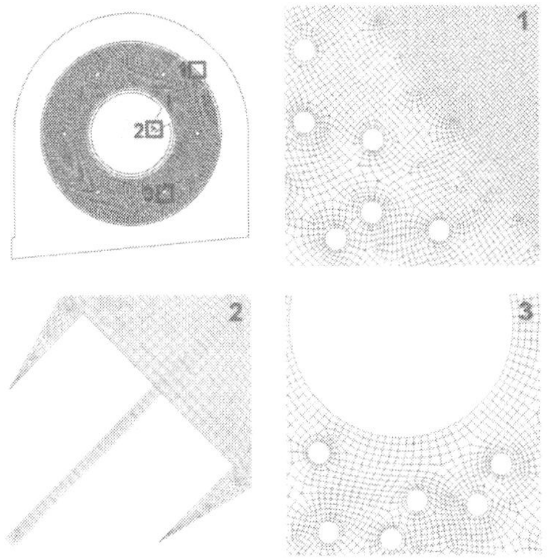

Fig. 4. Details of computational mesh

2.3.2 Physics

The numerical calculations were performed by the Siemens software, Star-CCM+ in version 2019.2 build 14.04.011. The physical setup is based on the assumption that the domain contains two immiscible newtonian fluids, water and air, both incompressible and mutually immiscible. The surface tension between the phases is 0.0728 N/m. The VOF method was used for the calculation, which solves the interactions between air (density = 1.18415 m3/kg; dynamic viscosity = 1.85508E-5 Pa/s) and water (density = 998.2 m3/kg; dynamic viscosity = 1.003E-3 Pa/s). The contact angle on the wires was set to 150° to simulate hydrophilic material, while the contact angle on the walls of RPB was prescribed to 30° (hydrophobic material) based on information from [3]. The calculation was performed on a two-dimensional domain. Turbulence was modelled using unsteady Reynolds Averaging of Navier-Stokes equations (RANS) approach with Realizable K-Epsilon Two-Layer model with all-y^+ wall treatment which calculates velocity profile of near-wall region for low values of wall-y^+ (>1) and model it for higher values. An implicit unsteady solver was set up with time step of 5×10^{-6} s with 2nd order of temporal discretization for time period of 0.55 second of phenomena. To decide whether the airflow field inside the RPB was stabilized, the liquid holdup values were monitored in all regions of the domain. The wired mesh motion was solved with a rigid body motion model with a rotation setting of 409.5 rpm in the direction corresponding to the experiment. Through the inlet (see Figure 3) in the inner region, 100% of the water fraction was injected into the domain at a velocity of 4.62 m/s corresponding to the flow rate during the experiment. The calculations were performed on 64 cores of Intel Xeon E5-2690 processors which are contained in the computing cluster located at Brno University of Technology. The run time to calculate 0.55 seconds of phenomena was 200 hours.

3 Results

With respect to intercomparison, the same area and time period was recorded in the numerical simulation as during the experiment. A comparison of the results obtained during the experiment and the numerical simulation is presented in Figure 5.

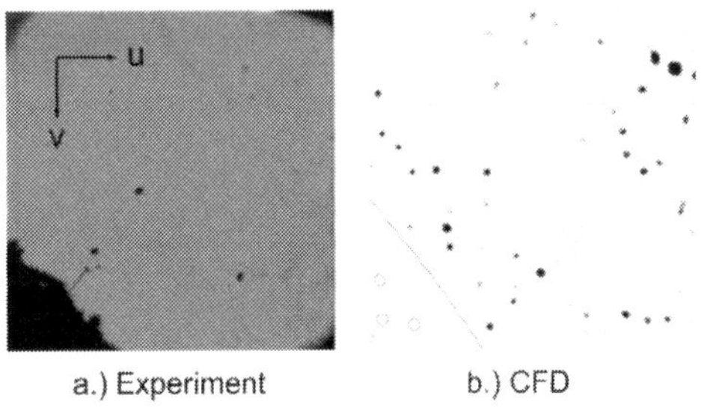

Fig. 5. Comparison of data obtained from experiment and CFD

The size and velocity of the passing water particles were then analyzed in MATLAB® (PIVlab addon) from the obtained images. In the experimental images a 37 particles were recorded and in the CFD images a 347 particles were recorded. The diameters of recorded particles were analyzed and compared in a histogram according to size classes in 0.1 mm increments (Figure 6).

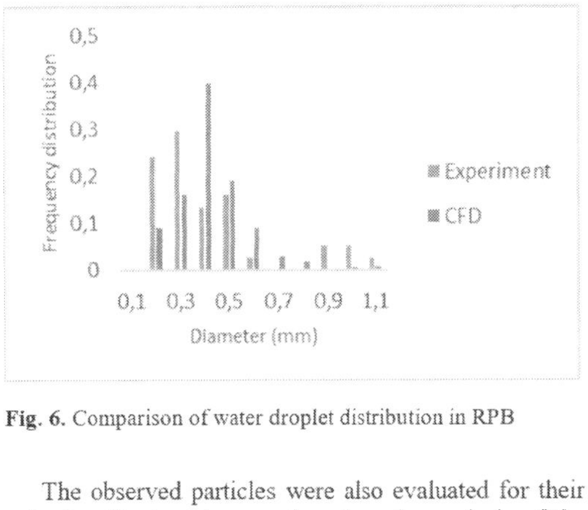

Fig. 6. Comparison of water droplet distribution in RPB

The observed particles were also evaluated for their velocity. The tracking was based on the analysis of the acquired images. The magnitudes and velocities in the direction of the individual components (see Fig. 4) were averaged and results are shown in Table 1. The analysis was based on the change in the position of each droplet within successive set of imaged obtained by high-speed camera and by numerical simulation.

Table 1. Velocity inside investigated region of RPB.

	Experiment	CFD
Velocity (mag)	*5.17 m/s*	*6,61 m/s*
Velocity (u)	*4,19 m/s*	*4,87 m/s*
Velocity (v)	*2,84 m/s*	*4,4 m/s*

The main difference between the compared cases is the number of samples analysed. Although the condition of the same time period and the same frame rate was observed, different amounts of fluid dispersed in examined space were realistically analysed. While in the case of the experiment, a thin slice through the domain of wire mesh, dependent on the depth of field of the camera, was observed and liquid droplets, located outside this depth, were not recorded, and thus could not be evaluated. In the case of CFD, which was solved on a 2D domain, an identical amount of fluid is concentrated in one plane and the amount of liquid droplets that can be analysed is consequently larger. This fact significantly affected the statistical sample size of the experimental data. Due to the density of the computational mesh, the VOF method is able to resolve droplets larger than the size of the control volume of the mesh, which in our case is 0.125 mm, while the high-speed camera should theoretically resolve droplets of one pixel size (0.023 mm).

4 Discussion

Although the numerical simulations were performed on a domain constrained in one dimension, thus neglecting the breakup due to the wires running in the RPB plane, the compared data show good agreement between experiments and numerical simulations. Disregarding the different number of liquid droplets captured during experiment and simulation, agreement in the capturing of the droplet size fraction is evident. The range of droplet sizes measured in both cases was between 0.2 - 1.1 mm. Both methods also capture the main size spectrum of particles moving in the investigated region in the range of 0.2 - 0.5 mm. These results suggest the suitability of using the VOF method to capture liquid dispersion on 2d geometry. Comparison of velocities in the measured region shows more than 20% difference in the case of magnitude and velocity in the v component direction. Since there is a significant turbulence inside the RPB due to the motion of the wire mesh, it is likely that the velocity component in the direction parallel to the axis of rotation of the RPB is non-zero and the liquid droplets captured by the high-speed camera do not move along the scanning plane and thus their velocity evaluated by the camera is distorted. The problem of the 2d numerical simulation lies mainly in the mass fraction of fluid present in the domain. Since there is no volumetric spread of the fluid injected by the inlet and a given velocity, the geometry will be overwhelmed and consequently the fluid will collide at the edge of the geometry, which does not correspond to reality.

Due to the smallest droplet size (0.1 mm in diameter) which needs to be solved using the VOF method, and the large range of length scales of the geometry (wire diameter vs. casing diameter), a full solution in 3D would require a minimum mesh size of about 80 million cells [4]. The number of elements in the mesh could be reduced by using an adaptive mesh refinement method or by prescribing a cyclic boundary condition and solving a slice of geometry [5], but the simulation demands due to the number of elements and the need to choose a short time step with respect to CFL < 1 [5] would still be too large. This comparison shows that the 2D approach, despite raising questions about the neglect of one dimension, is meaningful and may offer interesting results.

This work was financially supported by the project "Computer Simulations for Effective Low-Emission Energy" funded as project No. CZ.02.1.01/0.0/0.0/16_026/0008392 by Operational Programme Research, Development and Education, Priority axis 1: Strengthening capacity for high-quality research. And project Czech Grant Agency 21-45227L. The financial support is gratefully acknowledged.

References

1. J. Elcner, M. Jicha, MATEC Web Conf, *AEaNMiFMaE-2020*, **328**, 4 (2020)
2. Y.C. Yang, Y. Xiang, G.W. Chu, H.K. Zou, Y.Luo, M. Arowo, J.F. Chen, Chem. Eng. Sci. 138 (2015)
3. P. Xie, X. Lu, X. Yang, D. Ingham, L. Ma, M. Pourkashanian, Chem. Eng. Sci 172 (2017)
4. Y. Liu, Y. Luo, G. Chu et al., Chem. Eng. Sci 170 (2017)
5. W. Zhang, P. Xie, Y. Li et al., Chem. Eng. Journal 427 (2021)

Pollutant dispersion in a group of courtyard buildings

Simone Ferrari[1*]

[1]DICAAR - Dipartimento di Ingegneria Civile, Ambientale e Architettura, University of Cagliari, via Marengo 2, 09123 Cagliari, Italy

Abstract. As stated by the World Health Organization (WHO), the air pollution in the urban environment is the silent cause for around seven million death worldwide. This is due to the indoor and outdoor exposure to various pollutants emitted in the built environment: as the global trend is an increase of the population living in towns, this issue is predicted to become even worser. As a matter of fact, the built environment can cause the trapping of pollutants instead of their dispersion. In this work, the dispersion of a plume of a pollutant (carbon monoxide, CO), emitted from a chimneystack above the roof of courtyard in a group of courtyards, is investigated. This is achieved employing the ENVI-met software, able to model, among the others, the turbulence and pollutant dispersion in the built environment. Results show, among the others, how the pollutant emitted from an upstream building can harm also the downstream buildings.

1 Introduction

The 2021 report of the United Nations' World Health Organization (WHO) [1] has taken to the world attention the issue of the urban air pollution, with the majority of world population living in the most polluted areas, the cities (55% of the world population lives in urban areas, according to the UN, with a perspective of increase of this percentage to 68% by 2050 [2]). As a matter of fact, the urban air pollution has been recalled as the silent killer causing around seven million deaths worldwide, due to air pollution related causes, from polluted air over the cities to the smoke inside houses. Moreover, WHO estimates that almost all the global population (around 99%) is exposed to such a level of air pollution to encounter an increased risk for dangerous diseases as heart diseases, stroke, chronic obstructive pulmonary diseases, cancer and pneumonia. The built environment, in fact, alters the atmospheric air motion and can cause the trapping of pollutants inside the spaces between buildings instead of their dispersion (Kurppa et al., 2018 [3]).

The urban canopy can be classified in different ways, but usually the urban canyon (two linear buildings separated by a street) with its geometrical descriptors is the most commonly employed elementary block to describe the built environment, in particular now that individual urban canyons with their descriptors can be extracted on large extensions of real cities via free online available Geographic Information System (GIS) data (see, e.g., Salvadori et al., 2019 [4], Badas et al., 2019, [5], Salvadori et al., 2021-a [6], Salvadori et al., 2021-b [7]).

The study of the classification of elementary blocks of the boult environment can be called as a parametric study: the parametric study of the influence of urban blocks in pollutant dispersion has been a key element in the simulation via CFD (Computational Fluid Dynamics) for urban planning processes (da Silva et al., 2021-a [8], Costabile et al., 2006-a [9] and 2006-b [10]). Several authors investigated the pollutant dispersion in a single block or in a combination of blocks. Di Bernardino et al. (2017 [11]) investigated in a water channel the dispersion of a pollutant, emitted from three different point sources, around an isolated, cubic obstacle. Yassin (2011 [12]) have investigated via numerical simulations the impact of height and shape of building roof on the air quality in two-dimensional urban street canyons where the release of a pollutant was at the street level. Ferrari et al. (2019 [13]) performed, in the water tunnel, a similar experiment but with the release of the pollutant from a chimneystack. To study the pollutant dispersion, they employed a non-intrusive and quasi-continuous-in-space Digital Image Analysis technique (see Ferrari, 2017 [14], for a review) to measure the concentration, the Laser Induced Fluorescence technique employed in [15], [16] and [17]. Nosek et al. (2017 [18]) have investigated in a wind tunnel the impact of roof height non-uniformity on pollutant dispersion among groups of courtyard buildings and street intersections, with the pollutant released at the street level. Hall et al. (1999 [19]) studied in a laboratory wind tunnel the dispersion of a pollutant released from a point source at the courtyard floor level in courtyards and other enclosed buildings.

Among the blocks, the courtyard building (a building with an unroofed internal area) is one of the typical block establishing the urban environment shape, and it is also one of the most common, in particular in Europe (da Silva et al., 2021-b [20], Taleghani et al., 2014 [21]) but not only. As a matter of fact, the courtyard buildings are also very common in hot and arid climatic areas, because of their thermal efficiency (Chiri et al., 2020 [22], Al-Hafith et al., 2017 [23], Cantón et al., 2014 [24]).

* Corresponding author: ferraris@unica.it

As can be seen from the above quoted papers, most of the investigations of pollutant dispersion in the built environment are often carried on with the release of the pollutant from a street or a floor of the modelled building, in order to simulate traffic pollution. Less common is the release from a chimneystack, even if recent investigations highlight its relevance in the urban air quality issues (see, among the others, by Paunu et al., 2021 [25], Huy et al., 2021 [26] and Suryati et al., 2021 [27]). In none of those investigations, anyway, the chimneystacks were on a courtyard building.

For this reason, as, on the best of the author knowledge, the dispersion of a plume of pollutant emitted from a chimneystack above the roof of courtyard in a group of courtyards has been never investigated, the target of this paper is to fill this gap via a parametric study through numerical simulation with the software ENVI-met (see Section 2) on a group of four identical courtyards, where the length of the courtyard, i.e. the perpendicular-to-the-wind dimension of the internal open space of the building, is the variable parameter, while all the other parameters are kept constant. Among the others, the target here is also to evaluate if and how much the pollutant released form a chimneystack on a courtyard building can affect the internal courtyards of downstream buildings as well as the streets among the buildings.

2 Materials and methods

The ENVI-met software (Bruse and Fleer, 1998 [28]), able to model, among the others, the turbulence and pollutant dispersion in the built environment, has been employed in this work. ENVI-met has been widely employed for urban microclimate simulation (see the list of papers reported in Fabbri and Costanzo, 2020 [29]), as well as for the investigation of pollutant dispersion in the built environment. For instance, Nasrollahi et al (2017 [30]) employed ENVI-met to investigate the thermal performance of courtyards of different configurations, typical of Shiraz, in Iran. Among the others, also Wu et al., 2021 [31], Viecco et al., 2021 [32], Sun et al., 2021 [33], Jing and Liang 2021 [34] and Xing and Brimblecombe 2019 [35] employed ENVI-met to simulate the pollutant dispersion in several different urban configurations.

The modelled built environment is composed by (see Figure 1, where the names of the courtyards and of the streets are also reported) four identical courtyards buildings, with a constant height (10 m), a constant thickness (10 m), a constant courtyard width (W = 5 m where W is the courtyard dimension parallel to the wind direction), a constant width of the streets (i.e. a constant distance between the buildings, 10 m) and a variable courtyard length (L = 5 m, 10 m, 15 m, and 20 m, where L is the courtyard dimension perpendicular to the wind direction), and a 2m-height chimneystack on the roof of courtyard 1 (highlighted with the red dot on Figure 1) releasing a constant flow rate of carbon monoxide CO (that, consequently, is released form a point source). This arrangement was chosen as being typical of some family houses in the residential areas of many Italian towns. Anyway, it should be recalled that the target of this work is a parametric investigation on the influence of the relative dimensions of courtyard, building and road dimension on the capability to disperse the pollutant released from a chimneystack and not to simulate what happens in a portion of a real town. The configurations simulated where named W5L05, W5L10, W5L15 and W5L20.

The simulated wind is constant and from the East-direction, with an undisturbed velocity U_0 of 2.3 m/s and an air temperature of 19.8°C at the beginning of the simulation, i.e. at 6 a.m. of the 23/06/2018 (data taken from the meteorological station of Rome/Ciampino, Italy).

The domain is 250 cells-long x 150-180 cells-wide x 60 cells-height, with the single cell being a 1m-side cube. The difference in the width of the domain is due to the variable dimension of the courtyards. The domain size has been chosen to left enough space around the buildings for the flow to reattach after the detachment due to the presence of the building and to allow the stability to the numerical code (according to Blocken, 2015 [36]). The first cell close to the terrain is divided by the code into five smaller cells, in order to better simulate the typical boundary layer developing close to the terrain. Moreover, ENVI-met generates a logarithmic wind profile, typical of the Urban Boundary Layer, according to the given input data and the specific location and season. ENVI-met simulates the air turbulence via the so-called 2-equation Turbulence Kinetic Energy (TKE) Model, i.e. numerically solving the Kinetic Energy (KE) balance equation (describing the distribution of KE depending on its production, advection, diffusion and destruction) together with the Dissipation Rate of TKE (ε) equation.

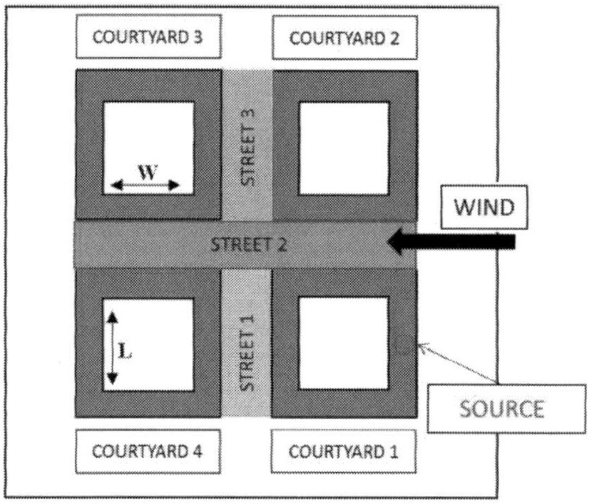

Fig. 1. Sketch of the group of courtyards, with the relevant dimensions, the pollutant source location, the wind direction, the names of the courtyards and the names of the streets.

3 Results

In Figure 2, the horizontal field of non-dimensional pollutant concentration C/C_0 (i.e., the pollutant concentration reduction, as C_0 is the CO concentration at the chimneystack outlet, i.e. the maximum concentration in the domain), extracted on the plane at the building roof height, is shown: colours close to the yellow point out where the concentration is higher, while colours close to the blue indicates where the concentration is almost zero. The white vectors represent the velocity direction (arrow) and magnitude (length). For the sake of clarity, not the whole domain is shown but only a zoom in the closest region to the courtyards where the flow is biased by the presence of the buildings. It is visible that the pollutant enters the courtyard C.1, the one upstream, but also the courtyard C.4, the one downstream, as well as the street 1, the one between C.1 and C.4, and downstream the courtyard C.4. On the contrary, in the courtyards C.2 e C.3 the pollutant does not enter. In particular, the pollutant concentration reduction C/C_0 in the courtyard C.1 is around 0.040, in the courtyard C.4 around 0.010 and in the street 1 around 0.015, with values close to the zero in the courtyards C.2 e C.3. The CO non-dimensional concentration in the street 1 is around 0.010, around 0.005 in the street 2 and almost zero in the street 3.

On Figure 3, the vertical field of non-dimensional pollutant concentration C/C_0, in colours, as well as the velocity vectors (white arrows) at the middle section of courtyards 1 and 4 are shown. Similar considerations as the ones for Figure 2 can be drawn.

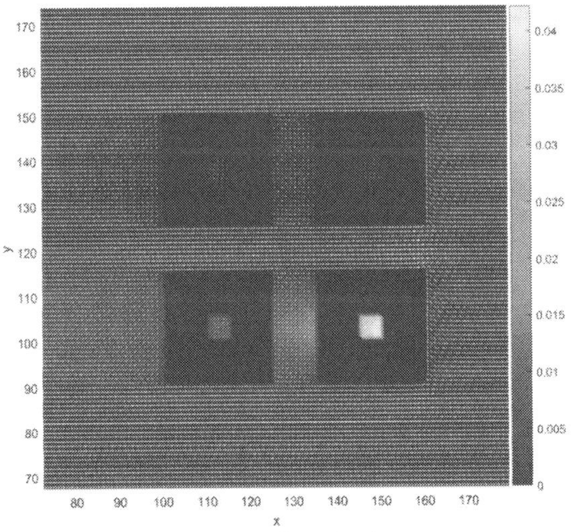

Fig. 2. Horizontal field of nondimensional CO concentration C/C_0 at the roof level (wind comes from the right).

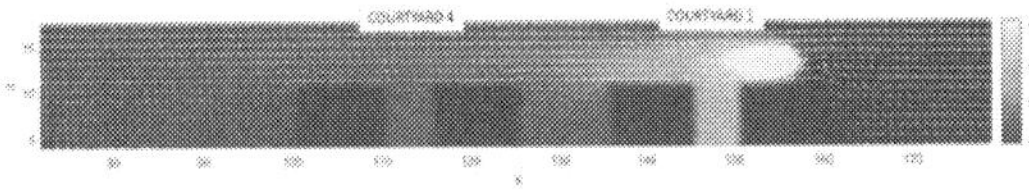

Fig. 3. Vertical field of nondimensional CO concentration C/C_0 at the middle section of courtyards 1 and 4 (wind comes from the right).

To quantify the effect of the simulated group of four courtyards and of their dimensions in reducing the pollutant penetration in the inner courtyards the non-dimensional maximum CO concentration C_{MAX}/C_0 inside the four courtyards, for the different simulated courtyard dimensions, in the volume inside the courtyard buildings is plotted, for each one of the four configurations, versus the hours of the day on Figures from 4 to 7. In the x-axis, the hours of the day are from 7 a.m. of the first simulated day to 6 a.m. of the following day. Similarly, on Figures from 8 to 10 the non-dimensional maximum CO concentration C_{MAX}/C_0 inside the three streets, for the different simulated courtyard dimensions, is plotted, for each one of the four configurations, versus the hours of the day; in this case, the concentration is measured at a 2 m height on the street level, i.e. at pedestrian level. In all the Figures from 4 to 10, the case W5L05 is represented by a continuous black line with black circles, the case W5L10 is represented by a dashed blue line with blue plus, the case W5L15 is represented by a continuous red line with red asterisks and the case W5L20 is represented by a dashed green line with green circles. The minimum non-dimensional concentration inside the courtyards is not shown, because it is always and everywhere almost zero.

Figures from 4 to 10 clearly show a time dependency due to the variable sunlight warming during the day which, in turn, increases the turbulent stirring during the warmest hours of the day.

On Figure 4 and 5, the non-dimensional maximum CO concentration C_{MAX}/C_0 inside the courtyards C.1 and C.4, for the different simulated courtyard dimensions, in the volume inside the courtyard buildings is plotted for each one of the four configurations, versus the hours of the day. The two courtyards show a similar trend, but with very different values, higher in the C.1 (as expected, because is the one where the chimneystack is located) and lower in the C.4. Moreover, for both the courtyards the concentration increases as the volume decreases and tends to have the highest value at 7 a.m., to then decrease until 1 p.m., increase until 9 p.m. and then remain almost constant until the end of the simulated day. This is due to the variation of the irradiation during the day, so during the warmer hours of the day when there is a higher temperature of the air volume close to the ground, which causes a reduction of the air density and an increase of the turbulent stirring which, in turns, favor the dilution of the pollutant.

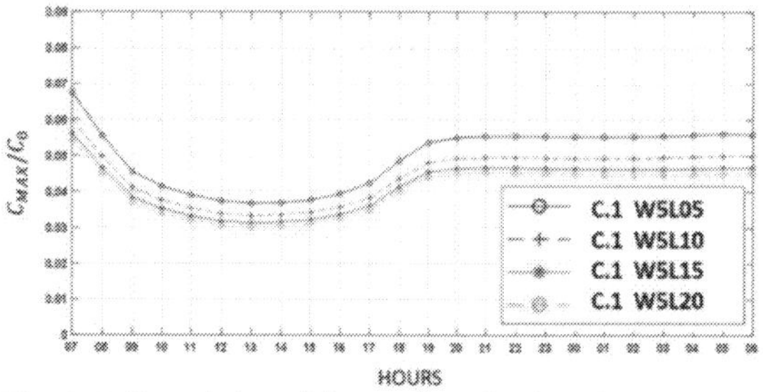

Fig. 4. Daily variation of the nondimensional maximum CO concentration C_{MAX}/C_0 inside the courtyard 1, for the different simulated courtyard dimensions.

On Figure 6 and 7, the non-dimensional maximum CO concentration C_{MAX}/C_0 inside the courtyards C.2 and C.3, for the different simulated courtyard dimensions, in the volume inside the courtyard buildings is plotted for each one of the four configurations, versus the hours of the day. As for the C.1 e C.4, the two courtyards show a similar trend, but with very different values, higher in the C.3 (that results downstream to the chimneystack) and lower in the C.4 (that results perpendicular to the chimneystack). Moreover, for both the courtyards the concentration increases as the volume decreases and it is almost zero from 6 p.m. to 8 a.m., to then increase until 1 p.m., where the highest value is found.

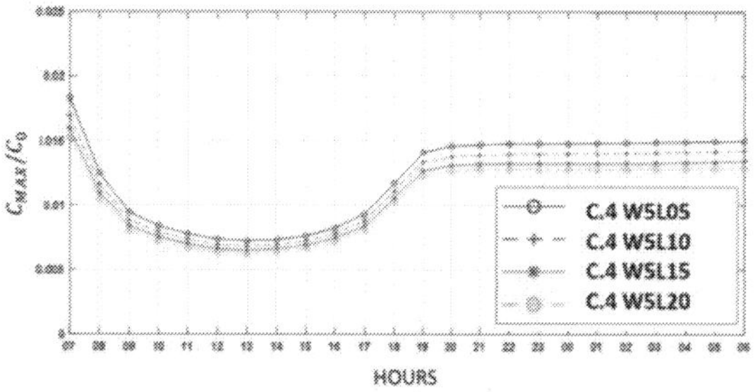

Fig. 5. Daily variation of the nondimensional maximum CO concentration C_{MAX}/C_0 inside the courtyard 4, for the different simulated courtyard dimensions.

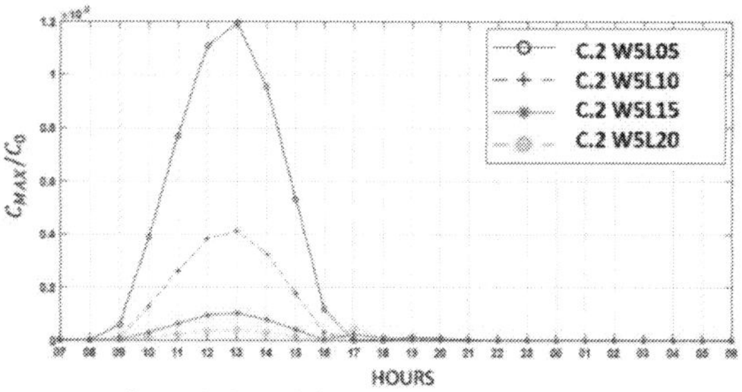

Fig. 6. Daily variation of the nondimensional maximum CO concentration C_{MAX}/C_0 inside the courtyard 2, for the different simulated courtyard dimensions.

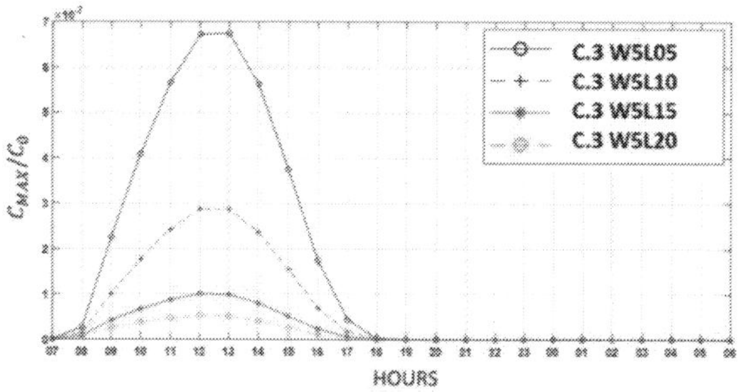

Fig. 7. Daily variation of the nondimensional maximum CO concentration C_{MAX}/C_0 inside the courtyard 3, for the different simulated courtyard dimensions.

On Figure 8, the daily variation of the nondimensional maximum CO concentration C_{MAX}/C_0 inside the street 1, for the different simulated courtyard dimensions, is reported. The pollutant trend is the same for all the simulated configurations, without a significative dependence on the courtyard dimension. The trend is very similar to the one of C.1 and C.4 and with values that are not negligible and similar to the one of the courtyard 4.

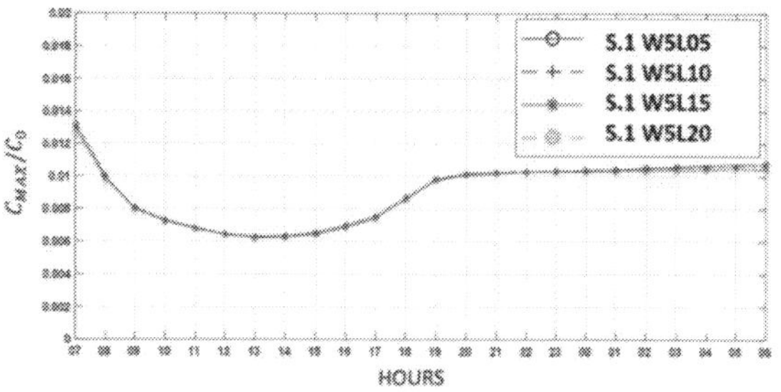

Fig. 8. Daily variation of the nondimensional maximum CO concentration C_{MAX}/C_0 inside the street 1, for the different simulated courtyard dimensions.

On Figure 9, the daily variation of the nondimensional maximum CO concentration C_{MAX}/C_0 inside the street 2, for the different simulated courtyard dimensions, is reported. The pollutant trend is the same for all the simulated configurations, with increasing values as the courtyard dimension decreases.
Eventually, on Figure 10, the daily variation of the nondimensional maximum CO concentration C_{MAX}/C_0 inside the street 3, for the different simulated courtyard dimensions, is reported. The trend is very similar to the one of C.2 and C.3, with values that higher than the ones of the quoted courtyards and with increasing values as the courtyard dimension decreases.

In summary, the results show, among the others, how the pollutant emitted from an upstream building can harm also the downstream buildings and the streets close to the buildings, with an increasing concentration as the courtyard volume decreases.

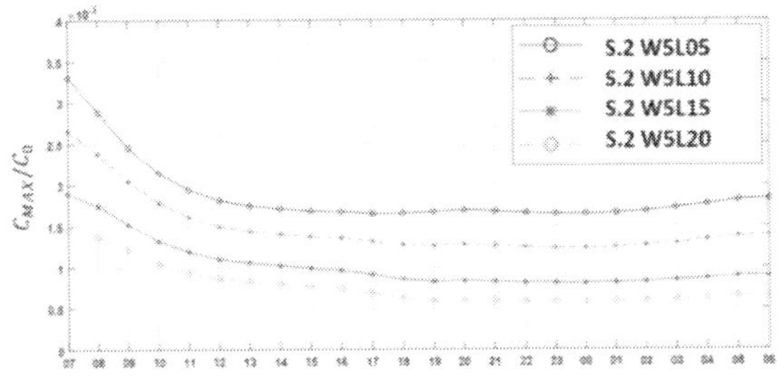

Fig. 9. Daily variation of the nondimensional maximum CO concentration C_{MAX}/C_0 inside the street 2, for the different simulated courtyard dimensions.

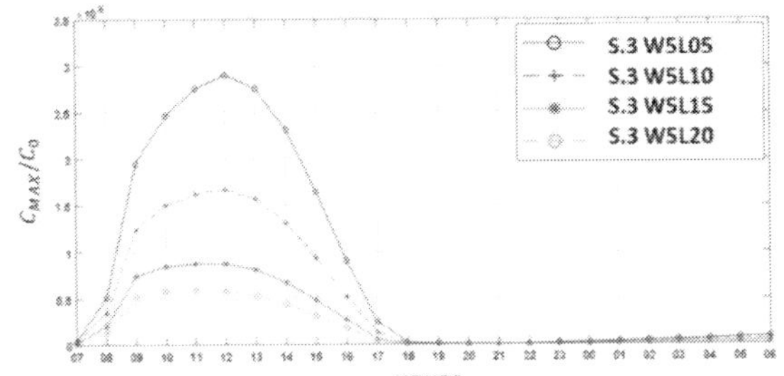

Fig. 10. Daily variation of the nondimensional maximum CO concentration C_{MAX}/C_0 inside the street 3, for the different simulated courtyard dimensions.

4 Conclusions

In this paper, the dispersion of a plume of pollutant emitted from a chimneystack above the roof of courtyard in a group of courtyards has been investigated, via a parametric study through numerical simulation with the software ENVI-met. The group consists of four identical courtyards, with a variable length of the courtyard, i.e. the perpendicular-to-the-wind dimension of the internal open space of the building, while all the other parameters are kept constant.

The main results are that the pollutant emitted from an upstream building can harm also the downstream buildings and the streets close to the buildings and, of course, the people living or working there. Moreover, the concentration tends to decrease as the courtyard volume increases, confirming the capability of the shape and dimensions of the courtyard building in contributing to the reduction of the pollutant concentration.

References

[1] World Health Organization, WHO global air quality guidelines: particulate matter (PM2.5 and PM10), ozone, nitrogen dioxide, sulfur dioxide and carbon monoxide. Executive summary. United nations, 2021.
[2] United Nations, '2018 Revision of World Urbanization Prospects'. 2018.
[3] M. Kurppa, A. Hellsten, M. Auvinen, S. Raasch, T. Vesala, and L. Järvi, 'Ventilation and Air Quality in City Blocks Using Large-Eddy Simulation—Urban Planning Perspective', Atmosphere, vol. 9, no. 2, p. 65, Feb. 2018, doi: 10.3390/atmos9020065.
[4] L. Salvadori et al., 'Similar urbanistic typologies and morpho-metric parametrization: Analysis of a possible date of construction based classification', presented at the 19th International Conference on Harmonisation within Atmospheric Dispersion Modelling for Regulatory Purposes, Harmo 2019; Bruges; Belgium; 3 June 2019 through 6 June 2019, 2019.
[5] M. G. Badas, L. Salvadori, M. Garau, G. Querzoli, and S. Ferrari, 'Urban areas parameterisation for CFD simulation and cities air quality analysis', IJEP, vol. 66, no. 1/2/3, p. 5, 2019, doi: 10.1504/IJEP.2019.104514.
[6] L. Salvadori, M. G. Badas, A. Di Bernardino, G. Querzoli, and S. Ferrari, 'A Street Graph-Based Morphometric Characterization of Two Large Urban Areas', Sustainability, vol. 13, no. 3, p. 1025, Jan. 2021, doi: 10.3390/su13031025.
[7] L. Salvadori, A. Di Bernardino, G. Querzoli, and S. Ferrari, 'A Novel Automatic Method for the Urban Canyon Parametrization Needed by Turbulence Numerical Simulations for Wind Energy Potential Assessment', Energies, vol. 14, no. 16, p. 4969, Aug. 2021, doi: 10.3390/en14164969.
[8] F. Trindade da Silva et al., 'Atmospheric dispersion and urban planning: An interdisciplinary approach to city modeling', Sustainable Cities and Society, vol. 70, p. 102882, Jul. 2021, doi: 10.1016/j.scs.2021.102882.
[9] F. Costabile, F. Wang, W. Hong, F. Liu, and I. Allegrini, 'CFD modelling of traffic-related air pollutants around an urban street-canyon in Suzhou', in Air Pollution XIV, The New Forest, UK, May 2006, vol. 1, pp. 297–306. doi: 10.2495/AIR06030.
[10] F. Costabile, F. Wang, W. Hong, F. Liu, and I. Allegrini, 'Spatial Distribution of Traffic Air Pollution and Evaluation of Transport Vehicle Emission Dispersion in Ambient Air in Urban Areas', JSME Int. J., Ser. B, vol. 49, no. 1, pp. 27–34, 2006, doi: 10.1299/jsmeb.49.27.
[11] A. D. Bernardino, P. Monti, G. Leuzzi, F. Sammartino, and S. Ferrari, 'Experimental investigation of turbulence and dispersion around an isolated cubic building', HARMO 2017 - 18th International Conference on Harmonisation within Atmospheric Dispersion Modelling for Regulatory Purposes, Proceedings, pp. 460–464, 2017.
[12] M. F. Yassin, 'Impact of height and shape of building roof on air quality in urban street canyons', Atmospheric Environment, vol. 45, no. 29, pp. 5220–5229, Sep. 2011, doi: 10.1016/j.atmosenv.2011.05.060.
[13] S. Ferrari, M. G. Badas, M. Garau, L. Salvadori, A. Seoni, and G. Querzoli, 'On The Effect Of The Shape Of Buildings And Chimneystacks On Ventilation And Pollutant Dispersion', EPJ Web Conf., vol. 213, p. 02017, 2019, doi: 10.1051/epjconf/201921302017.

[14] S. Ferrari, 'Image analysis techniques for the study of turbulent flows', EPJ Web of Conferences, vol. 143, p. 01001, 2017, doi: 10.1051/epjconf/201714301001.

[15] S. Ferrari and G. Querzoli, 'Laboratory experiments on the interaction between inclined negatively buoyant jets and regular waves', EPJ Web of Conferences, vol. 92, p. 02018, 2015, doi: 10.1051/epjconf/20159202018.

[16] S. Ferrari, M. Badas, and G. Querzoli, 'On the Effect of Regular Waves on Inclined Negatively Buoyant Jets', Water, vol. 10, no. 6, p. 726, Jun. 2018, doi: 10.3390/w10060726.

[17] S. Ferrari, M. G. Badas, and G. Querzoli, 'A non-intrusive and continuous-in-space technique to investigate the wave transformation and breaking over a breakwater', EPJ Web of Conferences, vol. 114, p. 02022, 2016, doi: 10.1051/epjconf/201611402022.

[18] Š. Nosek, L. Kukačka, K. Jurčáková, R. Kellnerová, and Z. Jaňour, 'Impact of roof height non-uniformity on pollutant transport between a street canyon and intersections', Environmental Pollution, vol. 227, pp. 125–138, Aug. 2017, doi: 10.1016/j.envpol.2017.03.073.

[19] D. J. Hall, S. Walker, and A. M. Spanton, 'Dispersion from courtyards and other enclosed spaces', Atmospheric Environment, vol. 33, no. 8, pp. 1187–1203, Apr. 1999, doi: 10.1016/S1352-2310(98)00284-2.

[20] F. Trindade da Silva, N. C. Reis, J. M. Santos, E. V. Goulart, and C. Engel de Alvarez, 'The impact of urban block typology on pollutant dispersion', Journal of Wind Engineering and Industrial Aerodynamics, vol. 210, p. 104524, Mar. 2021, doi: 10.1016/j.jweia.2021.104524.

[21] M. Taleghani, M. Tenpierik, and A. van den Dobbelsteen, 'Energy performance and thermal comfort of courtyard/atrium dwellings in the Netherlands in the light of climate change', Renewable Energy, vol. 63, pp. 486–497, Mar. 2014, doi: 10.1016/j.renene.2013.09.028.

[22] G. M. Chiri, M. Achenza, A. Cani, L. Neves, L. Tendas, and S. Ferrari, 'The Microclimate Design Process in Current African Development: The UEM Campus in Maputo, Mozambique', Energies, vol. 13, no. 9, p. 2316, May 2020, doi: 10.3390/en13092316.

[23] O. Al-Hafith, S. B K, S. Bradbury, and P. de Wilde, 'The Impact of Courtyard parameters on its shading level An experimental study in Baghdad, Iraq', Energy Procedia, vol. 134, pp. 99–109, Oct. 2017, doi: 10.1016/j.egypro.2017.09.539.

[24] M. A. Cantón, C. Ganem, G. Barea, and J. F. Llano, 'Courtyards as a passive strategy in semi dry areas. Assessment of summer energy and thermal conditions in a refurbished school building', Renewable Energy, vol. 69, pp. 437–446, Sep. 2014, doi: 10.1016/j.renene.2014.03.065.

[25] V.-V. Paunu et al., 'Spatial distribution of residential wood combustion emissions in the Nordic countries: How well national inventories represent local emissions?', Atmospheric Environment, vol. 264, p. 118712, Nov. 2021, doi: 10.1016/j.atmosenv.2021.118712.

[26] L. N. Huy, E. Winijkul, and N. T. Kim Oanh, 'Assessment of emissions from residential combustion in Southeast Asia and implications for climate forcing potential', Science of The Total Environment, vol. 785, p. 147311, Sep. 2021, doi: 10.1016/j.scitotenv.2021.147311.

[27] I. Suryati, R. Zulkarnain, and A. Pratama, 'Analysis of PM10 dispersion models from coal-fired power plant activities in North Sumatera by using CALPUFF', IOP Conf. Ser.: Earth Environ. Sci., vol. 802, no. 1, p. 012033, Jun. 2021, doi: 10.1088/1755-1315/802/1/012033.

[28] M. Bruse and H. Fleer, 'Simulating surface-plant-air interactions inside urban environments with a three dimensional numerical model', Environmental Modelling and Software, vol. 13, no. 3–4, pp. 373–384, 1998, doi: 10.1016/S1364-8152(98)00042-5.

[29] K. Fabbri and V. Costanzo, 'Drone-assisted infrared thermography for calibration of outdoor microclimate simulation models', Sustainable Cities and Society, vol. 52, p. 101855, Jan. 2020, doi: 10.1016/j.scs.2019.101855.

[30] N. Nasrollahi, M. Hatami, S. R. Khastar, and M. Taleghani, 'Numerical evaluation of thermal comfort in traditional courtyards to develop new microclimate design in a hot and dry climate', Sustainable Cities and Society, vol. 35, pp. 449–467, Nov. 2017, doi: 10.1016/j.scs.2017.08.017.

[31] J. Wu, K. Luo, Y. Wang, and Z. Wang, 'Urban road greenbelt configuration: The perspective of PM2.5 removal and air quality regulation', Environment International, vol. 157, p. 106786, Dec. 2021, doi: 10.1016/j.envint.2021.106786.

[32] M. Viecco, H. Jorquera, A. Sharma, W. Bustamante, H. J. S. Fernando, and S. Vera, 'Green roofs and green walls layouts for improved urban air quality by mitigating particulate matter', Building and Environment, vol. 204, p. 108120, Oct. 2021, doi: 10.1016/j.buildenv.2021.108120.

[33] D. (Jian) Sun, S. Wu, S. Shen, and T. Xu, 'Simulation and assessment of traffic pollutant dispersion at an urban signalized intersection using multiple platforms', Atmospheric Pollution Research, vol. 12, no. 7, p. 101087, Jul. 2021, doi: 10.1016/j.apr.2021.101087.

[34] L. Jing and Y. Liang, 'The impact of tree clusters on air circulation and pollutant diffusion-urban micro scale environmental simulation based on ENVI-met', IOP Conf. Ser.: Earth Environ. Sci., vol. 657, p. 012008, Feb. 2021, doi: 10.1088/1755-1315/657/1/012008.

[35] Y. Xing and P. Brimblecombe, 'Role of vegetation in deposition and dispersion of air pollution in urban parks', Atmospheric Environment, vol. 201, pp. 73–83, Mar. 2019, doi: 10.1016/j.atmosenv.2018.12.027.

[36] B. Blocken, 'Computational Fluid Dynamics for urban physics: Importance, scales, possibilities, limitations and ten tips and tricks towards accurate and reliable simulations', Building and Environment, vol. 91, pp. 219–245, Sep. 2015, doi: 10.1016/j.buildenv.2015.02.015.

Ventilation in a group of courtyard buildings

Simone Ferrari[1]*

[1] DICAAR - Dipartimento di Ingegneria Civile, Ambientale e Architettura, University of Cagliari, via Marengo 2, 09123 Cagliari, Italy

Abstract. In the contemporary urban planning, the outdoor comfort is more and more relevant. As a matter of fact, in some Nations the microclimate design, and so, among the others, a quantification of the outdoor comfort is already compulsory, while in many others it is recommended. Various methods to quantify the outdoor comfort can be adopted (e.g., among the others, the PMV-Predicted Mean Vote, or the PET-Physiological Equivalent Temperature), but in every formulation the quantification of the wind velocity, otherwise referred to as ventilation, close to the buildings under study is needed. In this paper, the ventilation inside and outside a group of courtyard buildings is studied via numerical simulations with ENVI-met. ENVI-met is a three-dimensional microclimate model able to simulate the ventilation in an urban environment and the interaction of air flows with surfaces of different materials, with plants and with other typical elements of the built environment in a given climate. Results highlight the relevance of the mutual influence of buildings and of their dimensions in modelling the ventilation inside and outside a courtyard.

1 Introduction

In the contemporary urban planning, the optimization of the outdoor comfort is becoming every day more and more relevant (see, e.g. Santamouris, 2001 [1]). As a matter of fact, in some Nations the microclimate design, and so, among the others, a quantification of the outdoor comfort is already compulsory, while in many others it is recommended (Eliasson, 2000 [2]).

Various methods to quantify the outdoor comfort can be adopted (e.g., among the others, the PMV- Predicted Mean Vote, or the PET- Physiological Equivalent Temperature; see the review of Coccolo et al., 2016 [3]), but in every formulation the quantification of the wind velocity, otherwise referred to as ventilation, close to the buildings under study is needed. In fact, the ventilation (i.e. the wind velocity and direction in the urban environment and its capability to exchange air in it) is the main driver for the air quality and the outdoor comfort ([4]).

Even if the main role of the built environment in the outdoor comfort has been confirmed by several scientific contributions and in various disciplines, most of the legislative guidelines and design activities are related to the optimization of single buildings, even if some studies (e.g. De Pascali, 2008 [5]) have demonstrated that this approach focused on the single building optimization leads to a reduced efficiency, as it is the interaction among buildings that plays the main role. As a consequence, the proper way to deal with the issue of the optimization of outdoor comfort in the urban built environment is to conceive it not as a collection of isolated buildings but as a unique environment, where the interaction among buildings is at least as important as each single building (see, e.g., the microclimate urban design process defined in Chiri et al, 2020 [6]).

The approach to the urban design can follow two main directions: a site-specific approach and a parametric one.

The first one allows to investigate complex built environment (see, e.g., Ouyang et al., 2022 [7], Sedagath and Sharif, 2022 [8], Piedra et al., 2020 [9], Lee et al., 2016 [10]), leading to realistic results on the specific case, but with less generalizable design advices.

On the opposite, in the second approach, the parametric investigation, simple general cases, such single buildings (e.g., Di Bernardino et al., 2017 [11]) or the classical urban canyons (see, among the others, Oke, 1988 [12] for a classification), are investigated via numerical simulations (e.g. Cui et al., 2021 [13], Sun et al., 2021 [14], Badas et al, 2017 [15]) or laboratory experiments (e.g. Badas et al., 2020 [16], Garau et al., 2018 [17], Ferrari et al, 2017 [18]), allowing to extract information about the influence of a single architectural aspect on the outdoor comfort, but without been immediately applicable to the complex nature of urban environment. Recent studies (Badas et al., 2019 [19], Salvadori et al., 2021-a [20] and 2021-b [21]) has allowed to automatically extract the geometrical features characterizing the urban canyons in real cities, via free online available Geographic Information System (GIS) data, so boosting the potentialities of this last approach to describe the urban canopy for atmospheric Fluid Dynamic purposes.

In addition to the urban canyon, the courtyard building is one of the most employed urban typologies. A courtyard building is a building with an unroofed

* Corresponding author: ferraris@unica.it

internal area, completely or partially enclosed by walls. Even if the ventilation in courtyard buildings have been extensively investigated, (see, e.g., Sun et al, 2021 [14], Piedra et al., 2020 [9], Subhashini and Thirumaran, 2020 [22]), as well as the ventilation in groups of buildings (see, e.g., Xuyi et al., 2020 [23], Ying et al., 2020, [24]), to the best of the author knowledge, a parametric study on the ventilation on a group of courtyard buildings has not yet been performed.

As a consequence, the target of this work is to perform a parametric investigation of the ventilation inside and outside a group of courtyard buildings, via numerical simulation with the widely employed ENVI-met numerical model (see Section 2). The variable parameter in this work is the length of the courtyard, i.e. the perpendicular-to-the-wind dimension of the internal open space of the building.

The possibility to model the ventilation through the shape, dimension and relative position is relevant for microclimate urban design purposes because, for instance, depending on the specific use of the internal courtyard, it could be more advisable to increase the ventilation (e.g., for recreative uses or in hot climates) or to decrease it (e.g. for storage uses or in windy climates).

2 Materials and methods

The group consists (see Figure 1) of four identical courtyard buildings, with a flat roof, a constant height (10 m), a constant thickness (10 m), a constant courtyard width (W = 5 m; W is the courtyard dimension parallel to the wind direction), a constant width of the streets (or, in other words, a constant distance between the buildings, equal to 10 m) and a variable courtyard length (L = 5 m, 10 m, 15 m, and 20 m; L is the courtyard dimension perpendicular to the wind direction). This configuration was chosen as being typical of some characteristic configurations of family houses in residential areas in many Italian towns. Anyway, it should be recalled that the target of this work is a parametric investigation on the influence of the relative dimensions of courtyard, building and road dimension on the ventilation and not the simulation of a portion of a real town. So, four different configurations were simulated, namely W5L05, W5L10, W5L15 and W5L20. The four courts are named, starting from the South-East one and in anti-clockwise direction, courtyard 1 (or C.1), courtyard 2 (C.2), courtyard 3 (C.3) and courtyard 4 (C.4), so C.1 and C.2 are upstream and C.3 and C.4 are downstream. The parallel-to-the wind street is street 2, while the two perpendicular ones are named street 1 and street 3.

The wind was constant, coming from the East-direction with an undisturbed velocity U_0 of 2.3 m/s, typical of the employed geographical location, the meteorological station of Rome/Ciampino (Italy). The other meteorological data where chosen in the same way, with an air temperature of 19.8°C at the beginning of the simulation, i.e. at 6 a.m. of the 23/06/2018.

The same 24 hours were simulated for each configuration, saving the data every hour.

The simulations ran on a personal computer with a 2,90 gigahertz Intel Core i7-7500U processor and 16GB of an 800 MHz RAM. The duration of the simulations was 78 hours and 27 minutes for W5L05, 82 hours and 26 minutes for W5L10, 84 hours and 22 minutes for W5L15 and 85 hours and 20 minutes for W5L20.

The single cell was a 1m-side cube, with a domain of 250 m in the wind direction, 60 m height and from 150 m (for the W5L5 configuration) to 180 m (for the W5L20 configuration) in the perpendicular-to-the-wind direction. The domain size was chosen, according to Blocken 2015 [25], to left enough space around the buildings to allow the flow reattachment after the detachment due to the presence of the buildings themselves and, consequently, to allow more stability to the numerical codes.

As stated in the Introduction, the numerical simulations were performed via the ENVI-met code (Bruse and Fleer, 1998 [26]), a widely employed software for simulating the urban outdoor microclimate (see, e.g., Ouyang et al., 2022 [7], Binarti et al., 2022 [27], Chiri t al., 2020 [6], Shinzato et al, 2019 [28], Tsoda et al., 2018 [29]). As reported in the review by Fabbri and Costanzo, 2020 [30], ENVI-met is the most widely employed microclimate model: in the quoted paper, a list of 59 recent (from 2013 to 2019) scientific papers where the ENVI-met software was employed to simulate the outdoor microclimate is reported.

ENVI-met is a three-dimensional microclimate model able to simulate the ventilation in an urban environment in a specific geographical location, and the interaction of surfaces of different materials, plants, solar radiation, energy exchanges, etc. with the atmospheric flow and its turbulent phenomena, in a given climate.

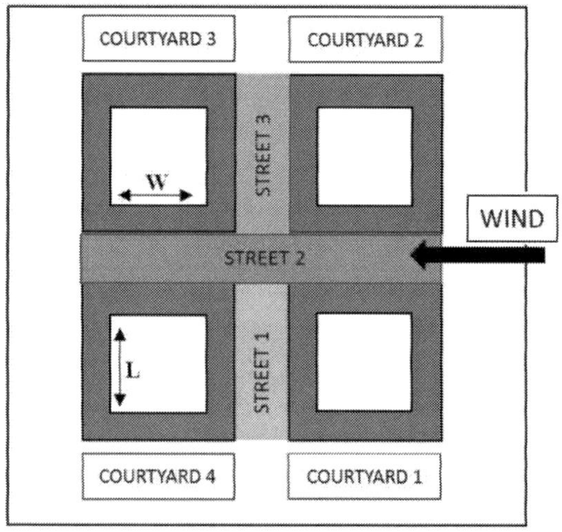

Fig. 1. Sketch of the group of courtyards, with the relevant dimensions, the wind direction, the names of the courtyards and the names of the streets.

3 Results

The data generated by the above-mentioned numerical simulations are in the form of three-dimensional matrices for each variable (e.g., wind magnitude and directions, humidity, temperature, pollutant concentrations, etc) for each hour of the simulated day.

In Figure 2, the horizontal field of velocity magnitude U in m/s, extracted on the plane at the building roof height, is shown: colours close to the yellow point out where the velocity magnitude is the highest, while colours close to the blue indicates that the velocity is almost zero. The white vectors represent the velocity direction (arrow) and magnitude (length). For the sake of clarity, only the closest region to the courtyards where the flow is more biased by the buildings is shown. As above stated, the wind comes from the right, so it is possible to spot the effect of the built environment on the flow. In particular, it is clear that the flow velocity increases in the street 2, the one parallel to the wind direction, due to the reduction of the section encountered by the wind, also called Venturi effect. Streets 1 and 3, the two perpendicular to the wind direction, are instead zones of relative calm, where the wind tends to enter from street 2, as well as the internal courts: these two streets could be consequently considered as a shelter from the wind when it should blow very fast. Moreover, the windward (i.e. upstream the building) recirculation zone and the leeward (i.e. downstream the building) wake are visible.

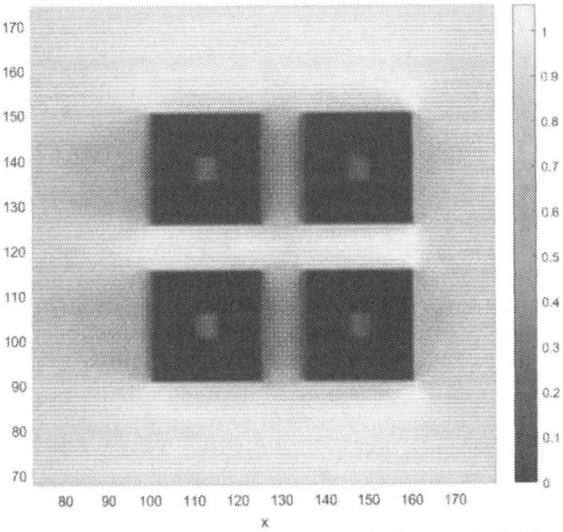

Fig. 2. Horizontal field (colour map) of velocity magnitude U [m/s] at the roof level (the undisturbed velocity U_0 is 2.3 m/s from the right), with the velocity vectors in white, for the configuration W5L05.

In Figure 3, the vertical field of velocity magnitude U [m/s], in colours, as well as the velocity vectors (white arrows) at the middle section of courtyards 2 and 3 are shown. As for Figure 2, it is possible to see that the influence of the built environment on the flow is much larger than the size of the buildings themselves, with the flow separation upstream (with the related recirculation vortex with a horizontal axis), the intrusion of the air inside the inner courts and in the perpendicular to the wind direction street, the downward wake (also here with a recirculation vortex with a horizontal axis), and the acceleration zone above the roof of the first building.

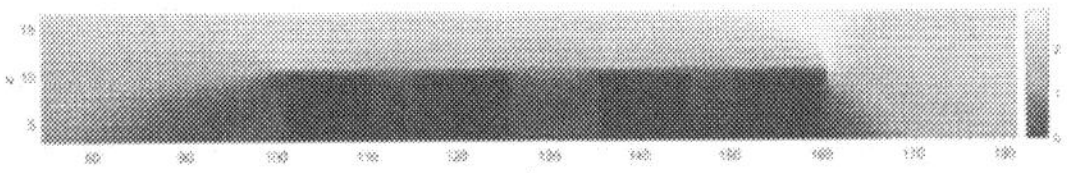

Fig. 3. Vertical field (colour map) of velocity magnitude U [m/s] at the middle section of courtyards 2 and 3 (the undisturbed velocity U_0 is 2.3 m/s from the right), with the velocity vectors in white, for the configuration W5L05.

In order to quantify the effect of the group of four courtyards and of their dimensions in reducing the wind velocity in the inner courtyards or, vice-versa, in allowing their ventilation, the maximum velocity U_{MAX}, as well as the minimum velocity U_{min}, in the volume inside the building have been divided by the wind undisturbed velocity U_0 and plotted for each one of the four configurations, versus the hours of the day on Figures 4 and 5. In this way, the velocity reduction (or nondimensional velocity) U/U_0 is determined. In the x-axis, the hours of the day are from 7 a.m. of the first simulated day to 6 a.m. of the following day. In Figures 4 and 5, the case W5L05 is represented by a continuous black line with black circles, the case W5L10 is represented by a dashed blue line with blue plus, the case W5L15 is represented by a continuous red line with red asterisks and the case W5L20 is represented by a dashed green line with green circles.

In Figure 4, the nondimensional maximum velocity U_{MAX}/U_0 inside the courtyard 1 during the simulated day is plotted. The time dependency, due to the varying sunlight warming during the day, which increases the forcing during the warmest hours, is evident. Moreover, on one hand, the velocity values tend to be the highest for the largest courtyard (W5L20) to then decrease for W515, w5L10 and to find the minimum values or the smallest courtyard (W5L05). On the other hand, the velocity trend and its values tend to be very similar for all the four simulated configurations. For all the courtyards, the lowest values are recorded during the night and the first hours of the morning, while the highest ones are encountered at 17:00 and 18:00. In particular, the absolute lowest value is around 0.17 and it is found at 7 a.m. in the configuration W5L05, while the highest one is around 0.26, found at 6 p.m. in the configuration W5L20.

The nondimensional maximum velocity U_{MAX}/U_0 inside the courtyard 2 during the day is not reported, because, as expected, as it has an identical behaviour as the same quantity in the courtyard 1.

In Figure 5, the nondimensional maximum velocity U_{MAX}/U_0 inside the courtyard 4 during the day is plotted. In a similar manner to the upstream courtyards 1 and 2, the velocity values tend to be the highest for the largest courtyard (W5L20) and then decrease for W515, W5L10 and to find the minimum values or the smallest courtyard (W5L05). Moreover, also in this case in a similar manner to courtyards 1 and 2, the velocity trend and its values tend to be very similar for all the four simulated configurations: as a matter of fact, in all the configurations, the lowest values are recorded during the night and the first hours of the morning, while the highest ones at 17:00 and 18:00. It has to be highlighted that the velocity values for the downstream courtyards 3 and 4 tend to be higher than the ones of the upwards courtyards 1 and 2. In fact, the absolute lowest value is around 0.23 and it is found at 7 a.m. in the configuration W5L05, while the highest one is around 0.27, found at 6 p.m. in the configuration W5L20. These highest values in the downstream courtyards can be explained with the different velocity direction in the two cases: as visible in Figure 3, in the case of the upstream courtyards the velocity is almost vertical due to the flow separation, while in the case of the downstream courtyards the velocity is almost horizontal, in a skimming flow manner (as described in Oke, 1988 [12]), so in this last case it has to modify less its direction to enter the internal volume of the courtyard and, in turns, a larger part of the velocity itself is able to contribute to the ventilation.

The nondimensional maximum velocity U_{MAX}/U_0 inside the courtyard 3 during the day is not reported, because, as expected, as it has an identical behaviour as the same quantity in the courtyard 4.

In all the cases, the maximum values are found close to horizontal open plane close to the building roof.

In summary, we can state that the increase of the internal volume of the courtyard through the increment of the dimension perpendicular to the wind direction contributes to increase the maximum velocity and so, in turns, to improve the ventilation.

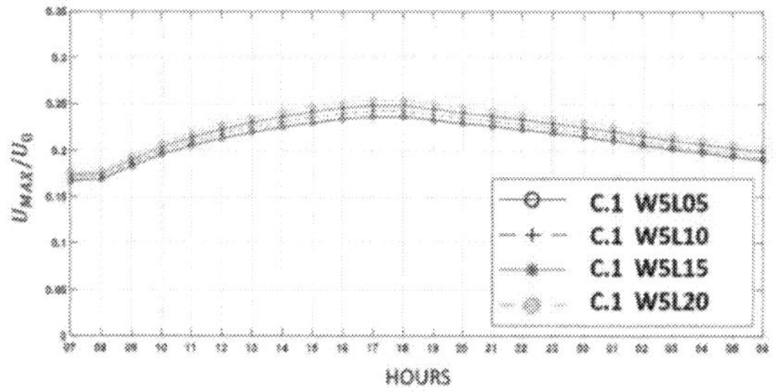

Fig. 4. Maximum nondimensional velocity U_{MAX}/U_0 in the internal court of courtyard 1 versus time (hours of the day), for the different simulated courtyard dimensions (in the legend); U_0 is the undisturbed wind velocity.

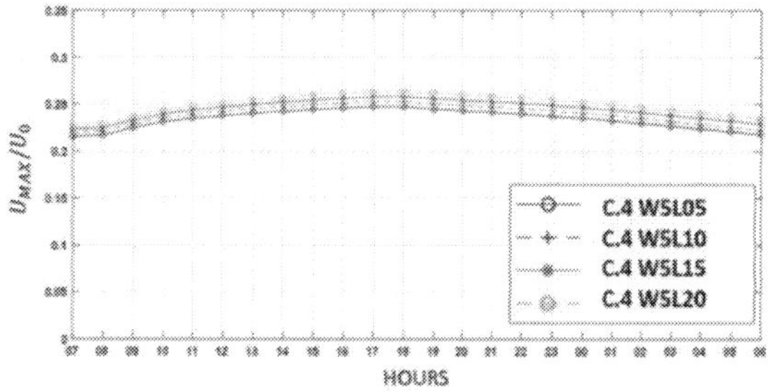

Fig. 5. Maximum nondimensional velocity U_{MAX}/U_0 in the internal court of courtyard 4 versus time (hours of the day), for the different simulated courtyard dimensions (in the legend); U_0 is the undisturbed wind velocity.

Finally, the nondimensional minimum velocity U_{mn}/U_0 inside all the courtyards during the day has always a very similar behaviour and, moreover, very low values, so it is not plotted here. Anyway, the minimum velocity inside the internal volume of the courtyards during the hours of the day has almost identical values and trends for all the four simulated configurations, with the lowest value (around 0.002) at 9 a.m. and the highest value (around 0.004) at 10 p.m, meaning, as just stated, that these values are almost negligible.

In summary, we can state that the increase of the internal volume of the courtyard through the increment of the dimension perpendicular to the wind direction does not contribute to increase the minimum velocity and so that all the courtyards have similar capability to isolate the internal open volume from the external wind.

4 Conclusions

In the presented work, the ventilation inside and outside a group of four courtyard buildings was investigated via numerical simulations with the ENVI-met code, a three-dimensional microclimate model able to simulate, among the others, the ventilation in an urban environment and its interaction with surfaces, plants and the atmosphere in a given climate. In this parametric study, the variable parameter is the length of the courtyard L, that is the perpendicular-to-the-wind dimension of the internal open space of the building, while all the other parameters are left constant.

Results highlight the relevance of the mutual influence of buildings and of their dimensions in modelling the ventilation inside and outside a courtyard.

In particular, the flow velocity increases in the street 2, the one parallel to the wind direction, due to the Venturi effect, while streets 1 and 3, the two perpendicular to the wind direction, are instead zones of relative calm, that could be consequently used as a shelter from the wind when it should blow very fast.

Regarding the ventilation in the internal courtyards, an increase of the internal volume of the courtyard through the increment of the dimension perpendicular to the wind direction contributes to increase the maximum velocity and so, in turns, to improve the ventilation. The velocities in the downstream courtyards C.3 and C.4 was found to be slightly higher than in the upstream courtyards C.1 and C.2: this can be explained with the different velocity direction in the two cases, because in the case of the upstream courtyards the wind is almost vertical (due to the flow separation), while in the case of the downstream courtyards the wind approaches the internal courtyard almost horizontally, so it is easier to enter the internal open volume of the building.

Moreover, the increase of the internal volume of the courtyard through the increment of the dimension perpendicular to the wind direction does not contribute much to increase the minimum velocity, meaning that all the simulated courtyard buildings have similar capability to isolate the internal open volume from the external wind.

Eventually, we recall that the presented results can help in modelling the ventilation through the shape, dimension and relative position, and this can be relevant for microclimate urban design purposes.

References

[1] M. Santamouris, *Energy and climate in the urban built environment*. London: James & James, 2001.
[2] I. Eliasson, 'The use of climate knowledge in urban planning', *Landscape and Urban Planning*, vol. 48, no. 1–2, pp. 31–44, Apr. 2000, doi: 10.1016/S0169-2046(00)00034-7.
[3] S. Coccolo, J. Kämpf, J.-L. Scartezzini, and D. Pearlmutter, 'Outdoor human comfort and thermal stress: A comprehensive review on models and standards', *Urban Climate*, vol. 18, pp. 33–57, Dec. 2016, doi: 10.1016/j.uclim.2016.08.004.
[4] Y. Tamura, *Advanced environmental wind engineering*. New York, NY: Springer Berlin Heidelberg, 2016.
[5] P. De Pascali, *Città ed energia. La valenza energetica dell'organizzazione insediativa*. Milano: Franco Angeli, 2008.
[6] G. M. Chiri, M. Achenza, A. Canì, L. Neves, L. Tendas, and S. Ferrari, 'The Microclimate Design Process in Current African Development: The UEM Campus in Maputo, Mozambique', *Energies*, vol. 13, no. 9, p. 2316, May 2020, doi: 10.3390/en13092316.
[7] W. Ouyang, T. Sinsel, H. Simon, T. E. Morakinyo, H. Liu, and E. Ng, 'Evaluating the thermal-radiative performance of ENVI-met model for green infrastructure typologies: Experience from a subtropical climate', *Building and Environment*, vol. 207, p. 108427, Jan. 2022, doi: 10.1016/j.buildenv.2021.108427.
[8] A. Sedaghat and M. Sharif, 'Mitigation of the impacts of heat islands on energy consumption in buildings: A case study of the city of Tehran, Iran', *Sustainable Cities and Society*, vol. 76, p. 103435, Jan. 2022, doi: 10.1016/j.scs.2021.103435.
[9] K. Piedra, W. Bustamante, and C. Schmitt, 'Living and Comfort Conditions in Heritage Housing: The Inner Courtyard as a Uniting Element in the Cités of Santiago Poniente', *IOP Conf. Ser.: Earth Environ. Sci.*, vol. 503, p. 012010, Jun. 2020, doi: 10.1088/1755-1315/503/1/012010.
[10] H. Lee, H. Mayer, and L. Chen, 'Contribution of trees and grasslands to the mitigation of human heat stress in a residential district of Freiburg, Southwest Germany', *Landscape and Urban Planning*, vol. 148, pp. 37–50, Apr. 2016, doi: 10.1016/j.landurbplan.2015.12.004.
[11] A. D. Bernardino, P. Monti, G. Leuzzi, F. Sammartino, and S. Ferrari, 'Experimental investigation of turbulence and dispersion around an isolated cubic building', in *HARMO 2017 - 18th International Conference on Harmonisation within Atmospheric Dispersion Modelling for Regulatory Purposes, Proceedings*, vol. 2017, Hungarian Meteorological Service, 2017, pp. 460–464.
[12] T. R. Oke, 'Street design and urban canopy layer climate', *Energy and Buildings*, vol. 11, no. 1, pp. 103–113, Mar. 1988, doi: 10.1016/0378-7788(88)90026-6.
[13] D. Cui *et al.*, 'Effects of building layouts and envelope features on wind flow and pollutant exposure in height-asymmetric street canyons', *Building and Environment*, vol. 205, p. 108177, Nov. 2021, doi: 10.1016/j.buildenv.2021.108177.
[14] H. Sun, C. Jimenez-Bescos, M. Mohammadi, F. Zhong, and J. K. Calautit, 'Numerical Investigation of the Influence of Vegetation on the Aero-Thermal Performance of Buildings with Courtyards in Hot Climates', *Energies*, vol. 14, no. 17, p. 5388, Aug. 2021, doi: 10.3390/en14175388.
[15] M. G. Badas, S. Ferrari, M. Garau, and G. Querzoli, 'On the effect of gable roof on natural ventilation in two-dimensional urban canyons', *Journal of Wind Engineering and Industrial Aerodynamics*, vol. 162, pp. 24–34, Mar. 2017, doi: 10.1016/j.jweia.2017.01.006.
[16] M. G. Badas, S. Ferrari, M. Garau, A. Seoni, and G. Querzoli, 'On the Flow Past an Array of Two-Dimensional Street Canyons Between Slender Buildings', *Boundary-Layer Meteorol*, vol. 174, no. 2, pp. 251–273, Feb. 2020, doi: 10.1007/s10546-019-00484-x.
[17] M. Garau, M. G. Badas, S. Ferrari, A. Seoni, and G. Querzoli, 'Turbulence and Air Exchange in a Two-Dimensional Urban Street Canyon Between Gable Roof Buildings', *Boundary-Layer Meteorology*, vol. 167, no. 1, pp. 123–143, Apr. 2018, doi: 10.1007/s10546-017-0324-4.
[18] S. Ferrari, M. G. Badas, M. Garau, A. Seoni, and G. Querzoli, 'The air quality in narrow two-dimensional urban canyons with pitched and flat roof buildings', *International Journal of Environment and Pollution*, vol. 62, no. 2/3/4, p. 22, 2017, doi: 0957-4352.
[19] M. G. Badas, M. Garau, G. Querzoli, L. Salvadori, and S. Ferrari, 'Urban areas parametrization for CFD simulation and cities air quality analysis', *International Journal of Environment and Pollution*, vol. in press, 2019, doi: 10.1504/IJEP.2020.10022368.
[20] L. Salvadori, M. G. Badas, A. Di Bernardino, G. Querzoli, and S. Ferrari, 'A Street Graph-Based Morphometric Characterization of Two Large Urban Areas', *Sustainability*, vol. 13, no. 3, p. 1025, Jan. 2021, doi: 10.3390/su13031025.
[21] L. Salvadori, A. Di Bernardino, G. Querzoli, and S. Ferrari, 'A Novel Automatic Method for the Urban Canyon Parametrization Needed by Turbulence Numerical Simulations for Wind Energy Potential Assessment', *Energies*, vol. 14, no. 16, p. 4969, Aug. 2021, doi: 10.3390/en14164969.
[22] S. Subhashini and K. Thirumaran, 'CFD simulations for examining natural ventilation in the learning spaces of an educational building with courtyards in

Madurai', *Building Services Engineering Research and Technology*, vol. 41, no. 4, pp. 466–479, Jul. 2020, doi: 10.1177/0143624419878798.

[23] G. Xuyi, L. Weicong, and D. Zhaoming, 'Research on Ventilation Optimization Design of Residence Community Based on Quantitative Analysis of PHOENICS : taking the Yunshan Poetry Courtyard as an example', in *2020 International Conference on Innovation Design and Digital Technology (ICIDDT)*, Zhenjing, China, Dec. 2020, pp. 92–97. doi: 10.1109/ICIDDT52279.2020.00025.

[24] X. Ying, Y. Wang, W. Li, Z. Liu, and G. Ding, 'Group Layout Pattern and Outdoor Wind Environment of Enclosed Office Buildings in Hangzhou', *Energies*, vol. 13, no. 2, p. 406, Jan. 2020, doi: 10.3390/en13020406.

[25] B. Blocken, 'Computational Fluid Dynamics for urban physics: Importance, scales, possibilities, limitations and ten tips and tricks towards accurate and reliable simulations', *Building and Environment*, vol. 91, pp. 219–245, Sep. 2015, doi: 10.1016/j.buildenv.2015.02.015.

[26] M. Bruse and H. Fleer, 'Simulating surface-plant-air interactions inside urban environments with a three dimensional numerical model', *Environmental Modelling and Software*, vol. 13, no. 3–4, pp. 373–384, 1998, doi: 10.1016/S1364-8152(98)00042-5.

[27] F. Binarti, M. D. Koerniawan, S. Triyadi, and A. Matzarakis, 'The predicted effectiveness of thermal condition mitigation strategies for a climate-resilient archaeological park', *Sustainable Cities and Society*, vol. 76, p. 103457, Jan. 2022, doi: 10.1016/j.scs.2021.103457.

[28] P. Shinzato, H. Simon, D. H. Silva Duarte, and M. Bruse, 'Calibration process and parametrization of tropical plants using ENVI-met V4 – Sao Paulo case study', *Architectural Science Review*, vol. 62, no. 2, pp. 112–125, Mar. 2019, doi: 10.1080/00038628.2018.1563522.

[29] S. Tsoka, A. Tsikaloudaki, and T. Theodosiou, 'Analyzing the ENVI-met microclimate model's performance and assessing cool materials and urban vegetation applications–A review', *Sustainable Cities and Society*, vol. 43, pp. 55–76, Nov. 2018, doi: 10.1016/j.scs.2018.08.009.

[30] K. Fabbri and V. Costanzo, 'Drone-assisted infrared thermography for calibration of outdoor microclimate simulation models', *Sustainable Cities and Society*, vol. 52, p. 101855, Jan. 2020, doi: 10.1016/j.scs.2019.101855.

 https://doi.org/10.1051/epjconf/202226401015

Analysis of velocity distribution in an air flow through a thin perforated plate

Adam Groma[1*], *Adam* Kazda[1], *Michal* Kotek[2], *Darina* Jašíková[2], and *Adam* Kuře[2]

[1]VÚTS, a.s., Department of Computation and Modelling, Svárovská 619, 460 01 Liberec, Czech Republic

[2]Technical University of Liberec, Faculty of Mechatronics, Informatics and Interdisciplinary Studies, Studentská 2, 461 17 Liberec, Czech Republic

Abstract. Modelling air flow in a curved fine perforated plate region was performed by porous zone substitution. This approach significantly simplifies the observed process, thus various limitations appeared. PIV measurements on an experimental test rig were performed and compared with the results from CFD computations.

1 Introduction

Perforated plates find their use throughout various industries. The most common perforated plate applications in fluid dynamics are with filtration, different particle size separation, flow homogenization or pressure drop control or throttling. Although a fluid flow through a perforated plate in a normal direction is often studied and well-described phenomenon, a flow hitting a plate under an acute angle is more complex and much less investigated.

An impact of a perforated plate on a fluid flow highly depends on both flow and plate properties. Particularly important are fluid viscosity, density and flow velocity magnitude and direction for the flow and orifices shape and plate thickness for the plate.

Porous Zone (PZ) model defines a source term in momentum equation composed of viscous loss taking into account viscosity and linear velocity dependence and inertial loss taking into account density and quadratic velocity dependence. Both terms include empirical coefficients defining specific properties of the porous zone itself. Thus, the model is widely used for perforated plate representation where the plate orifices properties are covered in the empirical coefficients.

To determine viscous and inertial loss coefficients of the perforated plate, various studies were executed on an experimental [1] and numerical [3-4] basis. Also, several analytical relations [4-5] and experimental methods were developed and described for this purpose. Nevertheless, the studies typically consider flow direction normal to the perforated plate.

In this study, an air flow through a curved perforated plate is investigated experimentally and numerically. First Particle Image Velocimetry (PIV) is used to visualize velocity distribution in the air flow. Further the results are compared with a numerical study using the porous zone model for the perforated plate with different means of porous zone properties evaluation. Suitability of the used porous zone properties evaluation represents a topic of further discussion and investigation.

2 Experimental setup

The experiment took place at the Technical University of Liberec where a test rig was designed and constructed. The rig scheme is shown in Fig. 1. The monitored geometry consists of an 80 x 80 mm square cross-section channel which is straight at the beginning to allow velocity profile development, before the end it is curved by 90° and it ends in a closed section. The outer (upper) wall of the curved part is made of a perforated plate and above it, a second channel is widening from the beginning of the curve to its end. The end of the second channel is connected to a suction device. Thus, all the air in the rig flows from the air inlet through the perforated plate to the suction device.

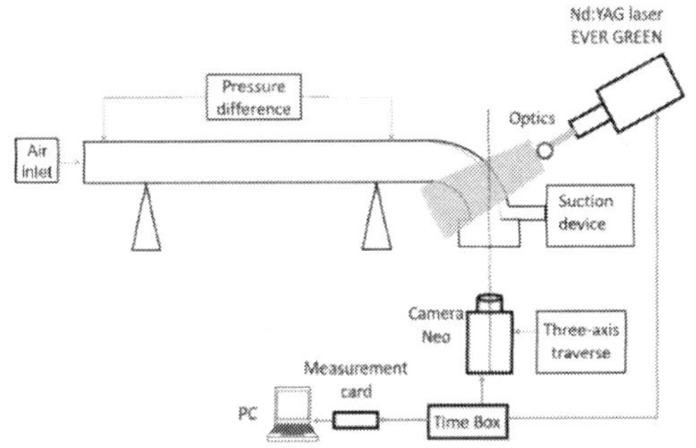

Fig. 1. Experimental setup for PIV measurements.

For the air flow behaviour and its velocity distribution also the perforated plate properties are important. The used plate has a constant thickness and is varnished. The perforation consists of constant radius circular holes pattern with specified axial spacing. The thickness of the plate including varnish is similar to hole diameter.

---
[*] Corresponding author: adam.groma@vuts.cz

3 PIV measurement results

Measurements were performed for three mean air flow velocities of 4 m/s, 5.7 m/s and 7 m/s. The mean air flow velocity was measured in the straight part of the channel.

The PIV method was used to evaluate the vector velocity distribution field in the curved channel area where the perforated plate is located. Velocity distributions for mean velocity 4 m/s, 5.7 m/s and 7 m/s are displayed in Fig. 2, Fig. 3 and Fig. 4 respectively. It is obvious that the distribution in the observed area has the same character in case of all three mean velocities. In the square channel, the velocity magnitude is steady along the curved region except for near the bottom wall. Here the flow velocity slightly decreases from the beginning of the curve and before the reach of the perforated plate end a significant drop occurs. Due to the immense air outtake at the end region, the flow velocity drops significantly in the whole cross-section. In the channel above the perforated plate, the velocity magnitude is steadily increasing along the curve and it reaches its maximum at the end of the curved part where the suction unit is connected.

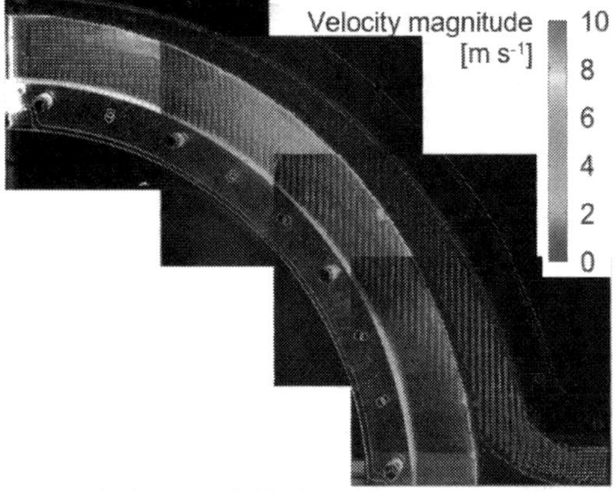

Fig. 2. Velocity vector field coloured by velocity magnitude for mean velocity of 4 m/s.

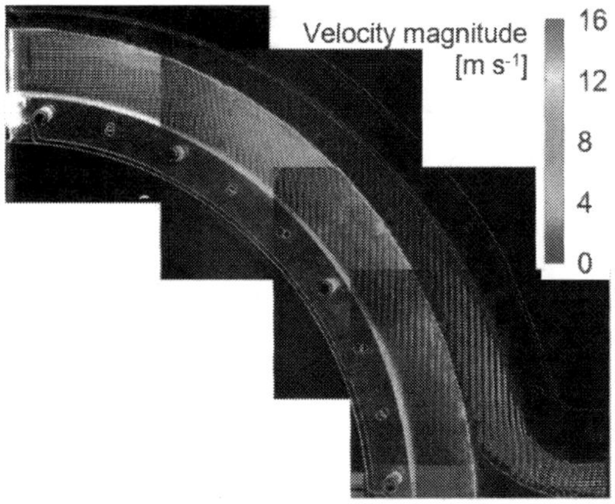

Fig. 3. Velocity vector field coloured by velocity magnitude for mean velocity of 5.7 m/s.

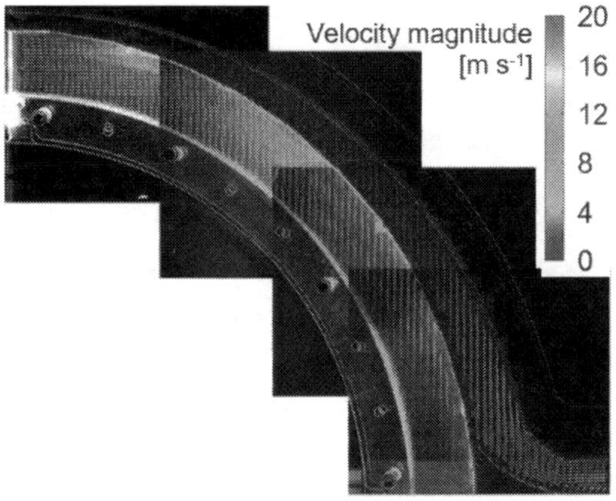

Fig. 4. Velocity vector field coloured by velocity magnitude for mean velocity of 7 m/s.

At the end of the perforated plate, the velocity is turning significantly more in the radial direction of the curve than anywhere else in the perforated plate region. Whereas the change in velocity magnitude in the perforated plate region does not change significantly, the volume flow rate through the plate is clearly highest at the perforated plate end.

4 Porous zone model

The air flow through the observed area was simulated by the CFD software ANSYS Fluent with the *k-ω* SST turbulence model. In numerical, studies the porous zone model is suitable for perforated plate simulations. Using the porous model a new source term is added in the momentum equation, [6]:

$$S_i = -\left(\sum_{j=1}^{3} D_{ij}\mu v_j + \sum_{j=1}^{3} C_{ij}\frac{1}{2}\rho|\boldsymbol{v}|v_j\right), \quad (1)$$

where S_i is the source term in coordinate direction i, $\boldsymbol{v}$ is velocity vector, v_i is velocity vector component in direction i, ρ is density, μ is dynamic viscosity and $\boldsymbol{D}$ and $\boldsymbol{C}$ are prescribed matrices of viscous loss coefficients D_{ij} and inertial loss coefficients C_{ij}. Appropriately the first term on the right-hand side of equation (1) represents viscous loss and comes out of Darcy's law and the second term on the right-hand side of equation (1) is inertial loss term. In equation (1) viscous and inertial loss coefficients are the only variables defining the porous zone properties.

4.1 Momentum source term

Various options are open on how to obtain perforated plate viscous and inertial loss coefficients for the momentum source term in the porous zone model. In this study, two different methods are used. First, the coefficients are obtained from an empirical equation, second, they are calculated by regression of pressure drop and velocity from the numerical simulation of normal flow through the defined perforated plate.

4.1.1 Empirical expression

The empirical expression presented by Kast [5] where the inertial loss coefficient in the normal direction is defined in terms of perforated plate porosity ϕ, thickness L, hydraulic (hole) diameter d_h and resistance coefficient k as follows

$$C_n = \frac{k}{\phi^2 L}\ ;\quad \frac{L}{d_h} \gg 0. \tag{2}$$

The porosity of the used perforated plate is $\phi = 0.3715$ and thickness is $L = 1.4$ mm. In terms of porosity and discharge coefficient

$$\alpha = 0.6 + 0.4\,\phi^2, \tag{3}$$

resistance coefficient is given for $\frac{L}{d_h} \gg 0$:

$$k = \left(\frac{1}{\alpha} - 1\right)^2 + (1-\phi)^2. \tag{4}$$

Thus, the inertial loss coefficient is fully defined. For this method, the viscous loss coefficient is set to 0.

4.1.2 Numerical data regression

For the second method, an air flow normal to the perforated plate in a square channel with periodic (or symmetry) boundaries in place of channel walls was simulated. For various mean flow velocities v_∞ at the channel inlet, the pressure drop ΔP between inlet and outlet was monitored. Further second-order regression curve was evaluated for the pressure-velocity data

$$\Delta P = -S_n L = l_n v_\infty + q_n v_\infty^2. \tag{5}$$

Using equation (1) viscous loss term in the normal direction D_n and inertial loss term in the normal direction C_n were calculated from linear coefficient l_n and quadratic coefficient q_n in equation (6) respectively.

Since the perforated plate used for experimental investigation is varnished the round holes in the plate are expected not to having sharp edges. To cover the possible case with sharp holes edges, two different geometries of perforation were used for the numerical data regression. One with simple sharp holes edges and the other with the edges rounded by the radius of 2 mm. The pressure-velocity data and regression curves for both cases are shown in Fig 5.

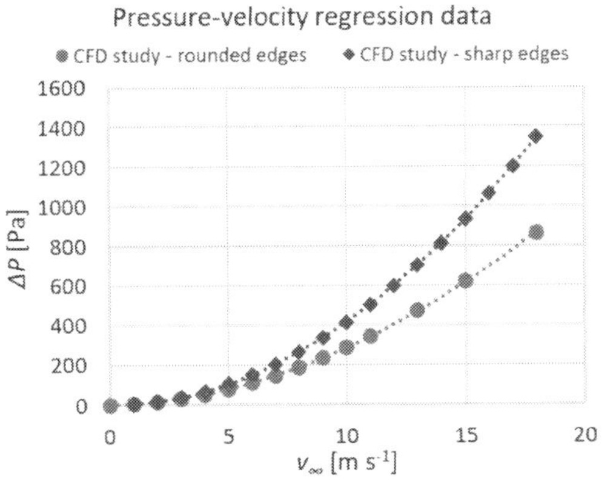

Fig. 5. Pressure drop ΔP on mean velocity v_∞ for normal air flow through perforated plate with sharp and round edged holes.

4.1.3 Tangential direction resistance coefficients

From the two above-described methods three sets of inertial and viscous loss coefficients in the normal direction to the perforated plate are evaluated, Table 1. For a proper numerical description of the porous zone, coefficients in the plate tangential direction should be used. Assuming the flow will not pass through the closed surface the tangential resistance should be higher than normal. Based on Oezcan [7] the coefficients in tangential direction were determined 10 times higher than in normal one. Thus, the porous zone model is fully defined in the local coordinate system associated with the perforated plate.

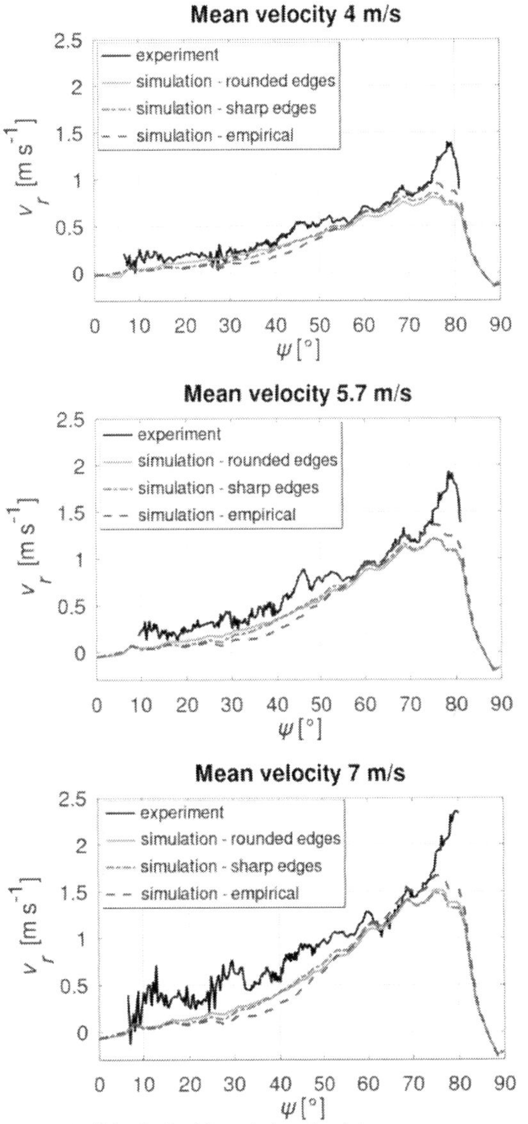

Fig. 6. Radial velocity 15 mm below the plate.

Table 1. Evaluated inertial and viscous loss coefficients in the direction normal to the plate from different methods.

	C_n [m^{-1}]	D_n [m^{-2}]
Empirical expression method	3478.80647	0
Regression method – sharp edges	4836.61808	3500394.63
Regression method – rounded edges	3059.70845	96811365.4

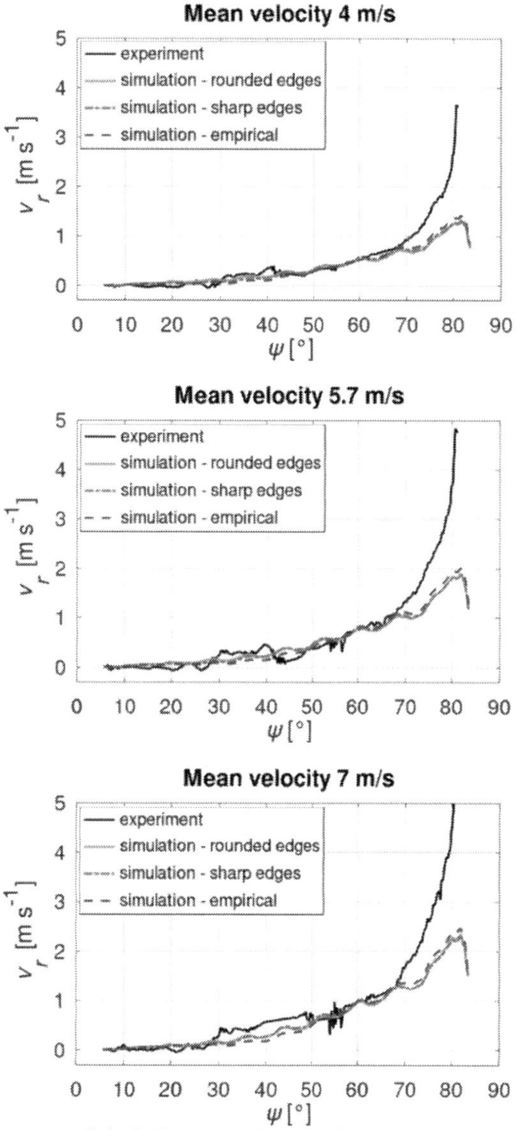

Fig. 7. Radial velocity 20 mm above the plate.

5 Numerical results

Simulations of air flow through the curved perforated plate were performed with the geometry and boundary conditions from the PIV measurements. Overall, three different mean (inlet) velocities 4 m/s, 5.7 m/s and 7 m/s and three different sets of porous zone coefficients (see Table 1) were used. Radial (normal to the perforated plate) and tangential velocity components to the channel curve were monitored 20 mm above and 15 mm below the plate.

Comparison of numerical and experimental results, for radial and tangential velocity components below and above the perforated plate, is shown for the mean velocity of 4 m/s, 5.7 m/s and 7 m/s in the Fig. 6 - Fig 9. The velocity is displayed depending on the angular coordinate ψ withthe origin at the beginning of the channel curve.

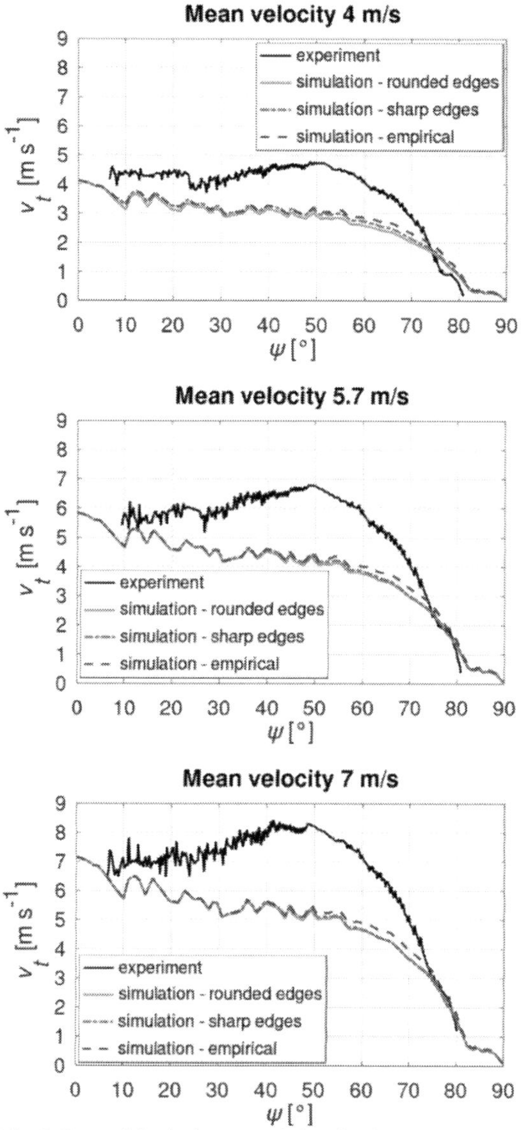

Fig. 8. Tangential velocity 15 mm below the plate.

The results show consistency for all used porous zone coefficient evaluation methods. The tangential velocity above the perforated plate, Fig. 9, shows a good agreement with the PIV experiment. The radial velocity below the plate in the Fig. 6 shows also a good agreement for the radial velocity component.

The experimental results for the radial velocity component above the perforated plate in Fig. 7, shows a peak at the end of the plate, which is significantly higher in comparison with the numerical results. There is a high chance that the difference is caused by measurement inaccuracy. In Fig. 8, the tangential velocity component below the perforated plate differs distinctly between simulation and experiment. The simplification of the porous model approach has probably the most influence here alongside the measurement inaccuracy.

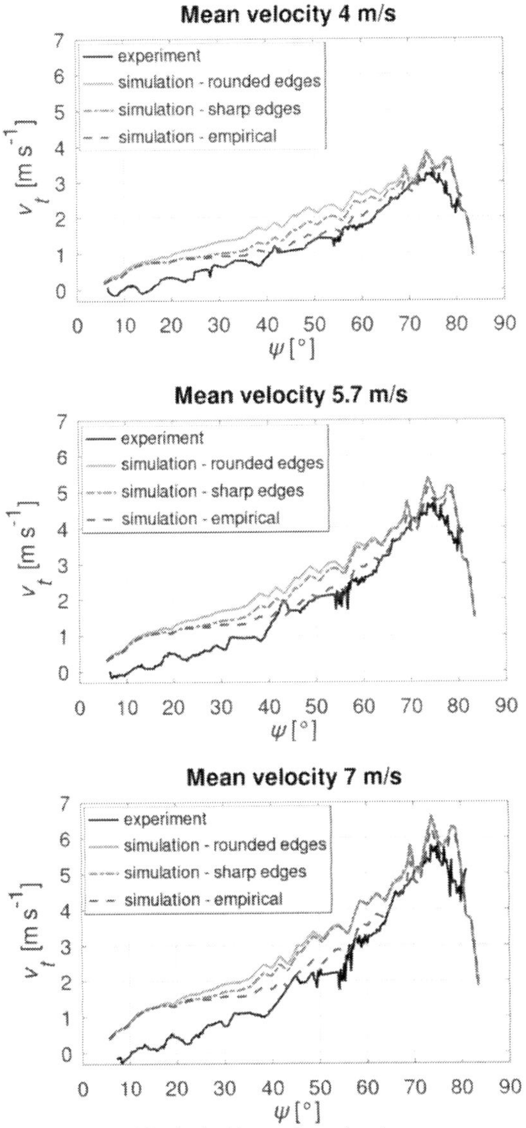

Fig. 9. Tangential velocity 20 mm above the plate.

6 Selected methods limitation

The definition of inertial and viscous loss coefficients by Oezcan [7] implies that the coefficients in the tangential direction should be higher, than in the normal direction. But there is no exact specification of how much higher they should be. It can differ based on $\frac{L}{d_h}$ ratio of the perforated plate. For the first simulations, the tangential coefficients were set 10 times higher than normal related by the example presented by Oezcan [7]. The resulting normal and tangential velocities below and above the plate are shown in Fig. 11. As a reference case, the mean velocity of 7 m/s and empirical porous zone coefficients were tested.

The numerical results in Fig. 10 are strongly dependent on porous zone tangential coefficients value and they are most consistent with the experimental results in the case of tangential coefficients 10 times higher than ones in the normal direction. It means that with the used perforated plate geometry, 10 times higher porous zone coefficients in the tangential direction (than in the normal direction) are the most sufficient, but with some other perforated plate geometries, it presumably would not be so.

To determine if there is a more appropriate way to define tangential porous zone coefficients a short study of pressure drop dependency on an air flow impact angle was carried out. A flat perforated plate geometry was used in an air flow with the mean flow velocity of 7 m/s and compared with the porous zone model with tangential coefficients 10 times higher than normal. The impact angle varied from 0° to 80° from normal with an increment of 10°.

The resulting pressure drop was consistent for the perforated plate and porous zone. But it was observed, that the porous zone redirects the air flow significantly more to the zone (or the plate) normal direction than the perforated plate. The flow redirection comparison of the perforated plate with rounded holes edges and the porous zone (coefficients from round edges regression) is shown in Fig. 10 for impact angle 40° from normal. It reveals that concerning the air flow velocity direction, the porous zone coefficients in the tangential direction should be lower.

The results of the study disagree with the original definition of the porous zone tangential coefficients. It suggests that tangential coefficients should be not only lower than normal coefficients but even that they should

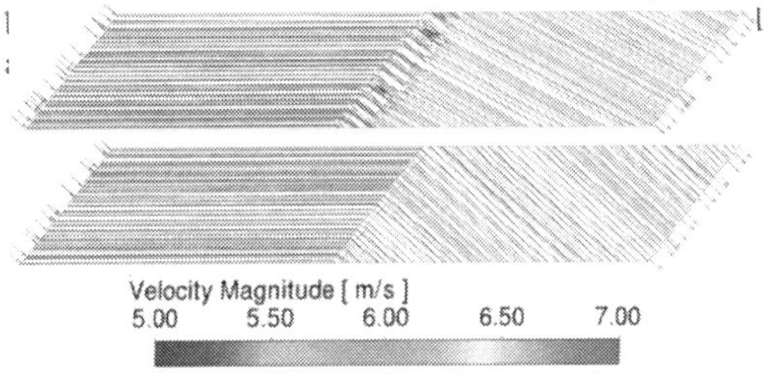

Fig. 10. Streamlines of air flow through perforated plate (top) and porous zone (bottom) coloured by pressure - impact angle 40°.

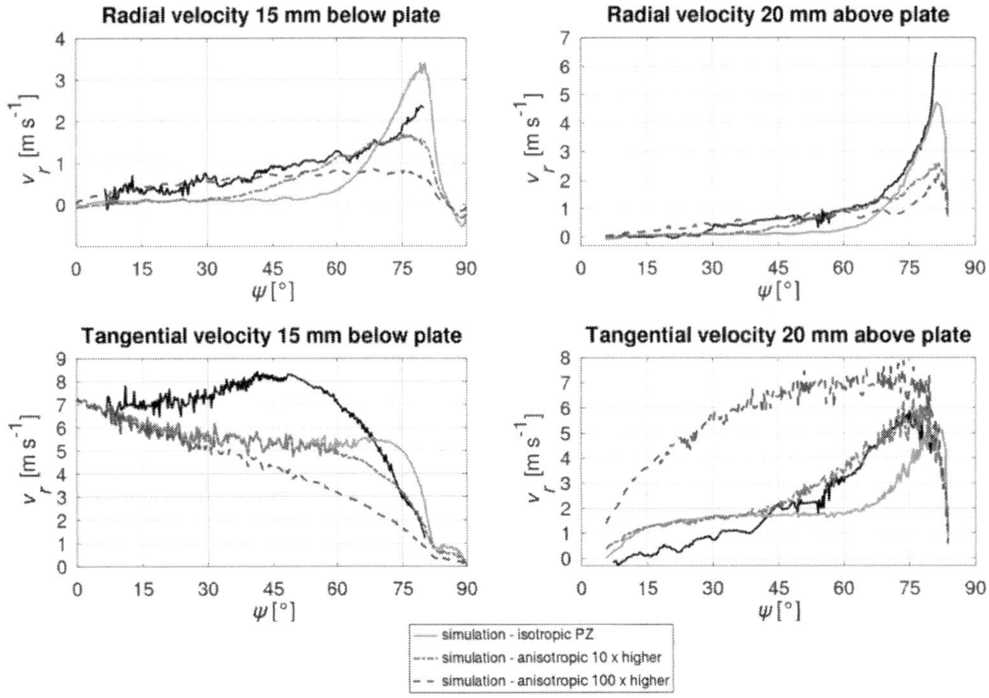

Fig. 11. Radial and tangential velocity for mean velocity 7m/s and empirical method porous zone coefficients.

7 Conclusion

Air flow through the curved perforated plate was investigated in this study. The first part was devoted to the experimental measurements of the flow velocity using the PIV method. Vector velocity fields were shown for three different air flow mean velocities.

Further porous zone model was employed to represent perforated plate in simulations. Three different sets of porous zone inertial and viscous loss coefficients were evaluated using two different methods. Normal and tangential velocities 15 mm below and 20 mm above the plate were compared for experimental results and numerical results with various porous zone coefficients.

The comparison showed satisfactory agreement between the experiment and simulation. But there are several deficiencies of the porous zone, especially in the used resistance coefficients evaluation methods. All the used methods are based on the porous zone coefficients definition in the direction normal to the perforated plate but do not explicitly define the coefficients in the tangential direction. For the investigated case of curved perforated plate, the most suitable porous zone coefficients in the plate tangential direction are 10 times higher than in the normal direction. So defined model performs well with the perforated plate normal to the flow, but not so well with an acute angle between the plate and the flow direction. It was shown, comparing perforated plate simulation with porous zone simulation for acute impact angle, that the flow is more redirected by the porous zone than by the perforated plate. For more corresponding results the porous zone coefficients in tangential direction should be set lower than they are. In a most precise definition, the viscous and inertial loss coefficients would presumably be functions of air flow impact angle on the plate and the mean flow velocity. This should be considered when using the porous zone substitution for perforated plate and might be a subject for further studies.

Acknowledgement

This publication was written at VÚTS, a.s. with the support of the project FV40025. Our acknowledgement belongs to all donors that helped us to reach the results of the work.

References

1. D. Allori, G. Bartoli, A. F. Miguel, Int. J. Fluid Mech. Res., 39, 136-148 (2012)
2. P. Tanner, J. Gorman, E. Sparrow, Int. J. Numer Method H., 29, 4310-4333 (2019)
3. A. F. Miguel, J. Porous Media, 22, 1439-1448 (2019)
4. I. E Idelchik, Handbook of hydraulic resistance, third edition, CRC Press Inc., Boca Raton, US (1960)
5. W. Kast, E.S. Gaddis, K. E.Wirth, J. Stichlmair, L1 Pressure Drop in Single-Phase Flow. VDI-Buch. Springer, Berlin, Heidelberg (2010)
6. Ansys® Fluent, Ansys Fluent User's Guide, 2021 R2, 1206-1242 (2021)
7. M. Oezcan, How to Use the Porous Media Feature to Define a Perforated Plate, SimScale (2020) https://www.simscale.com/

A dispersion of a droplet flow on crossing wires in an air counterflow

Ondřej Hájek[1*], *Ondřej* Cejpek[1], *Milan* Malý[1], *František* Prinz[1], and *Miroslav* Jícha[1]

[1]Energy Institute, Faculty of Mechanical Engineering, Brno University of Technology, Technická 2896/2, 616 69 Brno, Czech Republic

Abstract. Liquid dispersion on a wire mesh is a phenomenon that is utilized in many industrial applications, such as rotating packed beds. It is a very simple method of liquid atomization without a need for complex nozzles. This research focuses on an elementary case of a liquid dispersion on a crossing of two wires. Experiments were carried out in a wind tunnel to elucidate the influence of counterflow air velocity on a liquid sheet and droplets. High-speed camera was used to capture the impact of droplets on the crossing. Images were then processed using MATLAB® addon PIVlab. The effect of the input parameters, including a liquid flow rate in the range of 3.8 to 12 kg/h and air flow velocity varying from 0 to 9 m/s on the angle and velocity of dispersed droplets downstream of the crossing, was investigated. Finally, a qualitative description of the dispersion was evaluated. Results show that with an increasing liquid flow rate, the droplets dispersed in a wider angle. On the other hand, the influence of the air counterflow is significant only for low liquid flow rates. The atomization rate, determined by the number of small droplets, was better for higher liquid flow rates.

1 INTRODUCTION

Liquid atomization is common in all branches of industry. For example, paint is atomized into small droplets that impact the surface and create a thin layer to create coatings, or fuel is dispersed into tiny droplets that then evaporate and burn in engines. Dispersed droplets can be used for spray cooling. Finally, there are applications where liquid is used as an absorbent, and therefore the goal is to create a large interface between gas and liquid to enhance the mass transfer rate. Sprays and droplets can be created by several approaches, such as specially designed pressure swirl atomizers [1] which create spray due to interior design, two-fluid effervescent atomizers [2] or by impinging jet atomization where atomization is achieved by the impact of two liquid jets [3].

A liquid volume can also be dispersed by an impact on the solid surface, e.g., a wire or a wire mesh. This method presents a simple way to create a droplet cone or droplets without the need for complex geometries or pure liquids because it can disperse flows with particles. Possible applications of this atomization are oil-water separation [4], the delivery of pesticides and herbicides precisely above plants [5], or rotating packed beds (RPB). An RPB is a patented device [6] to increase the rate of mass transfer by adding centrifugal acceleration. A liquid jet is introduced on the inner surface of the rotating ring. The surface is formed by wire mesh [7], metal foam [8], or other shapes [9]. The jet is dispersed after an impact on the mesh. This creates liquid ligaments and droplets that further interact with the inner layers of the packing and create an interface. To improve the mass transfer rate, it is essential to understand how the first layer affects the dispersion with a consideration of the influence of the air counterflow.

The current state of the art concerning this atomization can be divided into two directions, both of which focus primarily on the impact of a single droplet on a wire or a wire mesh. This division is based on the ratio between the diameter of the mother drop before the impact and the dimensions of the wire or the mesh eye. For example, Soto et al. [5] studied millimetre droplets that impact on a fine mesh with a distance between the fibers in the order of hundreds of micrometres. This study concludes that this type of atomization seems useful for agricultural processes because it prevents the drift of pesticides to the environment. One conclusion is also that the liquid dispersion on the wire mesh increases the surface-to-volume ratio with low-energy loads. Sun et al. [10] studied different velocities of mother droplets that impact a fine superhydrophobic wire mesh. Such a liquid dispersion depends on the Weber number (We), where two types of sprays are described under the mesh for low and high We. Low We causes a droplet to bounce off the mesh, whereas droplets with high We pierce right through the mesh. The Weber number is a dimensionless number which describes the ratio between a kinetic energy and a surface energy of the drop. The impact of relatively large droplets on a fine mesh was described for other differing input parameters, such as several layers of wire meshes in Su et al. [11] or hydrophobic mesh in Zhang et al. [12]. A hydrophobic mesh shows a dispersion angle under the

* Corresponding author: Ondrej.Hajek3@vutbr.cz

mesh that is significantly greater than that of a hydrophilic mesh. Surface wettability was also studied in [13].

The other branch of these studies deals with a droplet that falls on a single wire or on the crossing of two wires. This phenomenon was studied by Tang et al. [14]. Their study focused on different velocities of mother droplets, droplet sizes, and shapes of the impacted objects. The authors found three typical dispersion modes of the liquid: one-drop mode, two-drop mode, and for some cases they described liquid ligaments. Haim et al. [15] published similar study where the authors study the amount of liquid captured on the wire, as well as the influence of the distance between the centre of gravity of the impacting droplets and the axis of the wire.

There are also several studies that describe liquid dispersion in RPBs, e.g., Lu et al. [16] or Wang et al. [17]. Both studies are affected by another input parameter, specifically rotations of the packing.

In real applications, such as RPBs, there is one parameter which seems to influence the angle of the dispersed droplets. This factor is the air counterflow. Thus, the goal of this study is to add the missing information about the influence of such a counterflow as well as to describe behaviour of the liquid jet dispersion on the wire.

2 MATERIALS AND METHODS

2.1 Wind tunnel

The whole experiment was carried out in an air tunnel described in Cejpek [18]. The test section of the tunnel was placed in the vertical position, so the gravitational acceleration is in the opposite direction to the air flow (Figure 1).

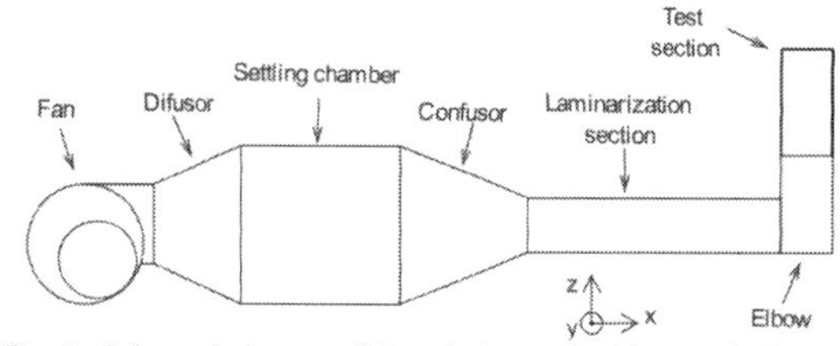

Fig. 1. Schematic layout of the wind tunnel with a vertical test section.

Phase Doppler Anemometry (PDA) was used to determine velocity profiles and turbulence intensity in the middle of the test section (Figure 2). Seeding particles with a diameter of approx. 1.5 μm were introduced in the suction of the fan and they followed the flow well due to their low Stokes number, thus providing results with a reasonable accuracy. The velocity profile and the intensity of the turbulence are shown in Figures 3 and 4. Counterflow velocities and turbulence intensities were measured in highlighted points in Figure 2 for three different fan velocities. The intensity of the turbulence was mostly around 2%, which can be considered a valid result.

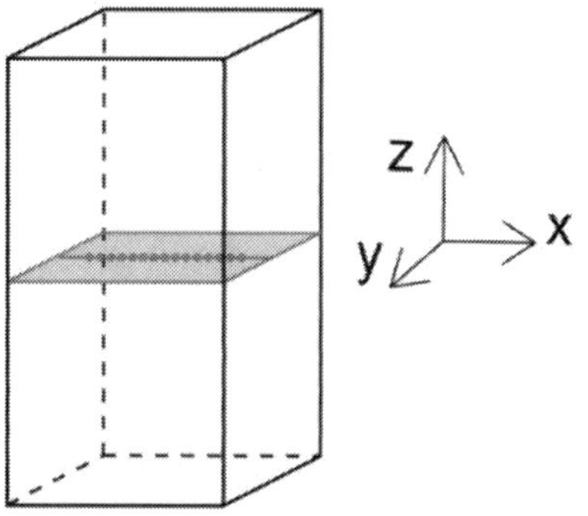

Fig. 2. Test section with with a highlighted cross-section.

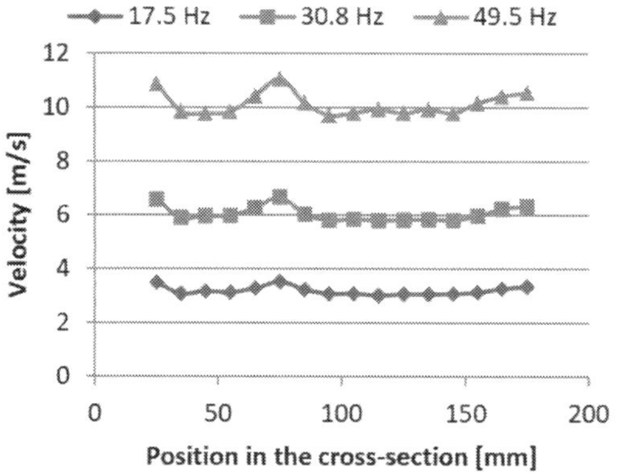

Fig. 3. Velocity profile of the test section measured in the middle of the cross-section.

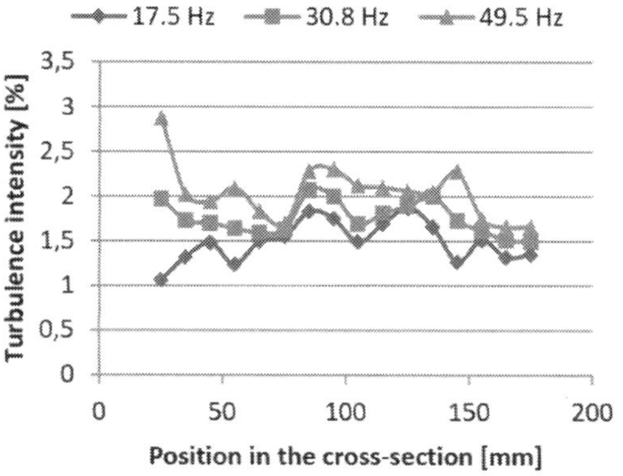

Fig. 4. Turbulence intensity.

2.2 Steel wire

The wires used in this study were taken from a knitted steel mesh, and therefore these wires were not straight. Both wires were attached in the corners of the test section (Figure 5). Due to the concave and convex shapes of the wires, there was a natural crossing on the axis of the test section. It was the same crossing as it is in the standard knitted wire mesh. The diameter of the wire was 0.8 mm and was made of AISI 304 stainless steel.

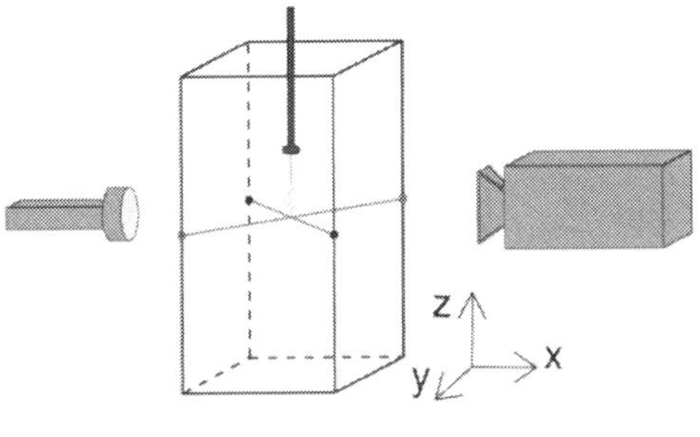

Fig. 5. Experimental setup.

2.3 Liquid

The working liquid was water at room temperature 20 °C. The liquid was pumped with a peripheral pump through a single orifice nozzle with a diameter of approximately 0.6 mm. This nozzle created a liquid jet. The orifice was placed 200 mm above the crossing. The liquid jet collapsed due to the Plateau-Rayleigh instability to droplets with diameters ranging from 0.8 to 1.2 mm. The following Table 1 shows the controlled input parameters i.e., the liquid flow rate and air flow velocity.

Table 1 Input parameters

Liquid flow rate [kg/h]			
3.8	8	12	
Air flow velocity [m/s]			
0	3	6	9

2.4 Camera and software

A high-speed camera Photron FASTCAM SA-Z type 2100K-M-16GB was used to capture images with a record rate of 20 000 fps. The shutter speed was 1.25 µs. The images were then processed with a MATLAB® code and the PIVlab add-on [19] to determine the velocity of the dispersed liquid. The angle of dispersed liquid and the atomization efficiency was analysed manually.

3 RESULTS AND DISCUSION

In the current study, three output parameters were monitored. The first one is the velocity of the dispersed liquid 12 mm downstream of the crossing. The second parameter is the angle at which the liquid is dispersed downstream of the crossing. The last one is the atomization rate determined by a qualitative description of the liquid dispersion.

The liquid flow rate was generalized by the Weber number We_d of the mother droplet before impact. This dimensionless number is defined in Equation 1:

$$We\_d = \frac{\rho \cdot u_d^2 \cdot D_o}{\sigma} \quad (1)$$

where ρ [kg/m³] is the density of water, u_d [m/s] corresponds to the mean velocity of mother droplets, D_o [m] is the mean diameter of mother droplets and σ [mN/m] is the liquid surface tension of water.

The second dimensionless number is the relative velocity re_v defined in Equation 2 as a velocity of the dispersed liquid u_{liq} divided by the mean velocity of mother droplets u_d:

$$re\_v = \frac{u_{liq}}{u_d} \quad (2)$$

3.1 Velocity

The velocity component of the dispersed liquid in the direction of the axis -z was derived from a rectangle at the bottom of each image (Figure 6). Figure 6 displays the impact of the droplets with We_d = 2217 on the crossing of two wires with the zero velocity of the air counterflow.

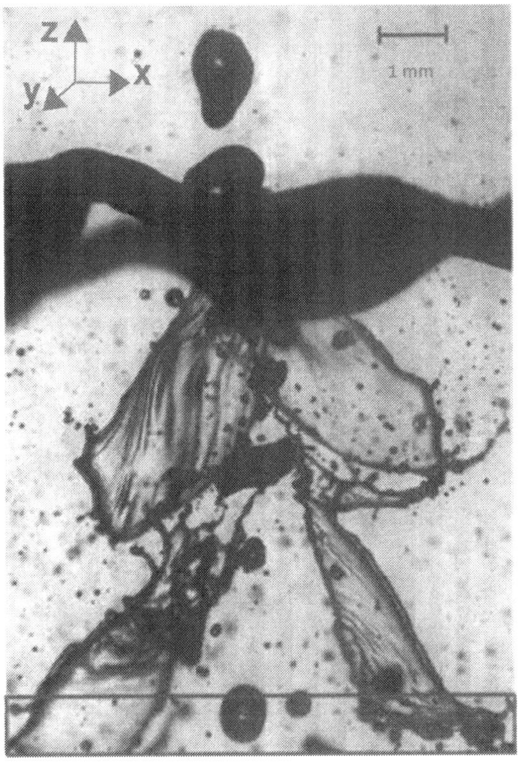

Fig. 6. The liquid dispersion on the crossing of two wires, We_d = 2217 and zero velocity of the air counterflow.

The velocity of mother droplets created from the jet before the impact was found to be approx. 3.5 m/s for the liquid flow rate of 3.8 kg/h. The velocity of droplets was 7.6 m/s for 8 kg/h, and the highest flow rate created droplets with the velocity of 12 m/s. Results can be seen in the Figure 7. The relative velocity of the dispersed liquid slows down with increasing velocity of the air counterflow but only for cases with We_d equal to 205 and 1073. For We_d = 2217, the relative velocity of the dispersed liquid increases. This phenomenon could be related to a feature of liquid jets. A liquid jet tends to drift with higher liquid flow rates and thus the stream of mother droplets created by the jet disintegration impacts on the crossing in various positions. Hence some droplets lost more kinetic energy than others which leads to differences in the velocity of dispersed liquid for We_d = 2217.

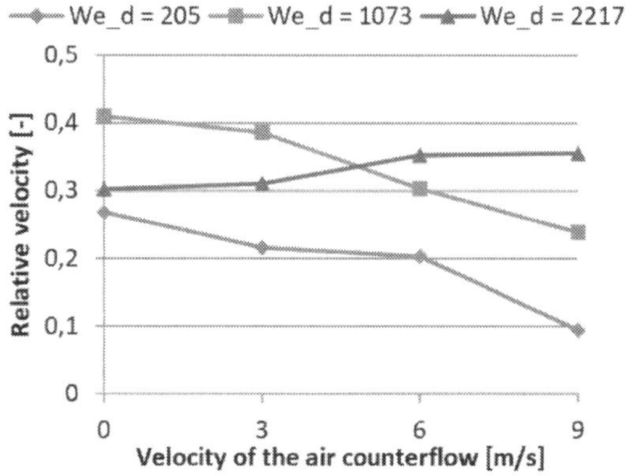

Fig. 7. Influence of the air velocity on the relative velocity of the dispersed liquid for different mother droplets' Weber numbers.

3.2 Angle

Figure 8 shows a method to determine the angle downstream the crossing. Overall results for the cone angle of the dispersed liquid can be seen in Figure 8. The angles were determined only for We_d equal to 1073 and 2217. For We_d = 205, the liquid generated only jets of

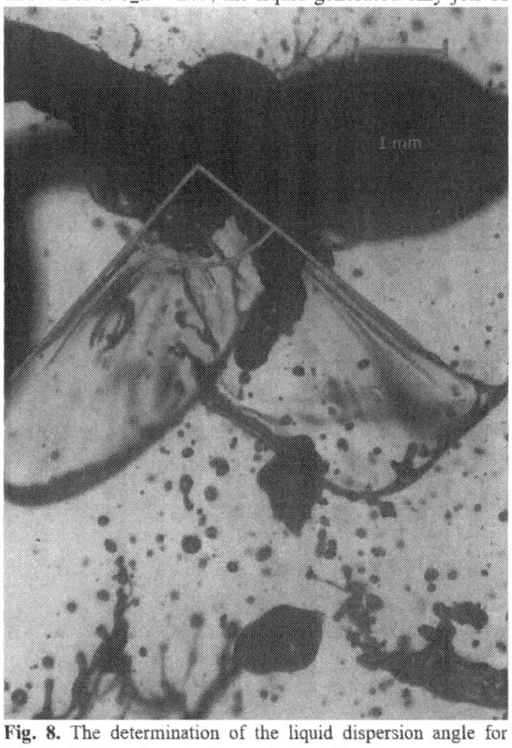

Fig. 8. The determination of the liquid dispersion angle for We_d = 2217 and counterflow velocity of 3 m/s.

Results in Figure 9 can be interpreted to mean that the air counterflow velocity has no influence on the cone angle. However, the liquid flow rate shows a significant impact on the cone angle. Therefore, the higher the liquid flow rate, the broader the cone angle. These results correspond with the research in [13].

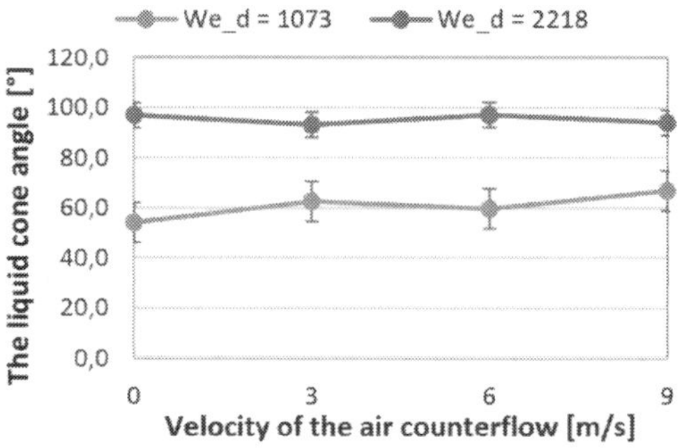

Fig. 9. Influence of the airflow velocity and mother droplets We_d on the cone angle of the liquid dispersion.

3.3 Dispersion

Atomization rate determined by a qualitative description of the liquid dispersion can be seen in Figures 10 (a), (b) and (c). Figure 10 (a) is for the lowest liquid flow rate defined by We_d = 205. This figure shows no liquid dispersion downstream the crossing. The liquid after the impact continues as a slow jet that will presumably disintegrate due to the Plateau-Rayleigh instability further downstream.

Figure 10 (b) shows mother droplets with We_d = 1073. Droplets in this mode disperse into liquid sheets with thick walls. These sheets quickly form jets.

The highest liquid flow rate with the Weber number of mother droplets We_d = 2218 forms more stable liquid sheets which disintegrate due to the Kelvin-Helmholtz instability (Figure 10 (c)). Dispersion for this We_d also presents the finest daughter droplets which arise from the dispersion of all observed input parameters. Similar conclusions were found in [14].

4 CONCLUSIONS

Research on the liquid dispersion on the crossing of two wires took place in the wind tunnel to determine the influence of the air counterflow rate on the liquid disintegration after the impact.

Velocity profiles and turbulence intensity in the wind tunnel were found to be reasonable for this study. Velocity component of the dispersed liquid in the direction of the axis -z, downstream the crossing, was evaluated with dimensionless numbers and important results are that the air counterflow velocity has a great impact on the liquid flow rates of 3.8 and 8 kg/h. Liquid flow rate of 12 kg/h shows some contrast to lower flow rates. This could be caused by the natural jet drifting. The next study will purposefully describe jet placed right above the crossing. Hence the influence of the drifting will be suppressed. The cone angle was described for liquid flow rates 8 and 12 kg/h and the conclusion is that the angle is only dependent on the liquid flow rate, not on the airflow velocity. The higher liquid flow rate has also an influence on the atomization rate since with a higher We_d dispersed mother droplets created a thinner wall of the liquid sheet.

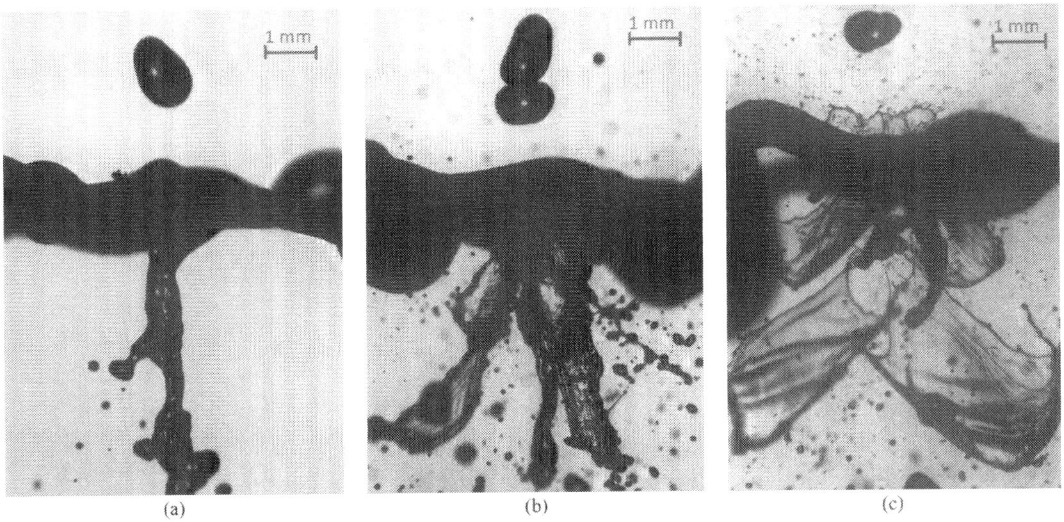

(a) (b) (c)

Fig. 10. Images of the liquid dispersion for different input parameters : (a) $We\_d = 205$, the air flow velocity 0 m/s, (b)$We\_d = 1073$, the air flow velocity 3 m/s, (c) $We\_d = 2218$, the air flow velocity 6 m/s.

Thus, daughter droplets from the disintegration of the liquid sheet were smaller, and smaller droplets would create greater interface for the mass transfer. A problem that arises with small droplets is their predisposition to be blown off with the air counterflow. In industry, this means that potentially dangerous chemicals could be released to nature. Hence, it is a direction for the next research.

ACKNOWLEDGEMENTS

The authors acknowledge the financial support from the grant The effect of ambient airflow on the quality of liquid dispersion on the wire mesh, which is realised within the project Quality Internal Grants of BUT (KInG BUT), Reg. No. CZ.02.2.69/0.0/0.0/19_073/0016948, which is financed from the OP RDE. This work was supported by the project no. GA 21-45227L funded by the Czech Science Foundation and the project FSI-S-20-6295 funded by the Brno University of Technology.

LIST OF SYMBOLS

u_d	Mean velocity of droplets
ρ	Density
D_o	Mean diameter of droplets
σ	Liquid surface tension
We	Weber number
re_v	Relative velocity
u_{liq}	Velocity of dispersed liquid

REFERENCES

1. M. Maly, J. Jedelsky, J. Slama, L. Janackova, M. Sapik, G. Wigley, M. Jicha, Int. J. Heat Mass Transf. **123** 805 (2018)
2. M. Zaremba, J. Kozák, M. Malý, L. Weiß, Rudolf, J. Jedelský, M. Jícha, Int. J. Multiph. Flow **103** 1 (2018)
3. M. Hafezi, M. Mozaffarian, M. Jafarikojour, M. Mohseni, B. Dabir, Photochem. Photobiol. A 389 112198 (2020)
4. D. Tian, X. Zhang, X. Wang, J. Zhai, L. Jiang, Phys. Chem **13** 14606 (2011)
5. D. Soto, H.L. Girard, A. le Helloco, T. Binder, D. Quéré, K.K. Varanasi, Phys. Rev. Fluids **3** 1 (2018)
6. C. Ramshaw, R.H. Mallinson, US4283255A, 1981
7. Y. Luo, J.Z. Luo, G.W. Chu, Z.Q. Zhao, M. Arowo, J.F. Chen, Chem. Eng. Sci. **170** 347 (2017)
8. M.-J. Su, Y. Luo, G.-W. Chu, W. Liu, X.-H. Zheng, J.-F. Chen, Ind. Eng. Chem. Res **57** 4743 (2018)
9. G.E. Cortes Garcia, J. van der Schaaf, A.A. Kiss, J. Chem. Technol **92** 1136 (2017)
10. L. Sun, S. Lin, B. Pang, Y. Wang, E. Li, X. Zu, K. Zhang, X. Xiang, L. Chen, Phys. Fluids **33** (2021)
11. M.-J. Su, Y. Le, G.-W. Chu, Y.-B. Li, L.-L. Zhang, Y. Luo, Ind. Eng. Chem. Res **59** 3584 (2020)
12. J.P. Zhang, Y. Luo, G.W. Chu, L. Sang, Y. Liu, L.L. Zhang, J.F. Chen, Chem. Eng. Sci. **170** 204 (2017)
13. M.J. Su, Y. Luo, G.W. Chu, Y. Cai, Y. Le, L.L. Zhang, J.F. Chen, Chem. Eng. Sci **219** 115593 (2020)
14. Y.Y. Tang, M.J. Su, G.W. Chu, Y. Luo, Y.Y. Wang, L.L. Zhang, J.F. Chen, Chem. Eng. Sci. **198** 144 (2019)
15. L.F. Haim, I. Sher, E. Sher, ICLASS 2012 - 12th ICLASS (2012)
16. Y.Z. Lu, W. Liu, Y.C. Xu, Y. Luo, G.W. Chu, J.F. Chen, Chem. Eng. Process **140** 136 (2019)
17. Y. Wang, Y. bin Li, M.J. Su, G.W. Chu, B.C. Sun, Y. Luo, Chem. Eng. Sci. **240** 116675 (2021)
18. O. Cejpek, Design and realization of an aerodynamic tunnel for spraying nozzles, Diploma thesis (2020)
19. W. Thielicke, R. Sonntag, J. Open Res. Softw. **9** (2021)

CFD Investigation of the test facility for forced blade flutter research

Jindřich Hála[1*], *Jan* Lepičovský[1], *Petr* Šidlof[2], and *David* Šimurda[1]

[1]Institute of Thermomechanics of the Czech Academy of Sciences, Dolejškova 1402/5, 182 00, Prague 8, Czech Republic
[2]Technical University of Liberec, Studentská 1402/2, 461 17, Liberec, Czech Republic

Abstract. With the increasing share of renewable power resources turbomachines need to be operated under a wider range of operating conditions including highly off-design regimes. Under such regimes an undesirable phenomenon of blade flutter might occur and possibly destroy the machine. To prevent this, intensive research is conducted by research teams worldwide. Blade flutter research program at the Institute of Thermomechanics of the Czech academy of sciences (IT CAS) mainly aims to advance experimental techniques for investigation of sonic and transonic blade flutter. For this purpose, the new sophisticated test facility was designed and manufactured. As part of the design process, the CFD computations were conducted in order to investigate the flow field in the test facility. This paper presents results of these computations with detailed analysis of flow structures occurring during the air flow through the stationary blade cascade.

1 Introduction

Recently a test section for investigation of forced oscillation of blades in a compressor blade cascade operating at transonic regimes has been designed in frame of a cooperation between the Institute of Thermomechanics of the Czech Academy of Sciences (IT CAS) and Technical University of Liberec (TUL) [1]. The test section is mounted on the modular in-draft wind tunnel (Fig. 1) stationed at Aerodynamic Laboratory in Nový Knín. The tunnel facility (Fig. 2) consists of a silica gel dryer (1), bell mouth inlet (2) with honeycomb flow straighteners (3), round-rectangular transition (4), flutter test facility with tested blade cascade (5), outlet part equipped with control nozzle (6) and fast acting gate valve (7) connected to a vacuum tank. The flow in the cascade is controlled using adjustable tailbords, throttling tailboard and regulating nozzle. The tested blades can be independently oscillated and a resulting flow field and mutual interaction between blades can be measured using numerous techniques including fast-response miniature pressure transducers, pressure sensitive paints, fibre optics interferometry, thermo-anemometer probes and high-speed torquemeter.Due to only slightly supersonic inlet velocity and limited space at the test section upper wall, it was decided that the inlet channel would be straight with no divergence. Preliminary tests would be run and according to the results of the first tests, a suction through the upper wall will be added to suppress reflection of shock waves back to the

Fig. 1. Test section with uncovered side wall and visible blade cascade (flow direction from right to left).

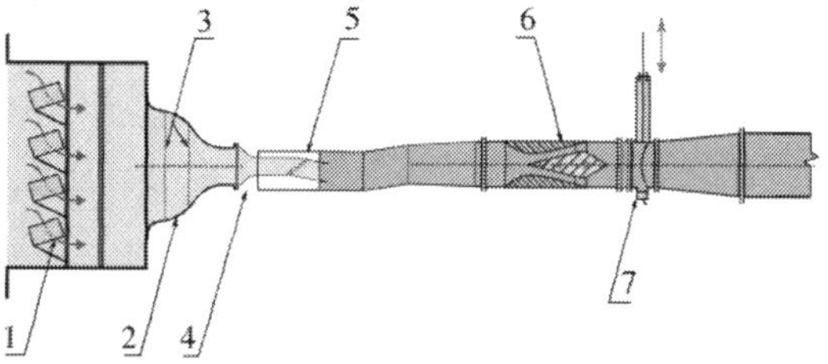

Fig. 2. Schematic depiction of the modular wind tunnel with the mounted flutter test facility.

*Corresponding author: hala@it.cas.cz

flow and also to allow for acceleration to higher inlet Mach numbers if necessary.

The aim of this work was to investigate the flow field in the test section using readily available numerical techniques represented by Ansys Fluent commercial code and provide numerical data for the initial stage of measurements. This work did not aim to investigate interaction between oscillating blades, thus, only stationary cascade was considered. Further sections briefly introduce numerical techniques used throughout this investigation including numerical grid and provide thorough analysis of the flow in the test section based on the obtained results.

2 Investigated blade cascade

Blade cascade represents a section of a fan rotor at 45% span. It has design inlet Mach number $M_1 = 1.09$, design inlet flow angle 148.5° and design flow turning 12.5°. Further details can be found in [2]. This cascade has been chosen since blades are reasonably thick to allow unchallenging instrumentation with Kulite fast response pressure transducers and also because data on its performance are available in the open literature.

3 Numerical simulation

The problem was numerically modelled in 3D using the system of time-averaged Navier-Stokes equations for compressible flows:

$$\frac{\partial \rho}{\partial t} + \frac{\partial (\rho u_j)}{\partial x_j} = 0, \tag{1}$$

$$\frac{\partial (\rho u_i)}{\partial t} + \frac{\partial (\rho u_i u_j)}{\partial x_j} = -\frac{\partial p}{\partial x_i} + \frac{\partial \tau_{ij}^{\text{eff.}}}{\partial x_j}, \tag{2}$$

$$\frac{\partial (\rho e_0)}{\partial t} + \frac{\partial (\rho h_0 u_j)}{\partial x_j} = \frac{\partial}{\partial x_j}\left(\tau_{ij}^{\text{eff.}} u_i + \rho\alpha^{\text{eff.}} \frac{\partial h}{\partial x_j}\right), \tag{3}$$

Since the flow medium is dry air, the equation of state of an ideal gas was considered. The Reynolds stress tensor and the turbulent thermal diffusivity were computed using $k-\omega$ SST turbulence model.

The inviscid fluxes were approximated using the Advection Upstream Splitting Method (AUSM) upwind scheme with linear reconstruction. The steady state solution was obtained using the density-based implicit solver in Ansys Fluent R2021 commercial code employing second order accuracy upwind schemes in space. The computations were conducted using the high performance computing cluster of the Institute of Thermomechanics using 16 cores of the *Intel Xeon E5-2683 v4* processor running at 2.10 GHz.

3.1 Computational domain and grid

The investigated cascade consists of five transonic profiles placed in the cascade with a chord length of $c = 120$ mm pitch $t = 74.52$ mm and stagger angle $\gamma = -41.49°$. To capture the effect of boundary layer growth inside the bell-mouth contraction and of the possible corner vortex formation in the round-rectangular transition upstream the cascade, the entire inlet part was included in the simulation. The computational domain can be seen in Fig. 3.

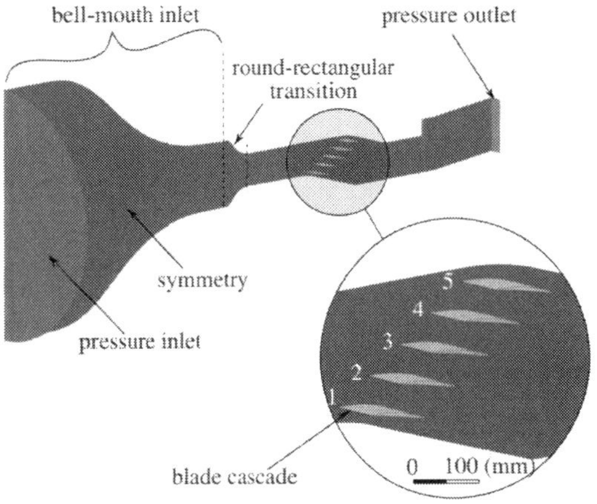

Fig. 3. Computational domain with the detail of the cas-cade and marked blade numbers.

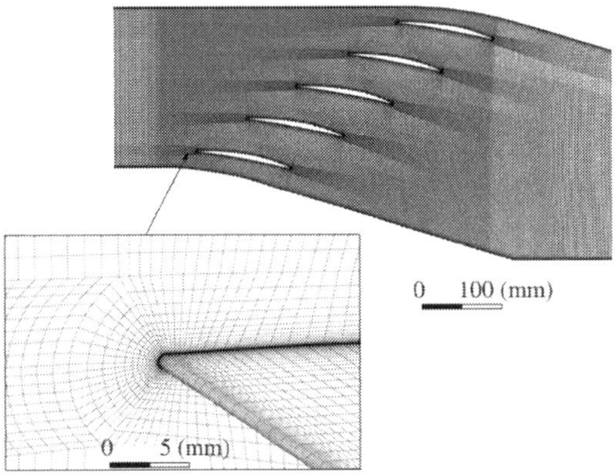

Fig. 4. Computational grid around the cascade and the detail of the blade – end wall transition close to a blade leading edge.

The computational grid consists of approximately $15 \cdot 10^6$ hexahedral cells. The dimensionless wall distance y^+ of the first cell at the blade surfaces and test section walls was less then 1 with the cell growth ratio 1.2. Fig. 4 shows the grid around the cascade and the detail of the blade – end wall transition close to a blade leading edge.

3.2 Boundary conditions

The domain is symmetric in the span-wise direction, therefore a symmetry boundary condition was used in the mid plane. At the outlet, the average static pressure of 20 kPa

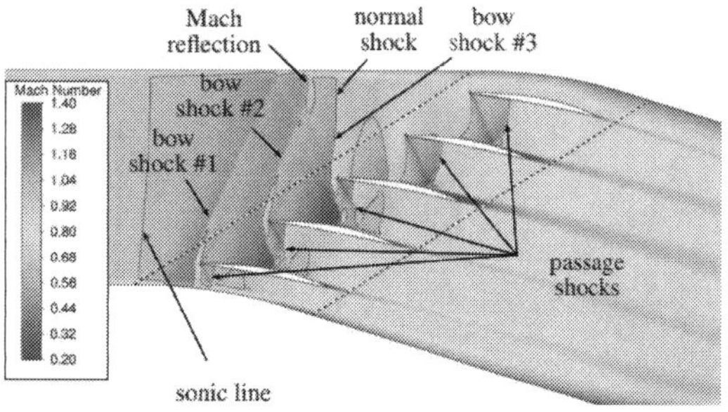

Fig. 5. Mach number contours of the region in the channel mid-plane around the cascade with the marked main flow features.

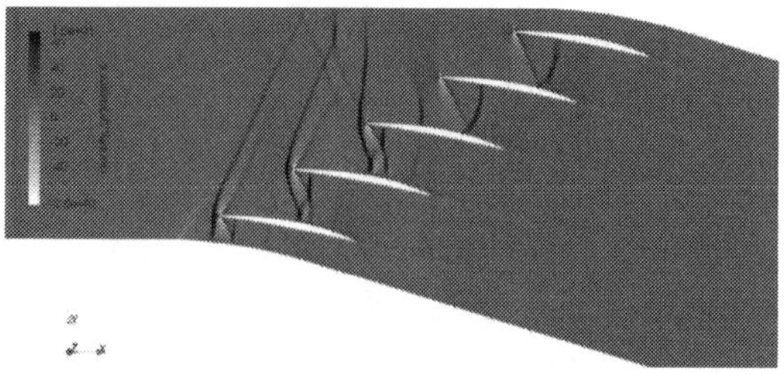

Fig. 6. Contours of the density gradient in x direction $\partial\rho/\partial x$ depicting shock waves in the flow field around the cascade.

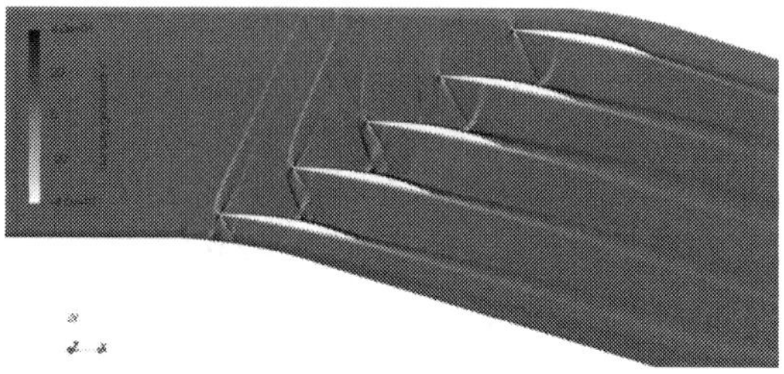

Fig. 7. Contours of the density gradient in y direction $\partial\rho/\partial y$ depicting shock waves and wakes behind blades.

was set. Following boundary conditions were set at the inlet:

1. constant total temperature $T_0 = 300\,\mathrm{K}$
2. uniform total pressure distribution $p_0 = 101.325\,\mathrm{kPa}$
3. constant turbulence intensity $Tu = 2\%$
4. constant turbulent viscosity ratio $\frac{\mu_t}{\mu} = 10$

4 Results and discussion

In reality, blading is annular and thus periodic. Therefore, a turbomachinery cascade modelling aims to model the flow as periodic as possible. This, however, appears to be one of the main difficulties with experimental modelling of transonic flow past finite linear compressor blade cascades. In the current investigation, owing to the setup, the flow is not periodic. Therefore one of the aims of the current investigation is to understand in detail sources of this aperiodicity in order to deal with it as best as possible. Figure 5 shows the Mach number contours around the cascade in the mid-plane with the sonic line in black and dashed lines upstream and downstream the cascade representing the locations in which the static pressure tapping will be located during the experiments.

As the contours of Mach number in Fig. 5 suggest, the flow is accelerated to supersonic velocity just before the cascade at the point where the lower wall becomes divergent (approximately 0.5 chord length before the leading edge of blade No. 1). The Sonic line stretches from this point across the whole inlet channel. This is an aerodynamic throat that determines mass flow through the blade cascade. Inlet flow is supersonic up to the blade No.3. Here, a normal shock knocks the velocity down to subsonic. This is due to interaction of detached bow shock waves #1 and #2 with the upper wall boundary layer. Upon the interaction with the shock wave #1, the boundary layer thickness starts to grow. This leads to such conditions that the interaction with shock wave #2 results in a Mach reflection. It can be seen that the normal shock of this interaction decelerates the flow to subsonic velocity and more importantly causes the boundary layer to rapidly grow in thickness. Subsonic region downstream the normal shock of the Mach reflection is not large and the flow accelerates back to supersonic velocity. Upon passing through the bow shock of the blade No. 3, however, supersonic velocity can not be maintained any more and flow upstream of blades No. 4 and No. 5 becomes subsonic with only a limited supersonic region on the suction surface of blade No. 3, i.e. upstream of the blade No. 4. Due to this configuration the passages in between blades No. 1, 2 and 3 operate at started supersonic inlet flow conditions whereas passages between blades No. 3, 4 and 5 operate at choked flow conditions [3]. This inlet flow aperiodicity is associated not only with the distribution of the inlet Mach number but also with that of the inlet flow angle (see Figs. 8a and 9a). In the region upstream of blades 4 and 5 where the flow is decelerated as described previously, the inlet flow angle drops to lower values towards slightly negative incidence. This results in the increased mass flow rate through these two interblade passages (Tab. 1).

The actual shock wave pattern is more apparent from Fig. 6 and 7 which shows the contours of density gradients in x and y direction respectively. Such contours mimic schlieren technique which is commonly used for shock wave visualisation in compressible flow research. The wakes behind the cascade which are nearly perpendicular to y axis can be better seen in Fig. 7. The darker regions then show compression in x and y direction, while the bright regions show expansion in these directions. It can be seen from Fig. 7 that thickness of the wakes is determined by the strength and position of the passage shock

Table 1: Mass flow rate through the individual inter-blade passages.

Passage number	Mass flow rate $\left(kg \cdot s^{-1}\right)$	Mass flux $\left(kg \cdot s^{-1} \cdot m^{-2}\right)$
6	0.223	77.81
5	0.761	129.05
4	0.769	130.44
3	0.732	124.12
2	0.742	125.84
1	0.381	126.41

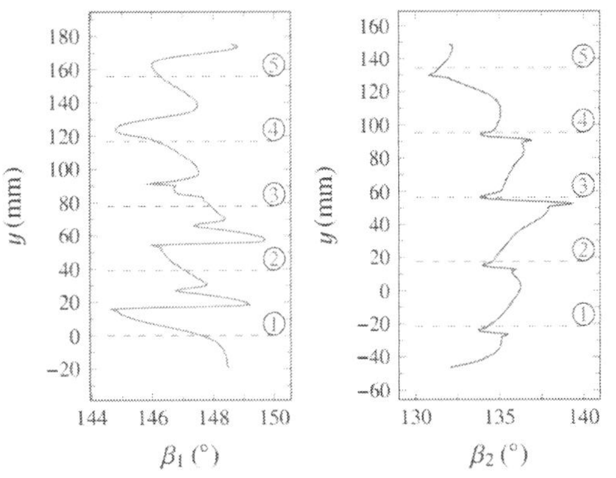

(a) (b)

Fig. 8. Flow angle β_1 in the plane in the normal distance of 15mm upstream the cascade (a) and β_2 in the plane 15mm downstream the cascade (b).

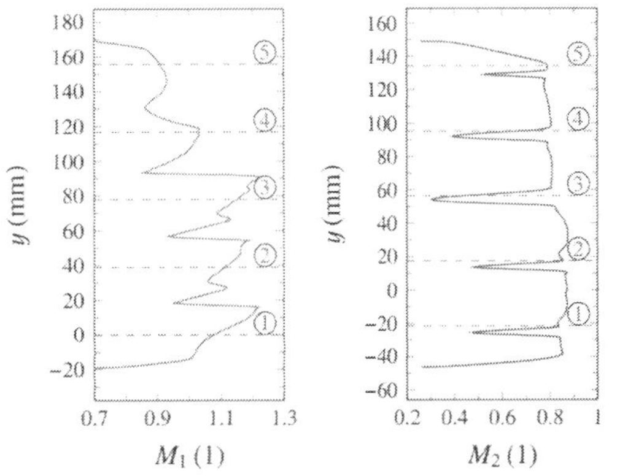

(a) (b)

Fig. 9. Mach number in the plane in the normal distance of 15mm upstream the cascade (a) and in the plane 15mm downstream the cascade (b).

wave in each inter-blade channel. The least favourable situation is in between blades 3-4 and 4-5. Here, the passage shock wave interacts with the suction side of blades 3 and 4 exactly at the point where the inter-blade channel starts to be divergent and the boundary layer is subject to adverse pressure gradient not mentioning the shock wave. Therefore Local flow separation takes place and the wake is wide. Situation in the inter-blade channel 3-4 is even more serious than in 4-5 since the passage shock wave is stronger there. This results also in distinct peaks in the distribution of the exit flow angle in F ig. 8b downstream of blades 3 and 4.

The qualitative difference between supersonic inlet flow conditions and choked flow conditions at the inlet to individual inter-blade passages is the principal cause of the flow field aperiodicity. As can be seen in Table 1, the mass flux through all the passages is nearly equal except the passage #6 (passage above the upper blade) in which case the mass flux is at approximately 60% compared to other passages. This is most likely due to the presence of the thick boundary layer at the upper wall of the test section. The increase in thickness is caused by already described successive interaction with the three shock waves.

5 Conclusions

The CFD computations of the new test section for investigation of the forced blade oscillations allowed to uncover the main flow features that might be expected during the experiments. The sources of flow aperiodicity, which represents one of the main concerns in the linear blade cascade investigation, were successfully identified in the form of shock wave-boundary layer interaction at the upper wall of the test section. It can be concluded that the planned installation of the upper wall suction should effectively mitigate the adverse effects of this interaction on flow periodicity and provide additional control to control the inlet Mach number. Further CFD simulations of the new test section will be focused on the simulation of the upper wall suction and the effect of the middle blade rotation on the forces acting on the other blades in the cascade.

This research was supported by the Ministry of Education, Youth and Sports of the Czech Republic - program Inter-Excellence, project No. LTAUSA19036. Institu-tional support RVO:61388998 of the Institute of Thermomechan-ics of the Czech Academy of Sciences is also acknowledged.

References

[1] J. Lepičovský, P. Šidlof, D. Šimurda, M. Štěpán, M. Luxa, AIP Conference Proceedings **2323**, 030001 (2021)

[2] H. A. Schreiber, H. Starken, ASME J. Eng. Gas Turbines Power **106(2)**, 288-294 (1984)

[3] H. J. Lichtfuss, H. Starken, Progress in Aerospace Sciences **15**, 37-149 (1974)

Fluid flow measurement in air ducts under unsteady flow conditions - fundamentals

Ludek Jancik[1*], and *Petra* Dancova[1]

[1]TU Liberec, Department of Power Engineering Equipment, Studentska 2, 46117 Liberec, Czech Republic

Abstract. Current methods for measuring fluid flow require a steady flow to achieve correct results. However, in practical applications, such a place is often not available and the measurement results obtained do not have the required reliability for further use. The project aims to design a methodology for measuring the flow in selected types of unsteady flow, focusing on the conditions in the ventilation ducts. The presented paper explains the state of the art in this field and identifies the most promising ways for further research. According to the results found in available sources, acceptable results with relative deviations of about 5 % can be obtained with an increased number of measurement points.

1 Introduction

This paper aims to present an overview of possible solutions for the measurement of unsteady fluid flow, especially in air ducts used for ventilation and air conditioning, in cases where it is impossible to find a measurement point with stabilized flow properties.

2 Motivation

Standard fluid flow measurement procedures require steady flow conditions with a stabilized velocity profile in order to obtain accurate measurement with acceptable levels of precision and uncertainty. This steady flow is mostly achieved by allowing the stream to stabilize along a relatively long path, usually defined as a number of relevant pipe hydraulic diameters.

For the field of industrial ventilation and air conditioning with characteristic dimensions of air ducts of more than 1 meter, it means finding a straight, clear and undistorted pipe segment of several meters in length before and even after the measurement point, which is mostly impossible in real applications.

Another reason for finding a feasible method of unsteady flow measurement in air ducts is the increased demand for energy savings, in this case, achieved by a more precise air flow rate control in the ventilation system.

3 Basic overview

As the first step, a literature review was performed with the following results:

1. Unsteady fluid flow measurement is mentioned and suitable methods are defined for specific cases in several technical standards.

2. Unsteady fluid flow measurement has got many practical applications (e.g., in medicine, combustion engines, natural gas transport).

3. Successful experiments with unsteady flow measurement in air ducts were carried out.

4 Valid standards

4.1 Air ducts (EN 14277)

European standard EN 14277 [1] gives the fundamental guidelines for air flow measurement in air ducts for ventilation and air conditioning. The most notifiable points are:

1. Pressure based methods mainly
2. For laboratory aerodynamic testing
3. Specification of basic measurement layouts
4. Specification of acceptable ranges of uncertainty

4.2 Asymmetric and swirling fluid flow (ISO 7194)

This standard (ISO 7194, [2]) defines measurement conditions for obtaining meaningful results for asymmetric (or drifting) and swirling fluid flow (Fig. 1.) in circular ducts using current -meters or Pitot static tubes.

As it states, even if it is advisable to choose a measurement point with a steady flow and fully developed velocity profile, there are cases where it is

* Corresponding author: ludek.jancik@tul.cz

impossible to find such a point, but it is still necessary to perform a successful measurement.

For this purpose, this standard describes methods for minimizing errors when measuring asymmetric or swirling flow, provides correction coefficients for certain types of errors, and defines procedures for determination of uncertainty increase in cases where it is not possible to compensate the measurement errors in another way.

Moreover, it provides a basic overview of the most common flow distortions in closed conduits: asymmetric flow velocities distribution occurs if all flow direction changes are in the same plane, whereas the swirling flow requires flow direction changes in two planes to be established.

Unlike the asymmetric flow, the swirling flow has a substantial effect on the reading of Pitot static tubes as well as current meters, and therefore, measurement in such cases should be avoided if possible.

If it is still needed to measure under such conditions, it should be noted that swirling flow lasts much longer than other forms of distortion and the required length for achieving a steady flow is several times higher when compared to other types of unsteady flow.

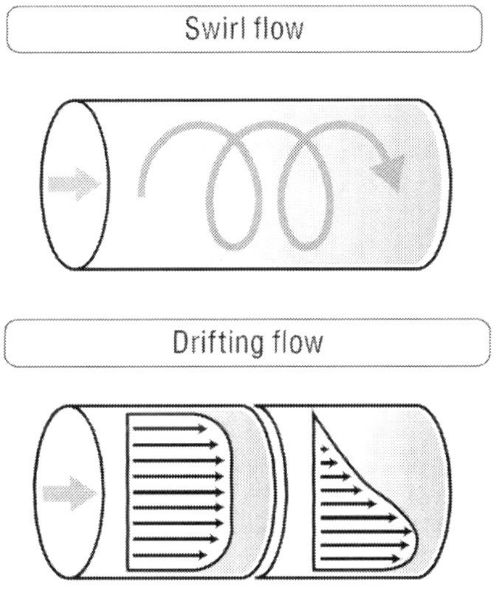

Fig. 1. Swirl flow and drifting flow – basic distortions in closed conduits [3].

5 Practical applications

5.1 A. Beaulieu et al., A flowmeter for unsteady liquid flow measurements

This paper describes a novel instrument for measuring pulsating flow in a special application using the Venturi tube principle (Fig. 2.) Unsteady flow conditions, in this case, are caused by a medical device called a liquid ventilator intended for human emergency external ventilation using a special liquid for O_2 and CO_2 transport. Medical devices for vital functions support are another example of systems working with unsteady (pulsating) fluid flow and thus requiring a proper measurement even under these non-standard conditions.

The proposed solution uses two symmetrical Venturi tubes oriented in opposite directions and equipped with three pressure sensors. As the pulsating stream flows through, the readings of the sensors are taken as inputs for solving a modified unsteady Bernoulli equation describing this system.

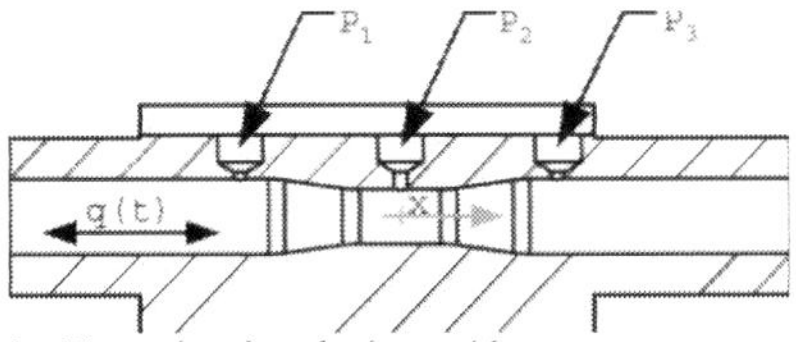

Fig. 2. Venturi tube device with pressure sensors for measurement of pulsating flow in a medical device [4].

5.2 F. Laurantzon, R. Örlü, N. Tillmark, P. H. Alfredsson, Response of common flowmeters to unsteady flow

This paper is focused on an experimental comparison of chosen fluid flow measuring devices investigating a pulsating compressible flow in a closed duct with mass flow rates from 80 to 130 $g \cdot s^{-1}$ for various pulsating frequencies (10 – 80 Hz), i.e., conditions similar to those in combustion engines.

Table 1. Relative deviations of mean mass flow from reference value of 80 g/s of various types of flow measurement devices for given flow pulsation frequencies [5].

Flowmeter	10 Hz	20 Hz	30 Hz	40 Hz
Venturi	1.02	1.06	1.09	1.05
Pitot	0.95	1.04	1.18	1.07
Hot-film	0.98	0.97	1.29	1.89
Vortex	0.98	0.99	1.08	1.01
Turbine	1.24	1.21	1.18	1.07

The significant deviation from the reference value of hot-film probe measurements for frequencies 30 and 40 Hz is notifiable compared to relatively close results for the frequency range 10 – 20 Hz (Table 1.)

5.3 F. Laurantzon, Flow Measuring Techniques in Steady and Pulsating Compressible Flows

This detailed work deals with a complex description of possible measuring techniques for applications with pulsating compressible flow found, e.g., in combustion engines.

Compared to the previously mentioned sources, in this one, another flow measurement technique was applied – thermal anemometry. A device called hot-wire mass flow meter was used to a determination of fluid flow rate in the pipes of the experimental set-up. When compared to a reference flow meter, it proved to be suitable for this purpose with a maximum relative deviation of ± 3 % (Table 2. and Fig. 3.)

A measuring sensor based on a hot-film probe was then used for the analysis of the pulsating flow with ambiguous results due to its inability to identify the backflow stream.

Table 2. Relative deviations of mean mass flow measured by hot-wire probe from reference value for given flow pulsation frequencies [6].

$\dot{m}$	$f_p = 0$ Hz	$f_p = 40$ Hz	$f_p = 60$ Hz	$f_p = 80$ Hz
80 g/s	1.01	0.997	1.01	-
130 g/s	0.968	0.979	0.988	1.02

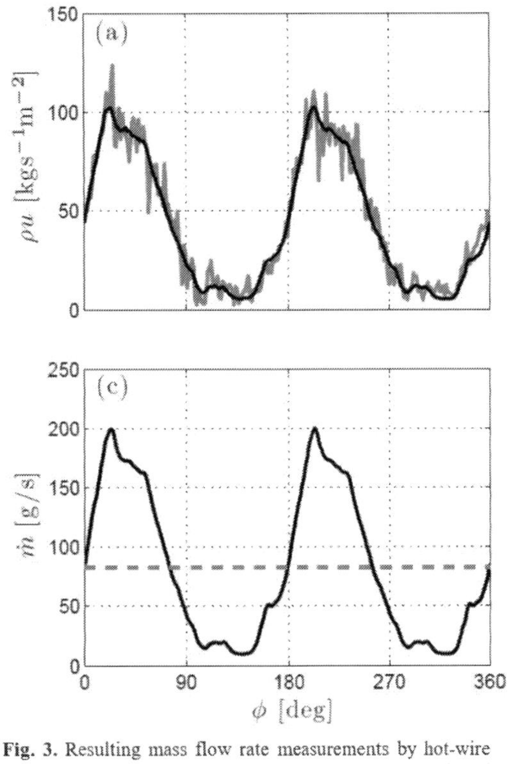

Fig. 3. Resulting mass flow rate measurements by hot-wire method (varying solid line, top chart) compared with phased averaged value (bold black solid line, both charts) and integrated mass flow rate (dashed line, bottom chart) for various pulse phases (horizontal axis) [6].

5.4 M. Kurth, P. Komp, Influence of installation and operating conditions on the measurement of high-pressure natural fluid flow, ultrasonic flow meters (in Czech)

The authors are focusing on flow measurement using ultrasonic flow meters in natural gas pipelines, i.e., closed circular ducts with large diameters and high pressures. Motivations for investigation of unsteady flow measurement are obvious – in the real pipelines, it is hard to find a place with the steady flow, but the measurement is still necessary. Lengths of about 28 diameters before and 13 diameters after the measurement point are mostly required for fluid flow stabilization in this field. Therefore, physical devices for a faster flow stabilization already mentioned, e.g., in ISO 7194, are widely used (Fig. 4.)

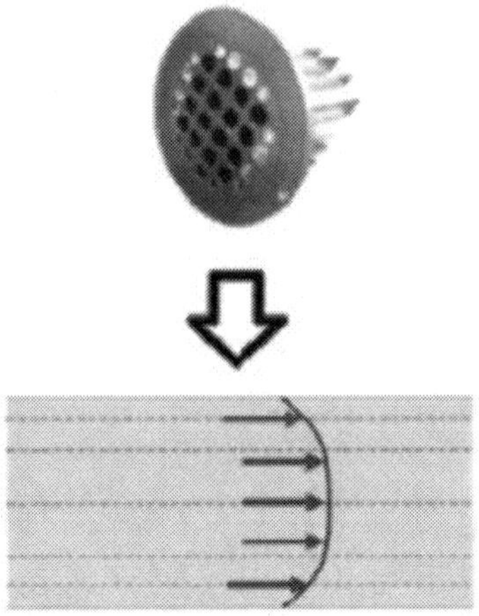

Fig. 4. Zanker-type flow conditioning plate used for fluid flow stabilization in natural gas pipelines [7].

As another solution, the authors are considering ultrasonic measurement in more planes or directions to overcome the errors caused by the unsteady flow and swirling (Fig. 5.)

Each of the presented configurations has its own advantages, drawbacks, and field of utilization. Generally, arrangements with more layers of ultrasonic probes produce a better approximation of the velocity profile, whereas configurations with one or more reflections succeed in better compensating both the symmetric and asymmetric swirls.

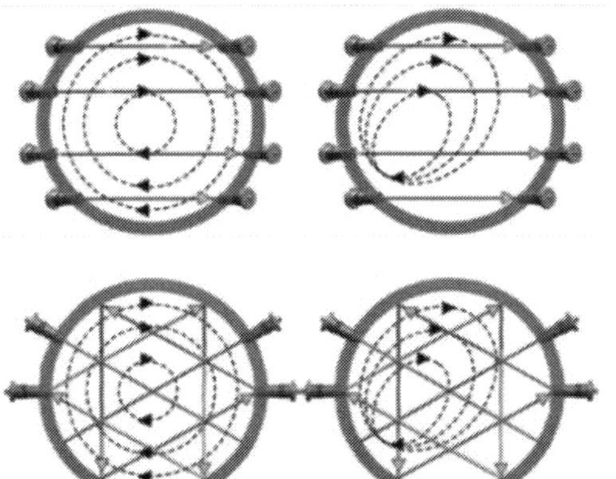

Fig. 5. Various configurations of ultrasonic flow meters with shown signal paths for compensation of symmetric and asymmetric swirls during measurement in natural gas pipelines [7].

5.5 I. Caré, F. Bonthoux, J.R. Fontaine, Measurement of air flow in duct by velocity measurements

In this paper, experiments with unsteady fluid flow in a circular air duct with a diameter of 200 mm are described. The measurement was performed behind a flow distortion in the form of a 90° bend; the measuring device used was a Pitot static probe.

Table 3. Relative maximum error for various configurations of measuring points and distance from the source of the flow distortion [8].

	E: maximum error in %							
	k	p	k*p	L/D				
				2	4	6	10	20
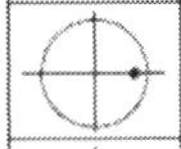	1	1	1	47	29	22	15	9
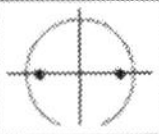	1	2	2	25	16	12	8	5
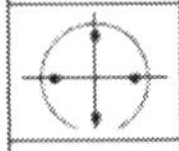	2	2	4	13	8	6	4	3
	1	6	6	11	7	5	4	2
	1	10	10	9	5	4	3	2
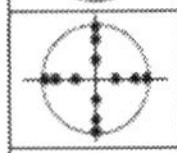	2	6	12	6	3	3	2	1
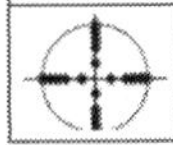	2	10	20	4	3	2	1	1

During the experiment, flow conditions defined with $Re > 2 \cdot 10^5$ were investigated. The method used for the compensation of increased measurement uncertainty due to the unsteady flow caused by the bend was based on an increased number of measurement points for a better approximation of the velocity profile. The results obtained are shown in the Table 3. As can be seen, a maximum error of 4 % can be achieved even in a very short distance from the bend (L/D = 2, where L is the distance between the probe and the bend, D is the diameter of the air duct) if we use 20 measurement points across the air duct diameter in two directions.

6 Conclusion

The general methodology for measuring fluid flow in closed profiles is defined, for example, by the relevant standards EN 14277 [1] and ISO 7194 [2]. In both documents, a sufficient distance of the measuring point from the sources of instability is required, which corresponds to several hydraulic diameters of the given pipeline.

The measurement under unstable conditions has some specific applications in natural gas supply or in pumping liquids, e.g., [7] and [4], respectively. A detailed description of the properties of measuring methods and instruments appropriate for the analysis of unsteady fluid flow is presented in [5] and [6]; among other things, an acceptable accuracy of thermal anemometric probes for a given purpose is documented here.

The key document for the specified purpose is [8]. In this work, the experiment performed deals with a ventilation air duct, and the first usable results are presented.

The presented measurements were taken with Pitot static tubes in a circular air duct with a diameter of 200 mm and for a Reynolds number of about 200,000, which corresponds to relatively high air velocities compared to typical values. The results show an acceptable relative error of 4 % even for measurements very close to the source of flow distortion (L/D = 2).

Based on the statements above, the research task can be specified in the following way: the measurement will be performed by hot-wire probes behind the 90° elbow bend in air ducts with a diameter of 300 - 400 mm, at air flow velocities from 0.5 to 5 $m \cdot s^{-1}$ (typical velocities in ventilation air ducts). Particle Image Velocimetry (PIV) visualization will be used to determine the most suitable places for measurement.

This work was supported by the Student Grant Competition of the Technical University of Liberec under the project No. SGS-2019-5073.

References

1. EN 14277 (European standard), Ventilation for buildings – air terminal devices – Method for airflow measurement by calibrated sensors in or close to ATD/plenum boxes (2006)
2. ČSN ISO 7194 (Czech national/international standard), Measurement of fluid flow in closed conduits. Velocity-area methods of flow measurement in swirling or asymmetric flow conditions in circular ducts by means of current-meters or Pitot static tubes (1994)
3. Keyence Corporation of America, *Flow Sensor Technical Guidebook*, (2011)
4. A. Beaulieu et al., *A flowmeter for unsteady liquid flow measurements,* Flow Measurement and Instrumentation, **22** (2011)
5. F. Laurantzon, R. Örlü, N. Tillmark, P. H. Alfredsson, *Response of common flowmeters to unsteady flow*, Technical report (2012)
6. F. Laurantzon, *Flow Measuring Techniques in Steady and Pulsating Compressible Flows*, Thesis (2010)
7. M. Kurth, P. Komp, *Influence of installation and operating conditions on the measurement of high-pressure natural fluid flow, ultrasonic flow meters (in Czech),* Techniky a technológie, **2** (2013)
8. I. Caré, F. Bonthoux, J.R. Fontaine, *Measurement of air flow in duct by velocity measurements,* EPJ Web of Conferences **77**, 00010 (2014)

Software-based processing system for phase Doppler systems

Jan Jedelsky[1], *Milan* Maly[1*], *Ondrej* Cejpek[1], *Graham* Wigley[2], and *James F.* Meyers[3]

[1]Mechanical Engineering, Brno University of Technology, Brno, Czech Republic
[2]Aero and Auto Engineering, Loughborough University, UK
[3]James F. Meyers, (retired), NASA - Langley Research Center, Hampton, VA, USA

Abstract. A Monte Carlo simulation of Phase Doppler systems has been developed. It consists of three sections, the droplet flow description, generation of the photomultiplier signals and then their processing to determine droplet velocities and the time shift between the signals from the three scattered light detection apertures. With highly realistic Doppler bursts being simulated and processed, the question arises as to whether the signal processing software could be used to process 'real-world' experimental signals. In a preliminary assessment of its capabilities in such a situation, actual spray Doppler signals (from a Dantec fibre-based PDA system with a BSA signal processor) were recorded and used as input to the software signal processor. The signals from the three photomultipliers were input first into a Picoscope and then into the BSA processor. In this way droplet velocities and size estimates would be available from the BSA as control data. The signal outputs were taken as csv files, and input directly into the software signal processor. Initially the software determined the time location of the centre of each signal burst envelop. This approach was shown to measure signal delays from single cycle to multiple cycles. For this experiment, the software was modified by adding a zero-crossing approach to measure the single cycle delays. The introduction of this method should establish the accuracy of the complete software package in the real world as the results from the preliminary experiment show good agreement between the two techniques.

1 Introduction

The Phase Doppler technique, PDA, is accepted as a robust optical method for point-wise simultaneous measurement of size and velocity of particles dispersed in a fluid. It is frequently used for the characterisation of sprays, aerosols and bubbles. This optical diagnostic technique is utilised widely for fluid mechanic experiments and is routinely used in many high-tech laboratories today.

The physical realisation is rather expensive. The instrumentation cost depends on the system specification, robustness, measurement capabilities, extensibility, and variability of the configuration [1]. It ranges from 40,000 €/channel for a basic system to more than 80,000 €/channel for a state-of-the-art system. A significant portion of that price is paid for the signal processor. It is also the hardware and software components that are superseded most often and, after a couple of years updated solutions are offered on the market.

The estimation of droplet or particle size by these commercial PDA systems is only dependent on the specification of the physical configuration of the laser transmitter and receiver systems and the assumption that the light scattering can be determined by geometric optics for either refraction or reflection. The size estimates are derived from determining the phase differences in the Doppler shifted signals, generated by the scattered light from each droplet, arriving at each of the two or three photodetectors in turn.

The size estimates can be extremely accurate if single droplets are probed under simple experimental conditions. As the measurement volume cross section has a Gaussian intensity distribution and that, there is a nominal square law relationship between droplet size and scattered intensity, then the scattered signal intensity becomes a significant factor in estimating droplet size distributions. Furthermore, the light scattering functions for a wide range of sizes are best considered by applying electromagnetic scattering, Mie theory, rather than the simplistic geometric scattering modes of refraction or reflection to determine the probability of measurements over a distribution of droplet sizes.

To understand this issue, a Monte Carlo type simulation program was developed to model phase Doppler systems [2]. The Monte Carlo technique consists of breaking systems down into small units and, using random number generators, develop a realistic approach to the real-world physics. The work is divided into three main sections, the development of a launch strategy of known droplet sizes, the generation of the photomultiplier signals, and then their processing to generate drop size estimates followed by the comparison of these data with the launched data. This provides a linear relationship between dropsize, and the signal time shifts.

The signal processing portion of the simulation measures the Doppler frequency in each signal to estimate

* Corresponding author: milan.maly@vutbr.cz

the droplet velocity. Rather than using measurements of the phase differences between the multiple Doppler-shifted signals the time shift between the arrival times of each signal is used to estimate the dropsize. The transmitter and receiver optical systems are defined as diffraction limited and the laser beams true Gaussian. Optical, and electronic noise sources can also be simulated while the signals generated, being dependant on droplet size, can have intensity levels ranging from photon resolved to photomultiplier saturation levels.

With highly realistic Doppler bursts being simulated and then processed the question arises as to whether the signal processing software could, on its own, be used to process 'real-world' experimental signals. The paper documents an application of the signal processing part to Doppler signals from an actual spray characterization by a Dantec fibre-based PDA system.

2 Material and Methods

In an assessment of the capabilities of the signal processing technique in such a situation, Doppler signals from an actual spray characterization were recorded and used as input to the software signal processor. A Dantec fibre-based PDA system, configured with a transmitter-receiver geometry for refractive scattering, was used with a BSA signal processor. The spray was provided by an artist's air brush.

Initially the software determined the location, in time, of the exact centre of each signal burst envelop. This approach was shown to measure signal delays from single cycle to multiple cycles. For this experiment, the signal processing software was modified by adding a zero-crossing approach to measure the single cycle delays. The introduction of this method should establish the accuracy of the complete software package in the real world as the results from the preliminary experiment show good agreement between the two techniques.

2.1 Principles of phase Doppler systems

The PDA technique (Phase Doppler Anemometry, or less frequently called phase Doppler interferometry) is a point-wise optical measuring method to provide Eulerian description of the flow field with embedded fluid markers. It gives statistical moments of the particle velocity and size. The principle of the technique is illustrated in Figure 1a, and it is detailed by [3]. A continuous-wave monochromatic laser beam is directed into a transmitting optics where it is split into two coherent beams of equal intensity that are intersected symmetrically at a small angle θ by a transmitting lens. The two in-phase beams interfere in the volume of their intersection (the measurement volume[a]) in a way to generate a series of parallel planes[b] of light and darkness (the so called fringes) as shown in Figure 1b. Light scattering from boundaries between two different phases (such as small inhomogeneities, particles or bubbles, present in the particle laden gas or liquid flows or artificially added tracer particles) generate a scattered light signal while passing through the small fringe zone. To eliminate the ambiguity of the direction which the particle passes through the measurement volume, the frequency of one of the beams is shifted by a fixed frequency amount f_0 using a Bragg cell so that the fringe pattern "scrolls" through the measuring volume with a known velocity. A receiver lens[c], placed in a separate receiving optical system, projects a portion of the light and dark interference bands (the fringe patterns) scattered by the particle on to a photodetector (receiver) which produces a "Doppler burst" signal (Fig. 1c) with a Doppler frequency[d], f_D, proportional to the velocity component u_n of particle motion normal to the fringe orientation divided by the fringe separation s:

$$f_D = f_0 + \frac{u_n}{s} = f_0 + u_n \frac{2\sin\theta/2}{\lambda} \qquad (1)$$

where λ is the wavelength of the laser light. The light signal is converted into a digital record and processed to provide information on velocity and arrival time of each individual particle.

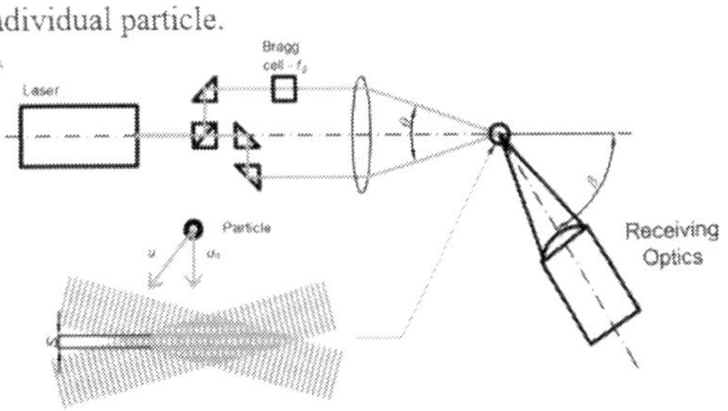

Fig. 1 The principle of the PDA: optical configuration of the system, fringe pattern formed by two coherent laser beams crossing with the fringe.

The Doppler signals are sensed by at least two detectors. The detector pair observes the light scattered by at the same scattering (the so called off-axis) angle, β. The detectors are located symmetrically about the observation plane at angles $\pm\,\psi$ with respect to the plane. A mask with two rectangular and horizontally elongated slits is used to define the elevation angles. A vertical slit can be used to reduce the effective length of the probe volume from where the scattered light is collected.

The Doppler signals collected at the different elevation angles ψ show a phase difference, $\Delta\varphi$, or a time shift with respect to each other. This shift depends on the particle radius R in the scattering direction and increases with increasing the elevation angles of the detectors. It may also depend on the particle/fluid refractive index n_p and may, to some, extent show fluctuations in the size diagram caused by partial wave resonances.

According to [3], the particle size can be determined based on:

[a] The measuring volume has a shape of 3D ellipsoid.
[b] The fringe planes are parallel to the bisector of the laser beams and perpendicular to the plane formed by the two beams
[c] which can be positioned in the forward, side, or backscatter direction
[d] often referred to as the Doppler frequency

• *either the instantaneous phase shift between modulated signal parts of the two detectors (phase Doppler technique)*

• *or the time shift between the signal maxima on the two detectors (time-shift technique)*

For the phase Doppler technique, a simple relationship between particle size and phase shift can be found using geometric optical ray traces based on refraction or reflection:

$$\Delta\varphi = \frac{4\pi R n_p}{\lambda} \Phi \tag{2}$$

The parameter Φ depends on the scattering mode and geometrical configuration of the PDA system [3]. The phase shift $\Delta\varphi$ is determined from the time lag of the two signals, Δt, with respect to the period of the Doppler burst $T = 1/f$ as indicated in Fig. 1b: $\Delta\varphi = 2\pi\Delta t/T$. The phase shift becomes ambiguous when it exceeds the 2π period. In this case a third detector is placed at a smaller elevation angle[e].

This provides two pairs of detectors with different phase relationships to help extend the dropsize range. Even so the maximum phase measurement will still not exceed $5\pi/2$. For the larger drop sizes the depth of the signal cycle modulation decreases making phase difference measurements uncertain. The dropsize measurement range is generally accepted to be no better than 45:1. However, these limitations are based only on the geometric calculations for the phase dropsize relationship and the signal visibility. The laser power, PM voltage and amplification levels determine the minimum detectable signals from small droplets and the PM saturation for large droplets.

Eqs. (1 and 2) show that the probed velocity component and particle size linearly depend on the measured frequency and phase shift respectively multiplied with a physical and geometrical constant.

The time shift approach determines the location in time, i.e. the mark, in each of the three sequential signal bursts that have the same signal characteristics. The first consideration would be to use the peak cycle signal to establish the mark location, but, with no synchronisation between the A/D sampling frequency and the Doppler signal frequency the marks would not have the same representative locations. This would also be the case for when the peak cycle is not at the mid-point of the signal burst.

The method used was to invert the signal burst, then apply a trigger level that was above the base noise. The samples, with amplitudes greater than the trigger level, are then accepted and used as the weight function in a weighted average process applied to the whole signal burst. The time shift delays are determined by the difference between the weighted average time of signal B subtracted by the weighted average time of signal A. Likewise, the same procedure is applied to the times between component C and component A. These two delays are then converted to two estimates of the droplet size based on equations (1) and (2).

The Doppler bursts from photo detectors are sent to a signal processor which extracts the velocity, transit and arrival time and possibly size data in digital form for further analysis. The opto-electronic part of LDA/PDA instruments is supplemented with a computer that controls the collection, manipulation, analysis, and data presentation.

2.2 Setup of phase Doppler system

The actual Phase Doppler system used a Spectra Physics 164 argon ion laser, operating in all lines mode, coupled with a Dantec fibre flow transmitter box, 60X41 and linked to a 112 mm fibre probe, 60X81 fitted with a 2X beam expander. For this work only the green, 514 nm, wavelength was selected. The output beam polarization from the probe was in the horizontal plane, i.e., parallel to the fringes in the measurement volume. The receiver was the Dantec 57X50. The essential optical parameters are given in Table 1.

Table 1. Optical set up parameters.

Transmitter		Receiver	
Beam diameter	2.70 mm	Aperture dia.	78 mm
Beam separation	75.24 mm	Lens aperture mask	A
Lens focal length	500 mm	Lens focal length	500 mm
Fringe spacing	3.429 µm	Spatial filter – slit	100 µm
Probe volume dia.	122.9 µm	Scattering angle	70 deg.

These settings were inherited from a previous experiment on a confined spray. There was no attempt whatsoever to optimise these settings to suit the measurement of the droplet size distributions contained in a spray generated by an artist's air brush as used in this study. However, this does not detract at all from the scientific integrity of the work here which was to use the Phase Doppler system purely as a tool to generate Doppler signals that could be then digitised and processed in software.

The air brush is a precision air blast atomizer. It was positioned 30 cm vertically above the measurement volume and operated with 2 bar compressed air. The working fluid was water.

The signals from the three photomultipliers were input first into a Picoscope 3405B with a 1 MOhm input impedance and then, via BNC T pieces, into the BSA processor. In this way droplet velocities and size estimates would be available from the BSA. The Picoscope is a very cost effective four channel A to D convertor with fast sampling, digital triggering, and a deep memory. In this preliminary experiment it was set up to record the three channels of photomultiplier signals with 668 sequential

[e] Present PDA systems use typically three detectors integrated into a single fiber-optic receiver.

Doppler bursts at a sampling rate of 0.25 GHz. The signal outputs were taken as csv files for ease of viewing and input into the software signal processor.

The photomultiplier, PM, voltage levels were determined using the signals displayed by the picoscope. The peak voltages were judged to be when signal limiting was first observed. The BSA PM output monitor signal should not be used as the output is non-linear making choice of the peak voltage levels for balanced Doppler bursts ambiguous. Furthermore, it adds noise and time/phase shifts to the signals when compared with the PM signals direct.

3 Results and Discussion

3.1 Doppler Bursts and Processing

Two Doppler bursts from apertures A and C for one droplet are shown in Fig. 2 as acquired by the PDA system.

The time shift approach, as with any measurement technique, has limits and problems. The time shift approach does not have the $5\pi/2$ phase limit of cycle-based methods, but it does have limits regarding the signal-to-noise in the signal bursts. Fig. 2 shows what looks like noise on the Doppler burst, but it is not. It is due to the interaction of the Picoscope sample frequency and the Doppler frequency. A peak of one frequency at the same occurrence of the other frequency may be a peak and peak, but it could be a positive in one and a negative in the other. This can be excessive if there are only a few of the amplitude steps within the signal burst. For the Doppler burst shown in Fig. 2 this is not a problem as the displayed voltage levels, 31, are considerably greater than the minimum resolution. However, if there are only 10 or less voltage steps in a Doppler burst then the burst centre could be far from the weighted average mark. This is particularly the case for small drop sizes. Also, a close up of the Channel A and Channel B signal edges, show that the small difference could be controlled any unseen characteristics in Channel A. This would also affect the location of the Channel A mark. Thus, the voltage level and voltage steps should be set to reduce the potential shift in the delay measurements.

The signal bursts captured by the Picoscope are transferred to the computer and stored. The signal bursts are then read by the software and the capture checked of a second signal burst occurred and the limits where there is no crossing of a second signal burst and/or noise levels. Once the signal burst limits are set, the pedestal within the signal burst is determined by using a first-in-first-out smooth technique. The pedestal is then subtracted from the signal burst leaving the amplitude centred at zero volts. The positive side is separated along with the inverted negative side. The centre of each side is determined by a weighted average. The two-time locations are averaged to obtain the mark of that signal burst. The time difference between the A and B marks yield the delay time between A and B.

In order to determine measurement accuracy, a small part of the program was added that would remove the delays on B and C signal bursts to zero when compared to A. A uniform random number generator would then pick a droplet size from the droplet range. The conversion of delay to droplet was inverted to yield the A:B and A:C delays for the selected droplet size. B and C signal bursts are then adjusted in time to set the calculated delays for the time between A mark and B mark, and between A mark and C mark. This approach was used on the signal bursts obtained by the PDA because the marks, especially B mark did not satisfy the A:C/A:B ratio set by the mask. The program was set to a 1-to-20- micron droplet size range, and the 668 array was processed. Since each droplet size was known, the comparison with the droplet size obtained by the program yields the measurement error for each droplet.

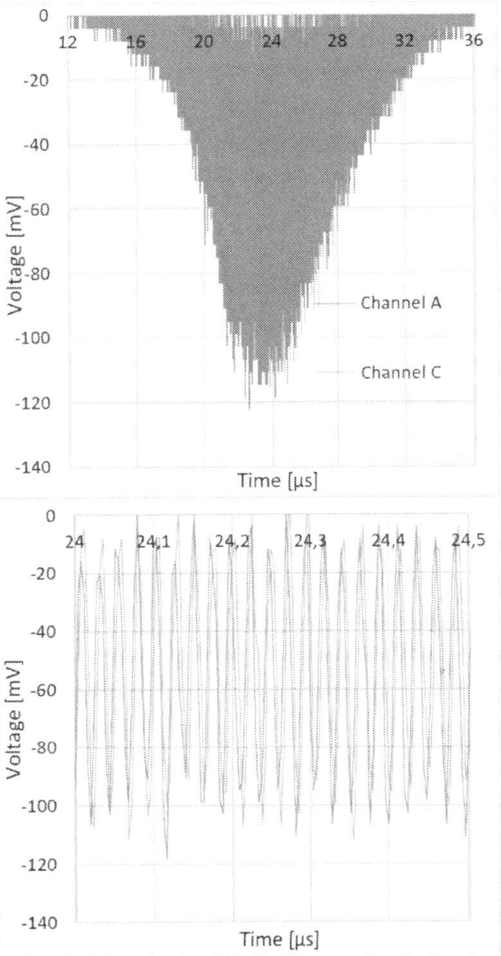

Fig. 2 Doppler Burst acquired by the PDA system. Top: The whole burst, bottom: detail in the time resolved signal.

The average error was +/-0.5 microns in both A:B and A:C droplet sizes. The processing time was 15 seconds to acquire the results and measurement errors for all 668 droplets. The computer was a standard PC with an i7 core processor, an M.2 PCIe SSD drive, and the Windows 10 operating system.

The processing of the Doppler burst of Channel C is shown in Fig. 3. In the first stage, a) the PM signal is simply inverted from the negative level to a positive level. In b) the pedestal has been determined and shown overlaying the signal burst as a red line from a 300-sample smooth. The view of the separated positive and negative halves as described above is shown in c). In this case the red pedestal plot was from a 100-sample smooth. Areas below 70% of the peak amps have been removed to reduce noise, and the peak amplitude has been set to 1.0. The change in amplitude makes it easier to match the characteristics of the signal bursts. The positive data are in black and the inverse negative data are green. The weighted average technique is used to determine the central time location for each, and the average of the two marks become the mark to be used with the A mark to determine the A:C delay.

The signal processing program has been recently increased by adding two different cycle techniques to determine what their measurement accuracies were. As mentioned above, the time technique has the advantage of no limits regarding delays, and the cycle technique has the advantage that noise, assuming high signal peak levels, does not cause problems. The objective was to use the time delay technique to determine the delays and the cycle techniques to verify the measurements. Running the same 668 array data, the measurement accuracy in all three techniques was +/- 0.5 microns over the range 1 to 20 microns.

3.2 Droplet size estimations

The time differential between the burst signals from apertures A and B, and A and C, yield the delays that are then converted to the A:B droplet size, and A:C droplet size. The conversion equations are determined by using the optical characteristics of the PDA system and the uniform random number generator to input a known two delays and droplet size into the signal bursts obtained from the spray. (The actual delay of the centroid is set to zero in the A, B, and C signal bursts. Then the B and C signal bursts shifted in time based on the random droplet size.) The measured delays are then plotted, one set for A:B and the other for A:C. The equalizations can be determined for droplet size versus delay. The signal processing software has two routes. The prime one is to take the signal bursts as they come and use the weighted average to determine where the mid-point is. If the three signal bursts are not identical, the location in time of the weighted average can change, yielding measurement errors – with no way to determine how much or why. In order to determine if the software is correct, was to add a second route. The signal bursts from the A/D convertor have the centroids based on the delays found (as in route 1). In order to control the measurement, the B and C signal bursts are shifted in time to match the centroid in A, i.e., no delays at all. A uniform random number selects a droplet size, and the A:B and A:C delays calculated. The B signal burst is now shifted to the A:B delay obtained between the A and B centroids. Likewise, the C signal burst is now shifted to the A:C delay obtained between the A and C centroids. The resulting signal bursts are then entered into the same signal processing path as route 1. The difference is that the delays are known and if the measurements yield the same delays, the software is working correctly, and processing errors determined over

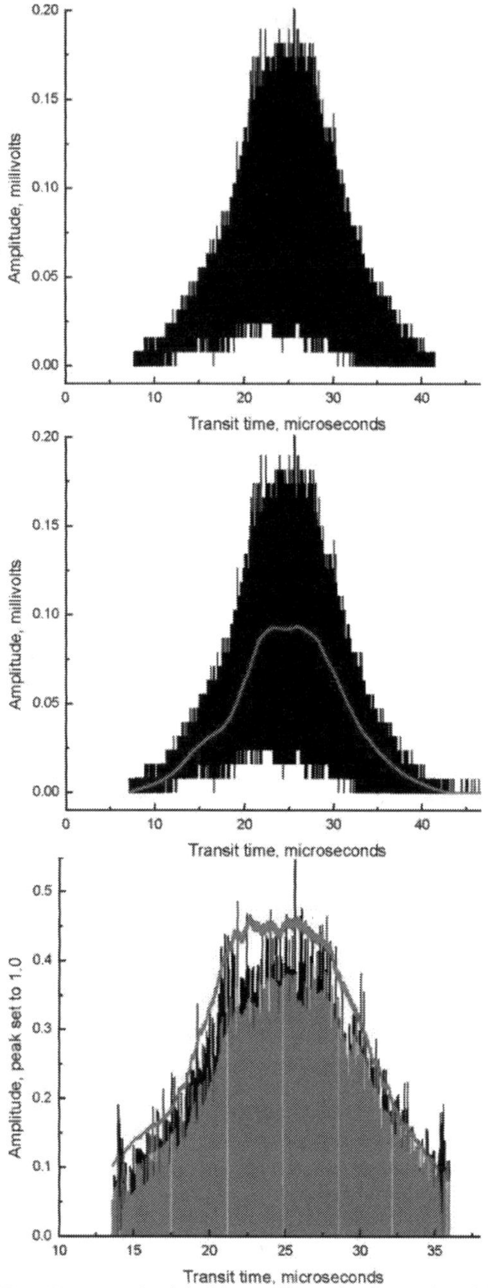

Fig. 3 From top: a) original inverted burst, b) The overlay of the pedestal on the signal burst, c) processed burst.

the entire 1–20 micron range in a droplet by droplet method. The software was run with synthetic signals – data set with uniform random droplet size was generated. Illustrative results in the form of dropsize histogram are shown in Fig. 4. Dropsize histogram of the measured data set that was acquired by the BSA processor is given in Fig. 5. Since the histogram in Fig. 4 was obtained for uniform random droplet size, it is not directly comparable to the BSA data.

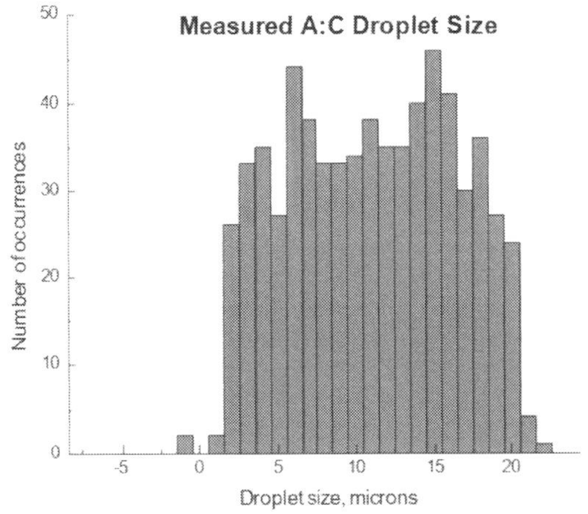

Fig. 4 The resulting size histogram.

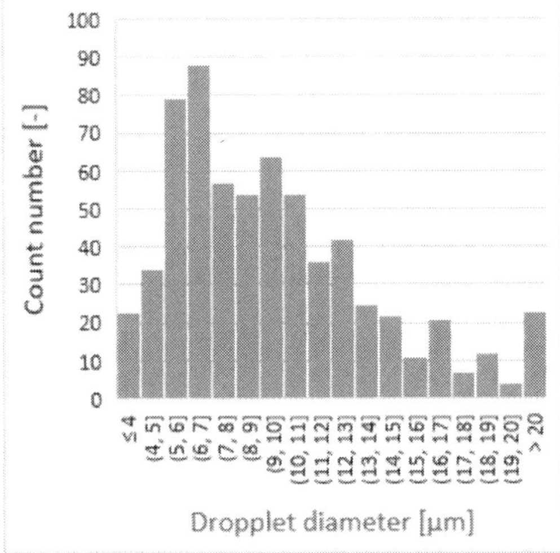

Fig. 5 Dropsize histogram as acquired by the BSA processor.

4 Conclusion

A processing part of a Monte Carlo simulation of Phase Doppler systems was used to determine droplet sizes from spray measurements. It was intended to replace the signal processing part of actual PDA system (Dantec fibre-based PDA system with a BSA signal processor and three channel receiver). The software was modified by adding a zero-crossing approach to measure the single cycle delays. The droplet sizes estimation was based on the time shift between signals from the three scattered light detection apertures.

Since the histogram obtained from the software was run with real signal bursts, with the delays set to known values as obtained from selected droplet sizes from a uniform random droplet size. The results are not comparable to the BSA data. It just illustrates that the signal processing software is working as expected and it could be used to process 'real-world' experimental signals. This is work in progress, and that signal problems should be removed with the upgraded A/D convertor. The alternative software-based signal processing approach will be widely usable with advantages of sizing non-transparent liquids (measurement in back-scatter mode), measurements above 2π for all scattering modes without a need to resolve between them and hypothetically also for multiple bursts.

Acknowledgement

This work was supported by the Ministry of Education, Youth and Sports of the Czech Republic under the project LTAIN19044 funded from the program INTER EXCELLENCE (INTER-ACTION).

List of Symbols

f_0	Frequency shift by Bragg cell [Hz]
f_D	Doppler frequency [Hz]
T	Period of Doppler burst [s]
u_n	velocity component of particle motion normal to the fringe orientation [m/s]
s	fringe separation [m]
β	Scattering angle [°]
λ	Wavelength of the laser light
ψ	Elevation angle [°]
Φ	Parameter dependent on scattering mode [-]
$\Delta\varphi$	Phase shift [°]
Δt	Time shift [s]
θ	Angle of intersecting laser beams [°]

References

1. F. Lizal, J. Jedelsky, K. Morgan, K. Bauer, J. Llop, U. Cossio, S. Kassinos, S. Verbanck, J. Ruiz-Cabello, A. Santos, E. Koch, and C. Schnabel, European Journal of Pharmaceutical Sciences 113, 95 (2018).
2. J. F. Meyers, Graham Wigley NASA/TM–2020-5006713 (2020)
3. H.-E. Albrecht, M. Borys, N. Damascke and C. Tropea, Laser Doppler and phase Doppler measurement techniques. (ISBN 978-3-642-08739-4 ISBN 978-3-662-05165-8, DOI 10.1007/978-3-662-05165-8, Springer, Berlin; New York, 2003).

High optical quality models for flow visualization and PIV measurement

Milos Kasparek[1*], and *Ludmila* Novakova[1]

[1]J.E. Purkyně University in Usti nad Labem, Institute of Machinery and Energetics, 40096 Usti nad Labem, Czech Republic

Abstract. Visualization of internal flow or PIV measurements in areas of real pipe elements is a method commonly used in flow studies. For these applications, transparent models with very complex internal geometries are required. Key features of the model are the high-quality optical access, which ideally allows to obtain image of the whole measured area without deformation and with noise suppression. The technology of 3d printing in combination with casting allows the production of almost any type of transparent models of piping elements. Models produced using this technique have almost ideal optical properties. By using index-matching fluid, optical distortions as well as reflections and refractions of light on the inner edges of the model are suppressed. The paper presents the results of visualization and 2D PIV measurements in the model of the globe valve with adjustable valve plug. The valve's inner geometry was rapid prototyped using a water-soluble material and casted with clear silicone Sylgard 184 under vacuum. The model exhibits excellent optical properties for PIV measurements, especially in the vicinity of walls.

1 Introduction

In fluid mechanics and in applications related to the flow of any transparent fluid, the PIV (Particle Image Velocimetry) method is often used. There are several preconditions for the successful application of the PIV method for internal flow of liquids [1]. It is necessary to use seeding particles that are able to track the fluid motion faithfully over a given range of velocities and accelerations. The illumination of the measuring plane must be sufficiently intensive and uniform. Optical access to the entire measured area is very important and the measuring area has to be captured by a camera with sufficient spatial resolution and light sensitivity. In summary, ideally a completely transparent model must be made without visible joints in the laser light path or camera's vision path. In addition to these requirements, we can add at least partial heat and chemical resistance and durability of the model material.

Currently, two methods of producing transparent models with complex geometry are commonly used [2]. The first is direct 3D printing of models from transparent materials, either by stereolithography or using PolyJet methods [3]. Stereolithography can achieve better surface quality but cannot print with a different material as support. PolyJet allows printing with a support material. The most commonly used materials for transparent models are Vero Clear resin and water-soluble support material. We have used the second technique, more labour-intensive, which brings the benefit of excellent optical quality of the models, affordability and the ability to produce any complex geometry [4,5]. The disadvantage of the presented technique is a slight inaccuracy in the final dimensions of the model due to the manual modification of the surface of the 3D print. For the purpose of visualization and measurement by PIV, a globe valve model was cast and completed with functional elements.

2 Manufacturing technique

2.1 Manufacture of the casting

The process of manufacturing of the presented transparent flow phantom of the globe valve is a combination of two technologies: casting and additive technology (3D printing). First, a mould of the valve's inner geometry is produced using 3D printing. This mould serves as the inner core during casting. Fused Filament Fabrication (FFF) 3D printing method was used with polyvinyl alcohol (PVA) as the filament. This material was chosen because of its water solubility. The printed surface of the inner core is not of sufficient quality and therefore the surface needs to be altered. Surface treatment is carried out by applying a solution of water and filament to smooth out surface irregularities. After surface preparation, the core of the valve is placed in the prepared container. This creates a mould for casting of the transparent phantom globe valve. Clear silicone resin is then poured into the mould and the mould is placed in a vacuum oven. After the clear silicone resin is solidified, the core is removed from the mould with water. The clear silicone resin used is a commercial variant of polydimethylsiloxane (PDMS)

* Corresponding author: milos.kasparek@ujep.cz

Sylgard 184. Sylgard 184 was used for its high-quality optical properties, in particular for its low refractive index (1.412 RI).

2.2 Functional components

The functional part of the globe valve model can be divided into five parts: the packing, the connecting shaft, the packing nut, the valve stem with the plug and hand wheel. The packing, the packing nut, connecting shaft and the hand wheel are made by 3D printing. The plug with the valve stem is made by a combination of 3D printing and casting technology - the stem is made by 3D printing, the plug by casting. As it subsequently turned out, the FFF 3D printing method is not suitable for the production of the functional parts of the valve, because liquid leakage between the printed layers. As a result, the functional parts were printed again, but using the stereolithography 3D printing method (SLA) to eliminate the leak problem. Thanks to the 3D printing quality of SLA, all functional parts can be printed in detail, including stem. To produce the plug with the valve stem, a PVA plug mould was printed using FFF 3D printing and its surface was modified. The printed stem was inserted into this mould. Silicone resin was poured into the mould and bubbles were removed from the resin using a vacuum oven. After the silicone resin was cured, the mould was dissolved using water. All the functional parts of the valve were assembled and inserted into the globe valve phantom. A schematic of the globe valve is shown in Figure 1.

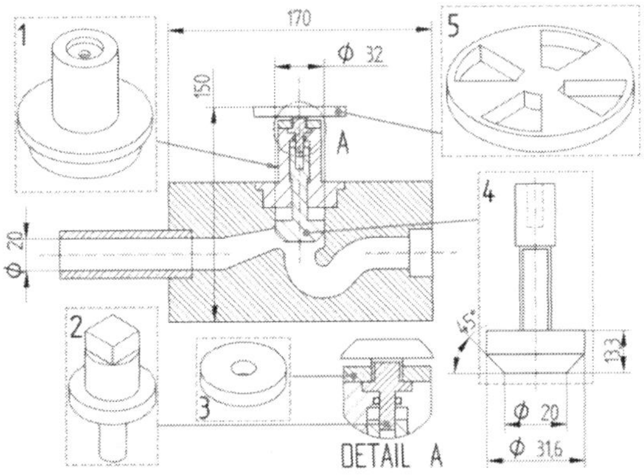

Fig. 1. Scheme of globe valve. The transparent body of the globe valve houses the functional parts of the model. The numbered positions represent the functional parts of the model: 1. Packing, 2. Connecting shaft, 3. Packing nut, 4. Stem with plug, 5. Hand wheel.

3 Experimental setup

Standard 2D PIV measurement and flow visualization were performed. The measurement plane for both PIV and visualization was placed in the axial plane of the globe valve model. For the measurement of velocity profiles by the PIV method, the fluid was seeded by glass hollow silver-coated particles of 10 μm, and a 532 nm camera filter was employed to minimize glare effects. For visualization, fluorescent particles (with fluorescent dye Rhodamine B:) of 50-100 micrometres were used and the signal was filtered with a 554 nm camera filter.

The aim of our measurements was to obtain the best quality data with the minimum optical distortion. For this reason, an index matching fluid was used [6,7]. It means that the liquid closely matches the refractive index of the model material. The magnitude of the optical deformation for water as a working fluid is shown in Fig. 2. The working fluid was a mixture of water and glycerine. The refractive index of Sylgard 184 is given by the manufacturer as 1.415 (Ri). This value can be influenced by curing time and temperature. The resulting optimum working fluid is mixed as a 40:60 weight ratio of water and glycerine. The resulting refractive index of the mixture is 1.4176 (Ri) [8]. The refractive index of the working fluid was examined refractometrically and visually. The refractive index measured by the refractometer was 1.412 (Ri). No optical distortion is evident from the visual inspection in Figure 3.

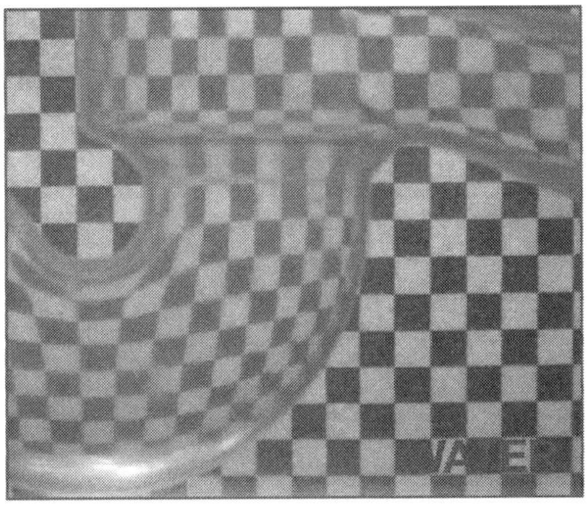

Fig. 2. Visual inspection of the optical deformation in the measured area using water (1.33 Ri) as the working fluid.

Fig. 3. Visual inspection of the optical deformation in the measured area using water-glycerine mixture (1.41 Ri) as the working fluid.

The optical properties of the model were tested by measuring of axial velocity profiles in a narrow section with laminar flow regime. The measurements were performed with both refractive matching fluid and water. This procedure allowed to quantify the effect of optical deformations when compared with the analytical result of the parabolic laminar profile. The results (Fig 4) show that the deformation of the optical profile near the wall is significantly suppressed in index matching fluid

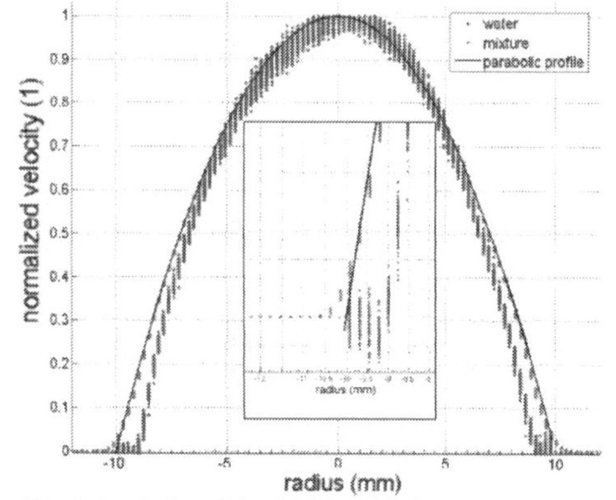

Fig. 4. Parabolic axial velocity profile in narrow section. The blue data represents data measured with water as working fluid, the red data with a mixture of water and glycerine. The black curve is obtained by analytical solution.

compared to water. The graph shows the result of the measurement of 30 velocity profiles along the narrow section. The velocity profiles of the index matching fluid are very close to the theoretical laminar profile. A significant deviation occurs at a distance of about 0.3 - 0.4 mm from the wall. At this distance, the deviation from the analytical solution is due to both optical deformation near the wall and particle-wall interaction. The results obtained with water show a more pronounced deviation from the theoretical profile. The typical "reduced" shape of velocity profile arises from optical deformation when the refractive index of the liquid is lower than the index of the model. The plot in Fig. 5 shows a histogram of the x-velocity uncertainties for both liquids. The histogram confirms the significantly smaller measurement uncertainties when using the water-glycerine solution as

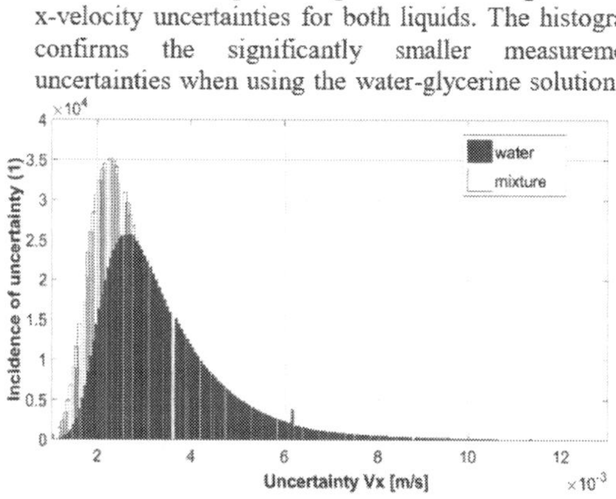

Fig. 5. Histogram of the x-velocity uncertainties. The blue columns represent the uncertainty distribution for water. Red columns represent the uncertainty distribution for index matching fluid.

the working fluid. The histogram for water shows the higher uncertainty values, which are probably associated with image distortion near the wall.

4 Results

The quality of the flow visualization in the axial plane of the globe valve model shows very good potential for flow studies. Visualization was performed again with both working fluids, water and a mixture of water and glycerine (Fig. 6 and 7).

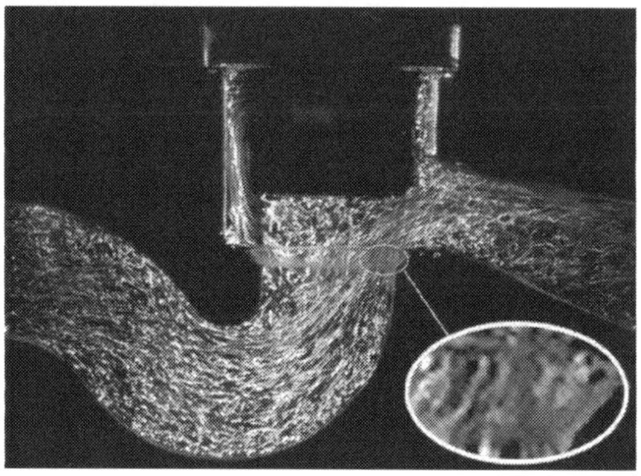

Fig. 6. Visualization in the axial plane of the model. Working fluid was the water (Ri 1.33).

The model showed remarkably good optical properties. The illumination of the laser sheet is homogeneous over the entire area under study. This indicates that the model material is optically homogeneous, with a minimum of bubbles, which are a common part of cast models. The detail in the images shows the reduction of particle visibility near the edge of the model. For water, this reduction in transparency is much more significant (Fig. 6). From the image of the mixture flow (Fig. 7), the flow can be very well analysed. From the flow pattern, regions of significant 3d flow as

Fig. 7. Visualization in the axial plane of the model. Working fluid was the mixture of water and glycerine (Ri 1.41).

well as vortices under the cone and in the outlet can be clearly identified. The visualization was performed on a model with an opaque plug. For the PIV measurements, this model was upgraded and the plug replaced with a transparent and tighter one. The new plug was made of Sylgard 184. PIV measurements confirmed the suitability of the model for optical methods. With a standard 2D PIV setup, it was possible to use adaptive correlation with an interrogated area of 8x8 pixels. For example, we obtained more than 80 values in the input velocity profile. Even at locations close to the edges of the model, the data were not significantly negatively affected. Despite the use of common non-fluorescent particles, no disturbing reflections were visible near the walls. Several regimes with different plug positions were measured. Fig. 8 shows the measurement results of two regimes. In Figure 8a, the detail of the leaking plug is clearly visible.

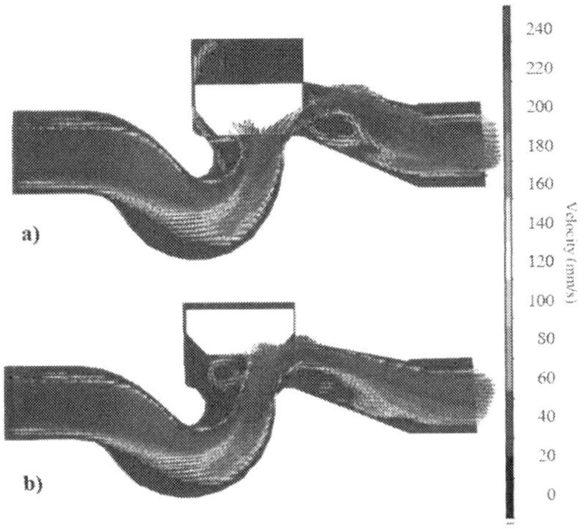

Fig. 8. Results of PIV measurements in the axial plane. Two regimes with the same flow rate but different plug position.

5 Discussion

The described method of manufacturing of transparent models ensures excellent optical access without significant optical distortion. The combination of 3d printing and casting enables to produce a cavity even with a very complex geometry. This technique is often used for medical applications, since data for 3D printing can be easily converted from medical imaging methods. As can be seen from our model, these castings can be supplemented with additional features that provide other functions of the model, such as the closing of the plug. These parts can be also made transparent to provide optical access to other parts of the model. We have verified that even in the case of a very complex geometry of the model, the optical access to the measured areas is very good. Our work has confirmed that 3D printing is an effective tool in producing models for PIV measurements, including functional parts of the models.

Acknowledgements

This work was supported by grant within student grant competition at UJEP - Jan Evangelista Purkyně University in Ústí nad Labem, project No.: UJEP-SGS-2020-48-003-2.
This work was supported by grant within student grant competition at UJEP - Jan Evangelista Purkyně University in Ústí nad Labem, project No.: UJEP-SGS-2021-48-003-2

References

1. Novotny, J., Machovska, I, Acta Polytech., **58**, (2018)
2. Yazdi, S., P. Geoghegan, P. Docherty, M. Jermy, A. Khanafer, Ann. Biomed. Eng. **46**(11), (2018)
3. W. H. Ho, I. J. Tshimanga, M. N. Ngoepe, M. C. Jermy & P. H. Geoghegan, Cardiovasc. Eng. Technol, **11** (2020)
4. Hopkins, L., J. Kelly, A. Wexler, and A. Prasad., Exp. Fluids, **29** (2000)
5. C. J. T. Spence, N. A. Buchmann, M. C. Jermy, S. M. Moore, Exp Fluids, **50** (2011)
6. Yousif, M.Y., Holdsworth, D.W., Poepping T.L. Exp Fluids, **50**, 2011
7. Miller P., Danielson K., Moody G., Slifka A., Drexler E., Hertyberg J., Exp Fluids, **41**, 2006.
8. Takamura K., Fisher H., Morrow N.R., J. Pet. Sci. Eng. **98–99** (2012)

Experimental investigation of homogeneous nucleation of water in various gases using an expansion chamber

Tetiana Lukianova[1], *Mykola* Lukianov[1*], and *Jan* Hrubý[1]

[1]Institute of Thermomechanics of the Czech Academy of Sciences, v. v. i. Dolejškova 5, CZ-182 00 Prague 8, Czech Republic

Abstract. Nucleation is an integral part of phase transitions and plays an important role in technology. The control of nucleation in devices such as turbines, rocket and jet engines, wind tunnels, and combustion processes is key to achieving efficient, ecologically sound operation. Our paper presents information about a modernized experimental setup used for homogeneous water nucleation research. The main goal of the study is an analysis of the influence of carrier gas on homogeneous nucleation. The experimental setup based on an expansion chamber has been used for research of homogeneous nucleation over several decades and has undergone various technical improvements. In the Institute of Thermomechanics, the setup was been completely disassembled and cleaned. Pressure transducers, laser, and the data acquisition system have been replaced with modern parts. After that, the chamber was assembled, and thoroughly tested and calibrated. Our investigations were carried with argon and nitrogen as carrier gases in a range of nucleation temperatures 220 – 260 K, pressure range 80-150 kPa, at several concentrations of water vapour. The results thus obtained are consistent with literature data. The findings suggest that this experimental approach is useful for homogeneous water nucleation research.

1 Introduction

The formation of the first microscopic fragments (small molecular clusters) of a new phase is called nucleation. In the absence of particles or walls, nucleation is homogeneous; otherwise, it is heterogeneous [7]. Nucleation is an integral part of phase transitions. Therefore, it is important for understanding many such common processes as the formation of rain, snow, crystals and smog, boiling of liquids. Playing an important role in technical processes, nucleation requires special control [1]. For example, in a nozzle flow, as well as in many other technological and environmental applications, droplet formation passes through nucleation, growth, and aging stages. The size distribution and polydispersity of the precipitate depend on the initial nucleation process. Therefore, considerable research efforts are undertaken to clarify the physicochemical properties of water. These are of interest for understanding nucleation and condensation of atmospherically relevant mixtures, e.g., in cloud formation [5]. Despite numerous studies, research of nucleation of water is an important kind of investigations to understand subtleties of the phase transitions processes. An analysis of the influence of carrier gas on homogeneous nucleation is a very important task of research in this direction. The paper presents a description of the experimental setup and the results of an experimental study of water homogeneous nucleation with argon and nitrogen as carrier gases. Our investigations were carried in a range of nucleation temperatures 220 – 260 K, initial pressure range 80 – 150 kPa, at several concentrations of water vapour.

* Corresponding author: lukianov@it.cas.cz

2 Experiment

Heist and He [1] assess the various experimental devices capable of quantitative nucleation rate measurements. The most significant review over the past five years was given by Wyslouzil and J. Wölk [8]. Authors described popular research methods, standard experimental techniques and conditions to make measurements during the nucleation process.

A "space to time" experimental method was described by Peng et al. [9]. This article presented tunable diode laser absorption spectroscopy (TDLAS) and light sheet technique which were integrated in the system to monitor the unsteady condensation process in a high speed expansion flow generated by the expansion system.

A novel design of a Pulse Expansion Wave Tube (PEWT) for the study of homogeneous nucleation in mixtures of vapors and gases was presented by Campagna et al. [10].

2.1 The nucleation pulse chamber

The nucleation pulse chamber (NPC) as one of the types of Expansion Cloud Chambers was developed by Wagner and Strey in 1984 [2]. Later it has been repeatedly used to investigate homogeneous nucleation of water [3-6] and has undergone various technical improvements [1-6]. In the Institute of Thermomechanics, the setup was been completely disassembled and cleaned. MKS Baratron 615A pressure transducers were installed for measuring the pressure of water vapor and the carrier gas during the preparation of the mixture. I/O Device PCI-6115 was installed as the data acquisition system. The laser and vacuum system were also replaced with modern parts. After assembling the setup, all the systems were checked and calibrated.

Figure 1 shows a schematic view of this experimental arrangement. Nucleation pulse chamber can be divided into three parts: nucleation reservoir (NR), recompression reservoir (RR) and expansion reservoir (ER). The mixture is prepared in the receptacle R.-Receptacle is evacuated to about $2\cdot10^{-5}$ mBar. Water vapor comes from the vaporizer V through the valves NV2 and V3 up to the required pressure. After recording the water pressure, the NV2 valve is closed and a carrier gas is filled up to a certain pressure through the NV1 valve. The pressure is recorded by pressure transducers P1 and P2. The mixture passes at a constant temperature. After preparing the mixture, the process of flushing the nucleation chamber begins. This is necessary to saturate the chamber walls by the mixture, so that in the experiment the mixture does not change due to the adsorption by the walls of the chamber. For expansion, Vrec and Vexp valves are closed and the required pressures are adjusted in the RR and ER nucleation chamber reservoirs using Pb1 and Pb2 volume pumps. In the NR reservoir, the prescribed pressure is maintained by the RV valve. After setting the pressures and finishing the flushing procedure, the expansion is performed. In the process of adiabatic expansion the Vexp valve opens, and after approximately 1 ms the Vrec valve opens also. This leads to pressure and temperature drop in the NR reservoir and the conditions under which homogeneous nucleation occurs are established. A following pressure increase (about 5%) leads to quenching the further nucleation of the droplets. The formed droplets are illuminated by a laser. The intensity of the transmitted laser beam is recorded by photodiode, and the intensity of the scattered light is registered with a photomultiplier. The scattered light is normalized to the transmitted light and we get series of peaks in good agreement with Lorenz-Mie theory[11]. These peaks allow us simultaneously determine drop radius and number concentration during the growth process.

To prepare for the next expansion, valves Vexp and Vrec are closed and valves V12-V16 are opened for flushing and setting of the required pressures.

In our experiments used gases from Messer manufacturer with purity for argon 99.999% and for nitrogen 99.996%. Water was cleaned in the laboratory to a purity level of 18.2 MΩ.

This experimental setup generally operate at low total pressures and the fluids formed by nucleation and growth are often sub-cooled with respect to the solid. Also, this setup does not imply retrieving particles, or characterizing them with respect to their composition or structure *in situ*. Taking this into account, comparison with the theory is sometimes difficult[8]. Nevertheless, the positive points described in the article[8] show the exceptional advantages of the complex of equipment we use.

The innovations introduced in the experimental setup allow obtaining more accurate data and made the work more convenient, as well as speeding up the experiment.

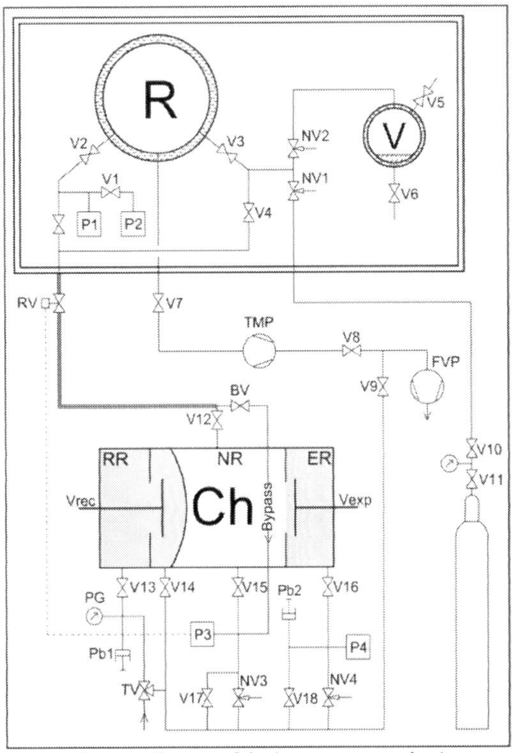

Fig.1. Schematic diagram of the homogenous nucleation setup: R – receptacle; V– vaporizer; TMP – turbomolecular pump; FVP – forevacuum pump; Ch – nucleation chamber; NR – nucleation reservoir; RR – recompression reservoir; ER – expansion reservoir; RV – electronically controlled regulation; Vx – valves; NVx – needle valves; BV - bypass valve; Vrec – recompression valve; Vexp – expansion valve; TV - three-way valve; PG – pressure gauge; Pbx – pump volume; Px – precision pressure transducers; dashed line – electric connection between regulating valve and pressure transducer in order to maintain the set pressure.

2.2 Data processing

The key parameter in quantifying the nucleation process is the nucleation rate J and supersaturation S. Nucleation rate is the number of droplets formed per unit of time and volume. For a vapor–gas mixture, supersaturation quantifies the current state deviation of the vapor

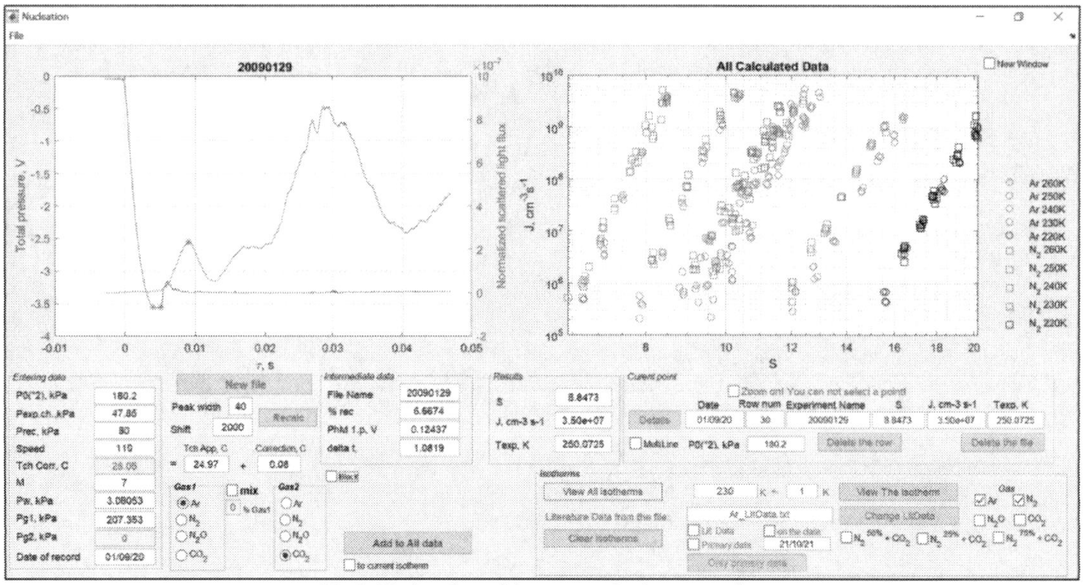

Fig. 2. The interface of the program for data processing

component from its corresponding (same *p* and *T*) phase equilibrium [10].

The classical nucleation theory (CNT) was first formulated by Volmer and Weber [12]. Later, numerous authors [13-19] improved the description, making it the most successful model for quantitatively predicting the phenomena of nucleation.

The temperature is calculated in accordance with Poisson's law

$$T_{EXP} = T_0\left(1 - \frac{\Delta P_{EXP}}{P_0}\right)^{\frac{K_m - 1}{K_m}}, \tag{1}$$

where T_0 and P_0 initial temperature and pressure. The ratio *Km* of the specific heats is obtained using the Richarz formula [20].

Supersaturation is calculated by the formula

$$S = \frac{\omega P_{EXP}}{P_{ve}(T)}, \tag{2}$$

where ω is the molar fraction of the water. P_{EXP} is expansion pressure and $P_{ve}(T)$ is saturation pressure of water at nucleation temperature.

Nucleation rate is evaluated as

$$J = \frac{C_{EXP}}{\Delta t_{EXP}}, \tag{3}$$

where C_{exp} is the number concentration of droplets is observed by light scattering, as described by Wagner [21], Δt_{exp} experimental duration of the nucleation pulse.

With the help of manual settings, we can get the parameters necessary for further calculations, such as the maximum of first peak of normalized scattered light, duration of the nucleation pulse and average pressure of nucleation pulse. This takes time and the accuracy of the data depends on the subjective opinion of the user. To automate this process, a special program was created in the MATLAB environment. The interface of this program shows on the Figure 2.

The experimental data is displayed in the form of a graph, which is subsequently analyzed. Figure 3 shows a normalized scattered light flux and pressure with dependence from time. The pressure signal shows expansion with duration a few milliseconds. A significant light scattering is observed only after the nucleation pulse has occurred. Nucleation rates can be calculated by a constant-angle Mie scattering method[21].

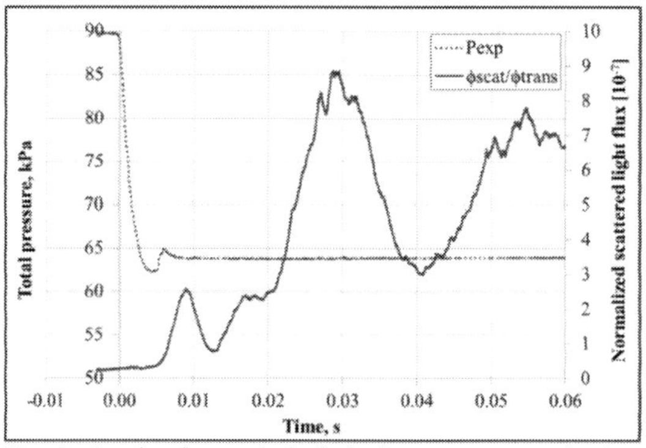

Fig. 3. Result of a typical single homogeneous nucleation experiment showing the total pressure and the light flux scattered at an angle of $\theta = 15°$ normalized with respect to the transmitted light flux as functions of time

3 Results

An experimental study was carried out at nucleation temperatures of 220 - 260 K and initial pressures of 80 - 150 kPa, with argon and nitrogen as carrier gases. Carrier gases were chosen in this way because they represent examples of monoatomic and diatomic gases. Results of the experiments are shown on Figures 4 and 5. In the figures, the lines show isotherms.

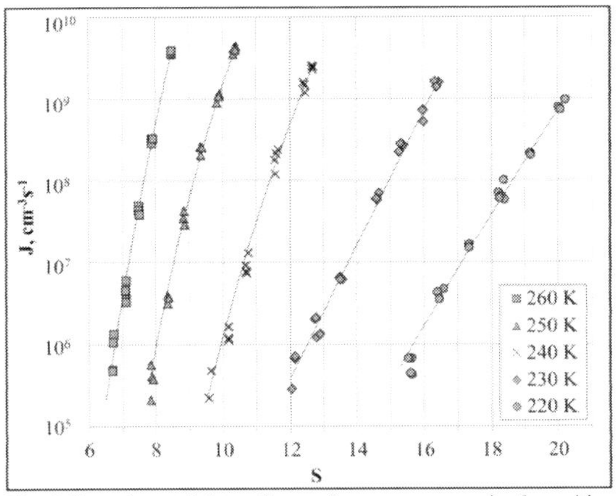

Fig. 4. Experimental data of water homogenous nucleation with argon as carrier gas.

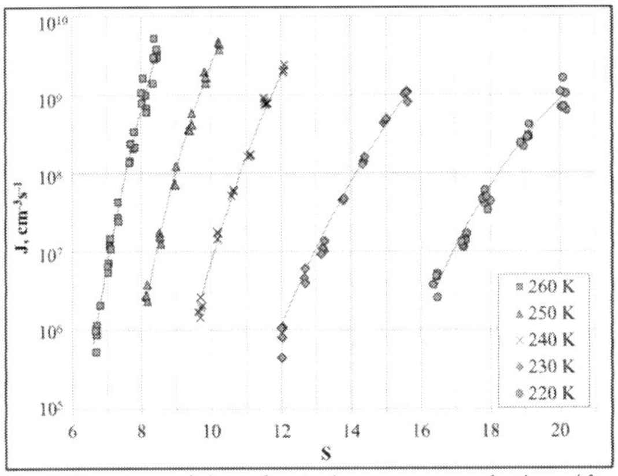

Fig. 5. Experimental data of water homogenous nucleation with nitrogen as carrier gas.

For a better visual comparison of the results, the experimental data were fitted by the equation

$$J = \exp\left(a + \frac{b}{S^c}\right), \quad (4)$$

where a, b, c are coefficients of equation.

Figure 6 shows a comparison of the experimental data fitted by the equation. As follows from Figure 6 on isotherms 220 – 250 K, the nucleation rate is higher at the same supersaturation with carrier gas nitrogen than with argon. This is due to the decrease in the surface tension of water because of adsorption effect of the carrier gas. As it was studied in[22] water surface tension reduction for nitrogen is grate than for argon.

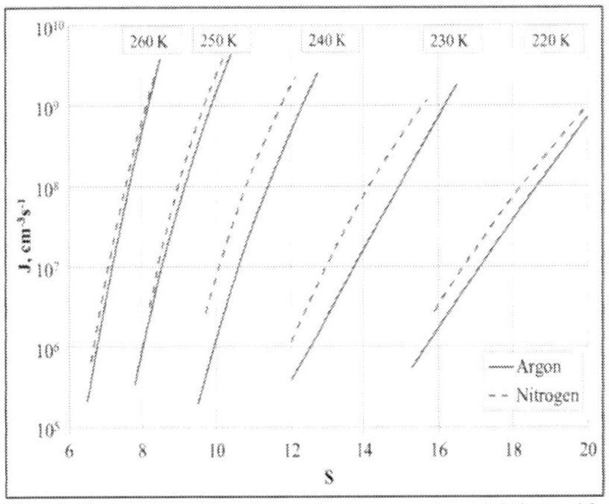

Fig. 6. Comparison of water nucleation experimental data with different carrier gases.

4 Conclusion

In the Institute of Thermomechanics, setup [2] has been completely disassembled and cleaned. Pressure transducers, laser, and the data acquisition system have been replaced with modern parts. After that, the chamber was assembled, and thoroughly tested and calibrated. To process the experimental data, a program was created in MATLAB environment, which greatly accelerated and facilitated the conduct of experiments and processing experimental data. The experiment conducted showed that the setup works properly and promising for further research. Investigation of the carrier gas influence on homogeneous water nucleation was also carried out. As a result, it was shown that due to the carrier gas adsorption effect, which leads to a decrease in the surface tension of the water [22], there is an increase in nucleation rate. This effect is more pronounced when using nitrogen as a carrier gas. Further studies of the influence of the carrier gas on the homogeneous water nucleation are planned.

Acknowledgments

We acknowledge a support from the Ministry of Education, Youth and Sports of the Czech Republic under OP RDE Grant No. CZ.02.1.01/0.0/0.0/16_019/0000753 "Research center for low carbon energy technologies" and institutional support RVO: 61388998

References

1. R.H. Heist and H. He, J. Phys. Chem. Ref. Data **23**, 781 (1994).
2. P.E. Wagner and R. Strey, J. Chem. Phys. **80**, 5266 (1984).
3. Y. Viisanen, R. Strey, and H. Reiss, J. Chem. Phys.

99, 4680 (1993).
4. R. Strey, P.E. Wagner, and Y. Viisanen, J. Phys. Chem. **98**, 7748 (1994).
5. J. Wölk and R. Strey, J. Phys. Chem. B **105**, 11683 (2001).
6. A. Fladerer and R. Strey, Atmos. Res. **65**, 161 (2003).
7. V. Holten, *Water Nucleation : Wave Tube Experiments and Theoretical Considerations* (2009).
8. B.E. Wyslouzil and J. Wölk, J. Chem. Phys. **145**, 211702 (2016).
9. D. Peng, Y. Jin, X.D. Fan, J.M. Yang, and C. Zhai, Rev. Sci. Instrum. **90**, (2019).
10. M.M. Campagna, M.E.H. van Dongen, and D.M.J. Smeulders, Exp. Fluids **61**, 108 (2020).
11. G. Mie, Ann. Phys. **330**, 377 (1908).
12. M. Volmer and A. Weber, Zeitschrift Für Phys. Chemie **119U**, 277 (1926).
13. L. Farkas, Z. Phys. Chem **125**, 236 (1927).
14. R. Becker and W. Döring, Ann. Phys **24**, 752 (1935).
15. J.B. Zeldovich, Kavitacia. Zh. Eksp. Teor. Fiz **12**, 525 (1942).
16. Y.B. Zeldovich, Acta Physicochem. USSR **18**, 1 (1943).
17. V.I. Kalikmanov and M.E.H. Van Dongen, EPL (Europhysics Lett. **21**, 645 (1993).
18. D. Kashchiev, J. Chem. Phys. **76**, 5098 (1981).
19. H. Reiss, W.K. Kegel, and J.L. Katz, Phys. Rev. Lett. **78**, 4506 (1997).
20. F. Richarz, Ann. Phys. **324**, 639 (1906).
21. P. Wagner, J. Colloid Interface Sci. **105**, 456 (1985).
22. M.M. Campagna, J. Hrubý, M.E.H. van Dongen, and D.M.J. Smeulders, J. Chem. Phys. **153**, 164303 (2020).

Optimization of the continuity of the aspherical production processes in asphericon s.r.o based on subsurface damage and microroughness analysis

Ondřej Matoušek[1, 2*], *Karolína* Sedláčková[1], and *Vladislav* Kuryshev[1, 3]

[1]asphericon s.r.o., 463 12 Jeřmanice, Czechia
[2]Technical university of Liberec, Institute of New Technologies and Applied Informatics, 461 17 Liberec, Czechia
[3]Technical university of Liberec, Dept. of Power Engineering Equipment, 461 17 Liberec, Czechia

Abstract In the paper is presented an analysis and optimization of the standardized sub-apertural grinding process used in the serial production in asphericon s.r.o. The monitored parameter was the depth of subsurface damage and surface microroughness. Tested were five grinding processes, which were automatically generated by the internal system, for five different diamond grit sizes (D151, D91, D64, D30, and D15). For evaluation of the depth of the defected layer was used modified wedge polishing method which is suitable for analysis of the rotationally symmetrical sub-apertural grinding processes [1]. For identifying the presence of the subsurface damage two methods were used. Defect detection using an optical microscope, as the broadly used and reliable method, and detection by standard ISO control to get the comparison with the method used in common serial production. The microroughness was measured using a white-light microscope concerning the used grinding tool and the amount of removed material. Within the experiment was found as the most effective two-step process uses D91 for rough grinding and D30 for fine grinding. D91 provides a very good removal characteristic with final subsurface damage of 44 µm which is possible to grind out using the D30 tool in two steps with final subsurface damage 22 µm in a total processing time of 137 minutes. This grinding process is timewise in best balance with 80 minutes long polishing process and therefore minimize the production cost. Result microroughness around 2 nm Sq in the fully polished zone is already limited by the polishing process. Using a finer grinding tool is not bringing improvement in the surface microroughness just shortening polishing time due to lower subsurface damage.

1 Introduction

Serial production of aspherical elements is becoming more widespread but compared to spherical optic is the productivity still way lower and due to that also price higher. In high volume production could even small optimization in the process significantly affect the final price. Extremely important is the proper continuity of every single processing step. These are in aspherical production much more interconnected compared to spherical production. One of the most important parameters to know for the proper planning of the whole production process is the level of the subsurface damage [2] after each machining step and the resulted surface microroughness [3] after applying the production chain.

The layer of the subsurface damage consists of the small cracks which are passing from the surface deeper into the material. In the optical industry is this layer generated mainly during the grinding process, when a glass is removed with a rigid diamond tool. Usually, multiple grinding steps with different diamond grit sizes, from rough to fine, are used in the production process. In the rough grinding step is removed a high amount of the material, but also high subsurface damage is caused. It leads to a very long polishing time. Therefore, the fine grinding step with constant removal is used to grind out the whole defected layer caused by the rough tool. After the fine grinding is the element not free of subsurface damage, but the depth of the damage is reduced which also reduces polishing time [2].

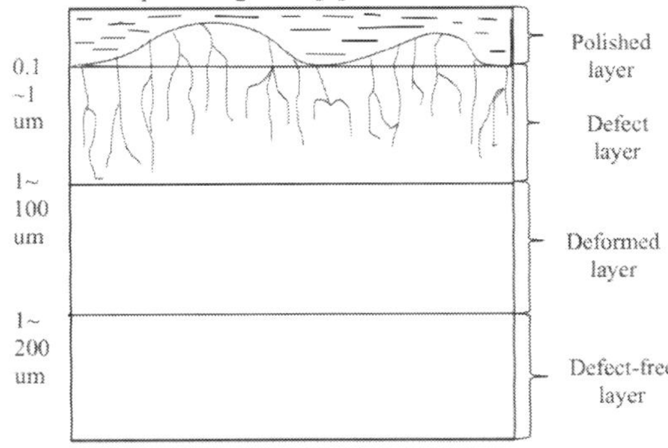

Fig. 1. Schematic view of the subsurface damage in brittal materials.

* Corresponding author: o.matousek@asphericon.cz

During the polishing process, the layer of the subsurface damage must be fully polished out. The necessary removal rate depends mainly on the depth of the defect layer, which must be removed. The deformed layer is in the case of the glass materials almost negligible. There is a lot of room for optimization of the processes. In general, the finer grinding tool is used, lower the subsurface damage is generated. But a fine diamond tool allows to remove only thinner layer of the material and the use a lower feed. It causes a longer processing time. In high volume serial production, as in asphericon s.r.o, is very important to balance polishing and grinding time to reach the most productive process. It requires setting up the grinding process time by choosing a proper tool with a certain diamond grit concerning the necessary time to polish out the defected layer.

The most often used method for subsurface damage identification is wedge polishing [4, 6]. Grinding is performed on a plano sample which is then polished under a certain angle. The angle is selected based on the expected deepness of the defects and the diameter of the sample. For optical materials such as glass, it is usually a few tenths of arc minutes. By the wedge polishing, three zones are formed on the surface. One zone is a ground surface that was not polished at all, the last zone is a fully polished surface. In between is the zone of subsurface damage, where the density of the defects is gradually decreasing from the ground zone to the fully polished zone (Figure 2). The depth position of the last defect is determined as the depth of the subsurface damage. The depth position or rather Z position can be measured using a tactile profilometer. The ground part of the sample is used as a reference and from it, the profile of the wedge, and from the lateral position of the last defect, the depth of subsurface damage is calculated.

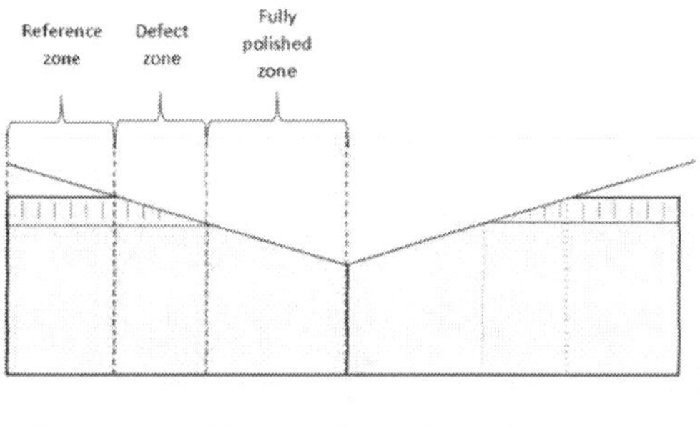

Fig. 2. V-shape polished surface with marked referenc (ground), defected, and fully polished zone.

This method can be modified in a way, to be more suitable for aspherical technology [1]. In this case, the wedge polishing is replaced by the polishing of a V-shaped hole into the surface using standard rotationally symmetrical aspherical polishing. As reference is left a ring of an unpolished surface at the edge of the sample. In this case, the ground and subsurface damage zones have the form of rings the fully polished zone is then the rest of the surface up to the centre of the sample. Because the profile measurement is done from edge to edge the reference part is on both sides. It makes it much easier and more precise to align the date and therefore also to calculate the depth of each defect compared to standard wedge polishing, where the reference is only on one side. The advantage is also that the form of the polished V-profile can be very easily driven and optimized by setting up the process parameters. The greatest benefit of using V-shape polishing is that for subsurface damage identification and the standard serial production is used the same polishing process. Therefore, the obtained values are in direct relation to the manufacturing process.

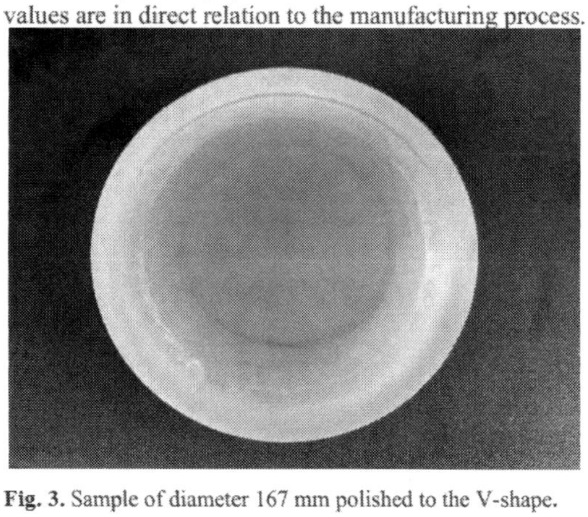

Fig. 3. Sample of diameter 167 mm polished to the V-shape.

The crucial step is the identification of the end of the subsurface damage zone. The most often used method is manual identification of the position of local defects at the edge of the zone using an optical microscope. This method is very time-consuming, but reliable. This method is usually used as a reference for newly developed methods and is used in the experiment as well. The second method used for defected zone localization is a control according to the ISO 14997 standard. In this procedure are used exactly defined illuminations lamps, magnification glasses, and comparison gauges to evaluate the final surface quality. It is worldwide the most common quality control (QC) method for optical elements and asphericon s.r.o uses it as well. An experienced controller can find even a single isolated defect of size close to one micrometer. Therefore, the method was chosen as a possible alternative for localization with a microscope.

But subsurface damage is not the only parameter taken into the account in the optimization. Important is also surface microroughness which is affected not only by the polishing process but also by the previous grinding steps. Too rough grinding increases polishing time not just by the necessity of removing the defected layer but also, because flexible tools are used for polishing, the rough structure is copied into the polished surface and takes a longer time to get rid of it. This effect negatively affects the surface microroughness and increases the total polishing time. Due to that is important to measure after polishing the surface microroughness on several radial positions concerning previous machining steps and the depth of the removed material. As standard for polished surfaces with microroughness deep under 10 nm Sq is used white-light interferometry [5]. The most common and reliable parameters used for 2D evaluation of the microroughness are Sa (1) and Sq (2) described below. Before its calculation, 4th order 2D polynomial is subtracted from the data to suppress the effect of the surface form.

Sa – average roughness:

$$S_a = \iint_a |Z(x,y)|\ dx\,dy \qquad (1)$$

Sq – Root mean square roughness

$$S_q = \sqrt{\iint_a (Z(x,y)^2)\ dx\,dy} \qquad (2)$$

2 Experiment

For the evaluation of the subsurface damage caused by the standard asphericon's sub-aperture grinding processes was chosen the most common glass Schott N-BK7. The diameter of the samples is 167 mm. The big diameter of the elements allows polishing gradual V-shape with a very low angle which helps to increase the precision of the method. The polishing process was held with standard tooling, polyurethane pad LP-35 from Universal Photonics, and cerium oxide-based polishing slurry AUERPOL PZ 110 with a mean particle size of 0.9 um. The parameters of the polishing process were at the beginning automatically generated by the internal technological system to ensure the relation with standard serial production. After that, the polishing times were linearly modified (shortened at the edge prolonged in the center) to get the required V-shape.

Table 1. Equipment used in the experiment.

Machining	Grinding	OptoTech ASM-100
	Polishing	OptoTech ASP-200b
Form measurement	Tactile form measurement	MarSurf LD 260 Y
	Non-contact form measurement	LuphoScan 260
Inspection	Ground surface roughness measurement	MarSurf LD 260 Y
	WLI for defects detection	MarSurf WM 100
	Optical microscope for defects detection	Zeiss Vario Axioscope

Asphericon s.r.o is mainly focused on serial production therefore all the processes are standardized and automatically generated based on parameters of the successful processes from the last 20 years. Input parameters for the creation of grinding processes are tool diameter, diamond grit size, type of glass, and diameter of the machined element. Based on this are generated tool speed, workpiece speed, feed, and removal in one step. Under the test were standard processes listed in the table below.

Table 2. Parameters of the grinding proceses for each tool.

Tool	Tool speed [u/min]	Lens rotation smooth grinding [mm/min]	Grinding track resolution [mm/point]	Z – removal in one step [mm]	Number of steps	Length of the one grinding step [minutes]
D151	-6131	7200	0,05	0,1	1	27
D91	-6131	7200	0,05	0,1	1	27
D64	-6131	7200	0,05	0,06	2	27
D30	-6131	4800	0,05	0,03	3	55
D15	-6131	4800	0,05	0,015	5	55

The used grinding tools from Lach Diamant have standard wheel construction of a diameter of 100 mm and a thickness of 10 mm. All the tools have the same metal bond and tested were five different diamond grit sizes (D151, D91, D64, D30, and D15). Dimond concentrations are C60 for D151 and D91, C100 for D64, and C50 for D30, and D15. Differences in the diamond grit size are visible in the below-presented microscope photos.

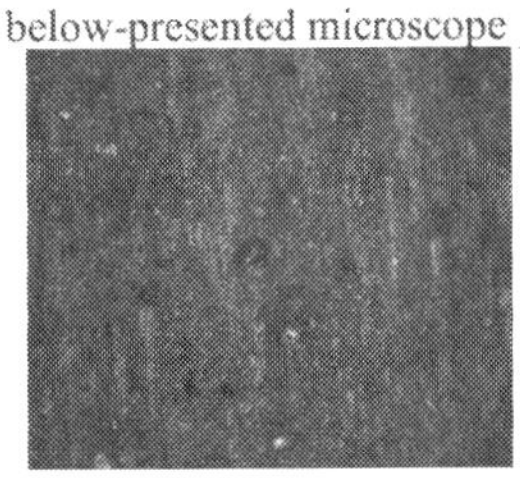

Fig. 4. Grinding tool with grit size D151.

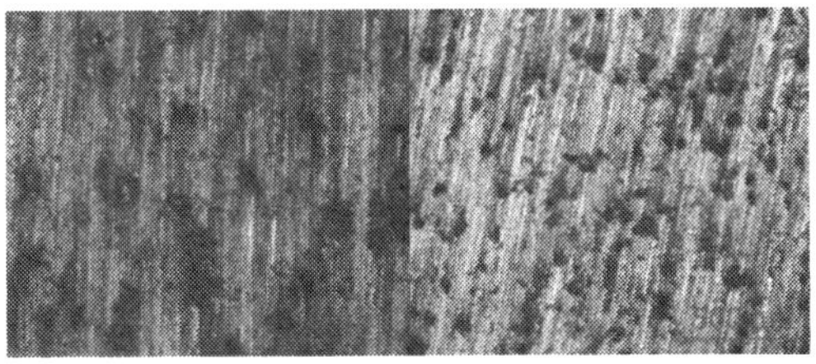

Fig. 5. Grinding tools with grit size D91 on the left and D64 on the right.

Fig. 6. Grinding tools with grit size D30 on the left and D15 on the right.

For polishing of the proper V-shape into the surface was at first generated standard polishing file for constant removal 30 µm all over the surface. The constant removal was then modified by doubling the polishing time in the center (position 0) of the lens and by shortening it by three at the edge (position 66) (Table 3 column "Position on the curvature"). In between the positions is polishing time linearly interpolated. To change the depth of the V-shaped hole for the different diamond grit sizes the polishing times in each position were simply multiplicated or divided, all by the same number (Table 3 column “Dwell time”). It of course changes the total polishing time (Table 3 column “Time”). All the other parameters were constant to reach the same polishing conditions and are listed in the table below.

Table 3. Parameter of the grinding processes for each tool.

Tool	Tool speed (outer/inner) [1/min]	Lens rotation polishing (outer/inner) [1/min]	Position on the curvature	Dwell time	Position of the tool	Time
D151	2/66	509/1499	0/33/66	60/36/12	5/27/49	2:31:06
D91	2/66	509/1499	0/33/66	20/12/4	5/27/49	0:50:22
D64	2/66	509/1499	0/33/66	18/10.8/3.6	5/27/49	0:45:20
D30	2/66	509/1499	0/28/66	16/9.5/3	5/27/49	0:37:06
D15	2/66	509/1499	0/28/66	9.5/4.5/1.5	5/27/49	0:18:33

Applying the grinding processes listed in the table above (Table 2) for grinding plan surfaces were reached PV form deviations under 5 µm on all test surfaces. It makes a good starting point for V-shape polishing because even without subtracting the form of the ground surface from the form of the polished one can be reached good precision. Despite that fact, the effect of the form deviation after grinding will be considered.

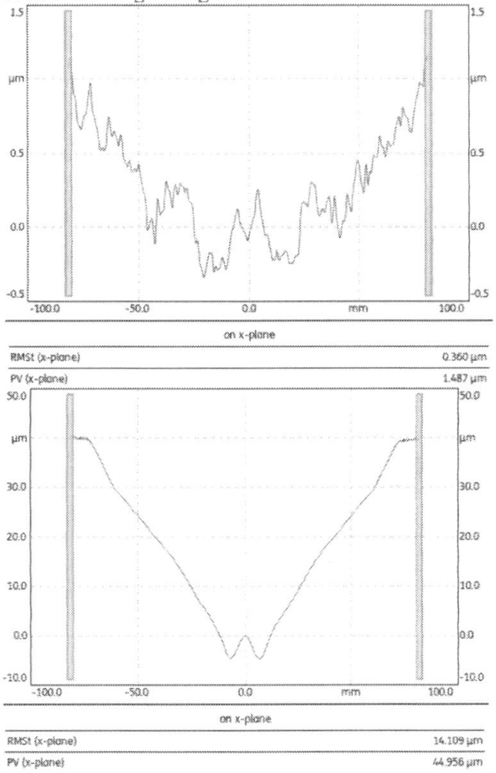

Fig. 7. Surface form deviation after grinding with the D30 tool on the top. On the bottom the same sample after the V-shape polishing.

Polishing with the parameters in the Table 3 led to a V-shaped form deviation on all the test elements. As can be seen in the measured data (Figure 7) in the centre of the elements is a small hill. It is not an issue because the fully polished surface was reached before the hill is rising on the surface. Therefore, the analysis of the subsurface damage is not affected by that. Only in microroughness measurement must be taken as a starting point the lowest point on the surface, not the center of the element. Polished was diameter 140 mm to preserve unpolished reference edges. The depth of the polished V-profiles including subtraction of the form of the ground surface is listed in the table below.

Table 4. Surface form parameters of each smaple after grinding and plolishing.

Tool grit size	PV after grinding [µm]	Depth of the V profile [µm]
D151	3.9	141.2
D91	5.3	53.3
D64	3.9	75.3
D30	1.5	45.0
D15	2.9	41.8

3 Results

After successful polishing it was possible to detect the edge of the subsurface damage using both methods, microscopic detection, and standardized ISO control. In the pictures below is presented the development of the surface quality depending on the amount of the material removed by the polishing process. With the increase of the depth are defects evolving from high density cracks through local defects with long grooves behind to the fully polished surface. The long grooves behind the defects are very typical for rotationally symmetric sub-apertural polishing processes and are given by the direction of the material removal. The major material removal direction is from the crack in the direction of the

Fig. 8. Microscopic view in depths 25 µm, 50 µm and 75 µm respectively of the V-shape polished surface grinded with the D151 tool.

Fig. 9. Microscopic view in depths 12 µm, 24 µm and 36 µm respectively of the V-shape polished surface grinded with the D91 tool.

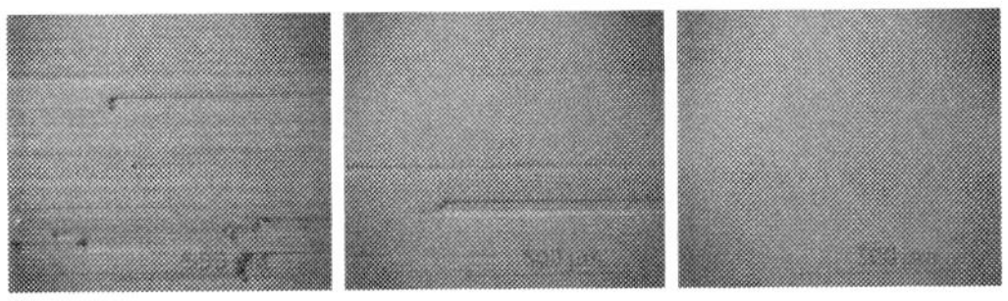

Fig. 10. Microscopic view in depths 10 µm, 20 µm and 30 µm respectively of the V-shape polished surface grinded with the D64 tool.

Fig. 11. Microscopic view in depths 7.5 µm, 15 µm and 22.5 µm respectively of the V-shape polished surface grinded with the D30 tool.

Fig. 12. Microscopic view in depths 4 µm, 8 µm and 12 µm respectively of the V-shape polished surface grinded with the D15 tool.

The depth of the subsurface damage is decreasing with decreasing diamond grit size, as it should be. It is an expected trend because the D number approximately defines the maximum size of the diamond particle in the bond. The ratio between the depth of the subsurface damage and the D number is 51 to 80 percent with an increasing trend from rougher to finer grit size (Table 5). Such dependence is typical for all grinding and even polishing processes of brittle materials using diamond particles. Using finer particles has a generally positive effect on subsurface damage or microroughness but the improvement is limited by any non-zero value. Therefore, is the ratio between quality and particle size more unfavourable for smaller particles.

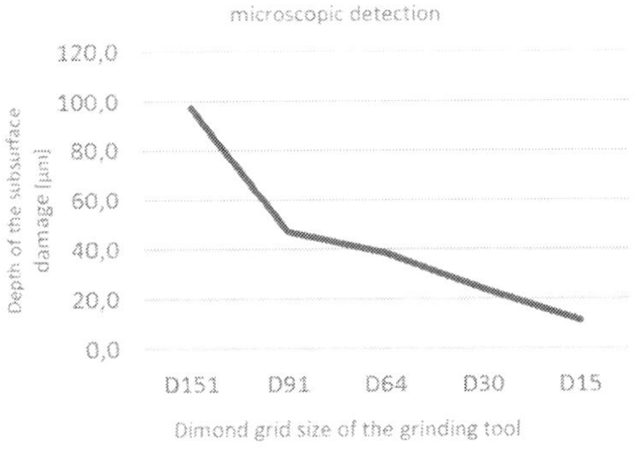

Fig. 13. Depth of the subsurface damage inspected by the referenc microscope based method.

As shown in the graph above and listed in the table below (Table 5), the detected subsurface damage is in the range from 97.4 µm for D151 to 11.2 µm for D15. A very positive result is that the ISO detection by the QC gives very similar results to detection by the microscope. Different values are only for D91 where the ISO control revealed the depth of the subsurface damage smaller by 0.8 µm and for D30 where is the difference 2 µm again smaller for ISO control. It gives us a good picture of that how very precise is experienced quality control. Important is to note that the detection using a microscope is at least ten times more time-consuming.

Table 5. Depths of the subsurface damege inspected by microscope and ISO controle and ration between grit size and subsurface damege.

Tool grit size	Depth of subsurface damage – microscope detection [µm]	Depth of subsurface damage – ISO detection [µm]	Ration D# / Depth of subsurface damage [%]
D151	97.4	97.4	64.5
D91	47.0	46.2	51.6
D64	38.5	38.5	60.2
D30	24.0	22.0	80.0
D15	11.2	11.2	74.7

From the grinding parameters, times and resulting subsurface damage was found the most optimal combination of the processes. The rough grinding using D91 and D30 for fine grinding.

Using D91 for rough grinding gives the best performance to quality ratio. Against the D151 allows removing only a thinner layer of the material. But maximal possible removal of 0.6 mm is suitable for most of the aspheres, which usually have smaller departures from the best-fit sphere. Therefore, can be ground in one step even with D91 with less than half the depth of the subsurface damage. Using D64 does not have many benefits against D91 mainly because is usually necessary to grind the surface in two steps and the benefit of lower subsurface damage does not sufficiently compensate the doubling of the processing time. But in specific cases can be used D64 as the only grinding tool (one level grinding). It is helpful for steep lenses with big diameters. In these cases, sometimes a second tool cannot be used due to the space in the machining chamber. The disadvantage of the one tool solution is worse final microroughness (Table 7). There are also processing issues such as high material removal with one grinding tool of intermediate grit size and long polishing time. Both those aspects cause high tool wear which leads to the lower long-term stability of the processes. Also, in general, is unwanted to have a longer polishing process than the grinding process, because there is no room for optimizing polishing, which is often necessary, without effect to total process time.

The fine tool with a grit size of D30 is the most suitable for serial production because allows reaching a better balance between the grinding and polishing times. The grinding time using the D15 tool is double compared to the D30 if all subsurface damage after rough grinding with D91 must be removed. With the D15 at least three but in real four cycles are needed, because removal 45 µm is too close to subsurface damage 44 µm, compared to two cycles with D30. The polishing time after the D15 is less than half compared to D30 but in high amount production, it only leads to longer standing times of the polishing machine because the total grinding time D91 + D15 247 minutes is at least four times longer compared to polishing. On the other hand, the benefit of the very low subsurface damage reached with the D15 tool can be used in special single-piece production or on very soft and chemically sensitive material in combination with unfavourable concave curvature which typically leads to long polishing times. Typical use is for grinding crystalline materials for example calcium fluoride where is the minimization of the subsurface damage critical for successful polishing.

Table 6. Possible production processes and the optimal process marked in bold.

Tool combination	Grinding time [minutes]	Polishing time [minutes]
D151 + D30	247	80
D151 + D15	412	40
D91 + D30	**137**	**80**
D91 + D15	247	40
D64 + D30	164	80
D64 + D15	219	40
D91	27	180
D64	54	140

Microroughness reached in the fully polished zone was relatively similar for all tested diamond grit sizes respectively only slightly lower for the two-fine tools D30 and D15. For the tool with grit sizes D151, D91 and D64 are final microroughness around 3 nm Sq and for the fine tools, D30 and D15 are microroughness under 2.5 nm Sq (Table 7, values are calculated as an average from three measurements in the fully polished zone). It means that the polishing process can remove quite well the grooves and micro defects and the surface microroughness is given mainly by the polishing process. But the benefit of the use of the fine grit size tools is not negligible and has a positive effect on the polishing time and on the final microroughness which is approximately 20% better too.

Table 7. Microroughness Sq in the fully polished zone.

Tool grit size	Sq microroughness in the fully polished zone [nm]
D151	2.9
D91	2.8
D64	3.1
D30	2.0
D15	2.4

In the measurements, below is showed a comparison of the not fully polished zones after tools D151 and D30 which were captured by the white-light interferometer. A different characteristic of the structure is clearly visible. The tool with grit size D151 creates much bigger local defects and due to that also polishing process produces much wider and deeper grooves on the surface behind them (Figure 14 - left) compared to the tool with the grit size D30 (Figure 14 - right).

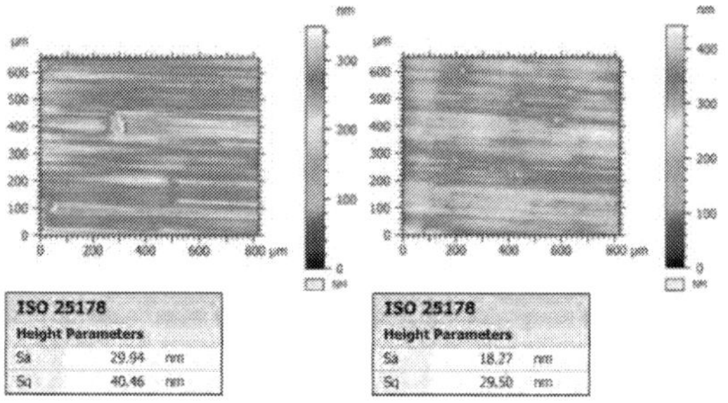

Fig. 14. Comparsion of the partly polished zones of the sample grinded with D151 (left) and D30 (right) tool measured with white light interferometer. Lateral resolution of both mesurement is 1.22 – 780 mm^{-1}.

The wider and deeper grooves tend to be copied by the flexible polishing tool for a longer time and therefore negatively affect the final microroughness. It could be well visible in the comparison below (Figure 15). Both pictures are from the fully polished zones. The left one is after grinding with D151 and the right one after D30. The surface structure after D30 is obviously finer compared to the structure after D151. As was already mentioned that the surface microroughness after fine tools D15 and D30 is approximately 20 % lower than after rough tools D151, D91, and D64. The higher value is coming from the rougher structure at the beginning of the polishing process and its copying by the flexible tool. It is different from the spherical or plano polishing where are used rigid full surface tools which do not tend to follow the waves on the surface. In the full-surface polishing is important to remove the whole defected layer and microroughness is then defined only by the polishing process and material properties.

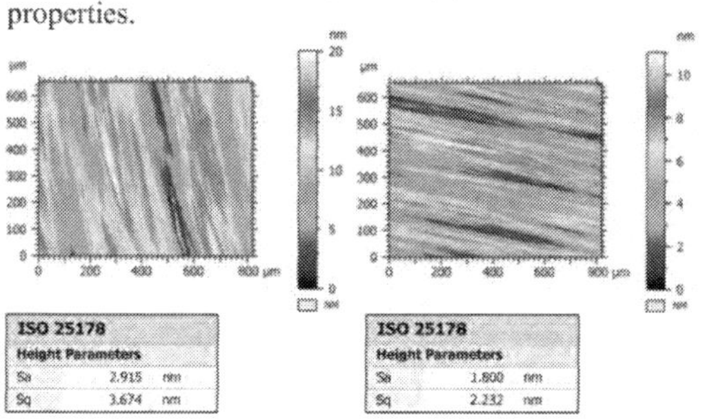

Fig. 15. Comparsion of the fully polished zones of the sample grinded with D151 (left) and D30 (right) tool measured with whitelight interferometer. Lateral resolution of both mesurement is 1.22 – 780 mm^{-1}.

4 Conclusions

In the paper was analysed five standard grinding processes used in asphericon s.r.o and their combinations to find the optimal production process for aspherical surfaces concerning the production time but also the required high surface quality. For analysis of the subsurface damage was used V-polishing method using the standard internal sub-apertural polishing process. It allows the analysis of the subsurface damage and microroughness related directly to the real production processes. For identification of the subsurface damage were used two methods. Localization by the microscope as a reference method and by ISO quality control. For microroughness measurement was used white-light interferometer with lateral resolution 1.22 – 780 mm-1.

In the experimental part was grinded five samples of a diameter of 167 mm with tools with 5 different grit sizes D151, D91, D64, D30, and D15. The resulted subsurface damage was identified very similarly with both mentioned methods with a maximal difference 2 μm which is less than 10%. It is a positive result of this comparison because it shows that the ISO control is precise enough to identify subsurface damage and way faster compared to the microscopic method. The measured depth of the subsurface damage is decreasing with the grit size of the tool from 97.4 μm for D151 to 11.2 μm for D15 (Figure 13). Based on the subsurface damage, removal rate, grinding time, polishing time, and final surface microroughness was chosen the most effective combination of the tools and processes. For rough grinding is the most suitable D91 grit size because allows removal of the required amount of the material in one step, which is beneficial compared to D64, and produces only 44 μm subsurface damage which is less than half compared to D151. For fine grinding, in combination with rough grinding using D91, was evaluated as the best the tool with a grit size of D30. The resulted subsurface damage is 24 μm. The removal of 0.03 μm allows doing only two steps of fine grinding which is the biggest advantage against the D15 tool which needs four cycles on the surface pre-grinded with D91. Even if the use of the D15 tool shortens the polishing time from 80 to 40 minutes is grinding with D30 more effective due to grinding the times of 137 minutes compared to 247 minutes. Also, there is no benefit of better surface microroughness after grinding with the D15 tool because the values show no significant differences. But the use of D15 has a reason for very specific materials, for example, soft crystals such as CaF2, for which machining are used very special and long polishing processes and therefore a limitation of the subsurface damage to the minimum is beneficial. An interesting alternative for D91 and D30 can be only one level grinding with D64. The overall processing time is very favourable. The two-step grinding process takes 54 minutes and the polishing time is 140 minutes (Table 6). The disadvantage of the process compared to two-level grinding is lower stability in grinding and polishing as well. In the grinding is problematic that the high amount of the material is removed by the tool with just intermediate grit size and in polishing very long processing time. It leads in both cases to high tool wear. Also as presented above (Table 7) the final microroughness of the polished surface is already slightly worsened by skipping the fine grinding step. But using only one grinding tool can be very helpful for big lenses with high curvature which often do not allow to have two tools in the processing chamber due to the limited space.

Acknowledgements

The research has been supported by the Ministry of Industry and Trade, project The Country for the Future, FX01030046.

References

1. D. Tomka, F. Procháska, O. Matoušek, J. Polák, "Development of methodology for evaluation of subsurface damage," Proc. SPIE 9442, Optics and Measurement Conference 2014, 94421B (7 January 2015); https://doi.org/10.1117/12.2176024
2. SHEN, Jian, Shouhua LIU, Kui YI, Hongbo HE, Jianda SHAO a Zhengxiu FAN „Subsurface damage in optical substrates“. Optik - International Journal for Light and Electron Optics. 2005, vol. 116, issue 6, s. 288 294. DOI: 10.1016/j.ijleo.2005.02.002.
3. GROSS, Herbert. Handbook of optical systems. Weinheim: Wiley-VCH, 2005. ISBN 3-527-40378-7.
4. Yaguo Li, Hao Huang, Ruiqing Xie, Haibo Li, Yan Deng, Xianhua Chen, Jian Wang, Qiao Xu, Wei Yang, and Yinbiao Guo, "A method for evaluating subsurface damage in optical glass," Opt. Express 18, 17180-17186 (2010)
5. MALACARA, Daniel a Brian J. THOMPSON. Fundamentals and basic optical Instruments ; Advanced optical instruments and techniques. Second edition. Boca Raton: CRC Press, 2017. ISBN 9781498720670.
6. Yu, Deping & Wong, Yoke & Hong, Geok-Soon. (2011). A novel method for determination of the subsurface damage depth in diamond turning of brittle materials. International Journal of Machine Tools & Manufacture - INT J MACH TOOL MANUF. 51. 918-927. 10.1016/j.ijmachtools.2011.08.007.

Nebulizer particle size distribution measured by various methods

Ondrej Misik[1*], *Andrea* Peštálová[2], *Miloslav* Belka[1] and *Frantisek* Lizal[1]

[1]Faculty of Mechanical Engineering, Brno University of Technology, Technicka 2896/2, 616 69 Brno
[2]Faculty of Pharmacy, Masaryk University, Palackeho trida 1946/1, 612 00 Brno

Abstract. Pari LC Sprint Star nebulizer is an inhaler device dedicated to delivering the liquid drug formulation into peripheral lung regions. The particle size is one of the critical parameters which determine the location of aerosol deposition within the human lungs. In this study, the particle size distribution of aerosol was measured by two different methods: Andersen Cascade Impactor (ACI) and Aerodynamic Particle Sizer (APS). Mass Median of Aerodynamic Distribution (MMAD), and Geometric Standard Deviation (GSD) were evaluated. MMAD gained by APS was approx. 7 μm, MMAD acquired by ACI was approx. 2.5 μm. According to the results, these two measuring methods are not very consistent, and the comparison of the particle size acquired by the different methods is not appropriate.

1. Introduction

Inhalation treatment became usable for an incredible range of diseases in the last decade. Most of the inhalers are dedicated to asthma or COPD (Chronic obstructive pulmonary disease), but there are also applications for the therapy of various other diseases like diabetes, cystic fibrosis, or bacterial lung infections [1–3]. Nebulizers are very attractive devices for inhalation drug delivery. In contrast with pMDIs (pressurized Metered Dose Inhalers) or DPIs (Dry Powder Inhalers), with nebulizers patients do not need to make a deep forced inhalation during the usage. The therapeutic aerosol from the nebulizers is inhaled during normal tidal breathing. This makes it much more comfortable and easier for patients and usable for weaker subjects like old patients or patients in a late stage of the disease.

There are four main categories of nebulizers: air-jet nebulizers, vibrating mesh nebulizers, ultrasonic nebulizers, and soft mist inhalers. Air-jet (AJN) nebulizers are probably the most common. The nebulization process works on the principle of the Venturi effect. Airstream is forced through a small jet placed in a tiny tube, therefore there is a small space between the outer wall of the jet and the inner wall of the tube. The bottom side of the tube is submerged in the liquid drug solution. When the air stream is passing through the jet orifice, it creates a vacuum in the space between jet and tube which sucks the liquid solution into the airstream. The solution is subsequently atomized into the aerosol due to shear forces caused by the airstream [4]. A baffle placed in front of the orifice makes the largest aerosol particles stay in the device and returns them back for the next nebulization [4]. Such devices are used very frequently, although, they need a compressed air supply and electricity connection.

The second group of nebulizers is vibrating mesh nebulizers (VMN). Such devices contain a small plate with a large number of small orifices (approx. 1 μm). This mesh vibrates in high frequencies. In contact with a liquid drug solution, small droplets of liquid are torn off by mesh orifices and emitted out of the device in form of an aerosol [4, 5]. Such aerosol particles have usually a narrow particle size distribution and particle size is affected by orifices size. This enables the determination of the aerosol particle size distribution (PSD) by manufacturing, which can be very helpful [5]. VMNs are usually quite small devices, which can be supplied by batteries as well. However, the biggest disadvantage of VMNs is probably their price. They are usually much more expensive in comparison with other nebulizer types.

The third category of nebulizers is ultrasonic nebulizers (USN). A piezoelectric element is placed at the bottom of the reservoir. This makes high-frequency vibrations and waves on the liquid level, which leads to liquid atomization [5]. However, these vibrations warm up the solution and for this reason, such nebulizers cannot be used for thermolabile drug solutions [4].

Soft Mist Inhalers (SMI) represent the last category. These devices are a combination of pressurized metered dose inhalers (pMDI) and nebulizers. Respimat SMI is the most popular and frequent member of this group. The liquid drug is stored in a reservoir in Respimat. Before the actuation, a dose of liquid is filled into the metering chamber. During the actuation, the drug solution is forced

* Corresponding author: Ondrej.Misik@vutbr.cz

through the twin nozzle, which makes a collision of two liquid streams, which leads to atomization [6]. Such aerosol is much slower than in the case of common pMDA and the time of atomization period is longer, which makes such a device suitable for patients who are not able to follow the proper inhalation technique as well [6].

The efficacy of drug delivery is affected by several factors. Particle size distribution and breathing pattern are probably the most significant. There are some commonly used abbreviations for particle size distribution and local lung deposition in pharmaceutical literature. Orally inhaled products are usually designed to have MMAD (Mass median aerodynamic diameter) under 5 μm. It is believed, that particles larger than 6 μm deposit preferentially in the oropharyngeal region, particles between 2 – 6 μm deposit in the central airways, and particles under 2 μm in the peripheral airways and alveoli[7]. Although, this is not completely true (explained in sources [8, 9][1]), the critical influence of particle size on lung deposition is obvious. Therefore, this study discusses the suitability and comparability of various methods for the measurement of aerosol PSD.

Lung deposition of aerosol particles is caused by the superposition of five deposition mechanisms: inertial impaction, interception, sedimentation, Brownian diffusion, and electrostatic precipitation. Each mechanism dominates in different conditions. Inertial impaction is significant for larger particles and occurs in the case of particles larger than 1 μm. Its rate grows with the particle mass and its velocity, in other words, it grows with Stokes number [9]. Interception is usually caused by the specific shape of particles and occurs in the case of prolonged particles like fibers [9]. The effect of sedimentation is significant only in the case of very low air velocities [9]. Brownian diffusion begins to be a significant cause of aerosol deposition for particles smaller than 0.3 μm when their movement starts to be affected by the collisions with ambient gas particles [9]. Electrostatic precipitation is caused by electrostatic charges of aerosol particles and walls [9]. According to the abovementioned, pharmaceutical aerosols are usually designed to have PSD between 2 – 6 μm, which is the reason why inertial impaction is considered to be the dominant deposition mechanism for pharmaceutical aerosols.

Methods for PSD measurements of inhaler aerosols are normalized by Ph. Eur. [10, 11]. Since inertial impaction is the main deposition mechanism for them, these methods separate the particle size fractions according to their inertial behavior. Ph. Eur. describes four such devices: Glass Impinger (GI), Multi-stage liquid impinger (MLI), Andersen Cascade Impactor (ACI), and Next Generation Impactor (NGI). These devices have one big advantage. They enable an assessment of an Aerodynamic Particle Size Distribution (APSD) directly for active pharmaceutical ingredients (API) since the measurements are evaluated by chemical analyses like HPLC (High-Performance Liquid Chromatography). However, such measurements are very time-consuming, and their uncertainty is very high. The list of variabilities and uncertainties of measurement on ACI is discussed in [12]. In the case of nebulizer measurements, an obstacle with the evaporation of droplets was found out, which should be solved by impactor cooling [13, 14]. Several studies showed significantly large particles while measured on a cooled impactor than at room temperature [13, 14].

However, there are some other methods for APSD measurements, which evaluate the sizes according to the particle aerodynamic and inertial behavior. These are time-of-flight measuring devices like an Aerodynamic Particle Sizer (APS, TSI Inc., USA) or previously Aerosizer (TSI Inc., USA). Such devices measure the same type of particle size as the impactors, the aerodynamic diameter. Moreover, this technique can be more precise and significantly less time-consuming than normalized methods, and even when it does not bring information directly about API, it can be suitable for some other applications. But are these devices comparable with ACI measurements and are they suitable for inhaler aerosol testing?

This study discusses the comparison of measurements of aerosol produced by nebulizers performed by different measuring methods.

2. Methods

Air-Jet nebulizer Pari LC Sprint Star was operating with a nebulization flowrate of 4.5 LPM. The atomized solution was distilled water with 100 μg/ml concentration (sodium salt) to enable the evaluation of ACI measurement. The vacuum pump Busch R5 PA0008C, control valve, and flowmeter TSI 4040 (TSI Inc., Minnesota, USA)

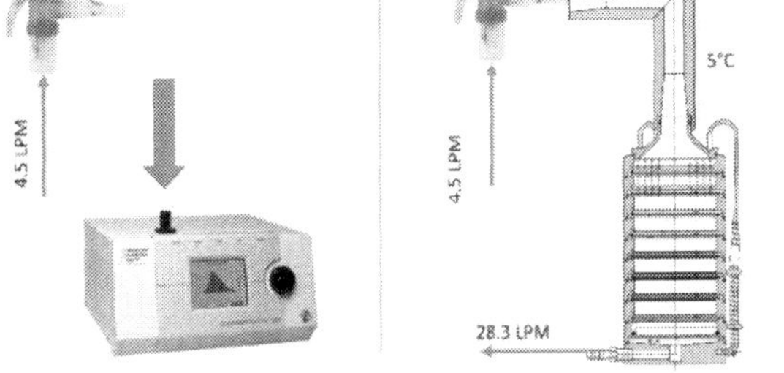

Fig. 1. Schema of the measurement rig: APS measurement (left), ACI measurement (right).

[1] Extra fine particles with MMAD < 0.3 μm use to deposit well in all regions of airways and lung deposition of particles > 6 μm can be effective as well, while it depends intensively on breathing pattern [8, 9].

Before the APS measurement, three samples of ambient air aerosol were measured. These were subtracted from the samples of nebulized aerosol. Aerosol was nebulized into the ambient environment above the APS sampler. 10 samples of aerosol were measured in this way.

Subsequently, aerosol from the same device was measured on ACI, as well. As is mentioned above, in this case, the nebulized liquid was dyed with fluorescein. ACI has been cooled in the refrigerator for at least 90 minutes at a temperature of approx. 5° C as is recommended in Ph. Eur. [11]. Dosing of aerosol into the impactor was performed within 5 minutes after taking ACI out of the refrigerator. ACI contained the kit for the flowrate 28.3 LPM which is the lowest flowrate ACI can be operated with. The nebulizer Pari LC Sprint Star was operating with a nebulization flowrate of 4.5 LPM and was connected to the ACI induction port through the modeled adapter. Before the measurements, the flow rates on the ACI inlet and outlet were checked to be the same and it was verified the ACI is leakage-free.

Dyed aerosol was dosed into the ACI continually for the time of 1 minute. Subsequently, ACI was disconnected and dismantled. Each segment of ACI was washed properly in a certain volume of distilled water to extract the deposited aerosol.

For analyses of deposited aerosol mass, a fluorimeter Quamtus™ Fluorometer (Promega Corp., USA) was employed. The calibration for the dye and solvent was performed. Concentrations of dye within the samples were determined and the mass of dye deposited in each segment was calculated.

Normalized mass aerodynamic particle size distribution was calculated from the obtained data. The deposited mass fraction was calculated as deposited mass at a certain impactor stage divided by the total deposited mass. Such results were also expressed by cumulative APSD. MMAD was gained by linear interpolation from the cumulative APSD and GSD was calculated by equation 1. according to the source [15]. The MMAD and GSD of APS results were evaluated similarly.

$$GSD = \frac{MMAD}{D_{16}} \quad [\mu m] \tag{1}$$

In ACI measurements it is appropriate to evaluate the Emitted Dose (ED) and Fine Particle Fraction (FPF; a fraction of the particles smaller than 5 μm as a percentage of emitted dose) as well. However, in the case of a nebulizer, it would be problematic to evaluate ED due to the principle of its operation. This is the reason why this study compares only APSD data, like MMAD and GSD.

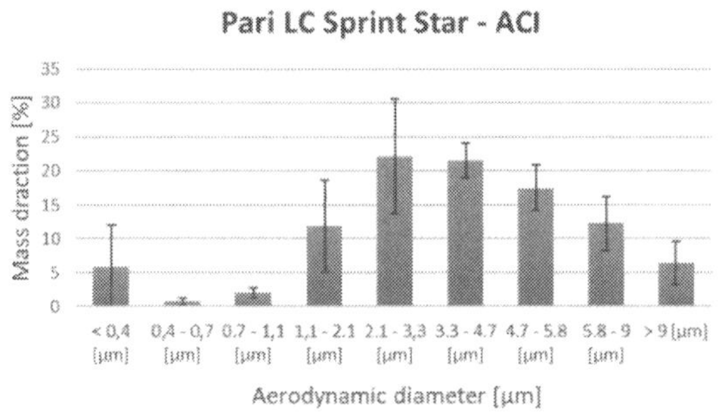

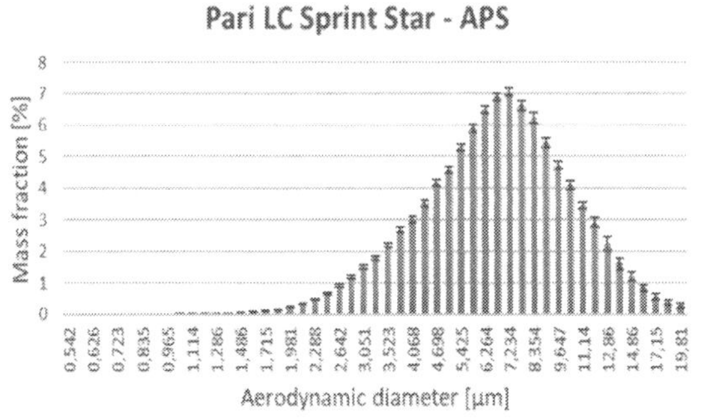

Fig. 2. Upper distribution - APSD of aerosol generated by PARI LC Sprint measured by APS; Lower distribution - APSD of aerosol generated by Pari LC Sprint Star measured by ACI.

3. Results and discussion

Table 1. APSD parameters measured by two different methods.

	ACI	APS
MMAD [μm]	2.53±0.56	6.89±0.27
GSD [μm]	2.76±0.89	1.60±0.17

According to the measured values, APS results show significantly larger particles than the results of ACI measurements. Parameters of APSD are shown in table 1.

Particle size distributions measured by both methods are shown in figure 1. From both, *table 1* and *figure 1* it is evident that the results measured by these two techniques differ significantly. However, data measured by ACI are in quite good agreement with the results of the same nebulizer measured on NGI reported in the study (MMAD approx.. 2.7 μm) [14]. A similar difference between ACI and other methods has been found in the literature as well. In the study [16], the aerosol generated by NE-U22V MicroAir (Omron Healthcare Inc. USA) was measured by ACI, NGI, and APS. Such results have shown significantly higher MMAD, approx. 6.9 μm, measured by APS than MMAD measured by ACI, respectively NGI (approx.. 4.1 μm, resp. 4.5 μm). However such differences can be found also in comparison of ACI with other techniques. McDermott et al [17] measured particle size distribution of vibrating mesh nebulizer Aerogen Solo by Phase Doppler Anemometry. The median of volumetric PSD (D_{v50}) of nebulized albuterol was 5.7 μm on average. The MMAD of albuterol aerosol nebulized by the very same VMN Aerogen Solo measured by cooled ACI in a different study [18], was 4.0 μm. Moreover, several studies [19–21] show MMAD of aerosol generated by Aerogen solo under 2 μm, but the aerosol was in these works dosed through the nasal cannula, which could hypothetically affect the APSD. On the other hand, measurements by Misik et al [22] showed good agreement between MMAD measured by APS and D_{v50} measured by PDA for Aerogen Solo nebulizer.

According to the abovementioned, it is probable, that in the case of aqueous aerosols, the ACI and APS (and maybe also other) measuring techniques are not very consistent and comparable, while ACI results tend to show smaller particle size. This should be investigated more. APS result can depends on the way of sampling. This should be tested and optimized in future steps.

4. Conclusion

The APSD of aerosol generated by AJN Pari LC Sprint Star was measured by two techniques: APS and cooled ACI. Our results confirmed the issue reported in [16] and APS showed MMAD approx. 6.9 μm while ACI is only 2.5 μm. Considering such results, it is obvious that methods are not very comparable. However, there has not been a very good explanation for this difference and more comparative measurements need to be done.

Acknowledgment. This work was supported by the Czech Science Foundation under the grant GA- 2027653S, FSI-S-20-6295, and FCH/FSI-J-21-7308.

1. Golshahi, L.; Seed, K. D.; Dennis, J. J.; Finlay, W. H. J. Aerosol Med. Pulm. Drug Deliv. 2008, 21 (4), 351–359.
2. White, S.; Bennett, D. B.; Cheu, S.; Conley, P. W.; Guzek, D. B.; Gray, S.; Howard, J.; Malcolmson, R.; Parker, J. M.; Roberts, P.; Sadrzadeh, N.; Schumacher, J. D.; Seshadri, S.; Sluggett, G. W.; Stevenson, C. L.; Harper, N. J. Diabetes Technol. Ther. 2005, 7 (6), 896–906.
3. Patton, J. S.; Byron, P. R. Nat. Rev. Drug Discov. 2007, 6 (1), 67–74. https://doi.org/10.1038/nrd2153.
4. NERBRINK, O. Inhalation 2016, 10.
5. Hickey, A. J. Pharmaceutical Inhalation Aerosol Technology; 2019; Vol. 112. https://doi.org/10.1201/9780429055201.
6. Dalby, R. N.; Eicher, J.; Zierenberg, B. Medical Devices: Evidence and Research. Dove Medical Press Ltd September 1, 2011.
7. Hillyer, E. V.; Price, D. B.; Chrystyn, H.; Martin, R. J.; Israel, E.; van Aalderen, W. M. C.; Papi, A.; Usmani, O. S.; Roche, N. J. Aerosol Med. Pulm. Drug Deliv. 2018, 31 (2), 111–113. https://doi.org/10.1089/jamp.2017.1396.
8. Cheng, Y. S. AAPS PharmSciTech 2014, 15 (3), 630–640.
9. Hinds, W. C. Aerosol Technology: Properties, Behavior, and Measurement of Airborne Particles, 2nd ed.; Wiley: New York, 1999.
10. Eur. Pharmacopocia 2005, No. 8, 2799–2811.
11. Eur. Pharmacopoeia 2008, 404–406.
12. Bonam, M.; Christopher, D.; Cipolla, D.; Donovan, B.; Goodwin, D.; Holmes, S.; Lyapustina, S.; Mitchell, J.; Nichols, S.; Pettersson, G.; Quale, C.; Rao, N.; Singh, D.; Tougas, T.; Van Oort, M.; Walther, B.; Wyka, B. AAPS PharmSciTech 2008, 9 (2), 404–413.
13. Zhou, Y.; Ahuja, A.; Irvin, C. M.; Kracko, D. A.; McDonald, J. D.; Cheng, Y. S. Medical, Respir. Care 2005, 50 (8), 1077–1082.
14. Schuschnig, U.; Heine, B.; Knoch, M. J. Aerosol Med. Pulm. Drug Deliv. 2021, No. June.
15. Finlay, W. H. The Mechanics of Inhaled Pharmaceutical Aerosols; 2001.
16. Waldrep, J. C.; Berlinski, A.; Dhand, R. C. J. aerosol Med. Off. J. Int. Soc. Aerosols Med. 2007, 20 (3), 310–319.
17. McDermott, K.; Oakley, J. G. Curr. Ther. Res. - Clin. Exp. 2021, 94, 100623.
18. Hassan, A.; Rabea, H.; Hussein, R. R. S.; Salah Eldin, R.; Abdelrahman, M. M.; Said, A. S. A.; Salem, H. F.; Abdelrahim, M. E. Pulm. Ther. 2016, 2 (1), 115–126.
19. Réminiac, F.; Vecellio, L.; Heuzé-Vourc'h, N.; Petitcollin, A.; Respaud, R.; Cabrera, M.; Le Pennec, D.; Diot, P.; Ehrmann, S. J. Aerosol Med. Pulm. Drug Deliv. 2016, 29 (2), 134–141.
20. Perry, S. A.; Kesser, K. C.; Geller, D. E.; Selhorst, D. M.; Rendle, J. K.; Hertzog, J. H. Pediatr. Crit. Care Med. 2013, 14 (5).
21. Madney, Y. M.; Fathy, M.; Elberry, A. A.; Rabea, H.; Abdelrahim, M. E. A. J. Drug Deliv. Sci. Technol. 2017, 39, 260–265. https://doi.org/10.1016/j.jddst.2017.04.014.
22. Misik, O.; Maly, M.; Cejpek, O.; Lizal, F. MATEC Web Conf. 2020, 328, 01006.

Dynamic behaviour of cavitation bubble close to a flexible wall

Miloš Müller[1*], *Jan* Hujer[1], and *Petra* Dančová[1]

[1]Department of power engineering equipment, Technical University of Liberec, Studentská 1402/2, 461 17 Liberec 1, Czech Republic

Abstract. The dynamic behaviour of cavitation bubble close to a flexible boundary is investigated experimentally. The cavitation bubble is produced as a consequence of an optical breakdown generated by the focus of 532 nm Nd-Yag laser beam. A gold mirror is used to focus the expanded laser beam into a spherically shaped plasma. A PVDF film sensor is used for the measurement of the interaction magnitude between the bubble and the flexible boundary. As the flexible boundary directly the PVFD film sensor is used. The sensor is flexibly mounted within a specially designed movable frame which enables to modify the elastic properties of the system. The bubble dynamics and the flexible wall movement are recorded by high-speed CCD camera and correlated with the acoustic signal obtained by the PVDF film. Different bubble collapse patters for different bubble wall distances and the corresponding acoustic signals are presented.

1 Introduction

In the last decades, the laser induced breakdown LIB [1] and its interaction with the biological tissue have been intensively studied [2], [3], [4], [5], [6]. This interest was motivated by the importance of the effects accompanying the LIB, the shock waves and cavitation bubbles. The determination of the type of interaction between the cavitation bubble and the flexible wall during the bubble collapse is important in order to estimate the damage potential of the collapsing bubble and also the range of the wall deformation.

For the solid wall it is considered that the bubble wall interaction is realised by three main effects [7]. The first effect is a fast liquid microjet formed during the bubble rebound after the collapse. In the case of solid wall this jet is directed towards the wall. This effect is produced by bubbles at medium distance from the wall. The second effect is so called splashing, which is generated by bubbles in direct contact with the solid wall. The shock wave [8], [7] is the last effect responsible for the cavitation erosion. The shock waves can be generated by the bubbles collapsing in the far field but also by bubbles collapsing directly on the wall. In the case of laser induced bubble, the shock wave is usually produced during the bubble initiation [9], [1] as a consequence of the rapid plasma expansion. When the bubble collapses close to a solid boundary, the bubble attraction to the solid wall can also be observed. This attraction is caused by so called "Bjerkens force" [7], [8].

In general, the cavitation bubble can collapse in a free field of liquid or close to a boundary. The magnitude of the bubble impact depends on several parameters, which include the bubble wall distance, the maximum bubble radius and the wall material.

The distance of the bubble inception (LIB) from the boundary s is usually normalized by maximum bubble radius R_{max} in form

$$\gamma = s/R_{max}. \qquad (1)$$

Dynamics of a spark generated cavitation bubble close to an elastic boundary was investigated e.g. by Gibson [10], [11]. The author investigated the bubble behaviour close to rigid boundaries with rubber coatings. An important result from their measurements was that for some range of the wall properties no bubble jet was generated during the bubble collapse and the bubble formed an hourglass shape. A cavitation bubble generated by laser near gelatine surface was studied in [12]. The authors observed that the cavitation bubbles inclined not to produce liquid jets and tended to migrate from the surface.

Shima showed in [13] that for some combinations of wall properties and the bubble wall distances no bubble migration occurs. The most comprehensive study of the dynamics of the laser induced cavitation bubble collapsing close to an elastic boundary is given in [14] by Brujan. The author uses polyacrylamide as the wall material. He studied the bubble collapse for large range of γ and identified the ranges where the bubble jet is directed towards the boundary and where the jet is directed away from the boundary. He also measured the jet highest velocities.

2 Experimental setups

The basic setup for the measurement of the interaction between the cavitation bubble and the flexible wall is given in Figure 1. The cavitation bubbles are generated by Nd-Yag laser at wavelength 532 nm focused into the

* Corresponding author: milos.muller@tul.cz

bath. To minimize the spherical abbreviation of the bubble the optical setup with gold spherical mirror was used [15]. The laser beam of diameter 5 mm is firstly expanded trough a Galilean beam expander (4x) and focused trough a gold mirror.

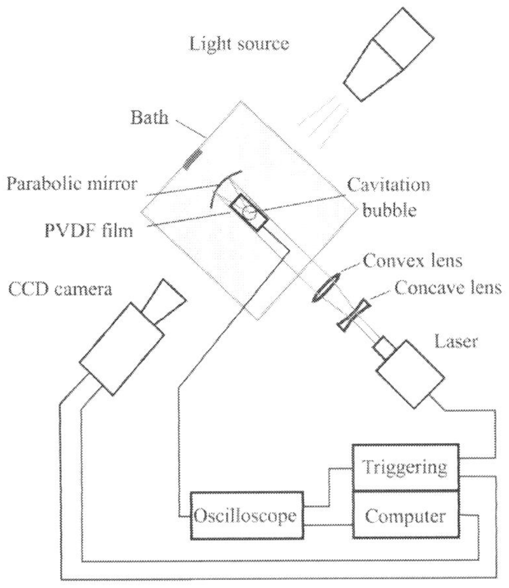

Fig. 1. Experimental setup used in the investigation of the cavitation bubble collapse.

Due to the arrangement of the experiment the LIB is generated out of optical axis to ensure no crossing between the expanded beam and the flexible wall. The cavitation bubble dynamics was investigated by synchronized acoustic and optical measurements.

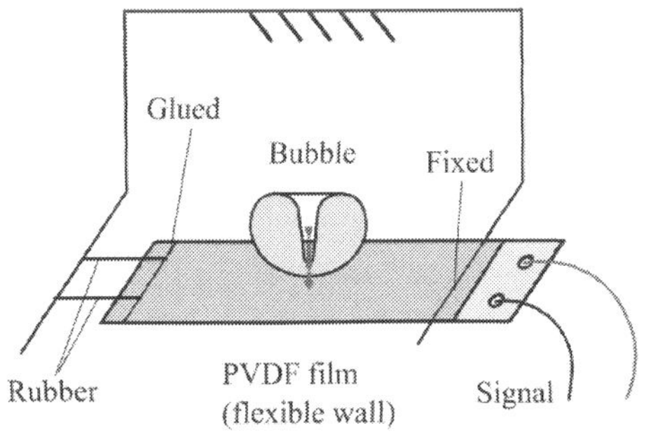

Fig. 2. Flexible wall arrangement.

The optical measurement was based on the high-speed photography using ultra-fast CCD cameras. The bubble collapse illumination was provided by the high-power flash lamp in continuous mode. The lamp was used with the collimating lens to provide the parallel illumination for visualization of the bubble surface. The acoustic measurement was realized with the high sensible piezoelectric polyvinylidene fluoride (PVDF) polymer transducer. The PVDF film was mounted on a movable frame submerged into the water bath and directly used also as the flexible wall. The bath walls were perpendicular to the optical axes. The placement of the PVDF film on the movable frame is shown in Figure 2. The film is mounted between two parallel supports. On one side of the support the film is glued directly on the supporting beam. To the second side of the support the film is connected by two rubber straps, which defines the wall support. To avoid the film deformation between the two rubber straps a wood beam is glued on the film. The film surface was covered with very thin black paint to protect the camera from the beam radiation during the wall deformation. The electrical contacts of the PVDF film are protected by a tape. The elastic modulus of the flexible wall was estimated based on a tensile test to 0.2 MPa.

The sampling frequency for the acoustic measurements was 60 MHz and for the optical measurement 170 kHz.

The PVDF film was used to measure the local short time exposure caused by the direct interaction of the PVDF film (flexible wall) and also to measure the long-time tensile load. The PVDF film is not calibrated so only the load character can be identified from the voltage signal. Optical measurements are synchronized by Dynamic Studio (Dantec) software. The acoustic signals are recorded by high-speed oscilloscopic card. NI LabVIEW Signal Express software is used for the acoustic data acquisition and processing.

3 Results

The experimental results are represented in the form of the figures of the time evaluation of the bubble radius and in the form of the PVDF film signals. The individual figure collections correspond to a single value of the relative bubble wall distance γ. Due to the bubble deformation close to the boundary the vertical and horizontal bubble radii were evaluated.

The bubble shape is deformed in the dependence on the distance between the initial position of the bubble centre and the initial wall position.

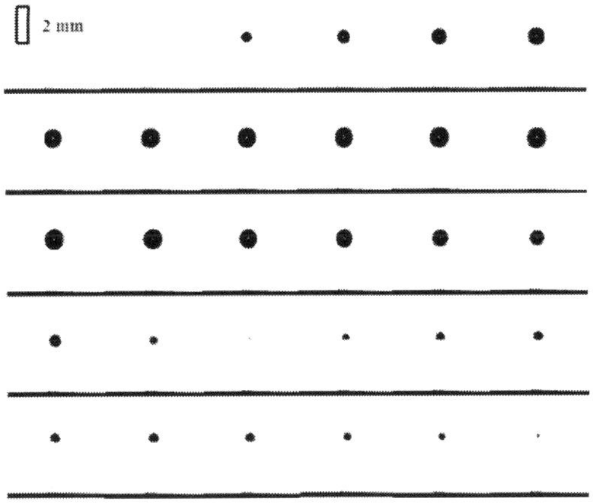

Fig. 3. Bubble dynamics for $\gamma = 4.6$. Time between the individual frames is 5.9 µs.

https://doi.org/10.1051/epjconf/202226401024

For the bubble wall distance γ = 4.6 (Figure 3) the bubble shape is not influenced by the wall presence. This is also evident in Figure 8 as the vertical and horizontal bubble radii are almost identical. For the bubble wall distance γ = 1.6 (Figure 4) the wall deformation and the bubble deformation are evident. However, the bubble jet in comparison to the rigid wall case is generated towards from the flexible wall.

Fig. 4. Bubble dynamics for γ = 1.6. Time between the individual frames is 5.9 μs.

For the bubble wall distance γ = 1.1 (Figure 5) the wall deformation and the bubble deformation in vertical direction are visible. The bubble jet is generated towards the wall but the bubble migrates towards from the wall after the rebound It can be also seen that the bubble is spitted during the jet formation.

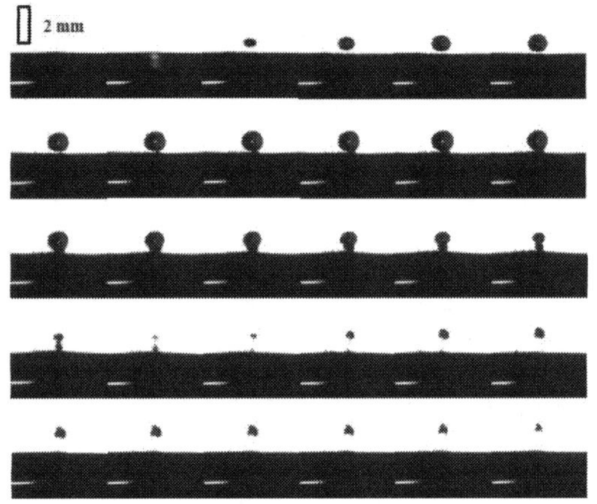

Fig. 5. Bubble dynamics for γ = 1.1. Time between the individual frames is 5.9 μs.

For the bubble wall distance γ = 0.9 (Figure 5) the bubble is in direct touch with the flexible wall. The vertical and horizontal bubble radii are significantly different. Due to the needle bubble shape after the rebound the vertical and horizontal radii are not evaluated in Figure 6.

The first and second bubble expansion and first and second bubble collapse are illustrated in Figure 6. It can be seen that the bubble shape is influenced by the wall presence mainly after the first collapse (the first bubble rebound). During the growth phase after the plasma expansion the bubble is rather spherical. This corresponds to the rigid wall case, where the bubble shape is deformed also during the first rebound.

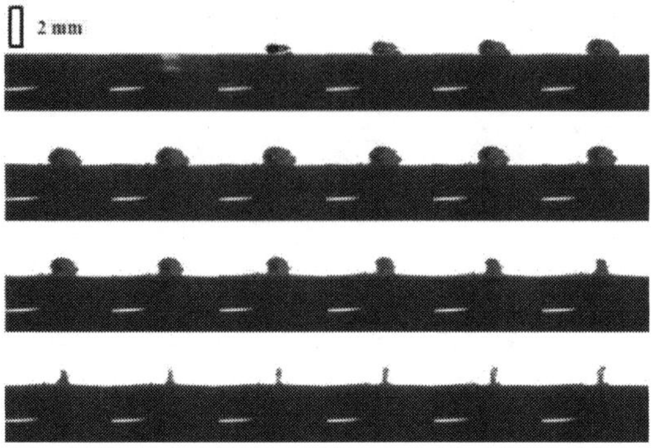

Fig. 6. Bubble dynamics for γ = 0.9. Time between the individual frames is 5.9 μs.

Figure 8 illustrates the voltage signals measured on the PVDF film sensor (flexible wall). The signal can be correlated with the horizontal and vertical bubble radii presented in Figure 7.

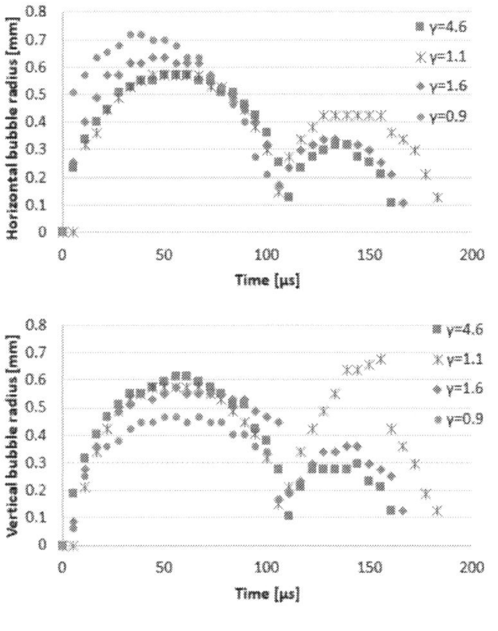

Fig. 7. Bubble radius in horizontal and vertical direction.

From the signal it is possible to identify several phenomena. In the first phase (up to 40 μs) the impact of the shock wave generated by the rapid plasma expansion is evident. The maximum bubble radius is reached about 70 μs. The bubble jet impact can be observed in the time after the first bubble collapse. In the time frame from 100 µs to 120 µs the first bubble collapse can be identified. The second bubble collapse can be identified in time about 180 µs.

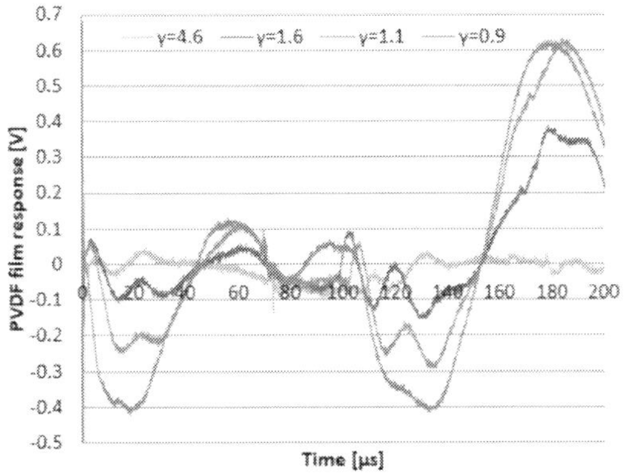

Fig. 8. Voltage signal measured on the PVDF film.

4 Conclusion

The cavitation bubble dynamic in the case of flexible wall exhibits completely different behaviour in the comparison to the rigid wall. The bubble jet generated during the bubble rebound can be generated for the flexible wall either towards the wall or from the wall. Also, the bubble is attracted either towards the wall or moves from the wall. The signal generated on the flexible wall indicated that the second bubble collapse is more intensive than the first bubble collapse.
Note that the cavitation bubble dynamic was investigated only for one set of elastic properties of the wall. It is known from the literature that the value of the wall elastic modulus can influence the bubble dynamic significantly. The investigation should be extended by the matrix of maximum bubble radii and the elastic moduli. The other part of the research should be focused to appropriate calibration method of the PVDF film sensor to obtain pressure of force value instead of the voltage signal.
The usage of the PVDF film sensor directly as the flexible wall represents a unique solution, which enables the possibility for the quantification of the bubble-flexible wall interaction without modification of the wall properties.

Acknowledgements

This research was supported by the Student Grant Competition of the Technical University of Liberec, grant number SGS-2019-5703 (21291) „ Experimental, numerical and theoretical research in fluid mechanics and thermomechanics", grant number SGS-2020-5028 (21397) „Investigation of cavitation collapse by PVDF sensors" and by the Ministry of Education, Youth and Sports of the Czech Republic, grant number 8J19FR018 „Assessment of the damage potential of cavitation for industrial hydraulic applications". Authors would like to acknowledge the help of Darina Jašíková and Michal Kotek from the Institute for Nanomaterials, Advanced Technologies and Innovation.

References

1. P. K. Kennedy, D. X. Hammer, and B. A. Rockwell, Prog. Quantum Elec-tron. **21** (3), pp. 155–248 (1997)
2. T. Juhasz, X.N. Hu, L. Turi, and Z. Bor, Lasers in Surgery and Medicine **15** (1), pp. 91-98 (1994)
3. E. A. Brujan, *Proceedings of the Romanian Academy-Series A: Mathematics, Physics, Technical Sciences, Information Science* **6**, pp. 155-160 (2005)
4. E. A. Brujan, Medical Engineering and Physics **31** (7), pp. 742-751 (2009)
5. A. Vogel, R. Engelhardt, U. Behnle and U. Parlitz, Appl Phys B **62**, pp. 173–182 (1996)
6. D. X. Hammer, R. J. Thomas, M. Frenz, E. Jansen, G. D. Nojin, S. J. Giggs, J. Noack, A. Vogel and B. A. Rockwell, *Proc. SPIE 2681 "Laser Tissue Interaction VII"*, pp. 437-448 (1996)
7. J. P. Franc and J. M. Michel, *Fundamentals of cavitation*. Kluwer Academic Publishers, Boston, ISBN 14-020-2232-8 (2004)
8. C. E. Brennen, *Cavitation and bubble dynamics*, New York: Oxford University Press, 282 p., ISBN 978-0195094091 (1995).
9. A. Vogel, J. Noack, K. Nahen, D. Theisen, S. Busch, U. Parlitz, D. X. Hammer, G. D. Noojin, B. A. Rockwell and R. Birngruber, Appl Phys B **68**, pp. 271–280 (1999)
10. D. C. Gibson and J. R. Blake, *Australasian Conference on Hydraulics and Fluid Mechanics*, pp. 283-286 (1980).
11. D. C. Gibson and J. R. Blake, Appl. Sci. Res. **38**, pp. 215–224 (1982).
12. T. Kodama and Y. Tomita, Appl Phys B: Lasers Opt. **70**, pp. 139–149 (2000)
13. A. Shima, Y. Tomita, D.C. Gibson and J.R. Blake, J. Fluid Mech. **203**, pp. 199–214 (1989)
14. E. A. Brujan, K. Nahen, P. Schmidt and A. Vogel, J Fluid Mech **433**, pp. 251–281 (2001)
15. D. Obreschkow, M. Tinguely, N. Dorsaz, P. Kobel, A. de Bosset, M. Farhat, Experiments in Fluids **54** (4), pp. 1-18

CFD analysis of the ultrasonic gas meter channel

Jan Novosád[1*], *Jaroslav* Pulec[1], *Petra* Dančová[1] and *Tomáš* Vít[1]

[1]Department of Power Engineering Equipment, Faculty of Mechanical Engineering, Technical University of Liberec, Studentská 2, 46117, Liberec, Czech Republic

Abstract. This paper describes a process of numerical testing and optimization of the ultrasonic gas meter channel. A conventional ultrasonic flowmeter requires a large installation space due to the quality and stability of the velocity profile. Previously, the shape of the inner gas meter channel was optimized to reach a suitable flow field for ultrasonic measurement. The goal of this work is to find an optimal design of reduced size gas meter to achieve a stable velocity profile with minimum disturbances regardless of the entry conditions in the smallest possible space. Flow characteristics for volume flow rate in the range (1 to 250) m^3/h were tested and the parameters of the gas meter were adjusted. Obtained results show good match with the requirements. Future work should aim to the measurement of transit time and the stability of measured values on the optimized design. Finally, it will enable to create the analogy between numerically obtained velocity fields and real transit time measurement.

1 Introduction

Ultrasonic (US) flow meters are devices for the fluid flow measurement using the ultrasonic principle. The flow rate is related to the fluid velocity. When the pressure disturbances are propagated at the speed of sound and the fluid also has a velocity, the absolute velocity of the pressure disturbance propagation is the algebraic sum of them [1]. A simple transit time flowmeter configuration is illustrated in Fig. 1.

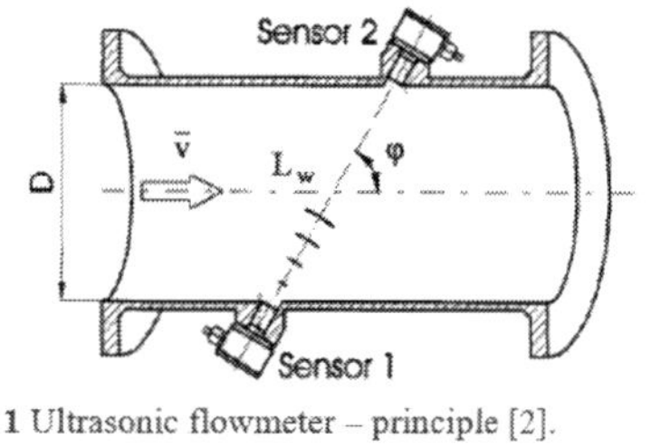

Fig. 1 Ultrasonic flowmeter – principle [2].

The transit time flowmeters evaluates the transit time following eq. (1) in the downstream (t_{12}) and upstream (t_{21}) direction [2]:

$$t_{12} = \frac{L_w}{c+v_a \cos\varphi} \text{ and } t_{21} = \frac{L_w}{c+v_a \cos\varphi}, \quad (1)$$

where L_w (m) is the distance in fluid between two transducers, c (m/s) the speed of sound at the operating conditions, φ the angle between pipe axis and the acoustic (US) path and v_a (m/s) is the axial velocity averaged along the acoustic path.

When the configuration with two transducers used both as transmitters and receivers, the time difference could be evaluated directly by the same pair. Then, the mean axial velocity along the US path in a circular pipe is given by

$$v_a = \frac{L_w}{2\cos\varphi}\left(\frac{1}{t_{21}} - \frac{1}{t_{12}}\right) = \frac{D}{2\cos\varphi\sin\varphi}\left(\frac{1}{t_{21}} - \frac{1}{t_{12}}\right) \quad (1)$$

where D (m) is a pipe diameter.

A disturbance element a deformation of the velocity field as wakes, vortex structures etc. Therefore, the sufficient length of a straight pipe, shown in Fig. 2, before/after the ultrasonic flowmeter is required to achieve optimal performance and accuracy [2].

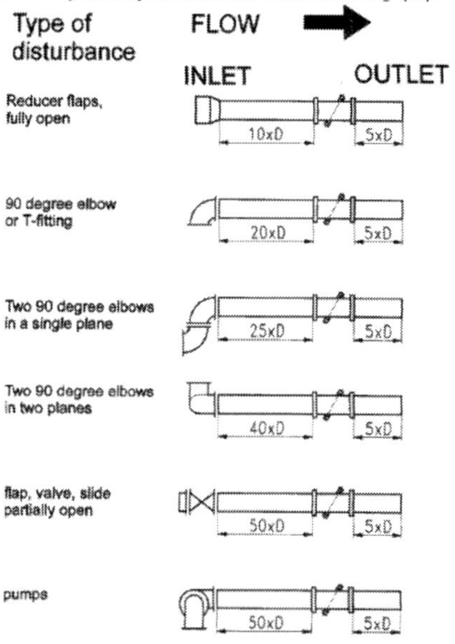

Fig. 2 Minimum straight pipe length for 1% accuracy of a single path transit time flowmeter [2].

* Corresponding author: jan.novosad@tul.cz

Technical applications of the gas meter follow the International recommendation OIML R137 [3]. These rules contain the specification of performance testing using different pipeline configuration in front of the flowmeter to test the stability of measured values. Fig. 3 illustrates two cases with a straight pipe as a reference case and disturbance OIML d+ as the worst case in terms of the flow disturbance.

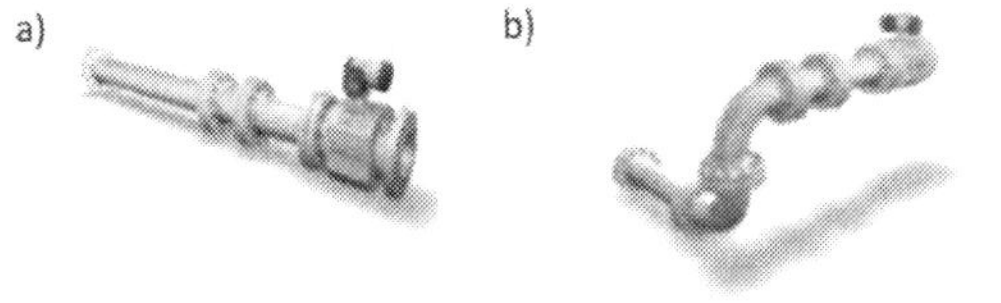

Fig. 3 Piping configurations for flow disturbances [3]: a) straight pipe (reference conditions), b) double elbow with a half-moon plate (OIML d+).

Our team investigates a new gas meter type with no required straight runs upstream. Previously, the disturbance testing of the gas meter with innovative channel shape [4] have been studied numerically. The results published in [5] shows that the disturbance does not affect the velocity field in the US measurement area. For the final prototype the dimensions were reduced so the inner shape had to be modified. These modifications have to be tested numerically to find the solution suitable for the US flow measurement.

2 CFD setup and results assessment

Based on the CAD model of the DN80 gas meter the computational grid and simulation setup was prepared in ANSYS Fluent [6]. Computational grid created in Fluent Meshing contains 2.3 million polyhedral elements. The near wall area is covered by the prism polyhedral elements with respect to the application of wall function models.

2.1 CFD setup

Boundary conditions were applied on the inlet and outlet boundaries. At the inlet cross section (see the green area with blue arrows in Fig. 4) the velocity corresponding to the flow rate Q01-Q06 was applied. Then, the pressure outlet boundary condition with p = 0 Pa was defined. Flow conditions are listed in Table 1.

Table 1 Simulated flow rates.

Label	Flow rate (m^3/h)	Velocity (m/s)	Re (1)
Q01	1	0.055	293
Q02	5	0.276	1463
Q03	10	0.553	2926
Q04	30	1.658	8778
Q05	65	3.592	19018
Q06	250	13.816	73146

Viscous forces are modelled using the two equation turbulence model SST k-ω model. Standard air considered as a gas with constant properties at 20°C and atmospheric pressure 10^5 Pa have been used as a working fluid.

Computation was set as steady state using the coupled scheme with the second order discretization for pressure, momentum and QUICK for turbulence quantities.

Fig. 4 Case with OIML d+ disturbance – BCs.

2.2 Results assessment

In Fig. 5 there is a sketch of the area of interest for results assessment. These area covers a space in which the ultrasonic measurement is realized. Two sets of cross-sections were defined and labelled as lineZ and lineY Depending on the x-coordinate, velocity profiles in both directions were observed (labelled as section01 to section09).

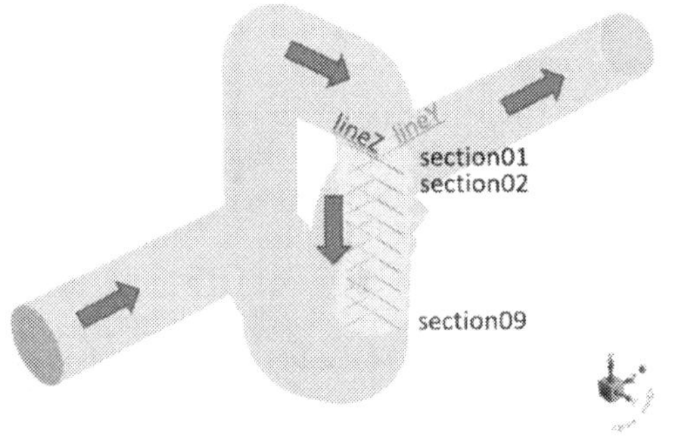

Fig. 5 Cross-section planes – sketch.

To compare the results for different flow rates, the dimensionless velocity U' have been evaluated as

$$U' = \frac{U}{U_{\max}} \text{ (1)}, \qquad (1)$$

where U $(\mathrm{m/s})$ is the velocity value in the axis direction at the specified point, U_max (m/s) the maximum velocity value on the evaluated line.

3 Results

The initial and optimized design have been analysed. The results shown the final design state. Fig. 6 shows results for the case with straight inlet pipe, Fig. 7 then the case with the disturbing element at the entrance. The horizontal axis symbolises the dimensionless coordinate where $\frac{r}{R} = 0$ means the axis of the pipe.

https://doi.org/10.1051/epjconf/202226401025

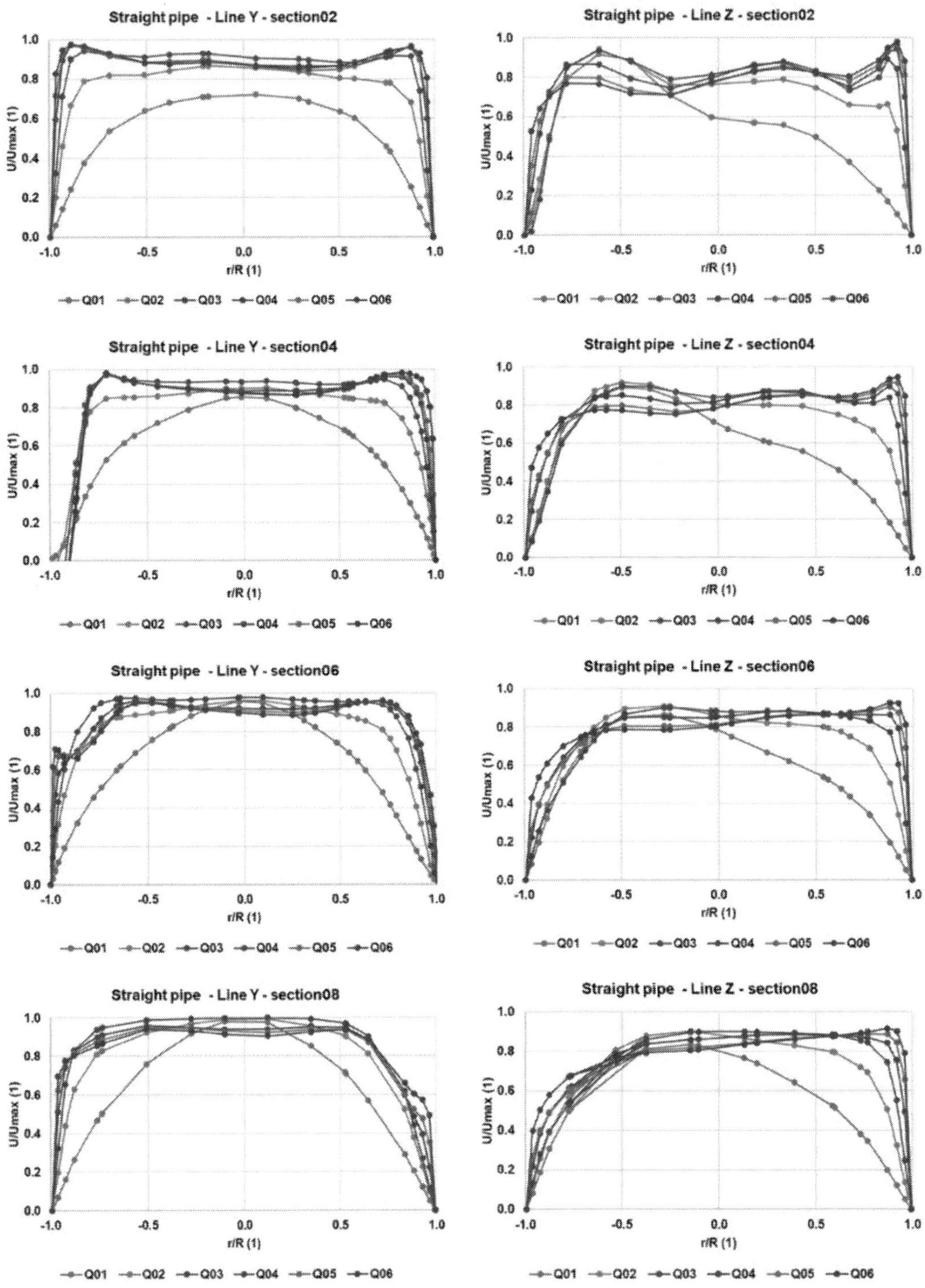

Fig. 6 Velocity profiles – straight pipe (no disturbance), selected sections.

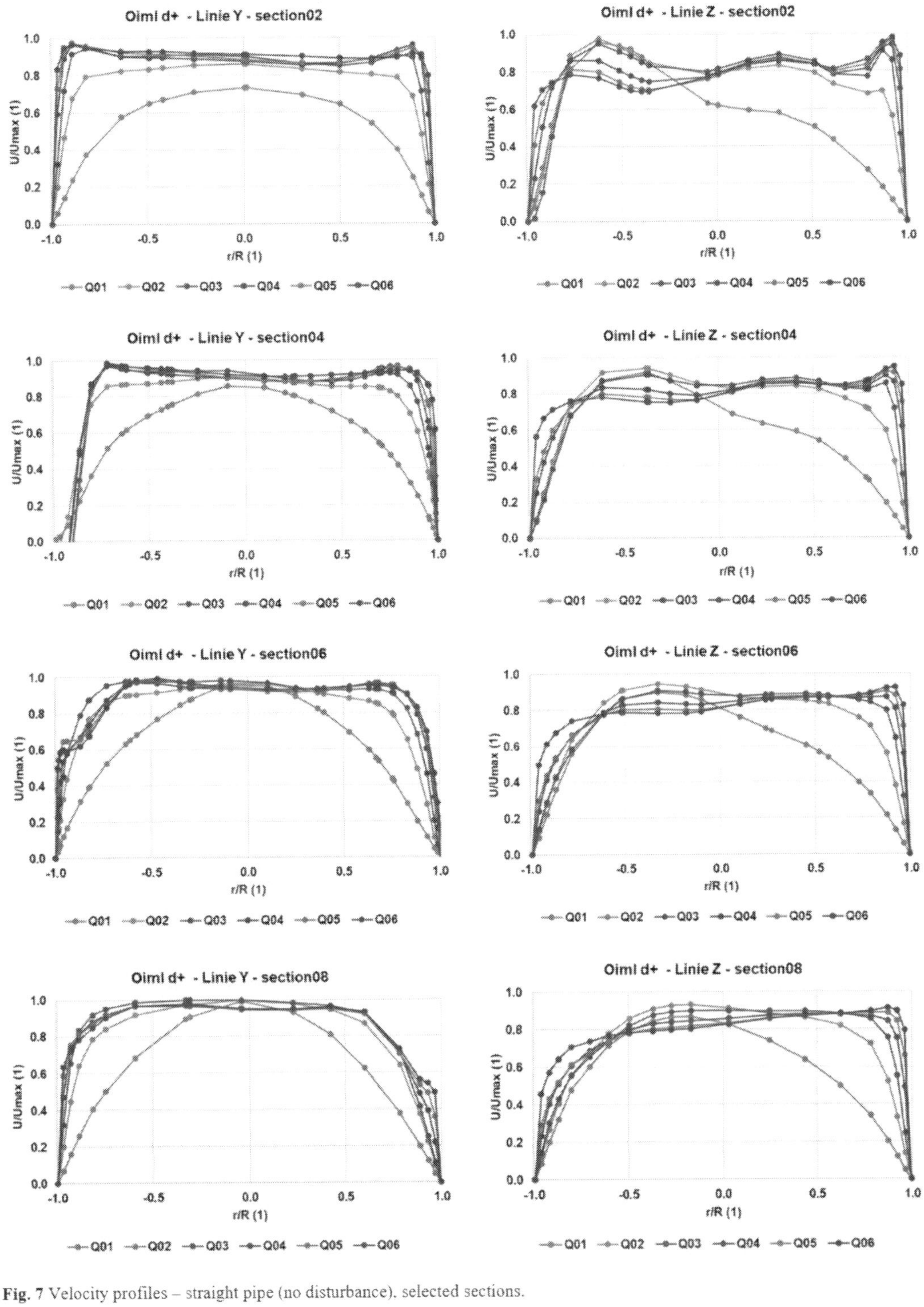

Fig. 7 Velocity profiles – straight pipe (no disturbance), selected sections.

The vertical axis follows the dimensionless velocity U/U_{max}. The graphs are plotted for flows Q01 to Q06. Except the lowest flow rate, the profiles in line Y direction are independent on the flow rate and the installed disturbance except the lowest flow rate QO1. The change in profile shows the different flow character in laminar (QO1, QO2) and turbulent flow. All the velocity profiles could be considered as symmetrical along the pipe axis. Finally, the change of the profile in the regions $|r/R| > 0.6$ for section08 shows the influence of the downstream channel shape.
The profiles in the second direction (lines Z) exhibit the same behaviour against the flow disturbance. No significant difference have been found between straight pipe and double bend cases. However, the profiles are not symmetrical, when higher velocity values are situated for $\frac{r}{R} > +0.8$. This part is affected by the 3D shape of the upstream channel where the flow is more adjacent to the outer wall of the gas meter for turbulent flow. In laminar regime the flow behaves in the opposite way, when the main flow stream is located closer to the gas meter centre.

4 Conclusion

The current research project aims to the innovative gas meter channel. The main goal is the design optimization of the gas meter with limited dimension, actually. Secondly, the optimized design have to meet the OIML requirements, especially the stability of the velocity fields in the US measuring area.

Based on the previously proposed methodology for a numerical testing using CFD methods, the disturbance effect on the flow field inside the optimized channel have been studied. Several flow regimes have been analysed to find the stability of the flow field across the whole measuring range of the gas meter. The results were presented as the velocity profiles in predefined cross sections. These curves show the change of the profile depending on the flow rate (inlet velocity), as well as the comparison of both disturbance cases which have been taken into account. Generally, main goal was reached, so the profiles match well for both cases.

On the other hand, a few significant differences was found for specific conditions. The velocity profiles evolve gradually from the lowest velocity corresponding to laminar flow, when the increasing of velocity, i.e. the transition to turbulence, leads to the stabilization of the profiles. Then the geometry of the channel causes the nonsymmetric profiles in line Z. The asymmetry is similar for different cases and flow rates, but it is necessary to analyse that phenomena further for the influence on the US measurement.

Finally, the presented results are limited to the CFD methods only. The velocity fields have been analysed as the results. However the ultrasonic measurement is based on the transit time, which is affected by the axial velocity and the working fluid parameters. The future work should be aimed to the experiments for validation of numerical results, also. Then, finding a suitable analogy between numerically obtained velocity fields, experimentally measured transit time and the stability of measured values will enable the interconnection of all partial activities in the design and optimization of the gas meter channel. It is planned to use the PIV (particle image velocimetry) method for validation of the numerical results.

Author gratefully acknowledges financial support by Czech Technological Agency under the project TACR FW02020048. This work was also partly supported by the Student Grant Competition of the Technical University of Liberec under the project No. SGS-2019-5073.

References

1. Doebelin E. O. Measurement Systems: Application and Design. McGraw-Hill Inc., 1990.
2. Webster, J. G. The Measurement, Instrumentation and Sensors Handbook. CRC Press, 1999. ISBN 0-8493-2145-X
3. International organization of legal metrology. Gas meters. OIML R 137-1 & 2 Edition 2012.
4. Mikan, J.: Compact Ultrasonic Flow Meter, Especially For Gas. Patent nr. 17/056784.
5. Vít, T., Novosád, J., Dančová, P.: International Journal on Engineering Technologies and Informatics, Volume 2 Issue 1- 2021.
6. Ansys Inc. Ansys Fluent User´s guide. Release R2021R2.

Optimization of air mass flow in a PEM fuel cell

Jan Novotný[1*], *Ludmila* Nováková[1], *Roman* Čížek[1], *Miloš* Kašpárek[1], and *Ilona* Machovská[1]

[1]Jan Evangelista Purkyně University in Ústí nad Labem, Faculty of Mechanical Engineering, Institute of Machinery and Power Engineering, Pasteurova 1, 400 96, Ústí nad Labem, Czech republic

Abstract. In our work, we focused on measuring the performance characteristics of a test hydrogen fuel cell with an open cathode. The aim of the work was to verify the effect of the mass flow through the fuel cell on the power output as a function of the current density and voltage on the fuel cell. A test bench capable of testing fuel cells up to 2000 W was used to measure the power characteristics. The test fuel cell was composed of three cells and its total membrane area was 100 cm2. A PIV measurement system was used to measure the mass flow. The mass flow measurement was performed at the inlet of the fuel cell. The measurements included the determination of the inlet temperature distribution using a thermal imaging camera. The presented measurements were performed for two configurations with original and optimized diffuser at the inlet of the fans. Finally, the effect of mass flow and overall thermal management on the fuel cell performance is discussed, including the effect of the newly designed diffuser.

1 Introduction

The aim of our work was to verify the influence of air flow rate and air distribution inside a hydrogen fuel cell with a proton exchange membrane PEM on the performance characteristics. In open cathode fuel cells, electrical and thermal energy is produced by the electrochemical reaction of hydrogen and oxygen contained in the air. This type of fuel cell operates optimally at temperatures typically below 60°C (for example, ([1], [2], [3]). An additional advantage of PEM cells is their short ramp-up time to maximum zero-emission performance. For these reasons, it is clear that the use of hydrogen fuel cells will experience rapid development in the coming decade. In open cathode fuel cells, the air flowing through the fuel cell carries oxygen which is consumed by the electrochemical reaction and, on the other hand, removes excess heat outside the fuel cell. This dual role of air in this type of fuel cell leads to a simpler PEM cell design but, due to the sensitivity of open cathode fuel cells to operating parameters, places great demands on the control of the input air parameters such as its temperature humidity and mass flux distribution across the bipolar plate. A critical parameter for open cathode PEM cell operation is the thermal balance of the water mass flux ([4], [5], [6]).. Excess water inside the fuel cell leads to flooding and reduces the mass transfer across the membrane, leading to a decrease in fuel cell performance. A small mass flux of water leads to drying of the membrane and a decrease in the electrochemical reaction. High or low fuel cell temperature then reduces fuel cell performance and higher temperatures lead to reduced lifetime and in limiting cases can lead to membrane destruction. It is clear that maintaining the optimum thermal management of an open cathode fuel cell is essential for its operation. For this reason, a number of studies have been carried out to investigate the effect of heat and mass transport within fuel cells on their operating characteristics ([7], [8])... Another parameter that significantly affects the performance characteristics of the fuel cell is the shape and distribution of the flow channels of the bipolar plate on both the hydrogen and air sides. Much attention is paid to the shape of the cooling channels for air flow. In these studies, both the effect of channel geometry on oxygen transport through the membrane and the effect of channel shape on water movement and droplet shape on the channel walls are investigated ([9]). The effect of the channel shape on the overall aerodynamic drag is also investigated. The effect of changing the channel cross-section across the fuel cell leading to improved reactant mixing, better water drainage and thus increased fuel cell performance is also investigated. The reaction at the anode and cathode of the PEM cell can be written using the following equations:

$$2H_2=4H^++4e^- \tag{1}$$

$$O_2+4e^-=2O_2^- \tag{2}$$

$$2O_2^-+4H^+=2H_2O \tag{3}$$

$$H_2+O_2=2H_2O \tag{4}$$

* Corresponding author: novotny@ujep.cz

2 Experimental set up

To measure the effect of the mass flow through the fuel cell, we used a LEANCAT test station with a maximum output of up to 2KW. This test station is capable of automatically recording and adjusting all provoyance parameters such as pressure and hydrogen flow, measuring the current and voltage on individual fuell cells, including temperature monitoring both inside the fuel cell and on its surface. The station is equipped with nitrogen for cleaning and inertia. The schematic of the test station and the circuitry is shown in Figure 1. Fuel cell test station is shown in Fig.2. Shape of the channels of the bipolar plate on the air side is rectangular with a width of 1 mm and a channel depth of 2 mm. The membrane inside the harness is MEA-7 from Altpolymer. The maximum provoyage temperature of the membrane is 65°C and the surface area is 100 cm2. The operating hydrogen overpressure during testing was set to 300 Pa and the purge duration was 200 ms. The fuel cell was operated in constant voltage mode at 2V, 1.8V and 1.6V. At each fuel cell load, the effect of mass flux magnitude was tested for both the original fan connections to the fuel cell and the optimized diffuser shape Figure 3. The optimized diffuser shape is designed with internal bars that distribute the mass flow across the fuel cell. The partitions are designed to increase the mass flow on the sides. 3D printer with FDM printer technology with a PTEG print filament was used to produce the optimized dofuzor. The temperature inside the fuel cell was measured with a type T thermocouple and was also monitored with a thermal camera Fig.6. The mass flow distribution at the inlet of the fuel cell was measured with a 2D PIV system from Lavision. The resolution of the CMOS camera was 5Mpixel; with a bit depth of 16 pit;. A 200 mJ/pulse Nd:YAG laser was used to illuminate the inlet plane. We used condensed oil droplets of about 1-2um as marker particles. The evaluation was performed by the classical multistem

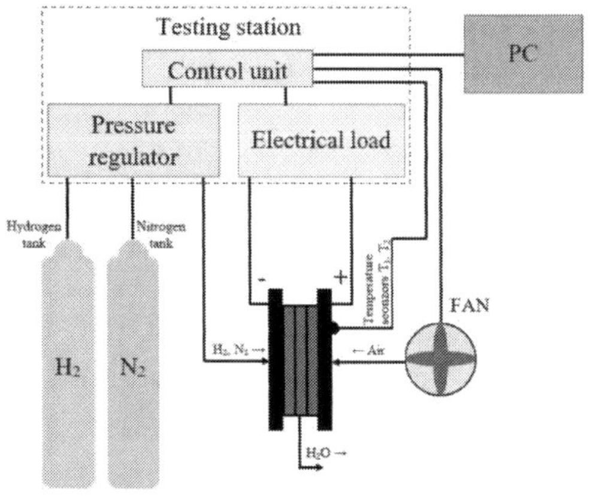

Fig. 1. Set up of the test bench.

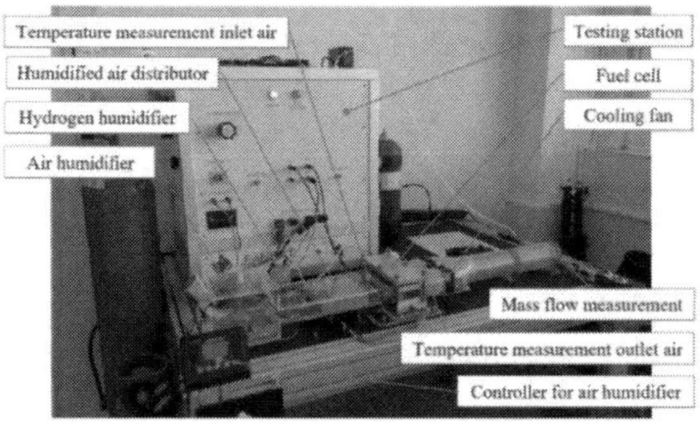

Fig. 2. Fuel cell test bench.

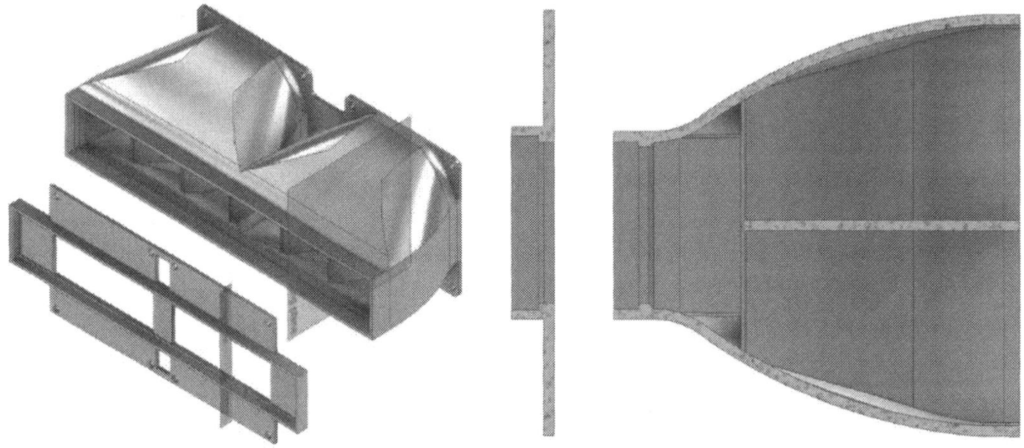

Fig. 3. Original and optimized diffuser shape for air outlet from the fuel cell with the indicated cutting plane on the left. Cross section of the original and optimized diffuser on the right.

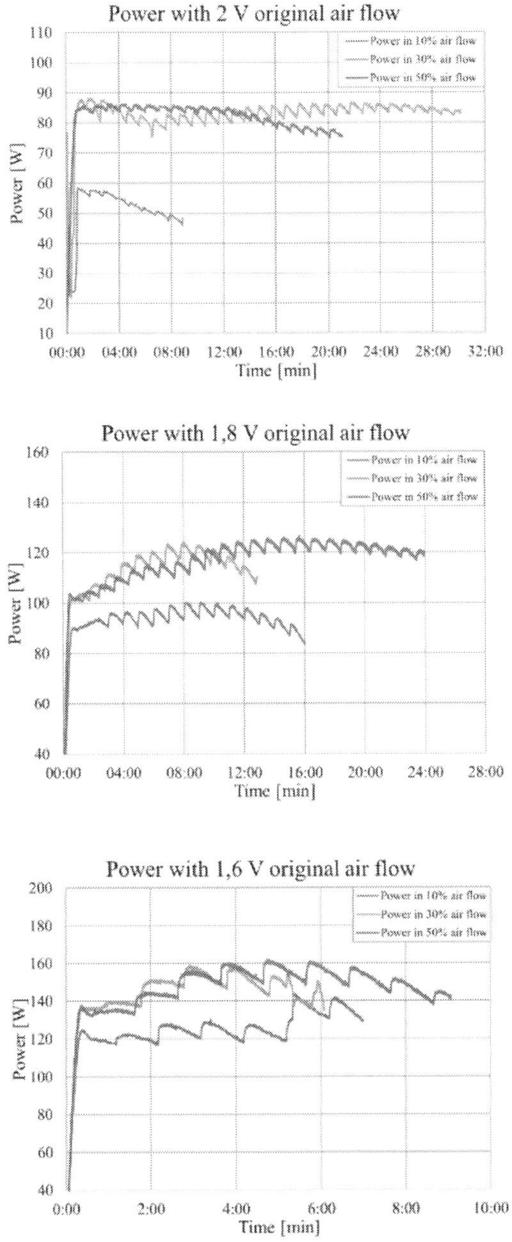

Fig. 4. Dependence of fuel cell performance on air flow and voltage - original adapter.

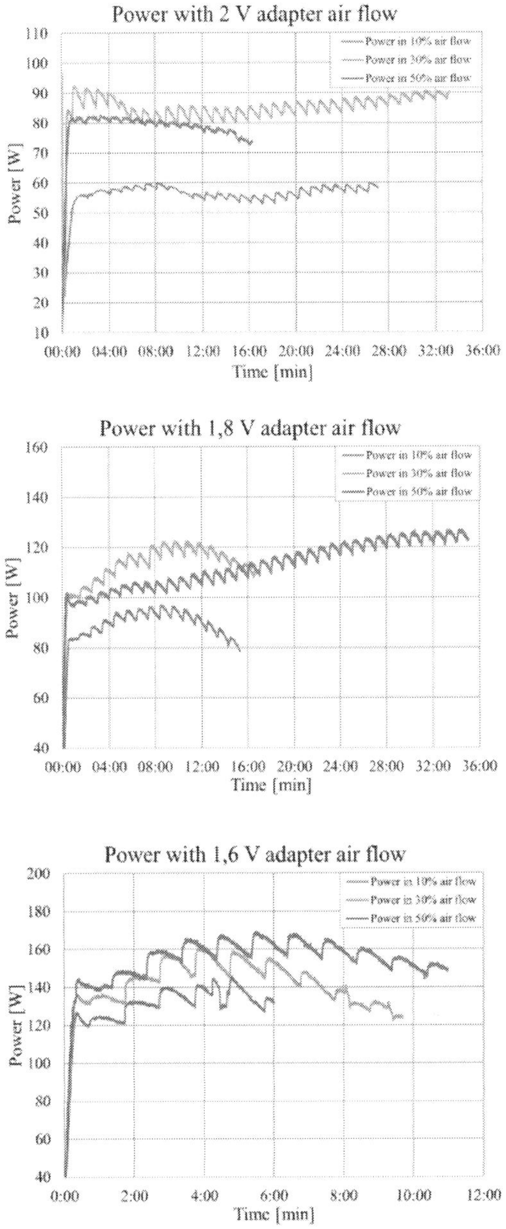

Fig. 5. Dependence of fuel cell performance on air flow and voltage – optimised adapter.

stndard cros corelation algorythm algorithm using the DaVis environment. The measurements were performed in the axis of symmetry of the fuel cell and for all three fan speed modes both in the original configuration and for the newly designed diffuser.

3 Results

In the experiments performed, we recorded the current density, temperature and power of the tested fuel cell stack for different flow rates and two diffuser configurations. The comparison of the time course of the fuel cell stack as a function of the tested parameters is summarized in Fig. 4. The original diffuser configuration is on the left and the adequate results for the optimized diffuser shape as shown in Fig. 5 are on the right. In addition to the power ratio measurements, a quantitative determination of the inlet temperature of the fuel cell and a measurement of the mass flux distribution were performed. From these results, it is clear that the temperature distribution of the bipolar plates at the inlet exhibits considerable non-uniformity. As far as the measured dependence of the fuel cell performance for each voltage level and mass flow is concerned, it is clear that the positive effect of increased air flow and thus increased cooling effect and oxygen supply has a positive effect on the fuel cell performance. This effect is particularly noticeable when the fan power is increased from 10% to 30%. In the original configuration, however, there is no increase in performance with further increases in fan power. For the optimized diffuser shape, the positive effect of uniform air distribution and thus also the increase in performance with increasing airflow through the fuel cell is evident. With this configuration, however, only one difference in the fuel cell behaviour is seen and that is at low load (2V). In this mode, the negative effect of maximum air flow on fuel cell performance is visible. At this low load, as the air flow rate increases, the diaphragm dries out resulting in a decrease in fuel cell performance.

Fig. 6. Visualization of the temperature distribution at the fuel cell inlet - original adapter.

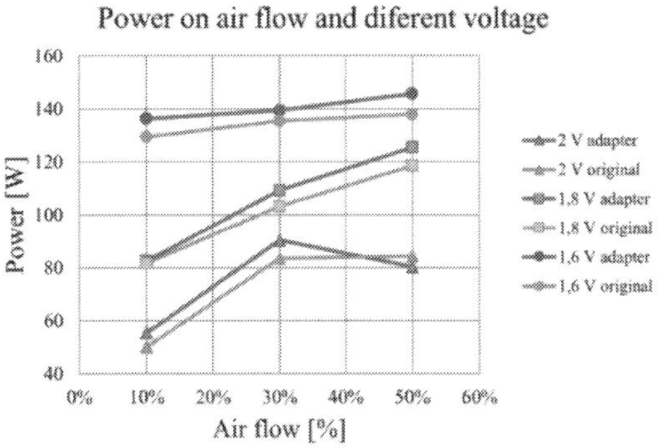

Fig. 7. Effect of air flow and fuel cell voltage on fuel cell performance.

4 Conclusion

When testing the influence of the diffuser shape and air flow through the fuel cell, we verified the positive influence of the appropriate shaping of the diffuser connecting the axial fans and the fuel cell stack Fig.7. However, it is clear from the presented results that for long-term operation of the fuel cell at maximum power for a given voltage, it will be necessary not only to control the air flow and its temperature, but it will be necessary to control the air flow depending on other parameters such as the temperature of the fuel cell and the humidity of the intake air. For these reasons, for the next experiment, it will be necessary to upgrade our measurement station and incorporate the fuel cell into an aerodynamic test bench with controlled inlet and outlet parameters to find a complete correlation between the operating conditions of the fuel cell and its performance characteristics. In the design of this new test bench, care will have to be taken to clearly determine the optimum balance of the mass flow of water through the fuel cell. In future work, we will take a detailed measurement of the mass flow uniformity at the fuel cell stage, which will then serve as a boundary condition for numerical simulations for further optimization tasks.

References

1. Barbir, Frano. PEM Fuel Cells: Theory and Practice. *Elsevier*, (2012)
2. Y. Wang, K.S. Chen, J. Mishler, S.C. Cho, X.C. Adroher, A review of polymer electrolyte membrane fuel cells: technology, applications, and needs on fundamental research, Appl Energy, **88** (2011),
3. Bvumbe TJ, Bujlo P, Tolj I, Mouton K, Swart G, Pasupathi S, et al. Review on Management, Mechanisms and Modelling of Thermal Processes in PEMFC. *Hydrogen and Fuel Cells, **20**,* (2016)
4. Rahgoshay SM, Ranjbar AA, Ramiar A, Alizadeh E. Thermal investigation of a PEM fuel cell with cooling Flow Field. Ene.**73**, (2017)

5. Shukui Liu, Zuchang Gao, Shunfeng Yang, Temperature Optimization and Control of Optimal Performance for a 300W Open Cathode Proton Exchange Membrane Fuel Cell, Procedia Engineering, **29**, (2012)
6. Bo Zhanga, Fei Lin Caizhi Zhang, Ruiyue Liao, Ya-Xiong Wang, Design and Implementation of Model Predictive Control for an Open-Cathode Fuel Cell Thermal Management System, Renewable Energy **154**, (2020)
7. Y. Song, C. Zhang, A. Deshpande, K. Tan, M. Han Fixed air flow-rate selection by considering the self-regulating function of low power air-cooled PEMFC stack *Int J Heat Mass Transf*, **158** (2020),
8. M. Hu, X. Zhu, M. Wang, A. Gu, L. Yu, Three dimensional, two phase flow mathematical model for PEM fuel cell: Part II. Analysis and discussion of the internal transport mechanisms, Energy Convers Manage, **45** (2004),
9. Jundika C. Kurnia Benitta A.Chaedir, Agus P.Sasmito, Tariq Shamim, Progress on open cathode proton exchange membrane fuel cell: Performance, designs, challenges and future directions, Appl Energy, **283**, (2021)

https://doi.org/10.1051/epjconf/202226401027

Experimental and numerical study of natural convection in 3D double horizontal annulus

Jan Pařez[1*], *Adam* Tater[2], *Jiří* Polanský[3], and *Tomáš* Vampola[1]

[1]Department of Mechanics, Biomechanics and Mechatronics, Faculty of Mechanical Engineering, Center of Advanced Aerospace Technology, Czech Technical University in Prague. Technicka 4, Praha 6, Czech Republic.
[2]Department of Technical Mathematics, Faculty of Mechanical Engineering, Center of Advanced Aerospace Technology, Czech Technical University in Prague. Technicka 4, Praha 6, Czech Republic.
[3]Department of Fluid Dynamics and Thermodynamics, Faculty of Mechanical Engineering, Czech Technical University in Prague. Technicka 4, Praha 6, Czech Republic.

Abstract. Presented paper is focused on numerical simulation of turboprop engine cooling in the section of a gas turbine. The method of heat transfer motor cooling by natural convection and radiation is described. The heat transfer in gas turbine section is represented on a simplified double annular geometry. Numerical data were compared with data that was gained from measuring on an experimental stand. The surface temperatures of the tubes were measured on the experiment stand during the cooling of the engine model. Numerical and experimental results reach a good agreement and will be further applied to the study of temperature-dependent deformation of aircraft engine parts, which have a significant effect on the safety and flawless operation of the aircraft engine.

1 Introduction

After the running turboprop engine is turned off, the engine is started to cooling down spontaneously. This cooling is not symmetrical due to natural convection. Due to the lower density, the hot gases rise upwards, resulting in a different temperature gradient between the upper and bottom side. The upper half of the engine parts are hotter and thus have lower heat transfer than the bottom part of the section (Neely and Smith [1]). Heat transfer due to natural convection in a simple 2D annulus has been described in numbers of papers with several approaches. (Yang and Kong [2], Zubkov and Narygin [3], Abu-Nada et al. [4], Francis et al. [5]). Natural convection in a circular enclosure with an internal polynomial cylinder geometry was studied in (Wang et al. [6]), on top of this, non-eccentric heat transfer due to natural convection was presented in (Prusa and Yao [7]).

A more complex phenomenon occurs when considering cooling in the generator turbine section of a turboprop engine. The turbine section of a turboprop engine consists of a double intermediate ring, where the inner intermediate ring corresponds to the flow path of the hot flue gases and the outer intermediate ring corresponds to the separation of the flow path from the engine case. The indicated scheme of the solved area is shown in Fig. 1.

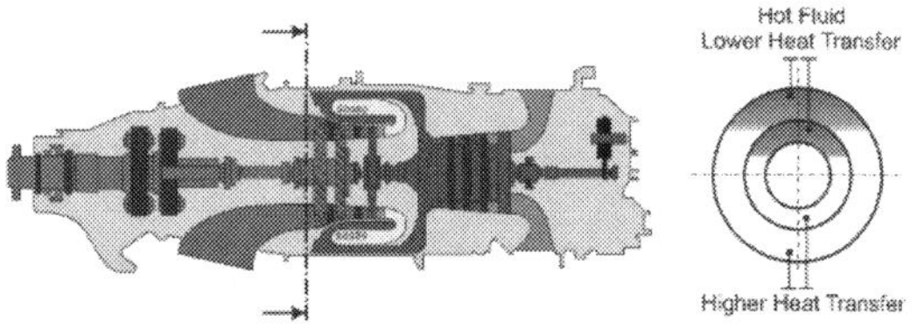

Fig. 1. Solving area of turboprop engine.

In the case of a double annulus, similarly to the simple model of the annulus, natural convection occurs when the engine is cooling down. The inner section of the flow path is asymmetrically cooled in the same time the outer section is asymmetrically heated from the hot gases of the inner section of the flow path, this asymmetry is caused by heat transfer of natural convection and radiation. In previous research, a 2D model of heat transfer caused by natural convection and radiation in a double annulus based on measuring data was presented [8].

This research has been simplified to a 2D model which does not include the effect of vorticity from natural convection and heat transfer by heat conduction in tubes. In (Padilla and Silveira-Neto [9], Krishnayatra et al. [10]) the 3D effect of natural convection is presented.

The presented paper presents a numerical study of cooling due to natural convection in a double annulus and heat transfer by radiation with an S2S model based on the geometry and boundary conditions of a turboprop engine. The geometry is showed in Fig. 2 and

* Corresponding author: jan.parez@fs.cvut.cz

corresponds to the simplification of the turbine section of the turboprop engine in Fig. 1. The initial conditions were chosen to be the constant temperature for the partially solved areas. Subsequently, there is the development of natural convection and heat transfer by radiation. Conditions are based on real measured temperature data on the running engine, which were presented in previous research (Solnař et al. [11]). Chosen models of S2S radiation and heat conduction by conduction through the tubes according to Fig. 2. Tubes thicknesses and tubes dimensions correspond to a simplification of the gas turbine. The values of the course of temperatures by the numerical model were further compared with experimental measurements on a measuring stand.

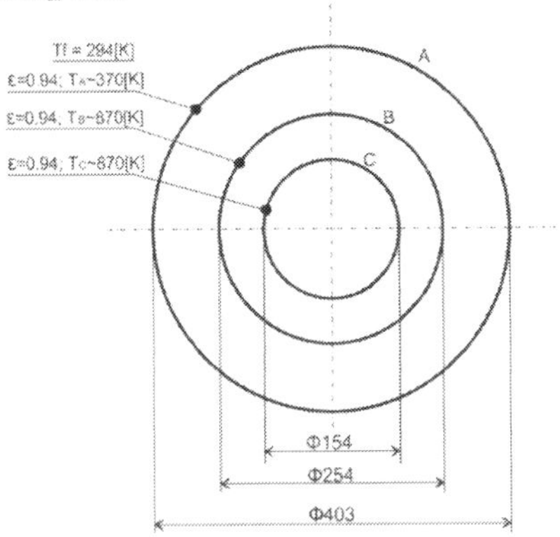

Fig. 2. Simplification sketch of generator turbine geometry.

2 Numerical model

Natural convection is very important in many engineering applications. With a suitable model it is possible to predict the flow caused by natural convection in a turbine engine. This makes it possible to influence temperatures and temperature fields when developing the engine.

The numerical model in Ansys Fluent software for natural convection and heat radiation are presented in this chapter. The conservation equations of mass, momentum and energy of fluid are considered.

A transient pressure-based coupled solver is used. Spatial discretization methods for pressure, momentum and turbulent properties are chosen as second order accurate. Time discretization is chosen as second order accurate implicit method with iterative time stepping. Air is modelled as incompressible-ideal-gas which density is defined as

$$\rho = \frac{p_{op}}{rT}, \tag{1}$$

where p_{op} is the operational pressure which is considered atmospheric, r is the specific gas constant and T is thermodynamic temperature. This model is recommended by Ansys for cases of natural convection with high temperature gradients. Moreover, specific heat is considered as piecewise-polynomial function. Thermal conductivity is assumed constant and viscosity is modelled by Sutherland's relation. Turbulence is modelled by k-omega SST model with additional transport equation for intermittency providing transition of laminar flow to turbulent. This turbulent flow solver must be further coupled with S2S radiation model with prescribed emissivity values described in Fig. 2. as radiation is the prevailing phenomenon in the beginning of the simulations. Initial values are constant temperatures for individual tubes. In fluid regions, gauge pressure is set to zero, temperature is considered standard atmospheric. Velocity in each fluid region is set to near zero value. Turbulence initial values correspond to previously mentioned values and characteristic length of computational domain. Boundary conditions on front and back faces are set as adiabatic no-slip walls. Walls that connect fluid regions with solid tubes are set to coupled no-slip walls. Heat transfer between the atmosphere and the outer surface of the largest tube is modelled by average heat transfer coefficient, defined in [12], which is variable throughout the simulation. Our transient simulations are performed until 45 minutes of real time from the start of the process. Time step length is constant and set to one second with 50 inner iterations. This prevents the Courant number to exceed 200 throughout the simulation. Computational mesh is a high quality, fully structured grid with nearly 2 million cells. The numerical method is then validated by comparing numerical results with experimental results from measurements.

The usage of turbulence models decides the „expected" flow regime in the solved geometry. In case of problems with natural convection, Rayleigh's number Eq. 2 determines the flow regime [13]. Let us informally estimate the value of the Rayleigh's number in the case of natural convection around the middle surface of the annulus.

$$Ra = \frac{\beta g \Delta T L^3}{\nu^2} Pr \tag{2}$$

where is used the middle diameter (ø254 mm) of the annulus as a characteristic dimension, see Fig. 2, and the air properties at the temperature difference of 580 [°C]. The calculated value of Rayleigh's number is 10^7. This value is in turbulent flow regime in natural convection. The showed numerical model is applied to the natural convection in a horizontal double concentric cylindrical tube with Rayleigh number in the range of 10^7 to 10^3 and Prandtl number in the range of 0.01 to 10. (Yang and Kong [14]).

In general, the flow is laminar at low Rayleigh number but turbulent at high Rayleigh number. The transition Rayleigh number from stable to unstable states is lower in the low Prandtl number cases than in the high Prandtl number cases.

In the following figures Fig. 3 to Fig. 8 is shown the temperature field affected by cooling due to natural convection and thermal radiation in a double concentric annulus. The first Fig. 3 is showed the temperature field after 15 minutes of cooling. The second Fig. 4 is shown after 30 minutes, Fig. 5 is after cooling for 45 minutes. There is a noticeable temperature difference between the top and bottom of the double annulus causing the temperature difference.

 https://doi.org/10.1051/epjconf/202226401027

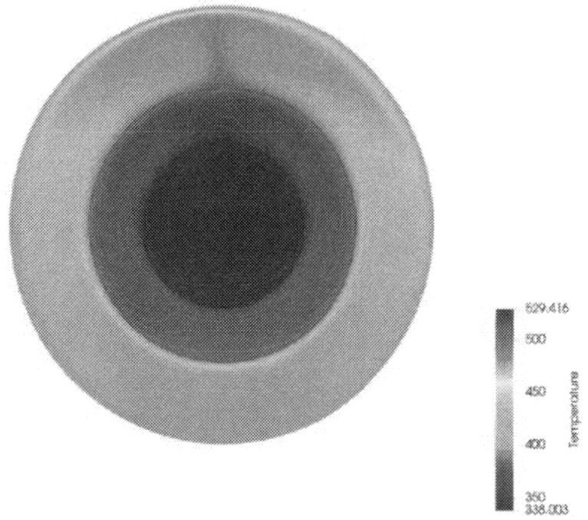

Fig. 3. Temperature field after 15 minutes.

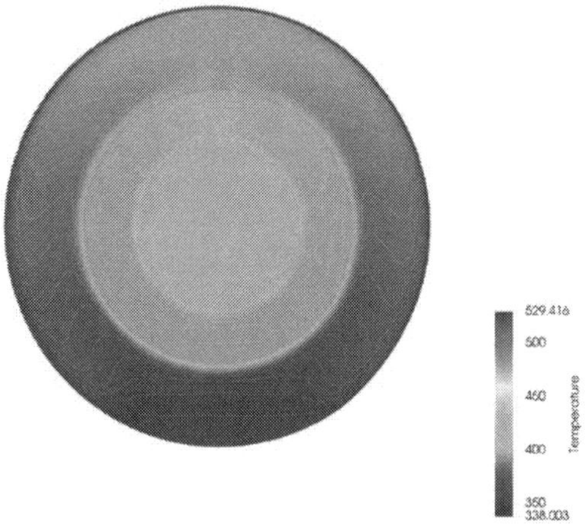

Fig. 4. Temperature field after 30 minutes.

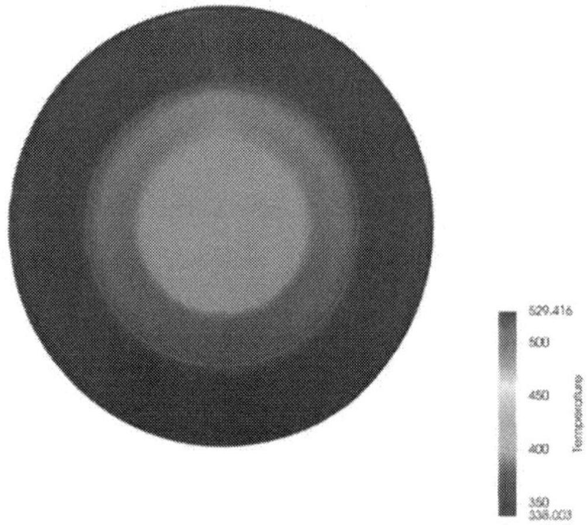

Fig. 5. Temperature field after 45 minutes.

On Fig. 6 to Fig. 8 is shown an identical temperature field in three perpendicular planes.

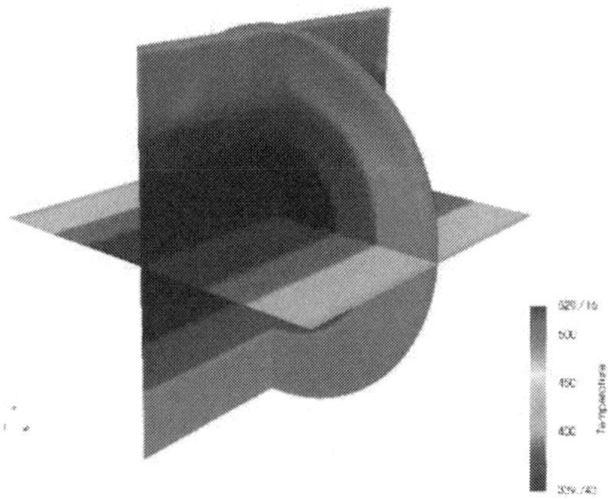

Fig. 6. Temperature field after 15 minutes.

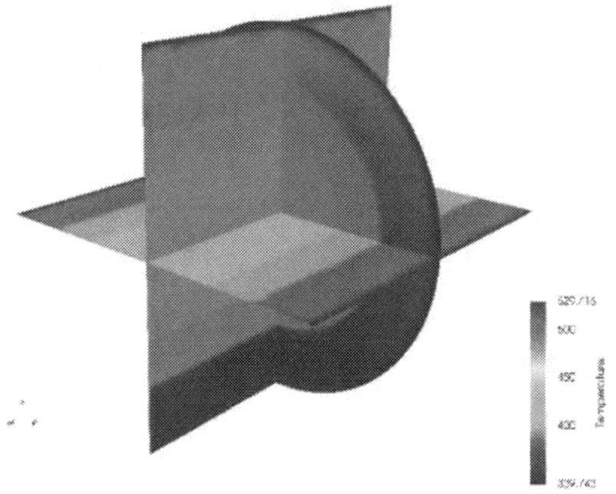

Fig. 7. Temperature field after 30 minutes.

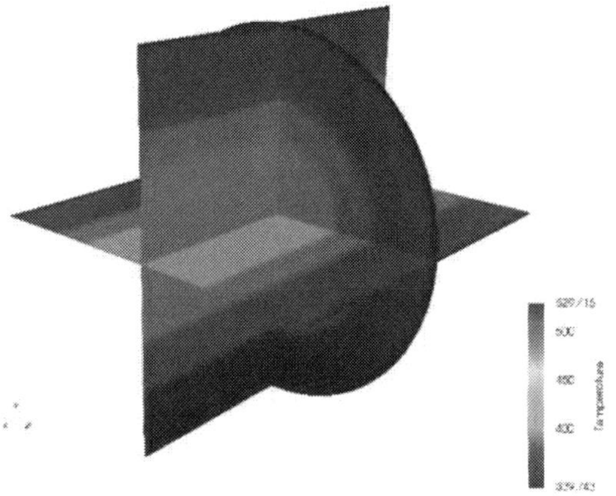

Fig. 8. Temperature field after 45 minutes.

The velocity field during natural convection due to the temperature-dependent change in density is shown in the following Fig. 9 to Fig. 11.

The flow is mainly at the walls of the tubes. This is due to the rotational circulation of air, where the hot air flows through the inner wall of the respective area of the annulus, where the air is further heated. The hot stream is then torn off the wall and forms a hot plumes. It flows upwards where it hits the upper wall and is pulled down by the cooler wall at the bottom of the wall.

The Rayleigh-Bénard Convection effect is noticeable in the flow. (King et al. [15]) studied this phenomenon in his works.

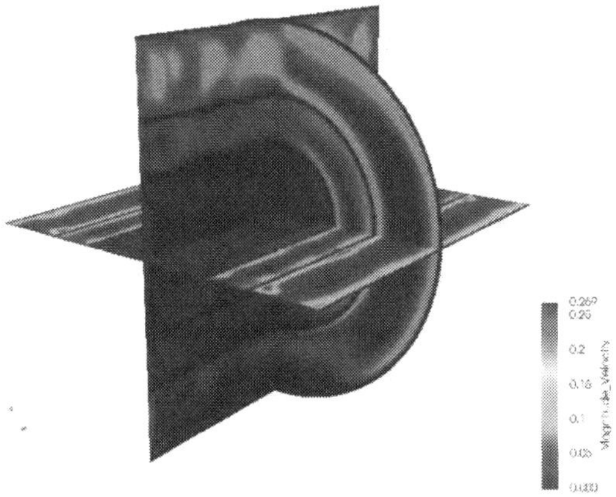

Fig. 9. Velocity field after 15 minutes.

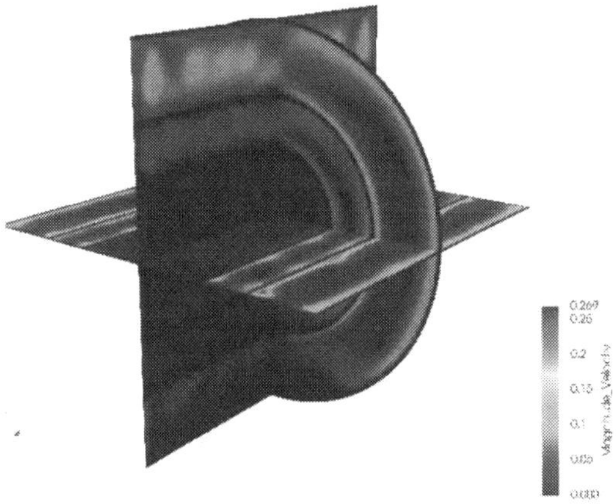

Fig. 10. Velocity field after 30 minutes.

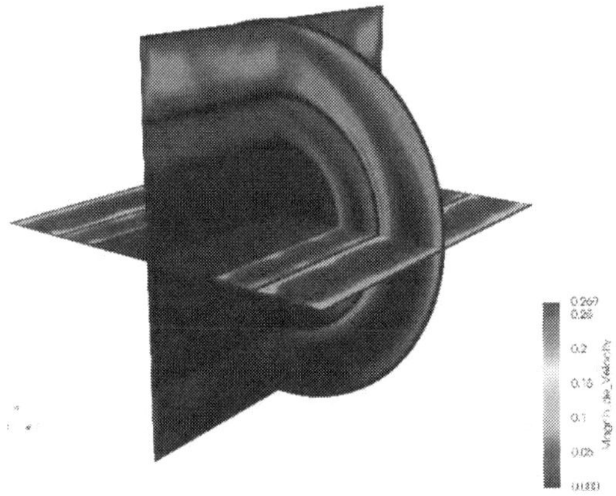

Fig. 11. Velocity field after 45 minutes.

The graph in Fig. 12 shows a convergence analysis of the solved mesh of the model. The trend of convergence shows that when using a solving mesh over 1 million elements, a sufficient deviation is achieved exceeding the range of thermocouple measurements on the experimental stand located according to Fig. 13.

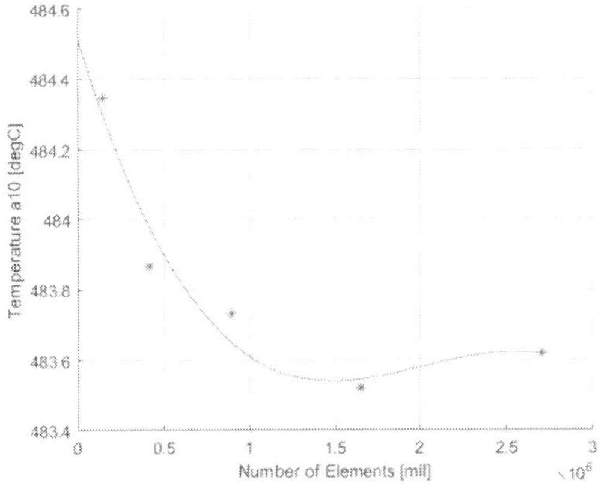

Fig. 12. Convergence analysis of the solved mesh of the model.

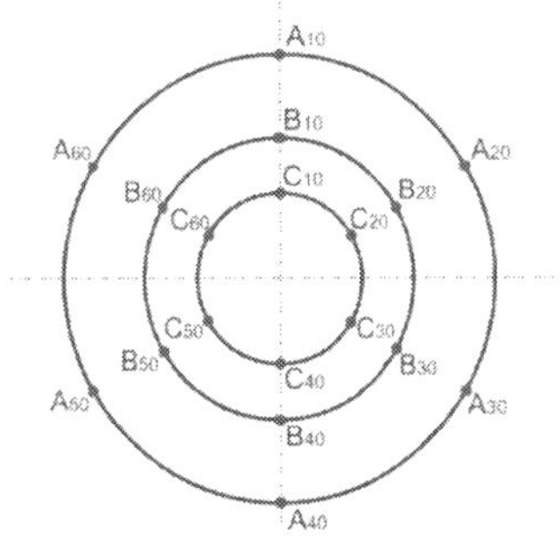

Fig. 13. Placement of K-Thermocouples on a double annulus.

3 Comparing of the numerical model on the measured data

3.1 Measuring stand Design

To verify the accuracy of the numerical model and its assumptions, an experimental stand was designed, where the visualization is showed on Fig. 14. The design of the stand is based on the geometry of the turboprop engine on Fig. 2 and is unified with a numerical model. The measuring section consisting of concentric tubes is mounted on a base frame. The front and rear sides are covered with isolated covers, in which the clearance is monitored and direct contact with the tubes is prevented. The measuring section is insulated from the covers and the base by ceramic high-temperature insulation Tyrasil made of ceramic fibres. This prevents heat transfer outside the measuring space.

Fig. 14. Visualization of measuring stand.

3.2 Instrumentation

The measuring stand is instrumented with heating elements for tempering to the required temperature and measuring thermocouples for measuring the temperature field. The flow path of the inner and middle tubes is heated by a Dawell resistance heater.

The temperature field was measured in three equidistant planes on each tube. Furthermore, the measurements in each plane were divided into 6 equivalent spaced K-Thermocouples on each tube shown in Fig. 13. This is a total of 54 K-Thermocouples for measuring the temperature field on tubes. Finally, 3 thermocouples were installed to record the course of heating temperatures. On the Fig. 15 and Fig. 16 is showed the measurement on the measuring stand. The Fig. 15 shows the analysis with a Flir thermal camera.

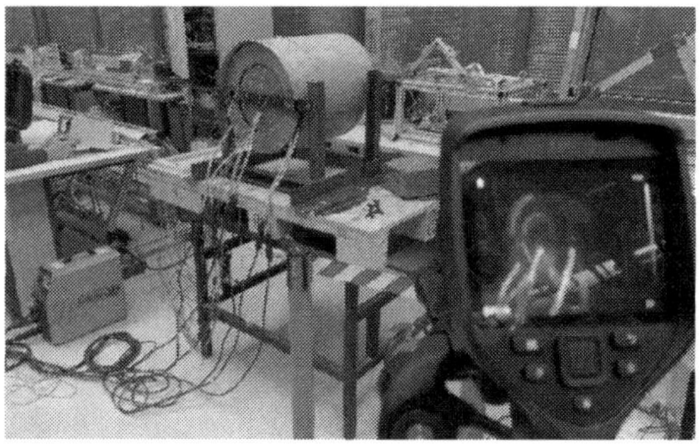

Fig. 15. Analysis by Flir thermal camera.

Fig. 16. Experimental measuring on the stand.

3.3 Validation

The numerical model was validated to verify they reach an agreement of the considered initial conditions and boundary conditions presented in the chapter of numerical calculation. By comparing the data obtained on the basis of experimental measurements with data from the numerical model, the waveforms and values of the temperature field during cooling of the computational model were compared as shown on Fig 17 to Fig. 19. Each curve corresponds to the location of the thermocouple in Fig. 13, where the highest temperature is at the upper point marked A10 and further descending to the point with the lowest temperature at A40. A similar temperature profile applies to other tubes on Fig. 18 and Fig. 19.

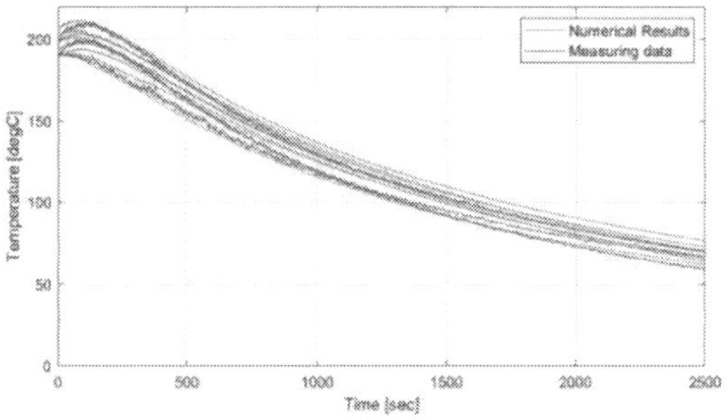

Fig. 17. Comparing the experimental and numerical results on case tube.

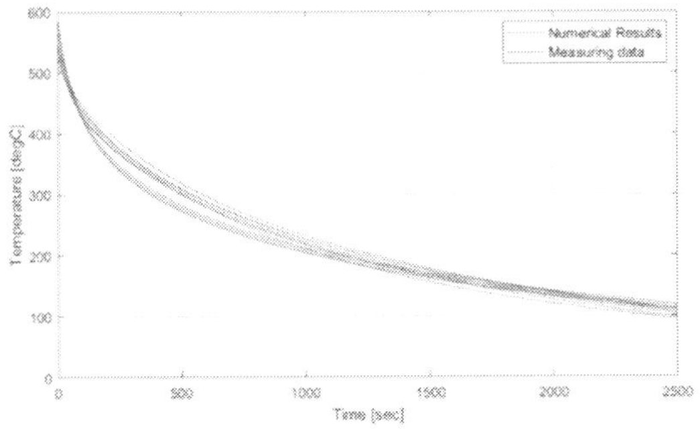

Fig. 18. Comparing the experimental and numerical results on middle tube.

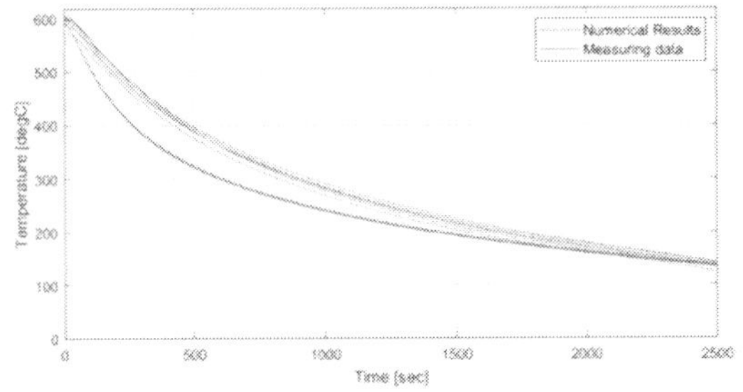

Fig. 19. Comparing the experimental and numerical results on shaft tube.

Based on the comparison, it can be concluded that the numerical and experimental results reach a good agreement.

The following Fig. 20 shows the percentage of heat transfer by radiation, convection and conduction from the outer tube. The individual heat transfer ratios are evaluated on the basis of a numerical calculation, where it can easily divide the partial heat transfer mechanism. The time dependence of the change in the mechanism of heat transfer from the dominant heat transfer by radiation to the stable method of heat transfer by radiation from 35% and heat transfer by convection 65% is shown. Heat transfer by conduction is negligible (less than 0.001%), as the computational model considers insulated walls and thus heat conduction is not significant. It is thus evident from the graphs on Fig. 20 that over time there is a greater effect of heat transfer by convection at the expense of radiation, this state stabilizes after 14 minutes at the 65%/35% ratio of radiation to convection. This is due to the temperature difference between the wall T_1 and the temperature of the surrounding air layer T_f seen in the equation Eq. 2,

$$q = \varepsilon \, \sigma_s \left(T_1 - T_f\right)^4. \qquad (2)$$

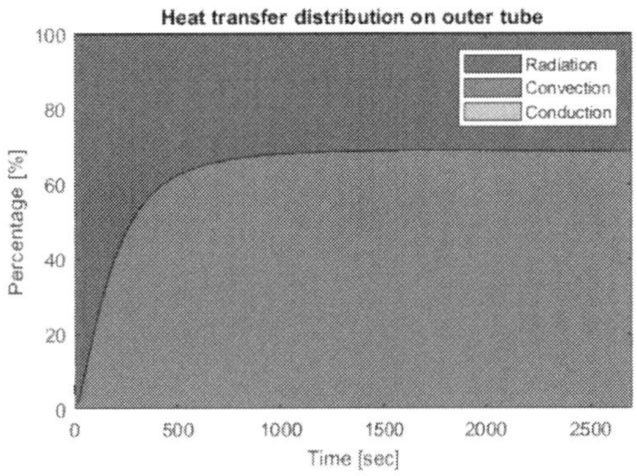

Fig. 20. Percentage of heat transfer on outer tube.

Finally, Fig. 21 shows the total heat flux through the outer tube where positive values indicate heat uptake from the environment and negative values indicate heat dissipation. The point of change of heat transfer from the direction of heat intake to heat outtake from the tube occurs after 4.5 min. At this point, heat transfer by radiation is no longer dominant, see Fig. 20.

This is followed by the point of maximum heat dissipation after 10.5 min. After this point, the system cools slowly until it cools down completely.

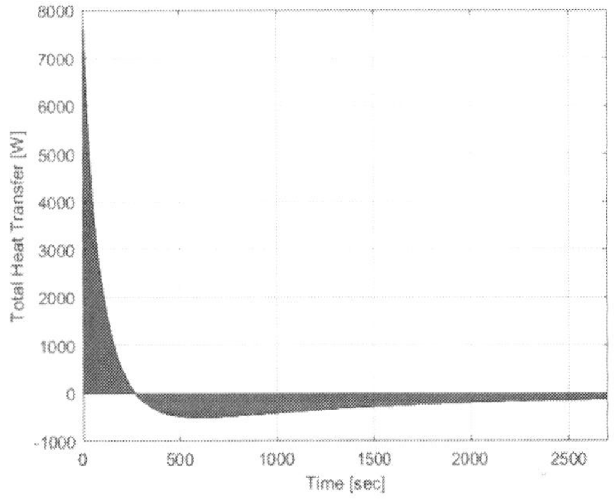

Fig. 21. Total heat transfer on outer tube.

4 Study of the middle tube influence

The verified numerical model was verified by experimental measurements on a specified middle tube. The following paragraph briefly describes the dependency of middle tube diameters on the temperature field parameters. The variable parameter of the middle tube diameter was determined and the solution for the selected steps of diameter changes was calculated. The variable parameter is related as the diameter of the middle tube to the diameter of the outer tube d / D = 0.5; 0.63 and 0.9. It is assumed that the initial conditions of the numerical model are identical to those when verifying the model on experimental data.

The aim of these calculations is to determine the change in the distribution of the temperature field of the system depending on the variable diameters of the middle tube. The velocity distribution and the vortices are shown on the following Fig. 22 to Fig. 24.

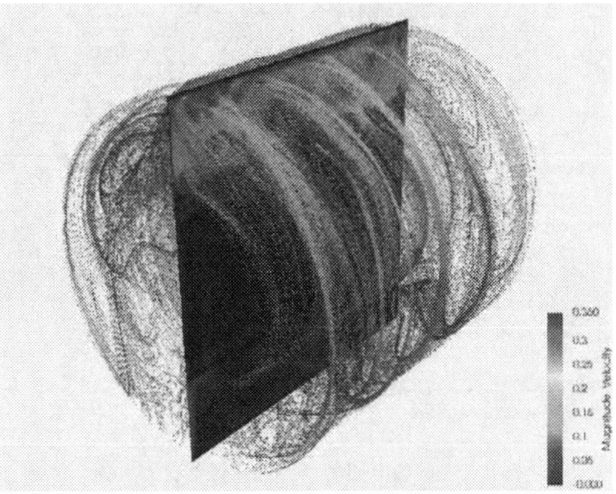

Fig. 22. The velocity field is plotted for the case d / D = 0.5.

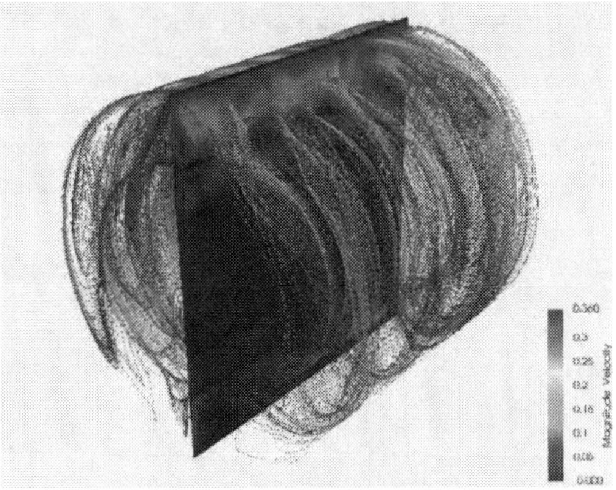

Fig. 23. The velocity field is plotted for the case d / D = 0.63.

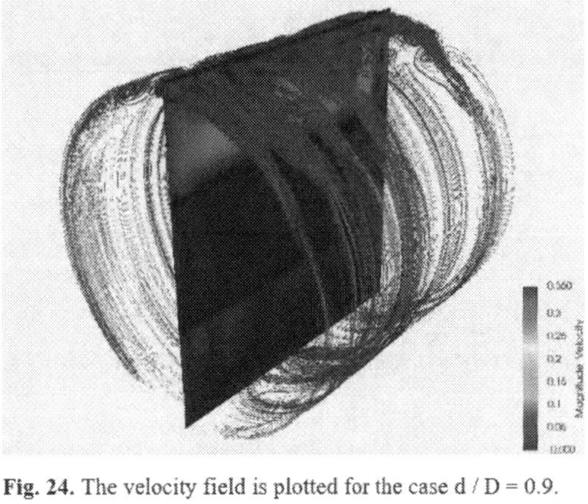

Fig. 24. The velocity field is plotted for the case d / D = 0.9.

Based on the figures shown above, it can be seen that the velocity field distribution changes with the changing size of the middle tube diameter. This results in a change in the temperature field and this affects the temperature difference between the top and bottom during cooling. When reducing the gap between the middle and outer tube, natural convection is not fully developed into one plume but into more plumes around the circumference. Furthermore, cooling with flowing hot air is not possible sufficiently rapidly, and thus cooling is slower.

Further, in Fig. 25 shows the iso-volumes of the velocity field on the initial case d / D = 0.63 from which is the steam the previously mentioned effect called the Rayleigh-Bénard Convection effect.

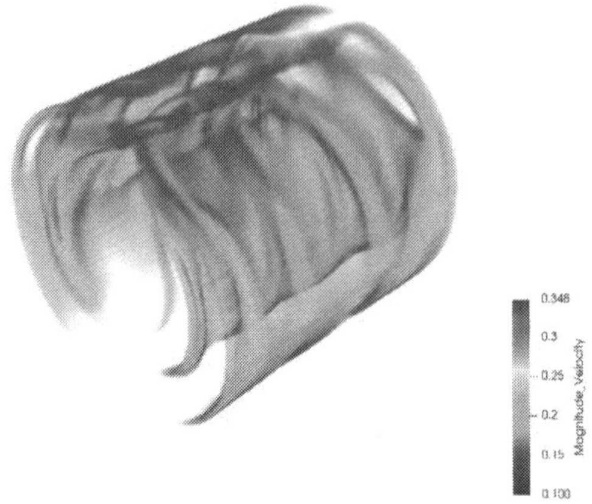

Fig. 25. Caption of the Figure 1. Below the figure.

From the above-described dependences of the velocity fields depending on the dimensions of the middle ring, it can be concluded that increasing the mean diameter has a positive effect on the consistency of the temperature field in the generator turbine section of the turboprop engine. By reducing the maximum values of the temperature difference between top and bottom, the effect of thermal expansion of the material is eliminated and thus the reduction of thermal deflection of the rotor parts of the motor is also called "Thermal Bow". This has a significant effect on the safety and trouble-free operation of the turbine engine during restart.

5 Conclusion

The results from the comparison of experimental data and the numerical model confirm the good accuracy of the numerical model. The suitability of the model used for the cooling analysis of the turboprop engine was confirmed and verified. This provides an effective tool for determining the non-stationary variable temperature field of a turboprop engine for stress analysis. The temperature-dependent deformation of a rotor assembly is called the "Rotor Thermal Bow" and will be studied in the following research. By predicting this phenomenon, it is possible to determine critical areas and conditions of thermal deflection. Contact between the stator and rotor part will be eliminated at the same time to maximize the efficiency of the turbine engine by reducing radial clearances.

Furthermore, the research will focus on the possibilities of influencing the temperature field. As already shown, the influence of the geometric characteristics of the middle tube or the influence of the flow will be studied.

Acknowledgment

Authors acknowledge support from the ESIF, EU Operational Programme Research, Development and Education, and from the Centre of Advanced Aerospace Technology (CZ.02.1.01/0.0/0.0/16_019/0000826), Faculty of Mechanical Engineering, Czech Technical University in Prague.

References

1. E. O. Smith and A. J. Neely. The impact of gas turbine compressor rotor bow on aircraft operations. The Aeronautical Journal. Royal Aeronautical Society, December 2017, **121**(1246), pp. 1808 - 1832. Available on: doi:10.1017/aer.2017.117
2. X. Yang and S. Kong. Numerical study of natural convection states in a horizontal concentric cylindrical annulus using SPH method. *Engineering Analysis with Boundary Elements*. December 2019, **2019**. Available on: doi:10.1016/j.enganabound.2019.02.007
3. P. T. Zubkov and E. I. Narygin. Numerical Study of Unsteady Natural Convection in a Horizontal Annular Channel. *Microgravity Science and Technology*. June 2020, **2020**(32), pp. 579 - 586. Available on: doi:10.1007/s12217-020-09791-2
4. E. Abu-Nada, Z. Masoud and A. Hijazi. Natural convection heat transfer enhancement in horizontal concentric annuli using nanofluids. *International Communications in Heat and Mass Transfer*. February 2008, **35**(2008), pp. 657 - 665. Available on: doi:10.1016/j.icheatmasstransfer.2007.11.004
5. N. Francis, S. Webb, M. Itamura and D. Janes. CFD Calculation of Internal Natural Convection in the Annulus Between Horizontal Concentric Cylinders. *International Communications in Heat and Mass Transfer*. Sandia National Laboratories, October 2002, (SAND2002-3132). Available on: doi:10.1115/HT2003-47515
6. Y. Wang, J. Chen and W. Zhang. Natural convection in a circular enclosure with an internal cylinder of regular polygon geometry. *AIP Advances 9*. June 2019, **065023**(2019). Available on: doi:10.1063/1.5100892
7. J. M. Prusa and L. Yao. Natural Convection Heat Transfer Between Eccentric Horizontal Cylinders. *Journal of Heat Transfer*. Lisbon, Portugal, February 1983, **105**(1), pp. 108 - 116. Available on: doi:10.1115/1.3245527
8. J. Pařez, P. Rohan and T. Vampola. Heat Transfer in Double Annular due to Natural Convection. *IOP Conference Series: Materials Science and Engineering*. Prague, Czech Republic, October 2021, **1190**(2021). Available on: doi:10.1088/1757-899X/1190/1/012002
9. E.L.M. Padilla and A. Silveira-Neto. Large-eddy simulation of transition to turbulence in natural convection in a horizontal annular cavity. International Journal of Heat and Mass Transfer. July 2008,**2008**(51). Available on: doi:10.1016/j.ijheatmasstransfer.2007.07.025
10. G. Krishnayatra, S. Tokas, R. Kumar and M. Zunaid. 3 Dimensional CFD analysis of Laminar flow Natural Convection of Hollow Cylinder with Annular Fins. *Proceedings of the 5th World Congress on Mechanical, Chemical, and Material Engineering*. Lisbon, Portugal, August 2019, (HTFF 181). Available on: doi:10.11159/htff19.181
11. S. Solnař, L. Popelka, M. Dostál, Cooling turboprop engine during shutdown. 13th International Conference on Experimental Fluid Mechanics (EFM 2018): EPJ Web of Conferences Volume 213, 2018. ISBN 9781510889668.
12. S. W. Churchill and H.S. Chu, Correlating equations for laminar and turbulent free convection from a horizontal cylinder, International Journal of Heat and Mass Transfer, 1975, Volume 18, Issue 9, pp- 1049-1053, ISSN 0017-9310,
13. R. B. Bird, W. E. Stewart and E. N. Ligthfoot. Transport Phenomena (John Wiley & Sons, Inc., New York / Chichester / Weinheim / Brisbane / Singapore / Toronto, 2002)
14. X. Yang and S. Kong. Numerical study of natural convection states in a horizontal concentric cylindrical annulus using SPH method. Engineering Analysis with Boundary Elements 2019. December 2019. Available on: doi:10.1016/j.enganabound.2019.02.007
15. M. P. King, M. Wilson and J. M. Owen. Rayleigh-Bénard Convection in Open and Closed Rotating Cavities. J. Eng. Gas Turbines Power [online]. April 2007, 129(2), pp. 305 - 311. Available on: doi:10.1115/1.2432898

Pool boiling of Novec-649 on minichannels filled with copper foam

Robert Pastuszko[1*], *Norbert* Dadas[1], and *Robert* Kaniowski[1]

[1]Kielce University of Technology, Faculty of Mechatronics and Mechanical Engineering, al. 1000-lecia Państwa Polskiego 7, PL-25-314 Kielce, Poland

Abstract. The article describes the experimental investigation of pool boiling heat transfer on minichannels with filling in the form of porous structure (copper foam). The results were compared with the data for a smooth surface and the minichannels without additional fillings. Tests were carried out for the boiling liquid Novec-649. Surfaces partially filled with porous structure were formed by inserting pieces of copper foam into the minichannels of 5 mm in depth and 1 mm width. Minichannels completely filled with copper foam formed the surface of MCC-F. The measurements were made with an increase in heat flux. The heat transfer coefficient obtained was four times higher than for the smooth surface. Additional foam fillings increased the heat transfer coefficient and reduced superheat for heat fluxes less than 100 kW/m^2. Visualisation was made using a high-speed camera which allowed to determine the diameters of the growing bubble.

1 Introduction

New surfaces that increase heat transfer rate (enhanced surfaces) are being extensively studied in many research centres in the United States, Europe, China, Japan, Iran, Brazil and South Korea.

Examples of surfaces with mini/micro channels and a porous structure in the form of metal foam, tested in the last 10 years, are presented in Table 1.

Table 1. Types of structures with micro, minichannels and metal foams for pool boiling enhanced heat transfer

Type	Reference	Configuration	Liquid	The highest HTC
Open micro/minichannels	Cooke and Kandlikar [1]	silicon microchannels 40 – 200 μm wide and 180 – 275 μm deep, etched in silicon plates	water	73 kW/m^{2}K
	Kalani and Kandlikar [2]	copper microchannels 245 – 470 μm deep and 194 – 406 μm wide	ethanol	72 kW/m^{2}K
	Gheitaghy et al. [3]	copper surface on 45° inclined microchannels (widths: 0.5—0.7 mm, depths: 0.5—1 mm)	water	about 120 kW/m^{2}K
	Kaniowski et al. [4]	microchannels 0.2-0.4 mm deep, 0.3 mm wide (HTC referred to the area of the heating cylinder - the tested area is 2 times smaller)	water Novec	64 kW/m^{2}K (water) and 8 kW/m^{2}K (Novec-649)
	Kaniowski and Pastuszko [5]	copper microchannels 0.3 mm wide, 0.2—0.5 mm deep (note as above)	water ethanol	63 kW/m^{2}K (water), 20 kW/m^{2}K (ethanol)
Plain metal foams/	Manetti et al. [6]	copper foams, porosity 0.9, thickness 1—3 mm	HFE-7100	about 20 kW/m^{2}K
	Zhu et al. [7]	copper foams, 10 and 20 ppi, porosity 90—98%, thickness 5 mm	R113	up to 4.7 kW/m^{2}K
	Yang et al. [8]	Cu foam 30-90 ppi, porosity 0.88-0.95, thickness 1—5 mm	water	up to 150 kW/m^{2}K
Gradient metal foams	Huang et al.[9]	40 ppi nickel foam, 10 ppi copper foam, thickness 4 mm, porosity 0.98	water	about 35 kW/m^{2}K
	Xu and Zhao [10]	copper and nickel foams, pore densities from 5 to 100 ppi, porosity of 0.98	water	up to 240 kW/m^{2}K
	Pratik et al. [11]	foam-like hierarchical hexagonal boron nitride (h-BN) nanomaterial	water	up to 37 kW/m^{2}K
	Xu et al. [12]	grooved array copper foam surfaces, porosity 0.9–0.98, 5–40 ppi, thickness 5–7 mm.	water	up to 90 kW/m^{2}K
	Qu et al [13]	copper V-shaped grooves, porosity of 0.95, 100-130 ppi, thickness 2 and 4 mm	water	about 93 kW/m^{2}K
	Xu and Zhao [14]	trapezoid-shaped copper-foam fins, pore densities 100–130 ppi, porosity 0.95, thickness 2—6 mm	water	about 90 kW/m^{2}K

* Corresponding author: tmprp@tu.kielce.pl

2 Experimental setup

The experimental setup to determine the heat transfer coefficient and diameter of the bubbles was presented in Fig. 1a. The heat supplied by the resistance heater to the main module can be adjusted using an autotransformer, and the power can be read on a wattmeter. The sealed sample formed the basis of a vessel with four rectangular glass walls, which made it possible observing bubble formation and growth. Photographs at 1000 fps were taken with a high-speed camera.

To improve the quality of the images, the main module was lighted with a SCHOTT KL 2500 LED. The linearity of the heat flux supplied by the heater to the specimen through the copper cylinder (Fig. 1b) was determined using four K-type thermocouples (T_5 – T_8) placed in the heating cylinder. Two additional thermocouples T_3 and T_4 (Fig. 1b,c) were placed directly below the sample. The specimen was soldered to the copper bar using a thin layer of tin. Differences in temperature indicated by the thermocouples T_3 and T_4, placed between the specimen and the cylinder, were less than 0.2 K. The main module worked as a thermosiphon – the condensed vapour returned to the vessel with the boiling liquid. Two thermocouples (T_1, T_2) were placed in the liquid above the specimen; they allowed us to determine the changes in temperature of the liquid during boiling.

The experiments were carried out at atmospheric pressure. The temperature difference between the thermocouples T_1 and T_2 was slightly greater than the measurement error, not exceeding 0.4 K.

The measurement uncertainties according to [15] were determined as follows:

- relative uncertainties in determining the heat transfer coefficient were about 24% – 3% for HTC in the range 8 – 24 kW/m^{2}K
- relative uncertainties in determining the heat flux varied between 22% to 4% for heat flux in the range 20 – 240 kW/m^2

Depending on the filling configuration, there were different lengths of the foam sections in the channels, as well as spaces between them. Figure 1c shows an arrangement of the porous fillings for different specimens, described in Table 2. The MC samples did not have porous fillings inside the minichannels, while the MCC-F specimen had a fully filled channel with porous structures. Foam with a porosity of 98% and pore densities of 100 ppi was used for the experiments.

Table 2. Specimen code and specification.

Code	p_p, mm	δ_p, mm	n_p	l_p, mm
MC	-	-	-	-
MCC-F	-	1	1	27
MCC-F-3.6-0.3-3	3.6	0.3	3	1.5
MCC-F-4.2-0.5-2	4.2	0.5	2	2

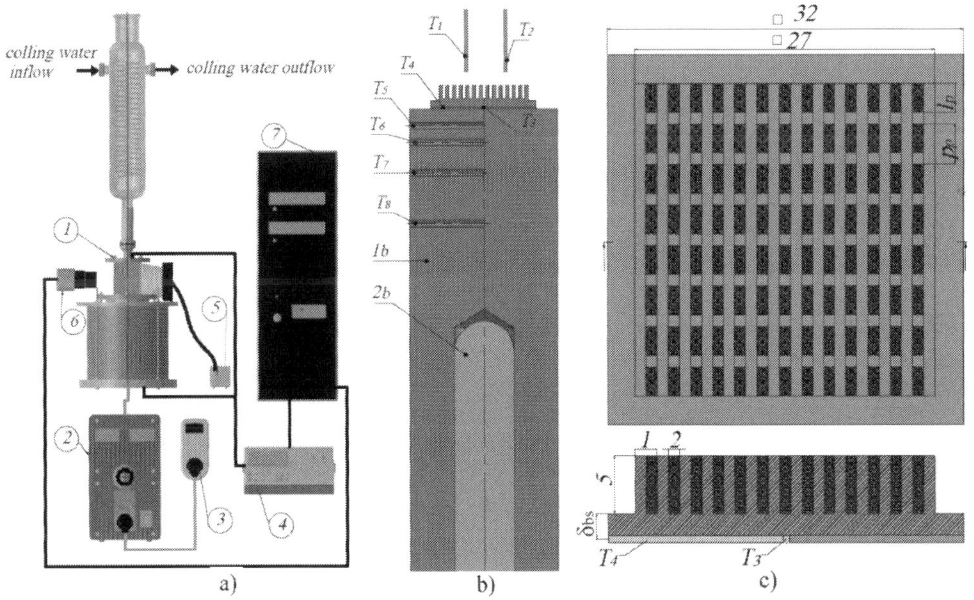

Fig. 1. a) Experimental setup, 1 – main module, 3 – autotransformer, 3 – wattmeter, 4 – data logger, 5 – light, 6 – high speed camera, 7 – PC, b) Arrangement of thermocouples, 1b – heating cylinder, 2b – cartridge heater, c) Specimen filled with porous foam structure.

3 Results

Figure 2 shows that at low heat fluxes the heat transfer coefficient (HTC) was higher when porous fillings were used. With an increase in the heat flux, a tendency toward a sharp decrease in HTC was observed along with a larger amount of foam filling the space in the channel. The minichannels with foam filling had higher superheating depending on the amount of filling (Fig. 3). Despite this, the heat transfer coefficient was several times higher than that of a plain smooth surface. MCC-F had one and a half times higher HTC than MC for a heat flux of about 50 kW/m^2. The superheat at the heat flux of about 50 kW/m^2 was lower for the MC by one and a half times than for the MCC-F. In the heat flux range 110 – 120 kW/m^2, the foam-filled specimens began to decrease the HTC compared to the minichannles without porous structure (MC). The highest difference in HTC for heat flux exceeding 120 kW/m^2 was observed for the MCC-F specimen, and the smallest was for the sample with longer distances between channel fillings (MCC-F-4.2-0.5-2). The highest HTC was reached for the MC specimen, the value was 25.4 kW/m^{2}K for the heat flux 382.5 kW/m^2. The performance at high heat flux for specimen without porous fillings was 20% better compared to MCC-F-4.2-0.5-2.

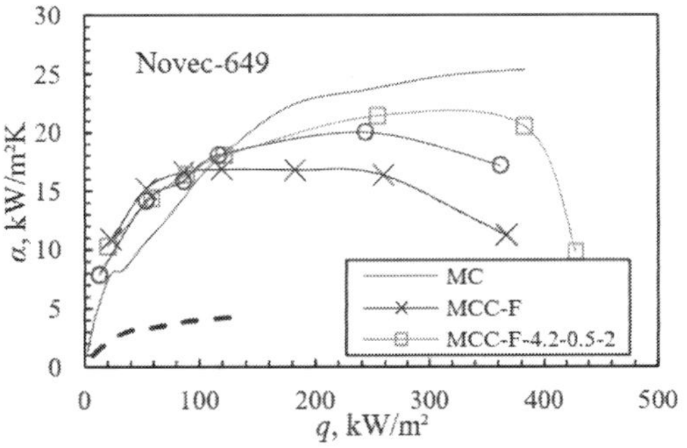

Fig. 2. Boiling curves for Novec-649, heat transfer coefficient versus heat flux.

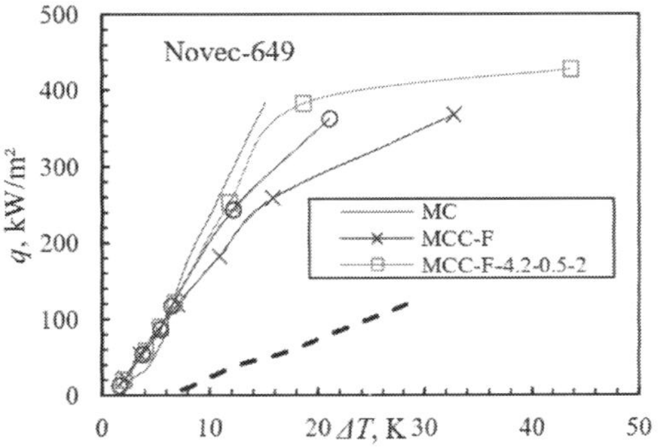

Fig. 3. Boiling curves for Novec-649, heat flux versus superheat.

The attempts to predict the diameter of the bubbles that grow during pool boiling were discussed in [16]. The authors mentioned the equation (1) for bubble growth in uniformly superheated liquid:

$$D = \phi_c \frac{4Ja\sqrt{a\tau}}{\sqrt{\pi}} \tag{1}$$

where τ is the time, a is the liquid thermal diffusivity and ϕ_c is sphericity correction factor, $\phi_c = \sqrt{3}$ according Plesset and Zwick. The Jakob number can be calculated from the equation:

$$Ja = \frac{\rho_l c_{pl} \Delta T}{\rho_v i_{lv}} \tag{2}$$

Liquid thermal diffusivity is described by the equation:

$$a = \frac{\lambda_l}{c_{pl}\rho_l} \tag{3}$$

Equation (1) was related to a plain surface. To correlate this equation for an extended surface with additional pieces of copper foam inside minichannels, the authors suggest the use of correction factor C_{mc}= 4.85, thus:

$$D_{mc} = C_{mc}\phi_c \frac{4Ja\sqrt{a\pi\tau}}{\pi} \tag{4}$$

The calculations according Eqs (1) and (4) were presented in Fig. 4 compared to the experimental data achieved during pool boiling visualization. The accuracy of bubble diameter measuring of 0.2 mm was assumed. The parameters of Novec-649 required for the calculations are shown in Table 3.

Table 3. Parameters of saturated Novec-649

T_{sat} °C	ρ_l kg/m^3	ρ_v kg/m^3	c_{pl} J/kgK	i_{lv} kJ/kg	λ_l W/mK	σ_l N/m
49	1532	13.4	1118	88	0.054	0.0092

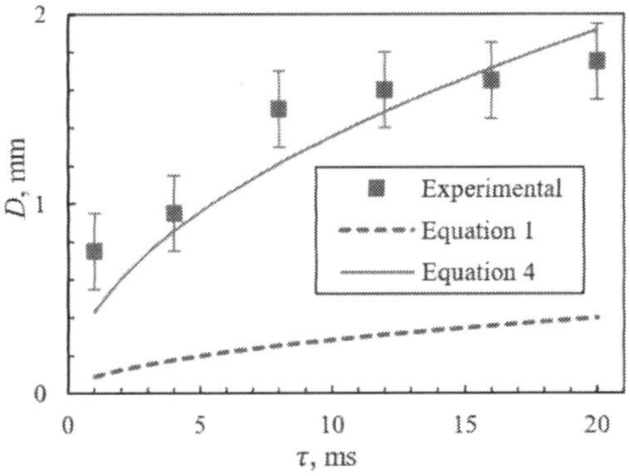

Fig. 4. Diameter of the growing bubble versus time.

4 Visualization of boiling

The bubbles formed at the minichannels outlets are shown in Fig. 5. By observing the places of growth of the bubbles and determining their diameters during growth, it is possible to better understand the mechanics that appear during the experiments. Bubbles were formed in the spaces between the porous structures and the liquid that was sucked by the porous fillings.

Fig. 5. Visualization of the pool boiling of Novec-649 on the surface of the minichannels surface MCC-M-4.2-0.5-2, q = 27.9 kW/m^2, a) t = 1 ms, b) t = 4 ms , c) t = 8 ms, d) t = 12 ms, e) t = 16 ms, f) t = 20 ms.

5 Conclusions

Based on the data presented in this paper, the following conclusions were drawn:

- For the Novec-649 working fluid, the highest heat transfer coefficients at heat fluxes less than 120 kW/m^2 were obtained using minichannels with additional porous filling. At larger heat fluxes, the largest HTC provided minichannels without fillings.
- The proposed surfaces with partially filled minichannels with porous structure are a new concept among structures designed to intensify the pool boiling heat transfer.
- The experiments carried out show that under certain conditions, fillings in the form of copper foam can improve the critical heat flux.
- The spaces between the porous fillings allowed for the efficient releasing of the vapour, and a significant enhancement was achieved both in CHF and in HTC.
- The proposed modified Plesset-Zwick relationship enables the presentation of changes in the diameter of the growing bubble as a function of time with satisfactory accuracy.

Nomenclature

a – liquid thermal diffusivity $[m^2/s]$,
c_p – specific heat at constant pressure $[J/kgK]$,
C – correction factor
D – diameter of the bubble $[mm]$,
i_{lv} – heat of vaporization $[kJ/kg]$,
Ja – Jakob number $[-]$,
l – length [mm],
p – pitch [mm]

Greek symbols

ΔT – superheat for specimen surface $[K]$,
δ – thickness [mm],
λ – thermal conductivity $[W/mK]$,
ρ – density $[kg/m^3]$,
σ – surface tension [N/m],
τ – time $[s]$,
ϕ_c – sphericity correction factor,

Subscripts

l – liquid phase,
mc – minichannels,
p – pore, porous,
v – vapour phase.

References

1. D. Cooke, S.G. Kandlikar, J. Heat Transfer, **133** (2011),
2. A. Kalani, S.G. Kandlikar, proc. ASME 10th Int. Conf. Nanochannels, Microchannels, and Minichannels **10** (2012),
3. A.M. Gheitaghy,. A. Samimi, H. Saffari, Appl. Therm. Eng. **126** (2017),
4. R. Kaniowski, R. Pastuszko, Ł. Nowakowski, EPJ Web Conf. **143** (2017),
5. R. Kaniowski, R. Pastuszko, EPJ Web Conf. **180** (2018),
6. L.L. Manetti, A.S.O.H. Moita, Int. J. Heat Mass Transf. **152** (2020),
7. Y. Zhu, H. Hu, G. Ding, H. Peng, X. Huang, D. Zhuang, J. Yu, Int. J. Refrigeration **34** (2011),
8. Y. Yang, X. Ji, J. Xu, Int. J. Therm. Sci. **49** (2010),
9. R.L. Huang, C.Y. Zhao, Z.G. Xu, Int. J. Multiph. Flow **103** (2018),
10. Z.G. Xu, C.Y. Zhao, Int. J. Heat Mass Transf., **85** (2015),
11. K.C. Pratik, A. Nammari, T. S. Ashton, A. L. Moore, Int. J. Heat Mass Transf. **95** (2016),
12. Z.G. Xu, Z.G. Qu, C.Y. Zhao, W.Q. Tao, Int. J. Heat Mass Transf. **77** (2014),
13. Z.G. Qu, Z.G. Xu, C.Y. Zhao, W.Q., Int. J. Multiph. Flow **41** (2012),
14. Z.G. Xu, C.Y. Zhao, Appl. Therm. Eng. **60** (2013)
15. R. Pastuszko, R. Kaniowski, N. Dadas, M. Bedla-Pawlusek, Int. J. Heat Mass Transf. **179** 2021
16. R. Cole, H.L. Shulman, Int. J. Heat Mass Transf. **9** (1966).

Advances in simulations of heat exchange in fluid flows by high-order methods

Jan Pech[1*]

[1]Institute of Thermomechanics of the CAS, Dolejškova 1402/5, 182 00 Praha 8, Czech Republic

Abstract. The article presents high-order simulations of heat transfer in fluid flow. Heat induced changes in material properties (density, viscosity and heat conductivity) substantially influence flow structures. The change of fluid density implies naturally the thermal expansion what results in non-homogeneous divergence of velocity and the model of evolutionary Navier-Stokes-Fourier equations is not strictly incompressible in this case. Proposed algorithm treats the model as a fully coupled system of equations. Applied method of operator splitting is promising to become a more general concept for numerical modelling of active quantities advected by incompressible fluids. This work follows the 2D results presented during EFM 2019 and extends to 3D together with advances of the code reimplementation to current version of the Nektar++ library (open-source framework for spectral/hp finite elements). Results for various settings for both the forced/mixed and natural convection are shown.

1 Introduction

In this contribution, we are going to revisit the problem of fluid flow coupled with a heat exchange, which was presented in previous EFM conference as a 2D study of natural and forced convection. The algorithm, unless based on the same idea, was reformulated now and implemented as a full 3D solver into the Nektar++ library [1], an open source library supporting the *spectral/hp* finite elements. The solver is expected to be available in the next release of the library. Mentioned 2D studies were based on early version (3.3.0) of Nektar++ and present reimplementation allows combination of new features of the library with the solver, namely stabilisation techniques (spectral vanishing viscosity), algebraic solvers, parallelisation or pre- and post-processing tools.

The set of equations to be discussed is the fully coupled Navier-Stokes-Fourier system with temperature dependent material properties (dynamic viscosity μ, thermal conductivity κ and density ρ).

$$\rho\left(\frac{\partial \mathbf{v}}{\partial t} + \mathbf{v}\cdot\nabla\mathbf{v}\right) = -\nabla p + \frac{1}{\mathrm{Re}}\nabla\cdot(2\mu\mathbb{S}) + \mathrm{Ri}\Theta\mathbf{g} \qquad (1a)$$

$$\frac{\partial \rho}{\partial t} + \nabla\cdot(\rho\mathbf{v}) = 0 \qquad (1b)$$

$$\rho\left(\frac{\partial T}{\partial t} + \mathbf{v}\cdot\nabla T\right) = \frac{1}{\mathrm{Re\,Pr}}\nabla\cdot(\kappa\nabla T)\,, \qquad (1c)$$

Table 1. Definitions of used nondimensional numbers. Note, that temperature T is normalised by farfield value T^+_∞ instead of widely used characteristic temperature difference ΔT^+.

Name	Symbol	Definition
Prandtl	Pr	$\dfrac{c^+_{p\,\infty}\mu^+_\infty}{\kappa^+_\infty}$
Rayleigh	Ra	RiPrRe^2
Reynolds	Re	$\dfrac{L^+\lvert\mathbf{v}^+_\infty\rvert\rho^+}{\mu^+_\infty}$
Richardson	Ri	$\dfrac{\Delta\rho^+ g^+ L^+}{\rho^+_\infty\lvert\mathbf{v}^+_\infty\rvert^2}$ or $\dfrac{g\beta\Delta T L}{\lvert\mathbf{v}_\infty\rvert^2}$

where

$$\mathbb{S} = \mathbb{D} - \frac{1}{3}(\nabla\cdot\mathbf{v})\mathbb{I} \qquad (2)$$

is related to the deviatoric part of the stress tensor (pure shear stress) and $\mathbb{D} = \frac{1}{2}\left[\nabla\mathbf{v} + (\nabla\mathbf{v})^{\mathrm{T}}\right]$. This dimensionless system of equations represents a model for flow of a heated fluid, while the fluid is supposed to be Newtonian, calorically perfect and obeying Fourier's law of heat conduction. The unknowns are velocity $\mathbf{v} = (u, v, w)^{\mathrm{T}}$, pressure p and temperature T, other symbols are listed in table of nomenclature. Used dimensionless numbers are summarized in table 1.

* Corresponding author: jpech@it.cas.cz

Nomenclature			
(We use dimensionless formulation, if a quantity acts as with physical units it is denoted by superscript "+")			
a, $\mathbf{a}$, $\mathbb{A}$	scalar, vector, tensor	*Greek letters*	
c_p^+	specific heat at constant pressure	β	thermal expansion coefficient
g^+	gravitational acceleration	κ	thermal conductivity
$\mathbf{g}$	unit vector in direction of gravitational force	μ	dynamic viscosity
$\mathbb{I}$	unit tensor	ρ	density
L^+	characteristic length	Ω	computational domain
$\mathbf{n}$	unit normal vector to a boundary		
p	pressure	*Symbols*	
t	time	$\lvert\cdot\rvert$	vector magnitude
T	temperature $T = \frac{T^+}{T_\infty^+}$, $T_r = \frac{T_W^+}{T_\infty^+}$	$(\cdot)^{\mathrm{T}}$	transposition
Θ	temperature $\Theta = \frac{T^+ - T_\infty^+}{T_W^+ - T_\infty^+}$	∇, $\nabla\cdot$, $\nabla\times$	gradient, divergence, curl
$\mathbf{v} = (u, v, w)$	velocity	∇^2	Laplace

Presented system considers the velocity to be divergent as a consequence of the thermal expansion unless the model originates in the incompressible formulation. This ambiguity follows from dependence of density purely on temperature, while the dependence on pressure is neglected. The system and algorithm get close to the border between incompressible and compressible formulations. Especially between numerical methods for incompressible and compressible flow have been substantial differences.

2 Algorithm

Proposed algorithm is derived from the high-order splitting scheme introduced for the incompressible Navier-Stokes system in Karniadakis [2] and splitting approach for variable viscosity of Karamanos [3]. Main aspects of its derivation stays in approximation of the time derivatives by Backward Difference formulas (BDF) and splitting the system to implicitly solved pressure-Poisson equation and a second order problems of form $(\nabla^2 - \lambda)u = f$. Idea for incorporating a time dependent material property (q) stays in its representation in form $q(\mathbf{x}, t) = \bar{q} + q_s$, where $\bar{q}$ is a time-independent part (constant or generally $\bar{q} = \bar{q}(\mathbf{x})$) scaling the implicitly treated operators, while $q_s = q_s(\mathbf{x}, t)$ acts in time-dependent terms evaluated within extrapolation (see Pech [4]). It is worthy to note, that we do not impose any restriction to the form of the functions describing temperature dependencies of material properties, they are expected to be fully nonlinear. The algorithm solves each step at once without any iterative process.

Its detail description is beyond the scope of this article. Denoting extrapolation of Q-th order

$$[u]^* = \sum_{q=0}^{Q} u_{n-q} \approx u_{n+1}$$

and BDF

$$\left[\frac{\partial u}{\partial t}\right]_{n+1}^{BDF} = \frac{\gamma u_{n+1} - \sum_{q=0}^{Q} \alpha_q u_{n-q}}{\Delta t} \approx \left.\frac{\partial u}{\partial t}\right|_{n+1},$$

the main steps may be sketched as follows:

1. Pressure-Poisson equation

$$\nabla^2 p = \frac{\gamma}{\Delta t}\left[\left[\frac{\partial \rho}{\partial t}\right]^{BDF}\right]^* + \nabla\cdot\left\{\frac{\gamma}{\Delta t}\rho^* \sum_{q=0}^{Q-1} \frac{\alpha_q}{\gamma}\mathbf{v}_{n-q} + [\mathcal{A}]^*\right\} \quad (3)$$

supplemented by the *High Order Pressure Boundary Condition* (HOPBC)

$$\frac{\partial p}{\partial \mathbf{n}} = \mathbf{n}\cdot\left[-\rho_n\left[\frac{\partial \mathbf{v}}{\partial t}\right]_n^{BDF} + \mathcal{A}\right]^* \quad (4)$$

2. elliptic problem for each of the velocity components

$$\nabla^2 \mathbf{v}_{n+1} - \frac{\gamma \mathrm{Re}}{\Delta t}\mathbf{v}_{n+1} = \mathcal{F}_V([\mathcal{A}]^*, [\nabla d - \nabla\times\nabla\times\mathbf{v}]^*). \quad (5)$$

3. the same type of elliptic equation, with specific source term, is solved finally for temperature

$$\left(\nabla^2 - \mathrm{RePr}\frac{\gamma}{\Delta t}\right) T_{n+1} = \ldots \quad (6)$$

where

$$\mathcal{A} = -\rho(\mathbf{v}\cdot\nabla\mathbf{v}) + \frac{1}{\mathrm{Re}}\left(-\mu\nabla\times\nabla\times\mathbf{v} + \nabla\mu\cdot\left[\nabla\mathbf{v} + (\nabla\mathbf{v})^{\mathrm{T}}\right] - \frac{2}{3}(\nabla\mu)\,d + \frac{4}{3}\mu\nabla d\right) + \mathbf{f}^{EX}. \quad (7)$$

and divergence is evaluated as

$$d = -\frac{1}{\rho}\left(\left[\frac{\partial \rho}{\partial t}\right]^{BDF} + \mathbf{v}\cdot\nabla\rho\right). \quad (8)$$

The correctness of the algorithm implementation and its convergence properties were tested using a smooth and bounded manufactured solution. For illustration, x-velocity component of the used exact solution was u `=2.0*cos(PI*x)*cos(PI*y)*cos(PI*z)*sin(t),`

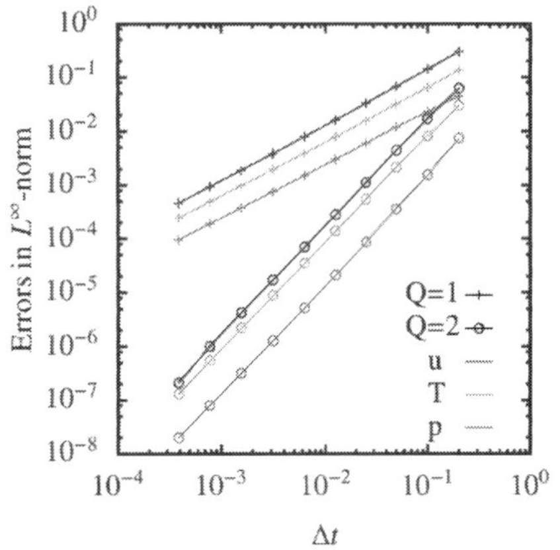

Fig. 1. First and second order, $Q = 1, 2$, convergence in time $\Delta t \to 0$.

other velocity components, temperature and pressure were in similar form. Functions describing temperature dependencies of material were set to identical form of power law $\rho = \mu = \kappa = (1 + 0.1T)^2$ for the convergence test. Error decay in u, T and p with shortening the timestep Δt is shown in L_∞-norm in fig. 2. Polynomial degree for this test was set to 14, sufficient for full resolution in spatial coordinates. Presented error decay therefore completely belongs to discretisation in time. Figure 2 shows the error decay of the scheme if the Dirichlet boundary condition is imposed, convergence with full HOPBC is slower in pressure component due to the error arising from boundary extrapolation and the fact, that pressure-Poisson step becomes fully Neumann.

The new implementation of the solver was applied in simulations of natural convection and forced convection/transient flow, studied previously using the same method in 2D.

3 Simulations

The temperature dependencies of material properties require knowledge of the farfield thermodynamic state. However, this information is often missing in published results. Materials have different properties at different temperatures and not only in their values, but also in sense of derivatives. Knowledge of values is sufficient for evaluation of the dimensionless numbers, but derivatives play an important role in our model and are unknown if the thermodynamic state of the experiment is not specified. Unfortunately, information about T_∞ was not found in none of the paper we followed (Kuehn [5], Ren [6]). For that reason, $T_\infty = 293,15$ was set, but it might be an artificial source of inaccuracy.

To check the scheme performance on a wider variety of cases, parameters for air were set in case of natural convection, while water flow was studied in case of transient

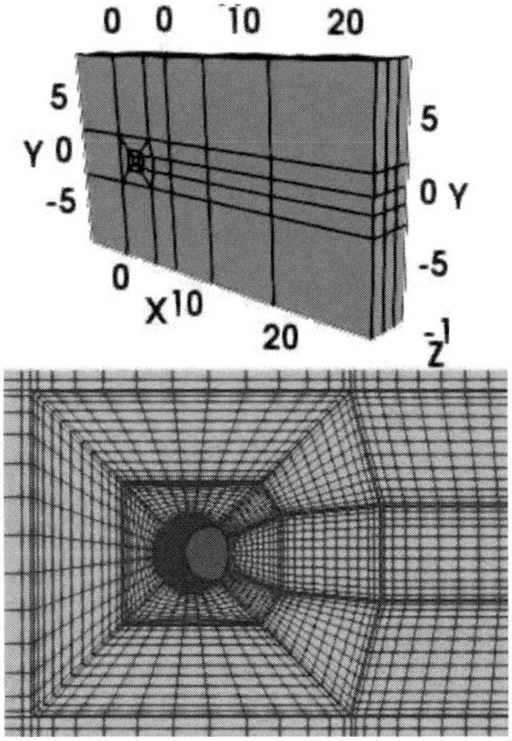

Fig. 2. Top: Computational mesh consisting of 111 elements. Bottom: Detail to cylinder area, abscissas connect the integration poins.

convection (in case of constant properties, models would be indistinguishable). The dependencies followed from fitting the data tabulated in [7] resulting as power laws or second order polynomial approximations.

Computational domain Ω for both tests was chosen to be of the same spatial extent 23x18x4 (cylinder diameter $L = 1$), the same as used in Ren [6]. Curved boundary of the cylinder was approximated by polynomial of order 15, which is sufficient for full resolution of this curvature. In both regimes the mesh consisting of 111 elements was used, see Fig. 3. Such a low number of elements in the mesh brings a limit for the number of cores over which the task may be parallelised.

3.1 Natural convection

Study of the natural convection was inspired by existence of data from Kuehn [5], which were already used for comparison with results obtained with the 2D Navier-Stokes-Fourier solver. The data are two dimensional, but geometrical setting of our 3D model is not expected to induce 3D patterns in vicinity of the source of heat.

The farfield/reference temperature was chosen to be $T_\infty = 293.15$, and ratio $T_r = \dfrac{T\_W}{T_\infty} = 1.01$ (T_W is the temperature of cylinder wall). Material parameters refered to air, therefore Pr = 0.71. The Reynolds number was set

Table 2. Boundary conditions used in the natural convection simulation. All the Neumann boundary conditions denoted $\frac{\partial\bullet}{\partial\mathbf{n}}$ were homogeneous, "HOPBC" denotes 4.

	$\mathbf{v}$	T	p
inflow	$(\frac{\partial u}{\partial\mathbf{n}}, \frac{\partial v}{\partial\mathbf{n}}, 0)$	1	0
top&bottom	$\left(\frac{\partial u}{\partial\mathbf{n}}, 0, \frac{\partial w}{\partial\mathbf{n}}\right)$	$\frac{\partial T}{\partial\mathbf{n}}$	HOPBC
outflow	$\left(\frac{\partial u}{\partial\mathbf{n}}, \frac{\partial v}{\partial\mathbf{n}}, \frac{\partial w}{\partial\mathbf{n}}\right)$	$\frac{\partial T}{\partial\mathbf{n}}$	0
cylinder	$(0, 0, 0)$	T_r	HOPBC
front&back	$\left(\frac{\partial u}{\partial\mathbf{n}}, \frac{\partial v}{\partial\mathbf{n}}, 0\right)$	$\frac{\partial T}{\partial\mathbf{n}}$	HOPBC

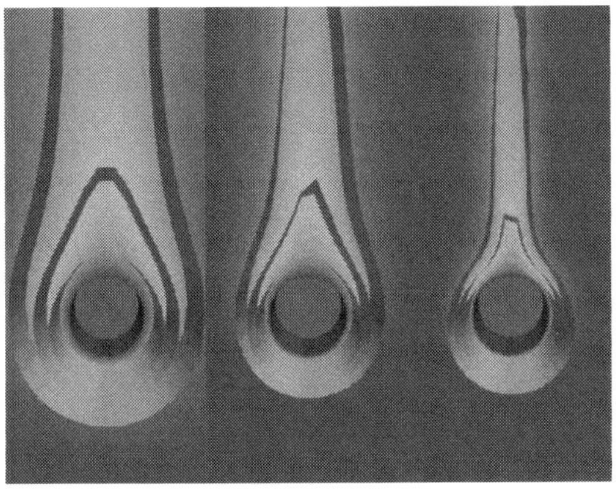

Fig. 3. Comparison of temperature distribution in vicinity of heated cylinder for Ra = 100 (left), Ra = 10^3 (center) and Ra = 10^4 (right).

to Re = 120. Computation was performed with timestep $\Delta = 0.005$ and polynomial approximation in spatial coordinates was of degree 10. Setting of the boundary conditions is summarized in table 3.1.

Only very basic outflow condition was available for the newly implemented solver and size of the computational domain was limited what resulted in instability of the computation at the time, when the wake structures touched the "outflow" boundary. However, the domain extent was sufficient to allow development of the field around the cylinder, see Fig. 3.1.

Contours of temperature in this buoyancy driven flow well coincide with results obtained for the same Rayleigh numbers in mentioned 2D simulations. More detail studies of 3D structures emerging later downstream are left for a future investigations.

3.2 Transient flow

The second simulation was an attempt to reconstruct results published in Ren[6] using currently implemented algorithm, which extends substantially the model (previous results were taken strictly with constant parameters and homogeneous divergence).

Table 3. Overview of boundary conditions used in the transient flow simulation. All the Neumann boundary conditions denoted $\frac{\partial\bullet}{\partial\mathbf{n}}$ were homogeneous, "HOPBC" denotes 4.

	$\mathbf{v}$	T	p
inflow	$(1, 0, 0)$	1	HOPBC
top&bottom	$\left(\frac{\partial u}{\partial\mathbf{n}}, 0, \frac{\partial w}{\partial\mathbf{n}}\right)$	$\frac{\partial T}{\partial\mathbf{n}}$	$\frac{\partial p}{\partial\mathbf{n}}$
outflow	$\left(\frac{\partial u}{\partial\mathbf{n}}, \frac{\partial v}{\partial\mathbf{n}}, \frac{\partial w}{\partial\mathbf{n}}\right)$	$\frac{\partial T}{\partial\mathbf{n}}$	0
cylinder	$(0, 0, 0)$	T_r	HOPBC
front&back	$\left(\frac{\partial u}{\partial\mathbf{n}}, \frac{\partial v}{\partial\mathbf{n}}, 0\right)$	$\frac{\partial T}{\partial\mathbf{n}}$	$\frac{\partial p}{\partial\mathbf{n}}$

Boundary conditions set in this case are summarized in table 3.2, $T_r = 1.01$.

The flow at Re = 85 and Ri = 1.0 stayed in the regime of parralel vortex shedding for very long time in current simulations, unless it is known it is unstable also in the spanwise direction. This may be a consequence of low level of numerical noise produced in spatial coordinates if high order approximation is used. Similar observation was reported also in 2D study comparing spectral/hp element method with the finite volumes in [8]. Therefore we imposed a small disturbance to velocity field in z-coordinate to speed up development of the final structures. Studied flow regime did not dump this impuls and slowly developed. In time interval $t \in [330 : 380]$ we can observe periodic evolution of the fluid forces acting to the cylinder, Fig. 3.2.

Accordingly to periodicity and amplitude in evolution of the aerodynamic forces (Fig. 3.2) we deduce, that the flow was fully developed after 350 time units. Periodic variation of small amplitude was observed also in the spanwise direction of velocity.

The final part of the simulations was done with polynomial degree 13 (in each spatial direction) and time step $\Delta t = 0.005$. Computation was parallelized up to 48 cores. Final part, where the full resolution was used, $t \in [300 : 380]$ (15000 steps), then took approx. 12000 s. To solve the global algebraic system, direct parallel XX^T solver with multilevel static condensation was used. Therefore a significant time interval of the computation was covered by evaluation of the inverse system, approx 1/4 of the computational time.

Unless the simulations were not pushed to extreme accuracy which high order methods offer, obtained results very well resemble structures described in Ren [6]. The three-dimensional structures, Fig. 3.2 are qualitatively comparable with [6] (pg. 28, fig. 2.13.). In Fig. 3.2, it is observable, that the pattern in y-component of vortic-

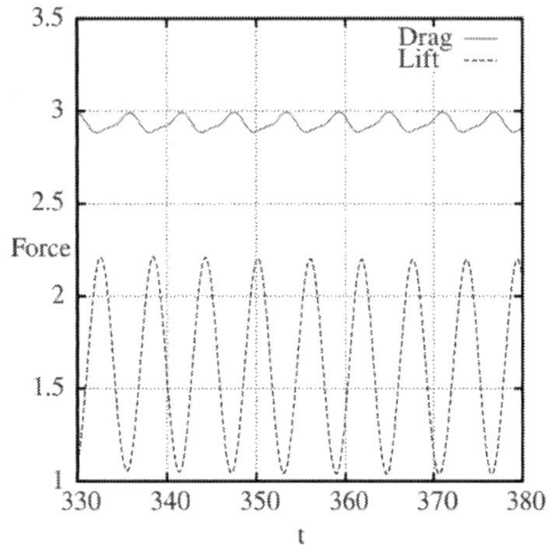

Fig. 4. Time evolution of drag and lift force for mixed convection case (Re = 85, Ri = 1.0).

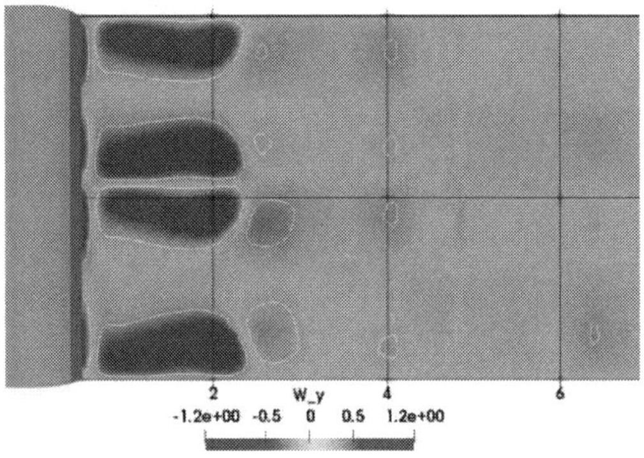

Fig. 5. Re = 85, Ri = 1.0, y-component of vorticity ω_y at $y = 0$.

ity is not positioned symmetrically at the presented time. This coincides with periodical evolution of low amplitude in mentioned spanwise component of force acting onto the cylinder.

Achieved data were processed at the same contour levels as in work of Ren [6]. Figure 3.2 shows contours of temperature and vorticity in x-direction. The polynomial space used in this simulation seems to be sufficient to represent tiny structures both in temperature and velocity in the wake behind the heated cylinder.

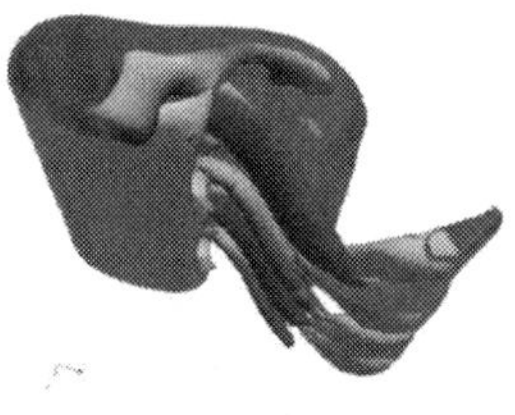

Fig. 6. Re = 85, Ri = 1.0, x-component of vorticity ω_x, contours

4 Conclusion

The main achievement presented in this contribution is in successful implementation of original scheme, which is suitable for simulations based on spectral/hp element method. The algorithm was implemented into the most current version of the open-source library Nektar++ and is supposed to be included in a next release version.

Present study showed, that the algorithm achieves expected convergence orders in time and is applicable to simulations of real physical processes.

Acknowledgements. The author wishes to acknowledge the support of the Czech Academy of Sciences, project MSM100761901.

References

[1] C. Cantwell, D. Moxey, A. Comerford, A. Bolis, G. Rocco, G. Mengaldo, D.D. Grazia, S. Yakovlev, J.E. Lombard, D. Ekelschot et al., Computer Physics Communications **192**, 205 (2015)

[2] G.E. Karniadakis, M. Israeli, S.A. Orszag, Journal of Computational Physics **97**, 414 (1991)

[3] G.S. Karamanos, S. Sherwin, Applied Numerical Mathematics **33**, 455 (2000)

[4] J. Pech, *Scheme for Evolutionary Navier-Stokes-Fourier System with Temperature Dependent Material Properties Based on Spectral/hp Elements*, in *Spectral and High Order Methods for Partial Differential Equations ICOSAHOM 2018*, edited by S.J. Sherwin, D. Moxey, J. Peiró, P.E. Vincent, C. Schwab (Springer International Publishing, Cham, 2020), pp. 465–475, ISBN 978-3-030-39647-3

[5] T. Kuehn, R. Goldstein, International Journal of Heat and Mass Transfer **23**, 971 (1980)

[6] M. Ren, Ph.D. thesis, Mechanical Engineering (2005)

[7] B. Gebhart, *Heat Conduction and Mass Diffusion*, McGraw-Hill series in mechanical engineering (McGraw-Hill, 1993), ISBN 9780071125147

[8] J. Pech, P. Louda, *Comparison of Finite Volume and Spectral/HP Methods on Navier - Stokes Equations for Unsteady Incompressible Flow*, in *Proceedings Topical Problems of Fluid Mechanics 2018* (2018), pp. 223–230

Lattice Boltzmann Simulation of the Air Flow Through the Children Respiratory Tract

František Prinz[1*], *Jan* Pokorný[1], *Jakub* Elcner[1], and *Miroslav* Jícha[1]

[1]Brno University of Technology, Faculty of Mechanical Engeneering, Energy Institute, Technická 2896/2, 61669 Brno, Czech Republic

Abstract. Aerosol particles like dust involved in the air can pass through the respiratory tract and can be deposited there. The long-term deposition of fine solid particles in the lower airways can stimulate lung diseases. In contrary, for the healing of certain lung diseases, the delivering of medicament's to the lower airways is important. Computational Fluid Dynamics (CFD) enables to simulate the flow field and the particles deposition in the lungs. By this method, it is possible to identify the probability of reaching particles into the lower airways through the tracheo-bronchial tree. In this paper, the flow field in the child lungs (5-year-old) in a constant sedentary breathing regime was simulated using the Lattice Boltzmann Method (LBM) with the LES Smagorinsky turbulence model. To verify the calculated velocities by LBM in OpenLB software, the finite volume RANS simulation in Star-CCM+ was performed. The results showed the good agreement in the upper part of the airways. Some discrepancies were found in the lower part. Nevertheless, LBM and OpenLB were proven to be a capable tool to solve the air flow through the respiratory tract and will be used for particle deposition.

Introduction

The last decades the research of aerosol particles and its deposition in the human lungs was carried out across the scientific community. Inhalation of these particles can cause diseases, e.g. lung cancer or pulmonary diseases. For instance, Raaschou-Nielsen et al. [1] revealed a correlation between long-term exposure to ambient air pollution and lung cancer incidence in European populations. They analysed data from 17 cohort studies in Europe countries and showed a statistical significant correlation between 10 μm particulate matter (PM_{10}) and lung cancer. On the other hand, S.P.Newmann in [2] discusses the importance of pulmonary drug delivery into the lungs which is successfully used for the treatment of diseases, i.e., asthma or chronic obstructive pulmonary disease (COPD). One possibility to reduce the effect of the behavioural mechanical barrier for preventing the inflow of the drugs into the lungs or airways is the usage of inhalator devices. To investigate the inhalation efficiency or aerosol spreading inside the respiratory tract the particle tracking method can be used. In the simulations it means to calculate the velocity field across the respiratory system and then simulate the particle trajectories. By these simulations, it is possible to evaluate phenomena, which are hard to measure on living humans. Therefore, an increased interest in the flow conditions inside the respiratory tract leads to a number of studies using computational fluid dynamics (CFD) methods.

To perform the CFD simulations of the airways, the 3D geometry model must be prepared. Earlier, simplified models were used (idealized symmetrical Weibel model [3], more sophisticated Horsfield model [4]) resulting in high error rates. In the last decades screening techniques, such as Computer Tomography (CT) and medical imaging methods, can be performed to obtain realistic 3D models of a respiratory tract of real human. These simulations on realistic geometry provide further information about local particle deposition inside the respiratory system. Breathing is a naturally transient process and could be distinguished into inspiratory and respiratory cycles and both are locally laminar and turbulent. Elcner et. al. [5] showed a good agreement between the velocity values obtained by RANS simulation and experimental data obtained by Phase Doppler Anemometry (PDA) at appropriate locations in the tracheobronchial airways. They investigated the temporal evolution during the respiration cycle and at distinct points in the geometry the findings resulted in numerical discrepancies. The particle deposition was not carried out in this study. With the improvement of the computational technologies, direct numerical simulation (DNS) gives even more accurate results. Elghobashi et. all ([6]) applied the Lattice Boltzmann method for DNS of the nose cavity and upper part of the trachea at the maximal inspiration mouth velocity with a tidal volume of 30 l/min. The geometry domain contained 208 millions voxels. Lintermann and Schröder [7] simulated the deposition of particles in the upper tracheobronchial tract. Two particle tracking simulation runs by the Lagrangian approach were

*Corresponding author: frantisek.prinz@vutbr.cz

computed with various particle radii. The analysis of the particle deposition showed that 32% of the heavier and larger particles deposit deeper in the airways, while for the particle mixture only 0.69% deposit in the primary, lobar, and segmental bronchi. This investigation reveals that small and light particles penetrate into the lung regions below the sixth generation. Nevertheless, the laryngeal tract was not assembled to the geometry which could influence the inflow conditions and thus impact the resulting deposition. Furthermore, they consider just the steady state solution and the time-dependent dynamic of the inspiration was not taken into account.

This paper describes the first step of our research, i.e., verification of the LBM in OpenLB to simulate velocities in the airways properly with the finite volume method using Star-CCM+. Particularly, these velocity profiles could be used to simulate the particle deposition (in the following periods of our study).

At first, the numerical method LBM is briefly introduced, then the simulation parameters and boundary and initial conditions are listed. In the results, the air velocities in predetermined points are compared between both numerical methods. Finally, relevant statements and ideas are discussed and the main findings are highlighted in the conclusion.

1 Methods

The Lattice Boltzmann method is a mesoscopic method based on the molecular dynamics. It is a statistical approach considering the probability of the position and velocities of particles in the domain. The governing equation is the Boltzmann equation [8] describing the time change of the particle probability distribution function f. The discretisation process derived by Chapman and Enskog [9] lead to a discretised set of Lattice-Boltzmann equations. It was proved the equivalence with the well-known macroscopic Navier-Stokes equations of fluid dynamics. Each time step of the equation is divided into a collision step and a streaming step. The whole algorithm is schematically demonstrated in the Figure 1.

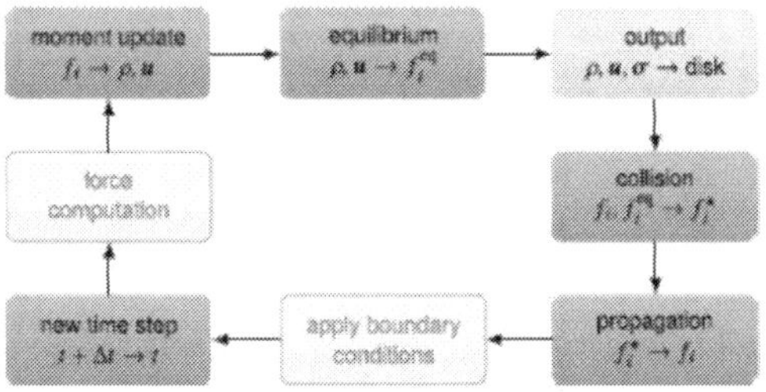

Fig. 1. Schematic illustration of the Lattice Boltzmann cycle, [10].

This approach is from the family of Direct Numerical Simulations (DNS) approaches which are very time expensive. Therefore, especially for turbulent flows, a combination with a Large Eddy Simulation (LES) model is efficient. The most common choice of a subgrid scale model is the Smagorinsky model [11]. In this model, only eddies larger than a defined grid length are simulated while the smaller ones are being modeled. Therefore, a modeled artificial turbulent viscosity is added to the kinematic viscosity. The filtering length is determined by the grid size length multiplied by the Smagorinsky constant C_f [12].

The resulting velocity field can be used for the simulation of particle motion (not part of this study). The aerosols can be represented by spherical solid particles and their motion is tracked by the one-way Lagrangian approach, i.e., the movement of the particles does not influence the air flow. If the value of particle Reynolds number is much less than the acting force on the particles, approximation by the Stokes force [7] is allowed.

2 Simulation setup

The 3D geometry of the airways used in this study was created by scaling of the adult airway model which was produced at Brno University of Technology and used for computational and experimental purposes [5]. The original model involves the mouth and nose cavity, nosopharynx, larynx, and the tracheobronchial tree up to the 7th branching generation. For the scaling properties, the information from [13] and [14] were used. In this paper, the trachea is determined as the distance between the carina of the first bifurcation and the glottis. The distance in our model is 136 mm and was reduced for the 5-year-old child to 73 mm. Furthermore, the tracheobronchial tree was simplified from seven to two branching generations (see Fig. 2). Respiration is considered as a combined

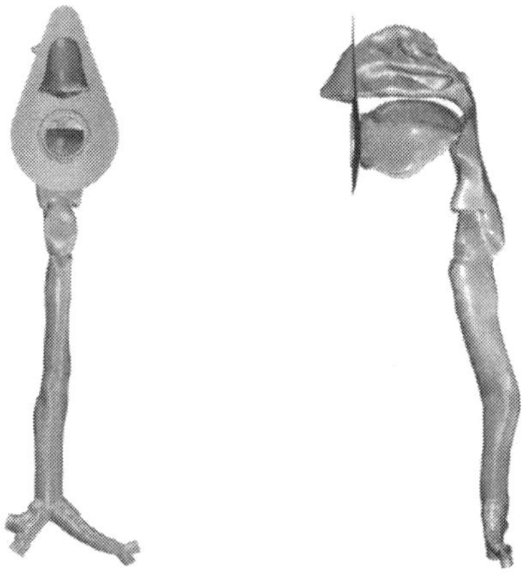

Fig. 2. Front view and side view of the rescaled STL geometry used in this simulation.

breathing, i. e., both through the oral and nasal airways, extended by a inhalator mask to investigate the efficiency of medical inhalation.

M. Roy and C. Courtey [15] measured the human respiration parameters for distinct age groups. According to them, the tidal volume for a 5-year child in a sedentary regime is 10.5 l/min with a respiration cycle period

 https://doi.org/10.1051/epjconf/202226401030

1.607 s. The data presented in [15] lead to specific flow rates through the inlet and outlets of each trachea branch in the state of averaged flow where constant piston profiles were considered.

The airway walls are assigned to a no-slip interpolated boundary condition for curved walls derived by Bouzidi et al. [16]. At the inhalator inlet a Dirichlet boundary condition with a constant velocity profile of 2.203 $\mathrm{m\,s^{-1}}$. This corresponds to the Reynolds number 1190 which is still in the laminar regime. However, deeper in the human airways a turbulent regime with vortical structure is developed. For the stabilisation of the flow we assigned a sinusoidal range up of the velocity from zero to the maximal value for a time period of 0.25 s. After that the velocity at the inlet boundary was hold at the maximal velocity which expects the averaged value of the sedentary respiration. The simulated fluid is air at normal atmospheric state, e.g., pressure of 101 325 Pa and constant temperature 20 °C, density of 1.3 $\mathrm{kg\,m^{-3}}$ and a kinematic viscosity of 1.67 $\mathrm{mm^2\,s^{-1}}$. At the starting time, the airways are occupied by the described air with zero velocity as the initial condition.

The created computational mesh was uniform with a grid length of 176 µm and time step 1.76471×10^{-6} s. This results in 5.7 millions voxels in the whole geometrical domain. The stability is governed by the relaxation time τ and the collision operator depends on this parameter. The value is given by the kinematic viscosity and time and spatial resolution. In this simulation the considered relaxation parameter results in 0.50314 which leads to a stable solution. Another important stability issue affecting the simulation is the initial condition. Dramatic velocity gradients in the domain reduces the stability and therefore a starting period at the velocity boundary is added. This period lasts 0.25 s in which the velocity magnitude grows sinusoidal from zero to the maximal value. The total simulation time was 1.5 s. We suppose that the stabilisation of the flow takes 1 s and then the last 0.5 s was determined for the resulting averaged velocity to outfilter the velocity disturbances produced by the turbulent character of the flow. The simulation is triggered with an applied LES subgrid scale shear-improved Smagorinsky model employed by Jafari et al. [11]. The Smagorinsky parameter was assigned to a value of 0.1. This significantly reduces the computational resources and still computes velocity field with a sufficient accuracy to perform particle tracking.

The simulations were obtained by OpenLB software, release 1.4-0 [17], [18]. This software has been developed at Karlsruhe Institute of Technology (KIT) by the Lattice Boltzmann Research Group (LBRG) of Dr. Krause. The C++ code of OpenLB software enables an efficient parallelisation of the solving problems, which divides the computation into more threads to reduce the simulation time significantly. The simulations were performed on the CPU cluster Barbora/Karolina in IT4Innovations in the project DD-21-19. For postprocessing and visualisation, the Gnuplot and Paraview software were used. For verification of the computed velocity field by LBM, another simulation was performed by the steady solution Reynolds Averaged Navier-Stokes (RANS) finite volume method (FVM) - Menter's modification of Wilcox's model – SST $k-\omega$ model [19] - in the multiphysics CFD software Star-CCM+. As the output, the velocity magnitudes were used for comparison with the LBM.

3 Results

In this chapter the comparison of the results for both method (FVM vs LBM) is presented. It was selected 67 distinct point probes, where the velocity magnitudes were compared. The resulting plots of LBM simulations were post-processed in Paraview. To study the influence of the whole geometry (e.g., wall treatment, mixing of the fluid at the branching area of the trachea or in the mouth and nose cavity), probe points were selected across the domain. The points were located on 6 cross-sectional planes through the domain: one in the mouth inlet, one each in the rare section of the nose and mouth chamber, one in the middle of the trachea and in the left and right bronchus after the first bifurcation. In every plane, a couple of points on two perpendicular lines were selected including the wall vicinity to study the effect of the wall boundary condition. A point set was evaluated at the tracheal axis, too. The positions of the cross sections used for comparison are depicted in Fig. 3. For the more accurate location of the point probes, the spatial directions are denoted by letters showing the given direction. For that purpose we used the coronal plane view of the airways from the back.

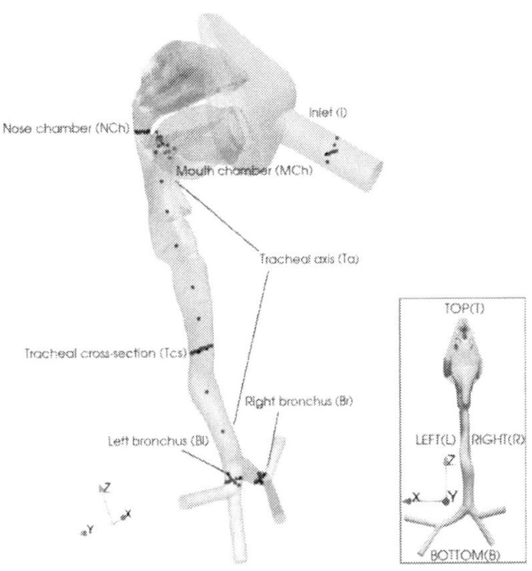

Fig. 3. Position of the investigated cross sections in the geometry matched by black spheres and direction description.

According to the applied system of coordinates, we used the left direction abbreviated by L corresponding to the X+ direction, the right direction R to the X- direction, the top side T by Z+, the bottom side by Z-, the front side F by Y- and the back side B by Y+. For an easier understanding of the position of the cross sections, these direc-

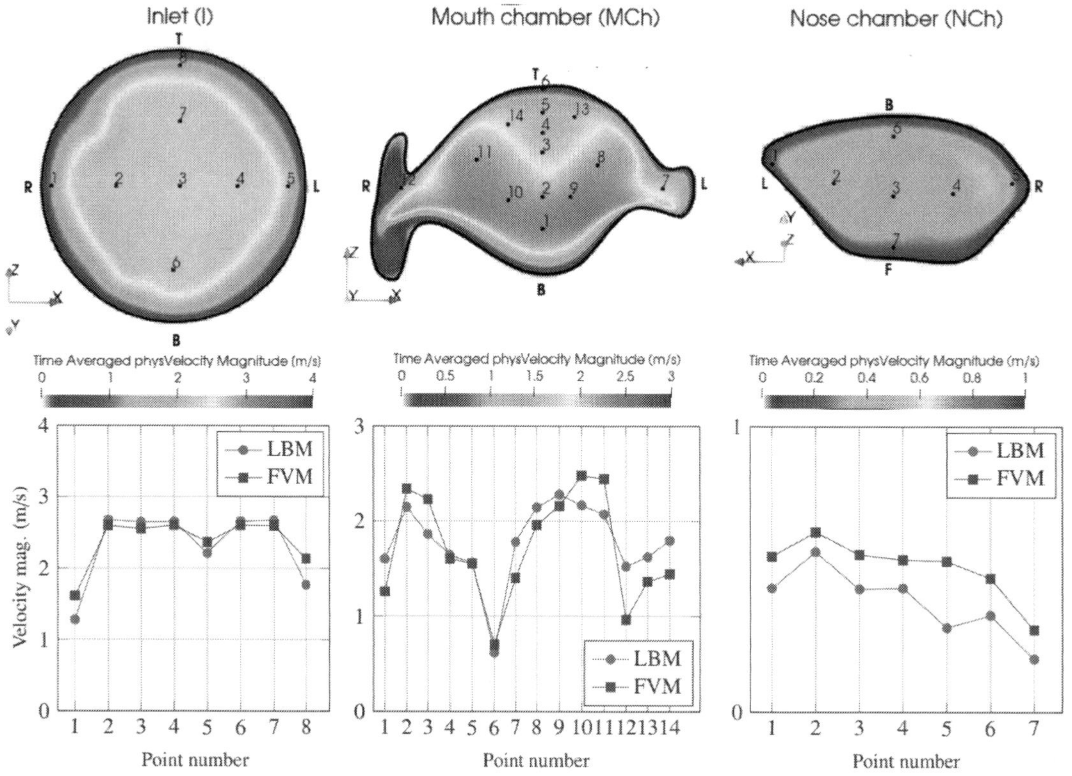

Fig. 4. Comparison of velocity fields at the investigated positions in the inlet and in the mouth and nose cavity.

tions are written down to the velocity fields in the following figures.

In Fig. 4, 5 and 6 we present the averaged velocity magnitudes on the desired cross sections. The point positions are shown on the cross section where the velocity field is depicted, too. The comparisons of each group of point probes are visualised in the charts below the pictures.

In Fig. 4 we can see the velocity field in the middle distance from the inlet tube and the outflow of the mouth and nose chamber as marked in Fig. 3. Due to the inlet Reynolds number of 1190, which corresponds to the laminar regime in the pipe, a symetrical flat profile evolves from the inflow piston profile at the inflow boundary which indicates a typical entrance region of a pipe with a further development of a fully parabolic profile. The velocity in the inlet channel shows a good agreement between both methods with a middle relative error rate of 8.4% with higher error rate near the walls. Moving to the cross sections in the mouth and nose cavity, we can observe slightly higher velocities in the mouth cavity. There are 14 probes determined through the domain in the mouth cavity and 7 in the nose cavity. The velocities with the Lattice Boltzmann method in the nasal cavity are lower than by the FVM method. The middle error rate in the mouth cavity is 15.4% and in the nose 37.4%.

In Fig. 5 the velocity field at a cross section in the middle part of the trachea and on the left and right bronchus after the main bifurcation is displayed. Due to the narrowing of the airways, the velocities are higher up to 8 $m\,s^{-1}$ than in the the upper airways. We can notice differences in the right and left bronchus cross section in the order of multiple of 10% relative, especially the velocity field in the left bronchus is predicted higher than from the FVM. Contrary, the velocity field in the tracheal cross section shows a good agreement in all points. Slightly greater difference is noticeable in the point number 5 which is in the wall vicinity. The relative error of all points in this cross section is 14.6 %.

In Fig. 6 we show the results at the axis of the trachea. The points were predetermined approximately in the middle of a cross section where every point is located. Here we can clearly distinguish a growing difference between the simulation values with a movement deeper into the airways towards the bifurcation. The highest discrepancy is in the point number 2 in the relative value of 10.6 % which is already acceptable. However, the relative error of this point sequence is 3.9 % which is showing a good agreement indeed.

Analysing the relative errors in the considered probes, some statements can be postulated. First, the values of the velocities in the upper airways above the bifurcation show a better agreement than under the bifurcation. We suppose some inaccuracy produced by the pressure boundary condition applied at the outlets. These are not aligned parallel

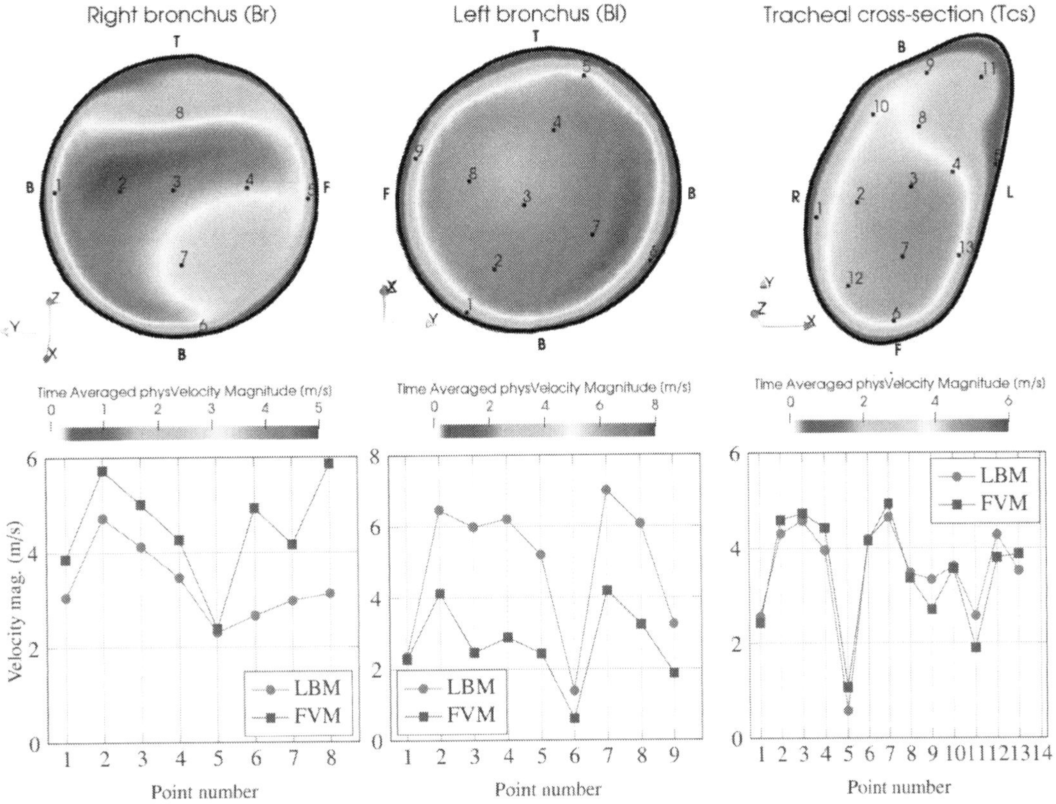

Fig. 5. Comparison of velocity fields at the investigated positions in the tracheal cross section and in the right and left bronchus.

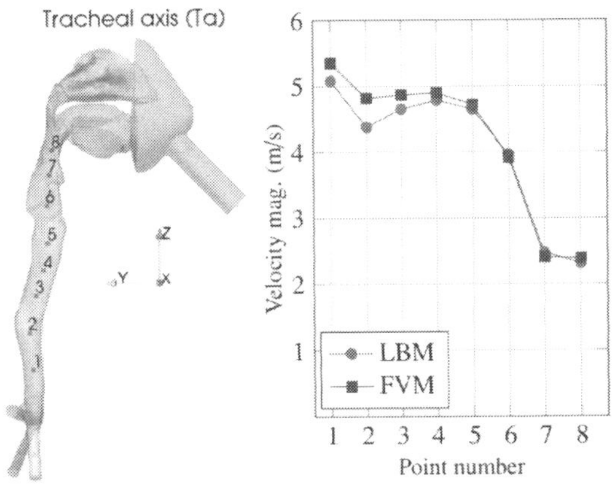

Fig. 6. Comparison of velocity fields at the investigated positions on the axis of trachea.

to the Cartesian axis and therefore cause some discrepancies acting on the bulk, too. Second, the flow through the nose cavity in the investigated probes predicted by the LBM indicates lower values than by the RANS FVM and therefore a lower mass flux through this part of geometry is simulated by LBM. Third, the relative errors near the walls are higher than in the core of the flow. This could be caused by the considered boundary wall condition. Eventually, a finer grid could also improve the results and could contribute to a more accurate distribution of the flow into the mouth and nose cavities. The middle relative error of all points in the domain is about 21.4 %.

To discuss the quality of the comparison between both methods, one could look at both approaches in more detail. Large eddy simulation ignores the smallest length scale by using the low pass filtering (LES filter) on both time and length scale. The RANS simulation is based on the Reynolds decomposition into its time average and fluctuating quantities. Therefore, the turbulences which are capable by the RANS approach are of a higher length scale than in the case of LES. Although the time averaging in the LES approach was obtained, some discrepancies could have been produced by the differences of both methods. To improve the confidence of the validation, a LES simulation in FVM should be performed and a comparison with experimental measurements, too, which is planned as a next step of our study. Another possibility could also be the effect of the time averaging period and a longer time for the relaxation of the fluid flow. Longer time averaging period and a longer range-up time could increase the accuracy, too.

Despite of several discrepancies, a relatively good agreement between both methods is verified. After some improvements, the simulated velocity field ought to be satisfactory for considering particle deposition simulation.

Conclusion

In this paper, a Lattice Boltzmann simulation of constant breathing of a 5 year-old child with a respiratory mask in the sedentary regime was obtained. The LBM results showed a good agreement with the FVM method, especially in the upper part of the airways. Some discrepancies were found in the lower part. Nevertheless, LBM and OpenLB were proven to be a capable tool to solve the air flow through the respiratory tract and it will be incorporated in the aerosol particles tracking to investigate the deposition. Furthermore, more realistic breathing regimes with inspiratory and expiratory cycles will be performed for 10-month infants. In this case, we expect a more detailed computational mesh to retain smaller-scale flow phenomena.

Acknowledgements. This work was supported by the Ministry of Education, Youth and Sports of the Czech Republic through the e-INFRA CZ (ID:90140), by the project of the Czech Science Foundation GA20-27653S and by the project FSI-S-20-6295 founded by Faculty of Mechanical Engineering, Brno University of Technology.

References

[1] O. Raaschou-Nielsen, Z.J. Andersen, R. Beelen, E. Samoli, M. Stafoggia, G. Weinmayr, B. Hoffmann, P. Fischer, M.J. Nieuwenhuijsen, B. Brunekreef et al., The Lancet Oncology **14**, 813 (2013)

[2] S.P. Newman, Therapeutic Delivery **8**, 647 (2017)

[3] R.J. Lorenz, Biometrische Zeitschrift **8**, 143 (1966)

[4] K. Horsfield, G. Dart, D.E. Olson, G.F. Filley, G. Cumming, Journal of Applied Physiology **31**, 207 (1971)

[5] J. Elcner, F. Lizal, J. Jedelsky, M. Jicha, M. Chovancova, Biomechanics and modeling in mechanobiology **15** (2015)

[6] Y. Wang, S. Elghobashi, Respiratory Physiology Neurobiology **193** (2014)

[7] A. Lintermann, W. Schröder, European Journal of Mechanics - B/Fluids **63**, 73 (2017)

[8] C. Cercignani, *The Boltzmann Equation* (Springer New York, New York, NY, 1988), pp. 40–103

[9] P. Rosenau, Phys. Rev. A **40**, 7193 (1989)

[10] T. Krüger, H. Kusumaatmaja, A. Kuzmin, O. Shardt, G. Silva, E.M. Viggen, *The Lattice Boltzmann Method - Principles and Practice* (2016), ISBN 978-3-319-44647-9

[11] S. Jafari, M. Rahnama, International Journal for Numerical Methods in Fluids **67**, 700 (2010)

[12] P. Nathen, D. Gaudlitz, J. Kratzke, M. Krause, 21st AIAA Computational Fluid Dynamics Conference (2013)

[13] M.H. Dave, M. Kemper, A.R. Schmidt, C.P. Both, M. Weiss, Paediatric anaesthesia **29**, 782 (2019)

[14] N. Griscom, M. Wohl, AJR. American journal of roentgenology **146**, 233—237 (1986)

[15] M. Roy, C. Courtay, Radiation Protection Dosimetry **35**, 179 (1991)

[16] M. Bouzidi, M. Firdaouss, P. Lallemand, Physics of Fluids **13**, 3452 (2001)

[17] M. Krause, S. Avis, H. Kusumaatmaja, D. Dapelo, M. Gaedtke, N. Hafen, M. Haußmann, J. Jeppener-Haltenhoff, L. Kronberg, A. Kummerländer et al., *OpenLB Release 1.4: Open Source Lattice Boltzmann Code* (2020)

[18] M. Krause, A. Kummerländer, S. Avis, H. Kusumaatmaja, D. Dapelo, F. Klemens, M. Gaedtke, N. Hafen, A. Mink, R. Trunk et al., Computers & Mathematics with Applications **81**, 258 (2021)

[19] F. Menter, AIAA Journal **32**, 1598 (1994)

EPJ Web of Conferences **264**, 01031 (2022) https://doi.org/10.1051/epjconf/202226401031
EFM 2021

Lift measurement of airfoil AH93-157 from wall pressure distribution

Pavel Procházka[1], *Vladislav* Skála[1*], *Pavel* Antoš[1], *Lukáš* Popelka[2], *Jiří* Fürst[2], and *Michal* Schmirler[2]

[1]Institute of Thermomechanics of the CAS, v. v. i. Dolejškova 1402/5 182 00 Praha 8, Czech Republic
[2]Czech Technical University in Prague, Faculty of Mechanical Engineering, Technická 4, 160 00 Praha 6, Czech Republic

Abstract. Measurement of the lift coefficient of the AH93-157 airfoil was performed by measuring the static pressure distribution on the wind tunnel walls along the test-section. A correlation was established between the lift coefficient value, determined by integrating the static pressure distribution on the wind tunnel walls, and the lift coefficient value, determined by integrating the static pressure distribution on the airfoil surface. This method is useful for easy and quick determination of the lift coefficient on a simple airfoil model without complicated static pressure tubing from the surface. The measurement was made within a closed test-section. The correlation relationship also eliminates the problem of the finite span and the effect of the side walls. Comparison of pressure distribution on airfoil with CFD was performed as well.

1 Introduction and motivation

Unmanned flying devices, drones (UAVs) are currently experiencing a fast rise. In order to improve the performance of these advanced vehicles, the propulsion and aerodynamic design needs to be optimized. These vehicles exhibit (compared to conventional manned aircraft) significantly smaller dimensions and achievable wing and propeller speed. UAV-wings and propellers operate at significantly lower Reynolds numbers than manned aircrafts.

In the available literature, there is minimal information on the characteristics of airfoil at so-called low and very low Reynolds numbers. Thus, it is difficult to perform UAV parameter optimization when there are not enough good baseline characteristics of airfoil at low Reynolds numbers.

This paper focuses on simplifying the method of experimental investigation of the lift from the static pressure distribution on the walls of the wind tunnel measurement section.

2 Measurement of the lift from the static pressure distribution on the walls

2.1 Method of measurement of the lift coefficient

The method of experimental investigation of the lift (lift coefficient) of the wing model from the static pressure distribution on the walls of the wind tunnel test-section is based on the integration of the static pressure difference on the opposite walls of the experimental channel. The problem is considered as two-dimensional. The basic principle of "action and reaction" is used, whereby the gradient of the static pressure field induced by the wing model in the test-section is reflected in the pressure field on test-sections walls. Direct measurement of the static pressure distribution on the airfoil is difficult and requires a wing model equipped with pressure taps, which is quite expensive. It is therefore not possible to use this procedure routinely. Another possibility is aerodynamical ballance and direct measurement of forces. Using this static pressure based method, the static pressure gradients are measured on the walls of the test-section, which are opposite to the suction and pressure side of the wing model. The measurement of the static pressure distribution along the walls of the test-section is carried out directly by means of a series of separate pressure taps. The observed static pressure waveforms are integrated and their difference corresponds to the lift acting on the airfoil. Since the test-space, or the area where the static pressure is measured, is limited, the relationship between the actual lift coefficient and the coefficient found by integration must be calibrated. For calibration purposes, it is necessary to know the actual lift coefficient of the wing model. An aerodynamic weight can be used to determine this or, as in our case, the static pressure distribution on the suction and pressure side of the wing model can be measured, since a model of the wing section with airfoil AH93-157 with static taps was available. By integrating the static pressure distribution, the actual lift coefficient is obtained. The advantage of this method is that it is based on a direct measurement of the local pressure field generated by the airfoil in the test-section. Calibration of the test-section by the wing model with pressure taps largely eliminates the effect of the finite size of the wing model, the effect of the model endings is very well eliminated, and the measurement is not affected by the flow in the gap between the model sides and the walls or the forces acting on the end-plates or by side wall effecst, as is the case for lift coefficient measurements using aerodynamic ballance. Second advantage of this method is that after calibration of the test-section it is possible to use wing models without special modifications. The end-plates are replaced directly by the test-section walls. The disadvantage of this method is the need to measure (ideally simultaneously) a large number of low pressures. It requires high-quality

* Corresponding author: skala@it.cas.cz

equipment for measuerement of a larger set of low pressures. [1]

2.2 Experimental setup

The experiments were carried out in the closed type wind tunnel cross section of the test-section of the wind tunnel was 0.9 m x 0.5 m. [2] The length of the test-section was 0.8 m. The free stream turbulence intensity *Iu* (at the leading edge of the model level) was naturally *Iu* = 0.009. The side walls of the test-section were each fitted with 16 static pressure taps (diameter 5E-4 m). All were connected to a common pressure connector (64 pressures).

2.3 Model

It was tested the AH93-157 airfoil, (Fig. 1). [3-5] The length of the model wing chord was 0.4 m, the span of the model was 0.5 m. The model was fitted with 45 static pressure taps (diameter 5E-4 m) on the suction side and with 45 static pressure taps on the pressure side. The static pressure taps were routed through the model interior and through the wall to the outside of the test section and connected via two multichannel pressure connectors (64 pressures) to the pressure scanner. The model of the wing was placed in the test section in a vertical position and was and was pivotally placed between the upper and lower wall of the test-section, see Fig. 2 and Fig. 3.

2.4 Measurement aparatus

Static pressure readings were taken using an Esterline Pressure Systems ESP 64HD pressure scanner. The pressure scanner was housed in a tempered thermobox (43°C). The pressure scanner was equipped with a pressure connector for easy connection of static pressure taps leadings from the walls of the test-section and from the suction and pressure side of the wing model. Measurement was made by connecting individual sets of static pressure taps for each case in turn. Unloaded pressure scanner readings were taken as a reference of zero signal.

The output signals from the pressure scanner were read using PCIe 6251 measurement card, and data acquisition was performed using dedicated software in the LabView software. A calibration of the pressure scanner with 4th order correction was applied and performed in the appropriate range. (range 1 kPa, accuracy ±0.015 % FS).The flow velocity reference was measured as the static pressure difference on the inlet difusser (contraction nozzle). The pressure signal and barometric pressure were connected to a Druck DPI 145 pressure transducer (range 7 kPa accuracy ±0.005 % FS).

Flow temperature was measured with a Pt100 thermometer. The turbullence intensity of the flow in the inlet was mneasured using single hot wire probe and Dantec Streamline unit. The output voltage was sensed by the HP 34970A data acquisition unit. Data ordering and graphs was made using MS Excel.

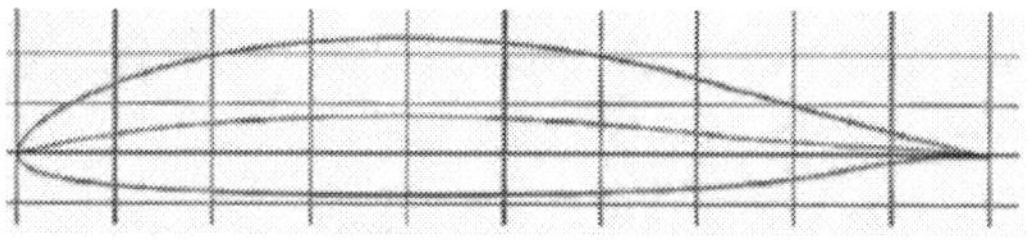

Fig. 1. Airfoil AH93-157 (http://airfoiltools.com/airfoil/details?airfoil=ah93157-il 10.11.2021).

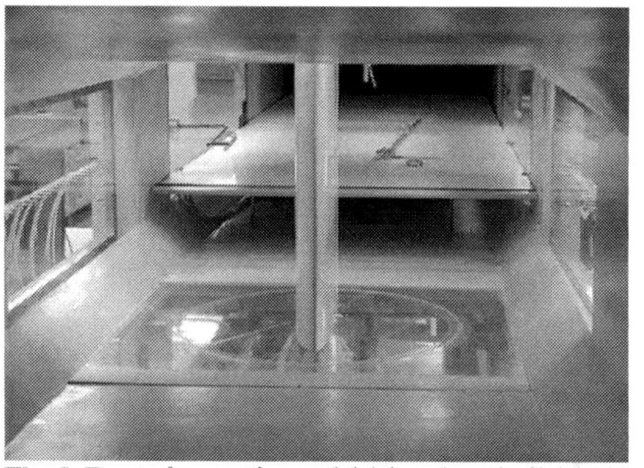

Fig. 2. Front view on the model (placed vertically) inside the test-section of the windtunnel.

Fig. 3. Side view on the model (placed vertically) inside the test-section of the windtunnel.

3 Results

3.1 Experimental results

The Fig. 4 and Fig. 5 show the static pressure distribution in dimensionless form of the local lift coefficient on the AH93-157 wing model and the static pressure distribution on the walls. On the Fig. 4 is evident different static pressure reference in comparison with CFD results.

The experiments were carried out by 14 m/s, corresponding *Re* = 4.1E5, turbulence intensity *Iu* in the inlet was measured *Iu* = 0.9 %. The static pressures were measured using pressure scanner, the time (duration) of data acquisition was t_a = 30 s.

3.2 CFD simulations

The experimental data were supported by numerical mathematical model. The *k-kl-omega* turbulence model was used, see example in the Figure 4 and Figure 5. The preliminary results was presented, simulation was performed for turbulence intensity Iu = 0.1 % and without test-sections walls modelling, so in the free stream. The simulation will be repeated for the case, where the walls are present in the future.

After sumarising of both parcial lift coefficients is evident, that the experiment is relatively in well agreement with CFS simulation. CFD simulation was made in the free flow field, so some correction of its results is necesarry.

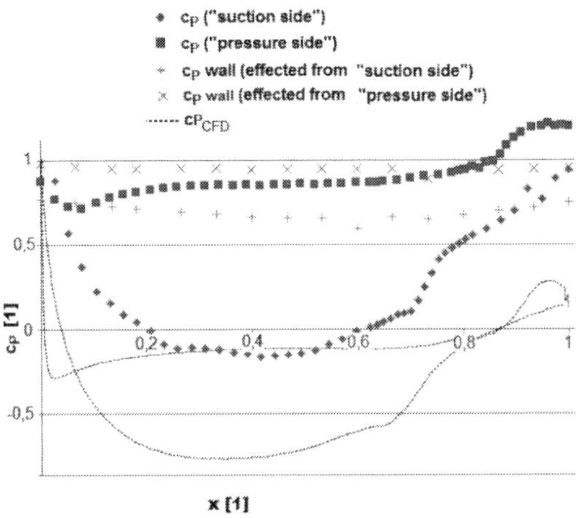

Fig. 4. Pressure coefficient c_P for suction side c_{PH} and pressure side c_{PD} of the airfoil AH93-157 and pressure coefficient c_{PW} on the corresponding walls. Experimental and preliminary CFD results, angle of attack α = 0 deg. The flow separation is good recognizable.

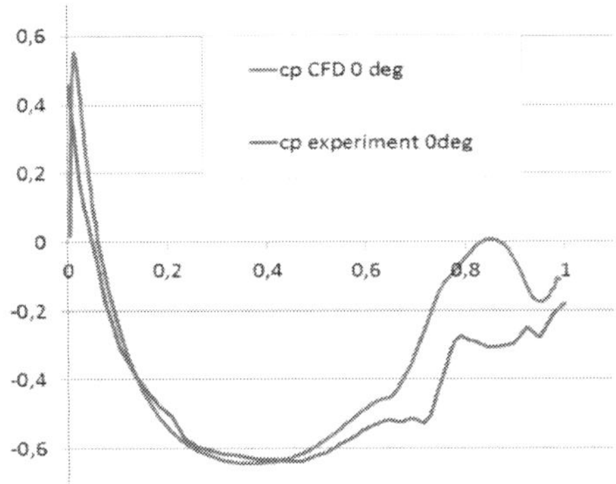

Fig. 5. Lift coefficient c_{PL} distribution along the chord of the AH93-157 airfoil.

3.3 Lift coefficient

It was possible to determine the lift coefficient for the AH93-157 airfoil by specified angle of attack α and Reynolds number Re.

$$c_{PL} = c_{PD} - c_{PH} \tag{1}$$

For example, for Reynolds number Re = 4.0E5 and angle of attack α = 0 deg, the lift coefficient was c_{PL} = 0.63. For presentet case (α = 0 deg) was detected the pressure difference on the walls of the test-section expressed by the pressure coefficient c_{PW} inducted by the model of the wing, corresponding pressure coefficient was c_{PW} = 0.27. The flow separation is very good recognizable, see Figure 4 and Figure 5.

3.4 Test-section calibration

Calibration of the test-section is presented in the Table 1 and Fig. 6. Calibration is expressed as the ratio of the lift coefficient and pressure wall coefficient c_{PL} / c_{PW}.

Table 1. Lift coefficient c_{PL} of the airfoil AH93-157 and pressure coefficient c_{PW} of the effected walls

α [°]	c_{PL} [1]	c_{PW} [1]	c_{PL} / c_{PW} [1]
-5	0.39	0.02	19.5
0	0.63	0,27	2.33
5	0.95	0.48	1.98
10	1.08	0.57	1.89
15	1.11	0.61	1.82
20	1.06	0.65	1.63
25	1.18	0.73	1.61

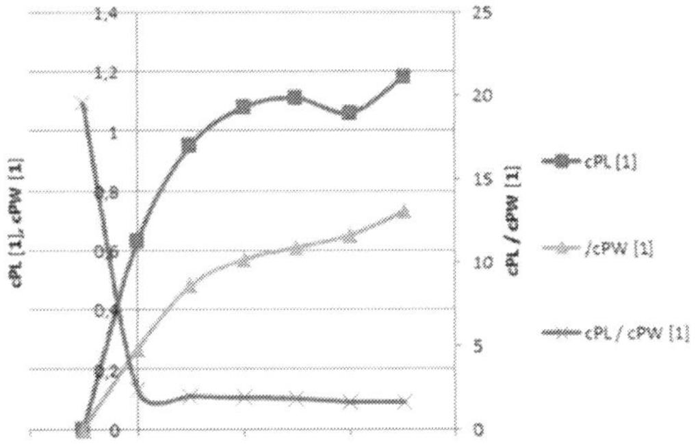

Fig. 6. Lift coefficient c_{PL} of the airfoil AH93-157 and pressure coefficient c_{PW} on the corresponding walls dependent on the angle of attack.

4 Conclusions

Experimental data on aerodynamic airfoil are rarely available for lower values of Reynolds number, examples can be found in [6-8].

At last a basic experimental determination of the most important characteristics of the airfoil can be performed using the procedure described above.

This method is capable to investigate large number of different airfoil variants, which can nowadays be easily realized by rapid prototyping (3D printing) methods. Other more experiments will be done in the next future.

Acknowledgement:

This work was supported by the research project of the Czech Science Foundation No. 20-26779S "Study of dynamic stall flutter instabilities and their consequences in turbomachinery application by mathematical, numerical and experimental methods".

References

1. L. Popelka. D. Šimurda, N. Součková : Experimerimental evaluation of pressure coefficient along upper and lower surface of wing-body configuration in the wind-tunnel 865 x 485mm. Institute of Thermomechanics AS CR v. v. i. Research report IT.D1.2010-T10-1 Prague, 2010

2. M. Píchal : Aerodynamický tunel 0.485 x 0.865 m2 se široce měnitelným pásmem turbulence. Research report Z 91/59 (Československá akademie věd, Ústav pro výzkum strojů Prague, 1959)

3. I.H. Abbott, A.E. Doenhoff, L.S. Stivers : Summary of Airfoil Data (NACA Rep. No. 824. Langley, 1945)

4. D. Althaus, F.X. Wortmann : Stuttgarter Profilkatalog I (Braunschweig, 1981) ISBN 3-528-08464-2

5. D. Althaus : Profilpolaren für den Modellug, Band 1 (Villingen-Schwenningen, 1980) ISBN 3-7883-0158-9

6. R.J.cMcGhee, W.D. Beasley : Low-speed aerodynamic characteristic of a 17-percent-thick airfoil section designed for general aviation applications (NASA TN D-7428, Langley, 1973)

7. J. H. M. Gooden : Experimental Low-Speed Aerodynamic Characteristics of the Wortmann FX 66-S-196 V1 Airfoil : Technical Soaring, Vol. 5, No. 3. , pp. 21-31 (1979) ISSN 07448996

8. M.D. Somers : Experimental and Theoretical LowSpeed Aerodynamic Characteristics of a Wortmann Airfoil as Manufactured on a Fiberglass Sailplane (NASA TN D-8324, Langley, 1977)

On the stall flutter occurrence in a blade cascade set to turbine and compressor geometry

Pavel Procházka[1*], *Pavel* Šnábl[1], *Sony* Chindada[1], *Ondřej* Bublík[2] and *Václav* Uruba[1,3]

[1]The Czech Academy of Sciences, Institute of Thermomechanics, Dolejškova 5, Praha 8, CR
[2]UWB, FAS, Department of Mechanics, Universitní 8, Plzeň, CR
[3]UWB, FME, Department of Power System Engineering, Universitní 8, Plzeň, CR

Abstract. A blade cascade allowing free pitch movement of five blades was developed in Institute of Thermomechanics. This model is introduced to study the phenomenon of stall flutter existing in rotary bladed wheels of steam turbines and other devices. This article describes how to induce the stall flutter for prescribed boundary conditions (as inlet velocity, various angles of attack, etc.) and gives survey about flow field differences around the cascade set to both the turbine and the compressor geometry. This experimental research utilized time-resolved Particle Image Velocimetry (PIV) to measure and to evaluate statistical quantities of the wake behind the NACA0010 profiles of the cascade. Also dynamical analysis was performed in the form of Fast-Fourier transform and also phase-locked measurement was applied. Gained knowledge will be utilized to design advanced model of the cascade allowing stall flutter examination without the use of ball bearings.

1 Introduction

A low structural and aeroelastic damping plays crucial role in the case of last stage blades (LPT) because the blades are very long and the profile of this tip region has very sharp edges and low curvature. More, these blades have no shroud to minimize the centrifugal forces. The (stall) flutter phenomenon can occur in the case of such rotary periodical bodies arising from aeroelastic coupling between the structural vibrations and aerodynamic instabilities. Especially stall flutter is very dangerous phenomenon since it would lead to high-cycle fatigue and consequently to fast failure of the whole device. This fact must be taken into account during the design of rotary bladed wheel as well as during the operation.

This article describes a partial goal of the project dealing with nonlinear stall flutter of flexible profiles witch pitch degree of freedom in subsonic flow to get more information about fluid-structure interaction (FSI) as flutter is very important part of aeroelasticity examined in turbomachinery [1]. The effect of aeroelastic coupling is commonly studied using three different approaches. The profiles can be subjected to flow instabilities and are self-oscillating [2] or profiles are exposed to kinematic excitation [3] or force excitation. This project combines the first and the last approach. This research follows on previous presented one [4, 5, 6] where the plunge degree of freedom of blades was utilized. However, the plunge motion is not able to cause such strong aeroelastic coupling because the flow separation does not occur when the blade vertical position is changing.

To prevent the stall flutter of the cascade, complex CFD models should be developed which could help during the design process. It is essential to model the velocity (pressure) field for proper modelling of aeroelastic coupling. However, CFD model could fail for some conditions as various angles of attacks (AoAs) and hence it should be validated using results from measurement in physical models. The model of such five-blade cascade is visible in figure 1 and its description will be present in the next chapter. There are some discrepancies between PIV data and two different CFD models (see figure 3a, b, c, d). The subjected quantity is velocity profiles plotted 100 mm behind the cascade (in the wake). As the comparison looks very promising (Fig. 2) for positive AoAs or zero value (compressor arrangement), for negative incidence angles situation differs. It seems that the process is more viscous in reality than in simulations. Obviously, the flow field is very complex in the vicinity of the cascade, there is a strong dependency on incoming angles and the dynamic activity is very important and it should be studied primarily by PIV method.

2 Experimental setup and CFD models

A two-dimensional model of rotational periodic body, which is a blade cascade, was built and placed inside a plexiglass channel connected to the blow-down wind tunnel facilities. This physical model was used to

* Corresponding author: prochap@it.cas.cz

perform time-resolved PIV measurement as well as force measurement. Five blades with profile NACA 0010 were used in the cascade geometry - profile chord length was 72.8 mm, stagger angle equal to 30° and vertical blade distance 16.14 mm. The first and last blade was auxiliary. Three middle blades were allowed to have pitch motion. The blade was supported by a shaft only at one end. The shaft has led through one side wall of the test section and necessary equipment (springs, bearings, encoders) were located aside the channel. A circular rotating table was used to allow AoA setting from -15° to + 15°. More, each blade could be compensated regarding static force affecting the AoA value within the cascade. The shakers were used to prescribe periodic harmonic motion of the pitch degree of freedom. The blade vibration was controlled with feedback loop to oscillate with the precise amplitude so we could get close to kinematic excitation. The forces (aerodynamic, inertia and elastic) were measured using the force transducer B&K type 8203. The blade position was acquired by miniature magnetic rotary encoder RLS RM08. These data were saved by Yokogawa ScopeCorder DL750. The transparent test section was connected to the tunnel orifice with cross-sectional dimension 100 x 250 mm. Although many regimes with different velocity were tested, this article will deal with the velocity equal to 20 m/s.

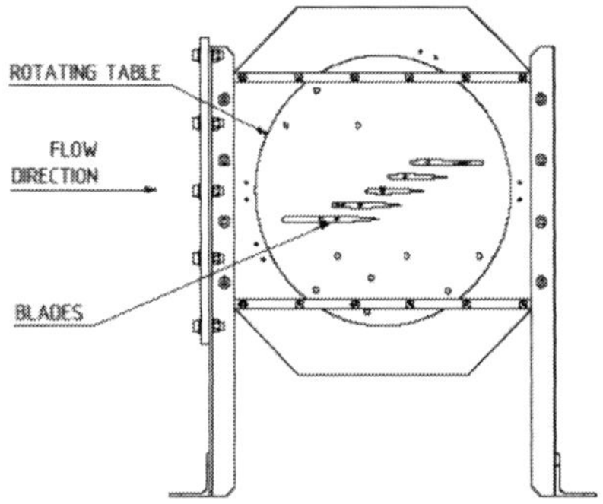

Fig. 1. Assembly of the blade cascade with a rotary disc.

The flow field was measured by PIV apparatus from Dantec company. To illuminate the tracing particles (oil droplets), laser New Wave Pegasus Nd:YLF with double-head was used (frequency 10 kHz, wavelength 527nm, energy in one pulse 10mJ for 1 kHz). Fast PIV camera Phantom V611 (up to 6000 double snaps per second, spatial resolution 1280x800 pix, memory 8 GB) was used to capture the particle movement. The system synchronization is performed via timer box. Dynamic Studio ver. 3.4 was utilized to capture the data and post-process data. Adaptive correlation with embedded validation was used to calculate velocity vector maps. An interrogation area of 16x16 pixels resulted in 159x99 vectors (maximal 5% invalid vectors). Further data analysis and plotting was performed in Matlab or in Tecplot software.

The acquisition frequency differs for individual purpose. The frequency for mean flow quantities evaluation and for higher statistical moments was set to 100 Hz during 10 seconds of recording. To detect and to analyse rapid flow fluctuation especially in the wake, the sampling frequency was set to 2 kHz. The motion of excited blade demonstrates periodic motion and the flow in the close vicinity has also pseudo-periodic character. To suppress the random component and to highlight the periodic component of the flow is very convenient to use a phase-locked measurement. The sampling frequency is set as a product of excitation frequency and the number of desired phases per one period. The dataset of each phase is then averaged over all periods resulting in mean phase velocity and its standard deviation (STD).

The regime with fixed blades is referred as base case. It is introduced here especially for comparison with CFD. One regime was studied with freely supported blades, but no free oscillation was observed. The periodic motion was prescribed by pitch amplitude equal to 3°, frequency of motion was 40 Hz and different inter-blade phase angle (0° or 180°) was set.

Numerical simulations were calculated using Ansys Fluent and FlowPro open-source CFD software. Ansys utilized compressible k-ω SST turbulence model. The fluid parameters were set to corresponding real values in the wind tunnel (viscosity ratio 1, inlet turbulence intensity 0.1% etc.). FlowPro is a multipurpose CFD software developed for complex fluid simulations as vibrating blades in a steam turbine. It utilized the incompressible fluid model together with Spalart-Allmaras turbulence model. The fluid parameters were set same as for Fluent solver.

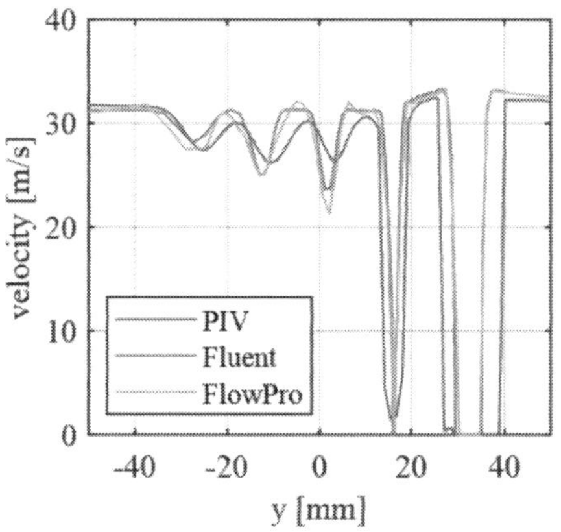

Fig. 2. Velocity profile comparison 100 mm behind for 0°.

3 Results

The coordinate system is set so that the origin is in the leading edge of the middle blade (3rd) in its hypothetic position with zero angle of attack. The vector field is depicted using black vector representing both velocity components in the plane of measurement. The color is used to plot the scalar value (mainly vector modulus). The flow is coming from the left side. A compressor arrangement of the cascade is represent by positive angle of attack 10°. Contrary, the turbine setup is characterized by negative value -12°.

3.1 CFD x PIV

There is a comparison for PIV data and data from Ansys and FlowPro for zero AoA in figure 2. The distribution of mean velocity across the channel behind the cascade can be simulated quite credibly. The position of wakes as well as the peak values corresponds for all three approaches. The situation is much more complicated when the AoA is negative (Fig. 3a). The velocity profiles do not corresponds at all. Nor the peak values either the wake position. It seems that the flow in CFD calculation is less viscous. The numerical model has big problem to detect the huge separation at the very first profile and also the wake itself (see figure 3b, c, d). It is obvious that the flow dynamics in the case of moveable profile will become much more complicated to predict. Further in the article, we will describe the flow topology and dynamics inside the cascade using PIV. There is a mean flow field distribution for compressor and turbine setup in the figure 4a,b. Compressor arrangement shows the detail of the 3rd, 4th and 5th blade. On the other hand, turbine arrangement shows the inter-blade channels between 1st, 2nd and 3rd profile. The FFT analysis is performed to detect frequencies present in the flow. The analyzed point coordinates are [80;-3] mm and [50;-3] mm for positive and for negative AoA, respectively. Both positions correspond to place of expected high activity of aeroelastic coupling. The point is chosen so it is between two moveable blades and in the vicinity of the trailing edge of preceding profile. It is obvious that the velocity value distribution differs completely for both cases. There is big separation of boundary layer at suction side of the first profile of the turbine blade cascade. The suction side is not fully covered by pressure side of adjacent blade in turbine setup and hence the flow separation is present more often. Also the mean velocity magnitude in inter-blade channel is higher than velocities present in compressor setup. However, this features is strongly influenced by the fact that blockage factor is much higher for negative values of AoA. Blades arranged in compressor cascade are located rather in a row behind each other and they are not distributed across channel height.

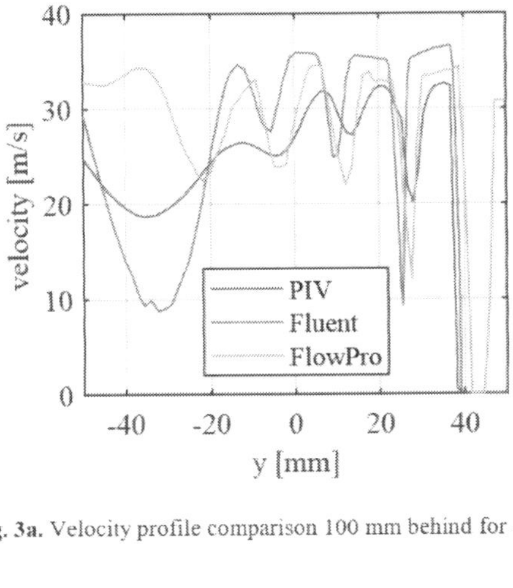

Fig. 3a. Velocity profile comparison 100 mm behind for -8°.

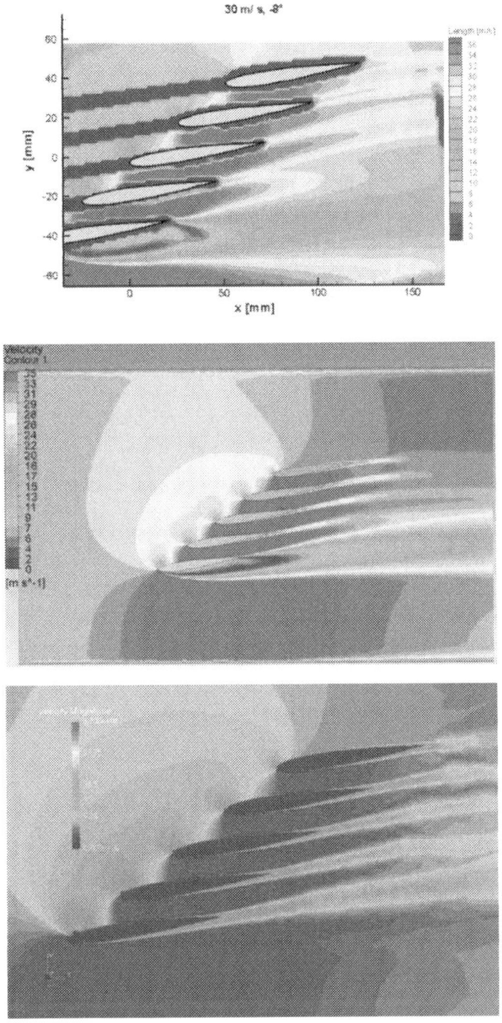

Fig. 3b, c, d. PIV, Fluent and FlowPro comparison, AoA -8°, velocity 30 m/s, distribution of mean flow velocities.

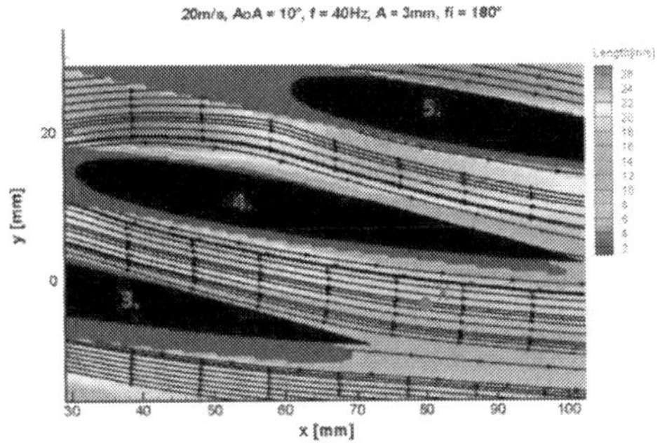

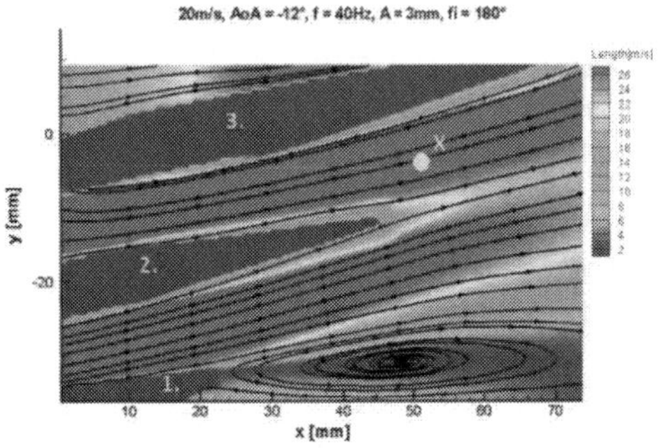

Fig. 4a, b. PIV measurement of compressor and turbine cascade, velocity distribution, vector lines, X-point location.

3.2 Flow dynamics

There are velocity variations in time, and frequency spectra for horizontal and vertical velocity component evaluated at point X for case with no excitation and negative AoA in figure 5. The blades are not fluttering, only insignificant vibrations were measured using sensitive encoders. There is no dominant frequency peak up to 50 Hz (plotted) and even no peak in higher frequencies. Further in article, we will search for PSD (power spectral density) value higher than one.

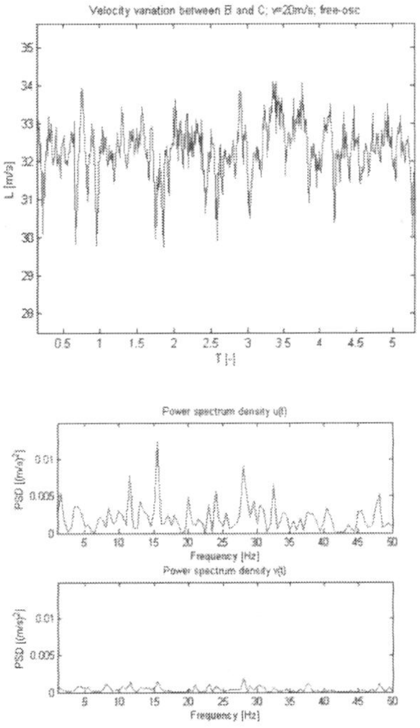

Fig. 5. Velocity variation and freq. spectra for both vel. components at point X for free-oscillating blades (turbine).

Further FFT analyses were performed for both type of cascades and for two values of inter-blade phase angle ϕ = 0° and 180° at point X. The dominant frequency 40 Hz was detected for all cases. It is evident that this frequency is embedded to the flow field by the excitation frequency. Both values are exact the same! There is an example (Fig. 6, 7) of instantaneous velocity in time and frequency spectra for both cascades and ϕ equal to 180°. The mean velocity value is about 19 m/s and 32 m/s for positive and negative AoA, respectively. The peak PSD value is about 1.5 for both cases and dominates for horizontal velocity component.

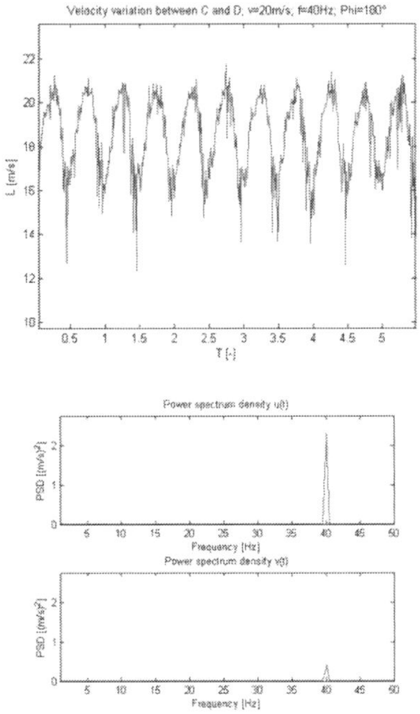

Fig. 6. Velocity variation and freq. spectra for both vel. components at point X for excited blades (compressor).

Next step is to plot the PSD value distribution in the whole plane of measurement for the most dominant frequency 40 Hz to search for the highest aeroelastic activity region. Only the result gained for phase angle 180° will be presented to maintain the article clarity. For this phase angle, the adjacent moveable blades are performing its movement in antiphase motion and hence the interaction between them is higher compared to case when blades go simultaneously. The PSD value distribution for horizontal and vertical component can be seen in figure 8 for positive AoA of cascade. The red area (PSD value higher than 5) can be detected as region with very strong aeroelastic activity. The horizontal velocity component dominates over vertical one for all regimes. The highest activity is present between 3rd and 4th blade below its leading edge. This is caused by the fact that the channel width variations are quite significant here – the blades rotate against each other. Not so high activity is also present in the wake behind the middle blade. Unfortunately, the arbitrary chosen point X is located in the region with rather small activity (blue bubble below 4th blade). The coupling is even stronger for case of turbine setup (Fig. 9). The area between the leading edge of the middle blade and adjacent blade has the same reason to exist as in former case. More, there is strong activity between the body of middle blade and both wakes from 1st and 2nd blade. The wake itself behind the 1st blade is huge (Fig. 4) and velocity variations are happening with exact frequency of 40 Hz despite the fact that this blade is fixed. Again the point X is outside the region with highest activity.

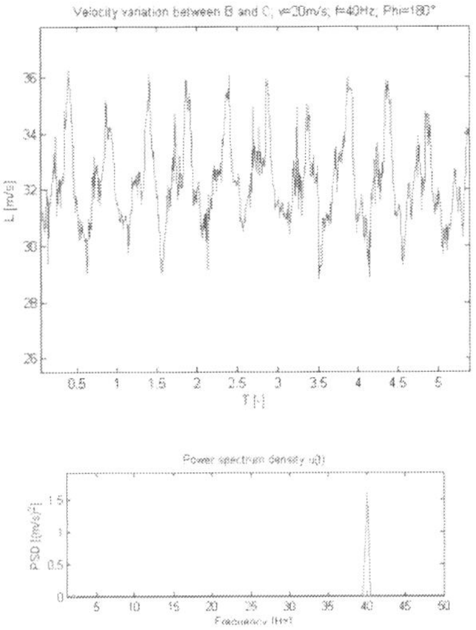

Fig. 7. Velocity variation and freq. spectra for one vel. component at point X for excited blades (turbine).

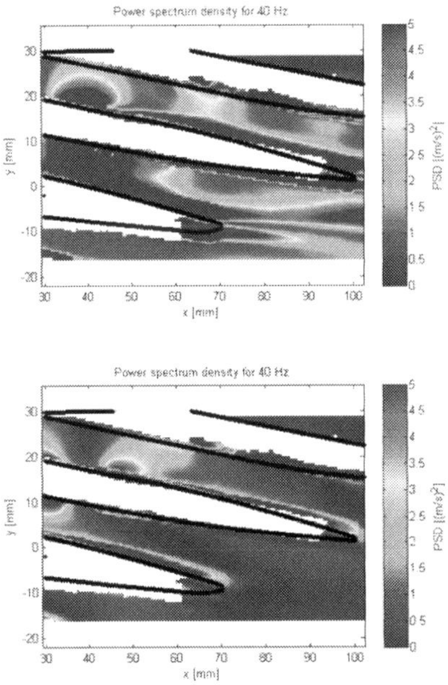

Fig. 8a, b. PSD value distribution for 40 Hz for both vel. components for compressor cascade, excited at φ = 180°.

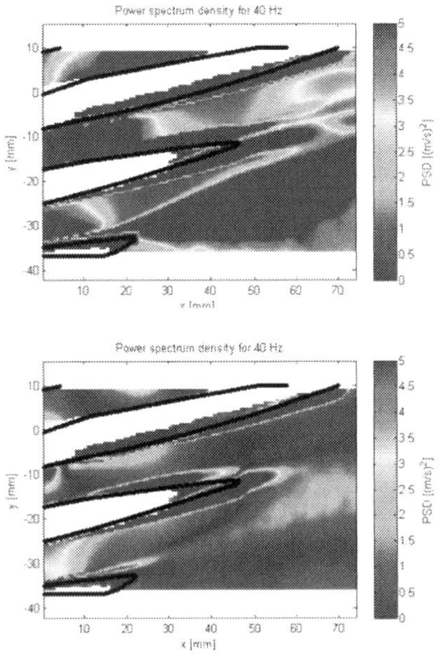

Fig. 9a, b. PSD value distribution for 40 Hz for both vel. components for turbine cascade, excited at φ = 180°.

3.3 Phase-locked measurement

The phase-locked measurements were conducted in distinct eight phases per period of harmonic motion prescribed by the shakers. The amplitude of this motion was set to 3°, so phase positions of the middle blade were set out in this order: 10°-11.5°-13°-11.5°-10°-8.5°-7°-8.5° for compressor setup. Notice, that the distribution is regular, e.g. the first and the fifth phase has the same position of the middle blade (and also adjacent blades). The phase positions were: 12°-10.5°-9°-10.5°-12°-13.5°-15°-13.5° with negative sign for turbine setup. The mean phase velocity in the plane of measurement was evaluated for all eight phases as well as standard deviation (STD) of this quantity.

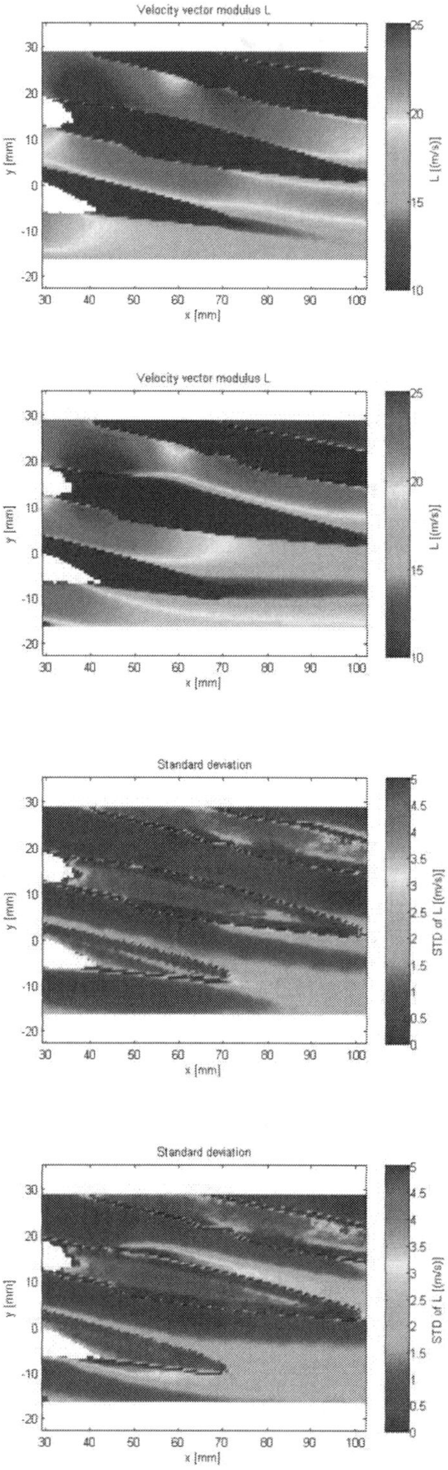

Fig. 10a, b, c, d. Mean phase velocity distribution and its STD value for excited regime (40 Hz and φ = 180°) for compressor cascade.

Although the starting position of the middle blade in the first phase is the same as the blade position in the fifth phase, we observed big differences in phase velocity distribution. There is a graph of this distribution for cascade arranged as the compressor in figure 10. Below middle blade (3rd), there is visible high increment of the velocity. On the other hand, the region between wakes from 3rd and 4th blades is affected by large loss of velocity value and the wake itself has totally bended shape in the fifth phase. Figures 10 c, d reveal the distribution of STD which can report the information about fluctuation activity of the flow. It can be seen that there is fluctuation value only in the wake in the first phase. Contrary, the fifth phase is characterized by separation region located below the middle blade and over the 4th blade. This separation is obviously a flow residual structure from the blade position in AoA of 13°. Figure 11 is plotted similarly for turbine setup. The wake interaction is clearly visible between 2nd and 3rd blade. Also the flow separation below middle blade can be seen, but only in the fifth phase. This is due to the fact that inter-blade channel at this place was more open during the previous phase position. The highest fluctuation activity is present below and behind the very first blade of this cascade. However, this one and the last one is introduced in the cascade as a periphery and they are actually boundary condition to help the flow to enter the cascade.

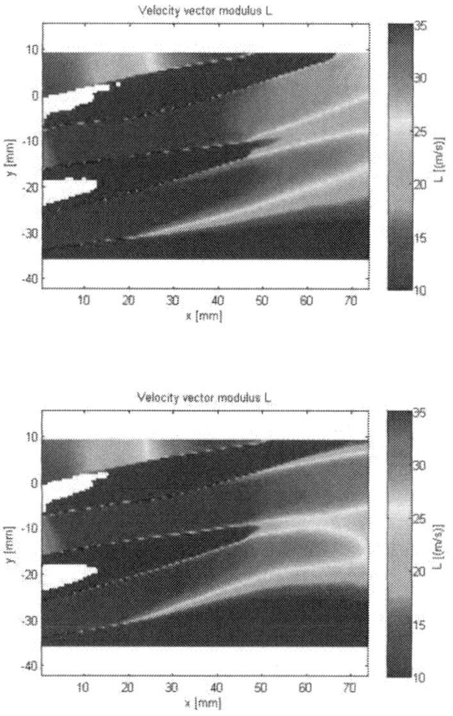

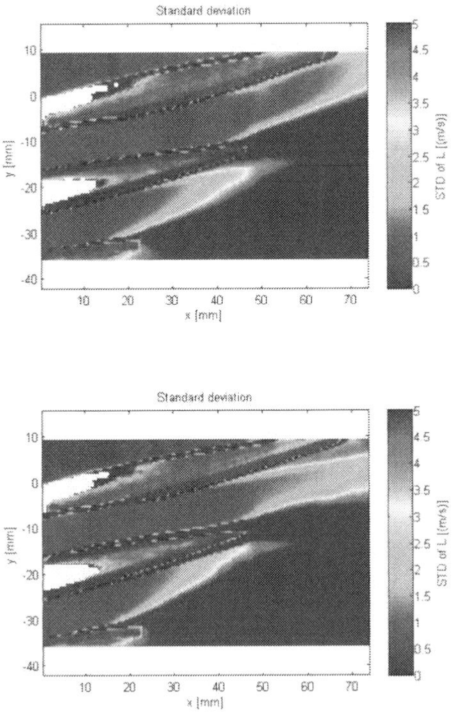

Fig. 11a, b, c, d. Mean phase velocity distribution and its STD value for excited regime (40 Hz and $\varphi = 180°$) for turbine cascade.

4 Conclusions

The stereoscopic TR-PIV measurement was conducted inside the model composed of five NACA 0010 profiles (blades) in the shape of a cascade with stagger angle equal to 30°. This assembly allowed to change the angle of attack from -15° to +15°. The important differences in flow field topology were observed for two standard cases; negative AoA simulating turbine arrangement and also positive AoA simulating a compressor. The comparison of statistical flow quantities with two CFD models was performed. Identified discrepancies will be resolved in the future.

During further investigation, the three middle blades were excited using three shakers, rotational harmonic motion with prescribed frequency 40 Hz and amplitude equal to 3° was introduced with different values of inter-blade phase angle. Dynamical analysis was utilized to search for regime and areas with important aeroelastic coupling. The phase angle equal to 180° is characterized by the existence of considerable FSI between blades motion and the flow properties as the channel between two adjacent blades is periodically expanding and narrowing which results in boundary layer separation.

Results evaluated from phase-locked measurements have shown that it is also very important if the channel between two blades is opening or closing. The flow topology is actually affected by previous state of cascade geometry quite significantly. Generally it can be said, that the highest interaction between two blades are within areas, where the inter-blade channel can get the narrowest which is close to the leading edge or more downstream in the wakes.

The obtained results will be used to design advanced model of the blade cascade allowing to investigation of stall flutter by means of self-induced vibrations.

This work was supported by the research project of the Czech Science Foundation No. 20-26779S "Study of dynamic stall flutter instabilities and their consequences in turbomachinery application by mathematical, numerical and experimental methods" An institutional support RVO61388998 is also gratefully acknowledged.

References

1. J. J. Waite, Physical Insight, Steady Aerodynamic Effects, and a Design Tool for Low-Pressure Turbine Flutter, AAT 10107521, **77**-09(E), Duke University, 2016
2. V. Tsymbalyuk, J. Linhart, XVII IMEKO World Congress, TC3 (2003)
3. J. Lepičovský, E. R. McFarland, V. R. Capece, T. A. Jett, R. G. Senyitko, NASA/TM-211894 (2002)
4. J. Vimmr, O. Bublík, A. Pecka, L. Pešek, The 13th ISAIF in Okinawa, ISAIF13-S-0089 (2017)
5. L. Půst, L. Pešek, International Journal of Bifurcation and Chaos, IJBC-D-16-00353R2 (2017)
6. P. Procházka, V. Uruba, L. Pešek, V. Bula, EPJ Web of Conf., **180** (2018) 02086

Design and testing of the supporting setup for the high-pressure vibrating tube densimeter

Olga Prokopová[1], *Miroslav* Čenský[1], *Aleš* Blahut[1], and *Václav* Vinš[1*]

[1]Institute of Thermomechanics of the Czech Academy of Sciences, Dolejškova 1402, 182 00 Prague 8, Czech Republic

Abstract. Vibrating tube densimeters (VTDs) have become relatively popular instruments for measuring the density of a large variety of gases and liquids since their development about fifty years ago. The apparatuses measure the characteristic frequencies of vibrating U-shaped tube filled with a fluid sample, when the fluid fill shifts the dynamic characteristics of the U-tube. This study describes design and testing of the in-house developed supporting setup for a commercial high-pressure VTD Anton Paar DMA HP. The instrument covers density range from 0 to 3000 kg/m^3 at pressures up to 700 bar and temperatures from -10 to 200 °C. Whereas the temperature of density measurement is fully controlled by the apparatus using a thermal block with Peltier unit, pressure has to be generated and monitored externally. Therefore, a high-pressure sampling system was designed and assembled. Special care is taken when measuring temperature with an external resistance thermometer connected to precise thermometry bridge. In order to avoid unfavorable condensation of air moisture during the measurement, VTD apparatus is protected by a dry box of own design. Technical details together with the calibrations of pressure and temperature probes are described in detail. A preliminary data for temperature dependence of density of the selected liquids (water, ethanol or ethylene glycol) obtained at the barometric pressure are also provided and compared to the reference density correlations.

1 Introduction

Vibrating tube densimetry (VTD) is a well - established experimental method often used for systematic studies of fluid density. Since its development about 50 years ago, the VTD method has been adopted by many scientists mainly for its relatively straightforward and fast measurements with good accuracy over wide ranges of temperatures and pressures. A commercial design of VTD offers reliability, small amount of sample required and convenient operation, when the density data can be obtained in the order of minutes or even seconds. The principle of VTD is based on measurement of resonance period τ of a vibrating, usually U-shaped, tube filled with the fluid sample, which can be related to the fluid density by means of relation derived from the spring mass model:

$$\rho = A \cdot \tau^2 - B \tag{1}$$

where

$$A = \frac{k}{4\pi^2 V}, \tag{2}$$

$$B = \frac{m}{V} \tag{3}$$

In equations (1) and (2), m stands for the mass of the cell, k denotes the spring constant, V is the inner volume of the cell, and ρ is the fluid density.

This work focuses on high pressure VTD equipped with a U-tube made of metal alloy and allowing measurements over a wide range of temperatures and pressures. In our laboratory, a commercial high-pressure VTD Anton Paar DMA HP is available. The device itself allows for temperature dependent measurements but for high pressure application it has to be equipped with an external filling and pressurizing system. Design, construction, technical details and testing of such in-house developed setup is described here.

Since most VTDs are operated in a relative measurement mode, it follows that in order to obtain accurate experimental density data a great attention must be paid to the instrument calibration including a suitable selection of the reference fluids. The calibration process is based on determination of parameters A and B, when the period of oscillation can be expressed as:

$$\tau = \sqrt{\frac{\rho + B}{A}} \tag{4}$$

The relation between the density and the period of oscillation is influenced by the temperature of the U-tube and the internal pressure. From this point of view, the calibration of high pressure VTDs becomes relatively complicated [1-9] compared to common atmospheric instruments. There have been developed several different approaches to obtain A and B parameters as described further.

* Corresponding author: vins@it.cas.cz

In this work, a preliminary density data of selected liquids (water, ethanol and ethylene glycol) for temperature dependence at atmospheric pressure compared to the reference density correlations are presented. The preliminary data were evaluated based on a polynomial model similar to that employed on a barometric VTD in our previous studies [10,11].

2 Temperature and pressure functions – Evaluation of parameters *A*, *B*

There are two main approaches to obtain instrument's parameters *A* and *B*. Measurements of two reference fluids made at almost exactly the same pressures and temperatures as the sample density required is one of the reliable way of calibration. Some authors call it a "classical" model. Using equation (1) at the measured points the parameters *A*, *B* are determined. To realize this linear calibration procedure, the following conditions must be fulfilled: (i) the densities of standards cover the whole range of pressures and temperatures, (ii) the density uncertainty of the standards is in full range sufficiently low, (iii) the standard densities do not differ much from each other and the sample density lies between the standard densities, (iv) viscosities of reference fluids are similar to that of the sample, (v) the reference substances under demanded temperatures and pressures are chemically stable. The uncertainty of density depends in this case on the uncertainties associated to each calibrating fluid densities. The main problem of such calibration lays in the finding of suitable calibration fluids because there is a lack of fluids with known densities as a function of temperature and pressure with sufficiently low uncertainty.

To solve the problem of unavailable reference fluids, alternative methods have been developed using just one calibration fluid (usually water) together with vacuum [3,4]. Vacuum can be realized by coupling a vacuum cylinder to a Hastelloy vibrating cell. When the air density is in the order of 1 kg/m^3 at 1 bar then the density of the vacuum at pressure below 100 Pa is lower than 0.001 kg/m^3, which is far below the resolution of most high-pressure VTDs and as such can be assumed as 0 kg/m^3. Using vacuum as the second standard solved the problem of its density uncertainty but arose a new one with the evaluation of the pressure dependence of parameters *A* and *B*. Bouchot et al. [5] showed that the assumption which parameters depend on the pressure (determination of boundary conditions) determines the shape of the calibration equation and subsequently relativizes the values of the constants *A* and *B*. This kind of problem does not occur when utilizing the classical calibration using two reference fluids along with eq. (4).

To figure out the way how to conserve the advantage of vacuum as one reference and remove the uncertainty arose from the classical discrete formula the physical signification of terms *m*, *V* and *k* has been evaluated independently [5,8,9]. To do this the equation (1) must be considered not as a discrete conversion tool but as a continuous transform [5]. Thus, calibration measurements can be made over a similar (not identical) range of conditions as the range of target conditions and parameters *A* and *B* can be expressed continuously as functions. This can be done either as a pure empirical function [1,8] or as a function in which correlation parameters have a physical meaning [5,9]. While maintaining the linear formalism of eq. (1) and preserving the mechanical spring-mass model, the constants *A* and *B* can be modeled as pressure functions using rigorous considerations on the stress and thermal behavior of the vibrating tubes.

Outcalt and McLinden [8] introduced a polynomial model, where 13 constants need to be evaluated from a calibration on a series of reference fluids at various temperatures and pressures.

$$\rho(T,p)=\begin{pmatrix}A_1+A_2T+A_3T^2+A_4T^3+\\+A_5p+A_6p^2+A_7Tp\end{pmatrix}\frac{\tau^2}{\tau_0^2}-\\-\left(B_1+B_2T+B_3T^3+B_4p+B_5p^2+B_6Tp\right) \quad (5)$$

In addition, a second-order polynomial for the fundamental period of an evacuated cell τ_0, has to be correlated to the experimental data for τ at varying temperature *T*.

$$\tau_0=c_0+c_1T+c_2T^2 \quad (6)$$

May et al. [9] developed a physically-based model having a general form similar to that of equation for atmospheric VTD.

$$\rho(T,p)=A(T,p)\tau^2-B(T,p) \quad (7)$$

where coefficients *A* and *B* are given as follows

$$A(T,p)=\frac{(1+\beta_\tau p)\left(\rho_M/S_{00}\tau_{00}^2\right)}{\left(1+\varepsilon_{\tau1}T+\varepsilon_{\tau2}T^2\right)^2\left(1+\alpha_VT+\beta_Vp\right)}, \quad (8)$$

$$B(T,p)=\frac{\rho_M}{S_{00}\left(1+\alpha_VT+\beta_Vp\right)}. \quad (9)$$

Quantities τ_{00}, $\varepsilon_{\tau1}$, and $\varepsilon_{\tau2}$ are determined from measurements of the evacuated cell at various temperatures. Other parameters S_{00}, α_V, β_V and β_τ are determined from measurements of a reference fluid. ρ_M is the density of the U-tube material, e.g., for Hastelloy ρ_M = 8890 kg/m^3.

3 Experimental setup

3.1 High-pressure vibrating tube densimeter

A commercial high-pressure vibration tube densimeter Anton Paar DMA HP was used for the measurements presented in this work. The instrument is designed to measure the density of liquids and gases at pressures up 700 bar in the temperature range from -10 °C to +200 °C. A simplified scheme of the experimental setup with DMA HP is shown in Fig. 1. Operation of the high-pressure cell in DMA HP is controlled via the DMA 5000 M densimeter designed for the atmospheric measurements. Both units are connected to a cooling water circuit with thermostatic bath for the measurements at temperatures below the ambient temperature. The DMA HP instrument has to be enclosed in a dry box connected to a drying circuit with air pump and a moisture absorber (silica gel column). The dry box allows measurements at temperatures below the dew point of the laboratory. A special filling and pressurizing system (FPS) for filling a liquid sample into the DMA HP U-tube had to be developed as described further.

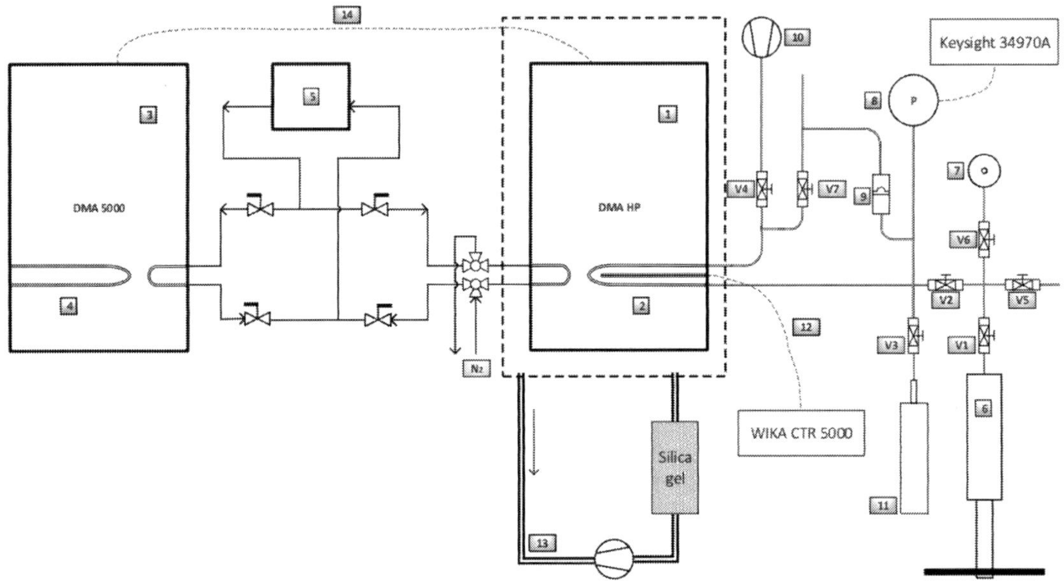

Fig. 1. Scheme of the high-pressure vibrating tube densimeter setup. 1 – DMA HP module, 2 – Hastelloy vibrating tube, 3 – DMA 5000 M atmospheric module, 4 – borosilicate glass vibrating tube, 5 – thermostat, 6 – hand piston pump, 7 – filling cylinder with sample, 8 – pressure transducer Omega, 9 – rupture disc, 10 – vacuum pump, 11 – 250 ml vacuum buffer or gas vessel (N_2), 12 – Pt100 with bridge WIKA CTR 5000, 13 – dry air circulation, 14 – data cable.

3.2 Filling and pressurizing system

A high pressure vibrating-tube densimeter measuring system was described in the literature [8,9,12]. A similar filling and pressurizing system (FPS) to the one used by Sampson et al. [12] was developed. Next to the DMA HP U-tube, FPS represents the core part of the apparatus as can be seen in Fig. 1. The FPS consists of high pressure needle valves (Parker 10V2081, USA), various fittings and adapters, safety rupture disc (Parker Autoclave Engineers, USA) and tubing (stainless steel 316L capillary 1/8" x .036" ~ 3.18 x 0.91mm), high pressure generator (manual pump HiP, USA, pressure range 0 to 10 000 PSI ~ 689 bar), pressure transducer (OMEGA, PX01S1 10KGI, USA), stainless steel sample vessel and a membrane vacuum pump (Vacuubrand MD1, max 1.2/1.4 m^3/h, 1.5 bar, Germany). FPS is operated in the following manner:

- Before the density measurement, FPS is cleaned and evacuated via valve V4 which is connected to the membrane vacuum pump.
- A sample is delivered into the first part of the system from the sample vessel via valve V6 while valves V2 and V5 are closed and sample enters via valve V1 also volume of the high pressure generator, which is open to its maximum.
- Then valve V6 is closed and the second part of the FPS is filled by opening valve V2, while valves V3, V4, V5, and V7 are closed.
- Whole system is then pressurized by the pressure generator to the required pressure and then the valve V1 is closed.
- Sample can be removed via valves V5 and V7 using either simple water vacuum pump or expelled by nitrogen via valve V3.
- For the calibration measurement under vacuum, a buffer volume of approximately 150 ml was connected via valve V3.

3.3 Dry box for VTD

A circuit for the humidity reduction in the dry box in which DMA HP is enclosed was developed. The drying box constructed from PMMA (Polymethyl methacrylate, plexiglass) is placed on an aluminum base. All necessary outputs and inputs were equipped with rubber grommets. During the operation, air is circulating in the closed circuit consisting of the inner space of the box, absorption column with inner diameter of 20 mm and a length of

65 cm filled with activated silica gel connected by polyurethane tubes. The level of humidity content was checked by thermohygrometer Testo 625 with its sensor inserted inside the dry box.

3.4 Pressure measurement

Pressure inside the system is measured using a transducer Omega PX01S1-10KG with the operating pressure range from 0 to 10 000 PSI (0 to 689 bar). DC output of the transducer corresponding to 4 to 20 mA is measured with data acquisition unit Keysight 34970A. According to manufacturer specification the linearity and hysteresis count ± 0.05 % of the transducer full scale.

Prior to the installation of transducer in the system, its accuracy was tested against a highly-accurate pressure transducer Paroscientific 415K-101. During the test measurement both transducers were installed simultaneously in the apparatus with gaseous nitrogen capable of stable pressure setting. Due to the fact that the transducer zero-point has a tendency to drift over time, the transducer is equipped with an adjusting screw. However, in our measurements we decided to adjust the zero-point numerically from the recorded data based on the reading at ambient pressure in the following manner

$$p_{\mathrm{offset}} = -I_0 p_{\mathrm{sl}} \quad (10)$$

where p_{offset} is pressure offset, I_0 is DC current reading at ambient pressure and p_{sl} denotes slope of pressure vs current I.

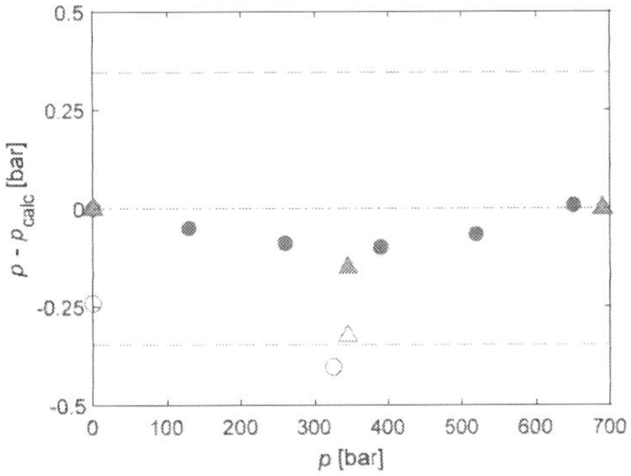

Fig. 2. Test of Omega PX01S1-10KG pressure transducer; deviations of the reference pressure measured with Paroscientific 415K-101 from the calculated values using linear pressure-current dependence. Triangles stand for calibration from the manufacturer, circles mark points from own measurements, when filled symbols denote measurements in the sense of increasing pressure, and empty symbols correspond to measurements with the descending pressure.

Comparison test was carried out at several points from ambient pressure up to 650 bar. Subsequently transducer was tested for hysteresis at two points in the sense of decreasing pressure. Test results are shown in Fig. 2 together with the data from original calibration from the manufacturer. Dashed lines correspond to ± 0.05 % tolerances. Value of transducer sensitivity, i.e. pressure vs DC current slope p_{sl}, stated in the calibration certificate from the manufacturer was confirmed within test measurements. Individual calibration points are compared in the figure with the values calculated using original slope p_{sl} and respective pressure offset calculated from eq. (10).

As can be seen, transducer exhibits some non-linearity, which can be better described in future, e.g., by quadratic equation. The transducer hysteresis was found lying within the tolerance level. For a reliable pressure measurement, it is recommended to carry out density measurements in the sense of increasing pressure. Also the initial point should be at ambient pressure to enable determination of pressure offset.

3.5 Temperature measurement

The DMA HPM instrument contains built-in Pt100 temperature sensor with an error < 0.1 °C stated by the manufacturer. In order to achieve more accurate temperature measurement and inspect its stability over time, an external Pt100 probe (Omega PR-10-3-100-M30-50-E-2M) with a length of 60 mm and a 3 mm diameter was used. The external thermometer can be placed in a hole located in the center of the DMA HP U-tube. Originally, the Pt100 probe had a specified length of 50 mm but approximately 10 mm of a plastic ring between the probe and the wiring had to be removed as it hindered inserting the probe deep enough. Probe is equipped with 2 m PFA cable resistant to the temperatures up to 200 °C. The calibration was done in the Czech Metrology Institute using a secondary temperature standard at 6 points in the range from -20 °C to 200 °C. In order to imitate condition in the hole for the external temperature probe in the DMA HPM instrument, the Pt100 thermometer was calibrated while placed inside a stainless steel tube 1/4" x 0.036" (6.35 x 0.91 mm) with a length of 400 mm. Expanded uncertainty based on the calibration was estimated to be below 0.048 °C over the whole temperature range. For the measurement of the resistance of the Pt100 probe a WIKA CTR 5000 thermometry bridge is used in our measurements, when the Callendar-Van Dusen equation correlated to the calibration data is used.

Both instruments WIKA CTR 5000 bridge and the Keysight 34970A data acquisition unit are connected to PC and data are collected by the LabVIEW program.

4 Water and dry air calibration

In order to prevent eventual temperature oscillations, the instrument and other accessories (pressure gauge and external resistance thermometer) were stabilized for 2 hours before each measurement campaign. The stabilization included (i) water circulation through the DMA HP cooling kit, (ii) drying the dry box below a required dew point value to prevent humidity condensation in the cell block inside the instrument, (iii) the air conditioning in the laboratory set to the constant temperature of 22 °C.

A preliminary high-pressure VTD calibration at atmospheric pressure was based on accurate measurements of water + air pairs over wide temperature range. Water from 2 °C to 90 °C and air from -5 °C to 90 °C with 5 °C step. In the calibration measurements, an ambient air applied into the U-tube was dried using a silica gel cartridge. To determine its density, the humidity was checked using a portable hygrometer (Testo 625, Testo AG, Germany). The relative humidity of dried air varied typically from 3.5 % to 6.0 %. The laboratory pressure was recorded at each temperature point by a high-precision external pressure gauge (Druck DPI 142, GE, USA). The pressure gauge was repeatedly compared to the stationary mercury barometer B 3 type from the VEB Werk für Technisches Glas Ilmenau manufacturer. The agreement of the measured pressures was typically better than 0.2 hPa. Reference density values of dry air were obtained from IAPWS guideline G10-8 for humid air [14].

Ultrapure water obtained from a purification system consisting of a reverse osmosis unit Rowapur purifying Prague tap water and of Purite Neptune Analytical unit was used in all measurements. The ultrapure water had a resistivity from 17.4 to 18.2 MΩ·cm, was free of particles larger than 0.2 μm and had an organic carbon content below $5 \cdot 10^{-9}$.

The different isotopic composition between the laboratory tap water and the Vienna Standard Mean Ocean Water (VSMOW) [15,16], to which the IAPWS-95 equation of state is correlated, was considered. Nevertheless, a slight correction of the molar mass of water that has been applied was not in the case of Hastelloy U-tube perceptible as it is one order of magnitude below the actual resolution of the DMA HP instrument.

4.1 VTD output quantities

VTD Anton Paar DMA HP with the Hastelloy C-276 cell measures the fluid density in the range from 0 to 3000 kg/m³ with a resolution of the period of oscillation 0.01 μs. The density error is declared to be ± 1.0 kg/m³ for higher pressures and even ± 0.1 kg/m³ for the optimum conditions. However, the final uncertainty of the measured density can be quite different as it depends on the considered temperature and pressure ranges, quality of the instrument calibration and the employed reference fluids, viscosity of the samples and other effects. We note that careful uncertainty analysis of the determined densities will be in the interest of our future research and is beyond the scope of this preliminary work.

Figures 3 and 4 show period of oscillation τ for water and dry air measured from 2 to 90 °C and from -5 to 90 °C, respectively, during three individual measurement campaigns. As it can be seen, the period is gradually increasing with increasing temperature. Periods measured at the same temperature on different days do not differ from each other by more than 0.01 μs both for water and dry air. These results confirm good reproducibility of the period of oscillation, i.e. a primary experimental output quantity. Any slight systematic offset between the individual air measurements caused by a different atmospheric pressure during individual measurements was not observed unlike in case of highly sensitive atmospheric VTD Anton Paar DMA 5000 M with a borosilicate glass U-tube [10,11]. This follows from the technical capabilities of the device. While the DMA 5000 M instrument has density resolution of 0.001 kg/m³ (per 0.001 μs oscillation period), the DMA HP cell can distinguish the difference in density of 0.05 kg/m³ (per 0.01 μs – device resolution). DMA 5000 M was found to achieve the uncertainty of water density better than 0.03 kg/m³ for the 2 to 90 °C temperature range [11].

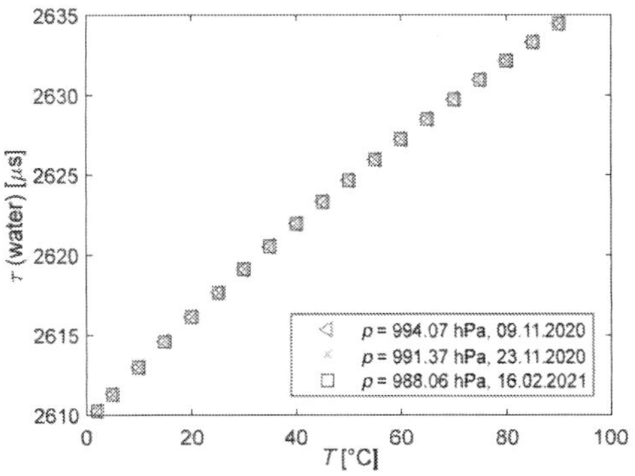

Fig. 3. Temperature dependence of period of oscillation τ for water at atmospheric pressure.

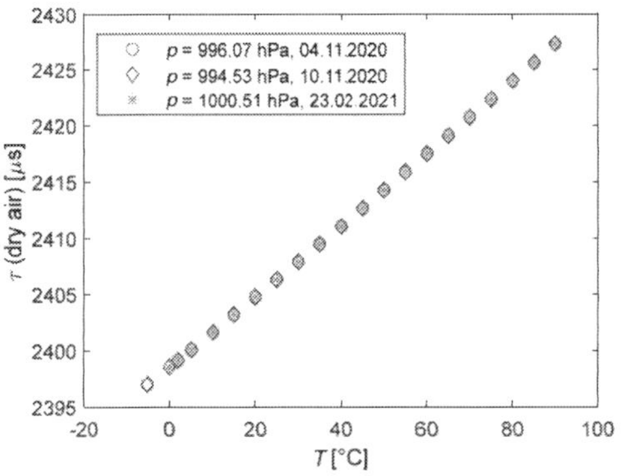

Fig. 4. Temperature dependence of dry air period of oscillation at atmospheric pressure.

4.2 Water and air measurements

The densities of water and dry air were determined from eq. (1). Temperature dependencies of instrument parameters A and B were fitted with cubic polynomials providing a sufficient accuracy. Figs. 5 and 6 show temperature dependencies of A and B reproduced with the following correlations:

$$A(T) = -1.5584 \cdot 10^{-5} T_r^3 + 3.582 \cdot 10^{-5} T_r^2 - \\ -1.3915 \cdot 10^{-4} T_r + 0.001 , \quad (11)$$

$$B(T) = -117.9996T_r^3 + 311.8838T_r^2 - \\ -345.149T_r + 5563.2 \quad , \qquad (12)$$

where T_r is the reduced temperature defined as $T_r = T/400$ with T in Kelvin.

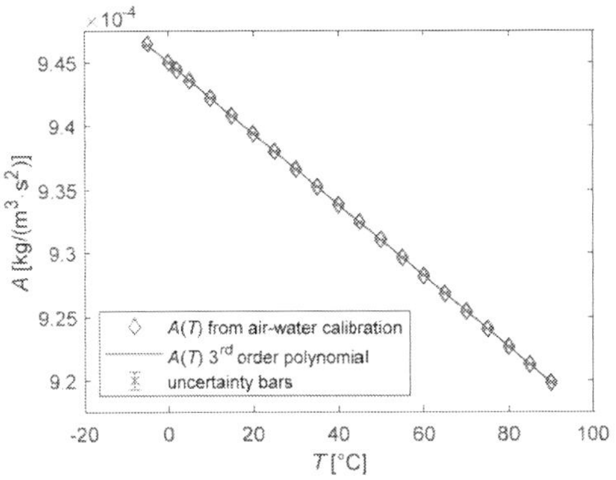

Fig. 5. Calibration parameter $A(T)$ correlated with 3rd order polynomial at atmospheric pressure.

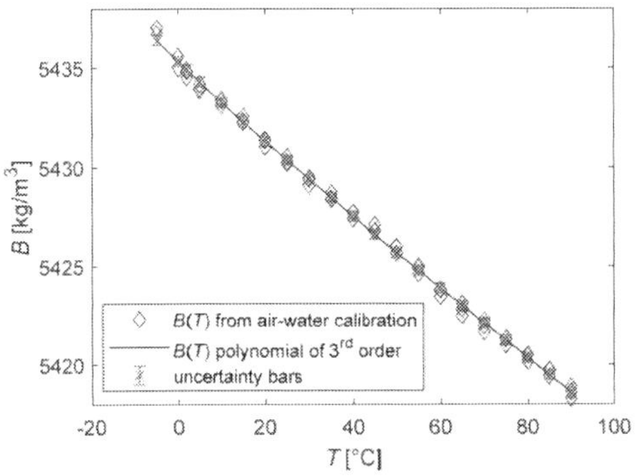

Fig. 6. Calibration parameter $B(T)$ correlated with 3rd order polynomial at atmospheric pressure.

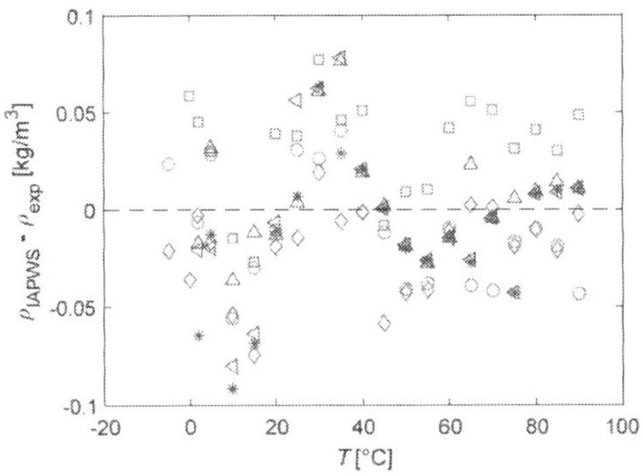

Fig. 7. Density of water (blue symbols) and air (red symbols) exp measured at atmospheric pressure compared to the reference models IAPWS – IAPWS-95 equation of state for water [16] and IAPWS guideline G10-8 for humid air [14].

Fig. 7 shows differences between the density of air and water from the reference values depending on temperature. The reference values were obtained from the models approved by IAPWS [14,16]. The preliminary density data determined from eq. (1) considering the polynomial correlations (11) and (12) led to the deviation from the reference values lying within ± 0.1 kg/m³.

4.3 Density of ethanol and ethylene glycol

In order to test the accuracy of the DMA HP Hastelloy vibrating tube, two commercially available chemicals, one with the density lower and one with the density higher than water, were selected as the test fluids. Ethanol from Supelco Merck batch number I1055390948 with the declared purity ≥ 99.90 % and ethylene glycol from Sigma Aldrich batch number SHBL0616 with the declared purity ≥ 99.96 % were used. Before and after each density measurement, the water content in the sample was determined using the coulometric Karl Fisher titrator Mettler Toledo C30. No considerable increase of the water content in the samples was detected during our experiments.

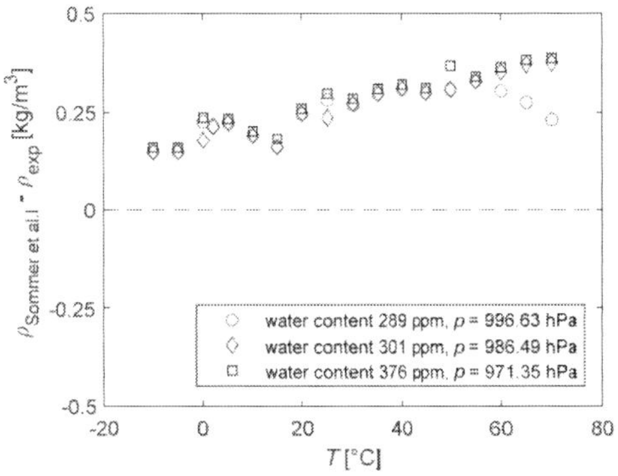

Fig. 8. Difference of the experimental density of three independent measurements of ethanol with similar amount of water impurities from the correlation by Sommer et al. [17].

Three individual measurements of ethanol and two individual measurements of ethylene glycol were carried out. In Fig. 8, it can be seen that the deviations from the reference correlation for liquid density of ethanol by Sommer et al. [17] does not exceed 0.40 kg/m³. In the case of ethylene glycol shown in Fig. 9, the mutual difference between the experimental density and the VDI correlation [18] taken as a reference lies in the interval from −1.1 to 0.37 kg/m³ in the temperature range from −10 °C to +140 °C. It should be noted that the ethylene glycol densities for temperatures below 2 °C and above 90 °C were determined based on an extrapolation of A and B polynomials given by eqs. (11) and (12). The standard uncertainty of the ethanol and ethylene glycol densities determined from the calibration are 0.36 and 0.37 kg/m³, respectively.

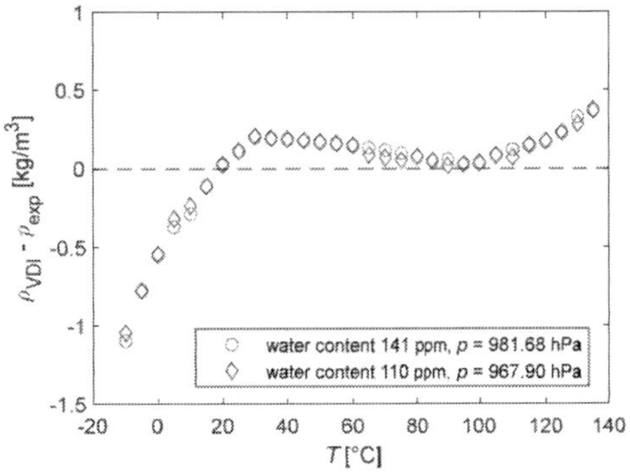

Fig. 9. Difference of experimental density of two independent measurements of ethylene glycol with similar amount of water impurities from the VDI [18] correlation.

5 Conclusions

Team of our laboratory is operating an atmospheric VTD Anton Paar DMA 5000 M with a borosilicate glass cell for more than three years. When properly treated and calibrated, the instrument provides high-accuracy density data at barometric pressure over the temperature range from 0 to 100 °C. The instrument can be extended with an independent high pressure vibrating cell DMA HP made of Hastelloy allowing the density measurements at pressures up to 700 bar. Our intention is to use these systems for accurate density measurements of large variety of technically interesting fluids over wide pressure ranges at temperatures from –10 ° C to +150 °C.

The effective use of DMA HP VTD linked with the atmospheric instrument DMA 5000 M required development of several supporting systems. In this work, the design, manufacture and operation of the new setup consisting of the filling and pressurizing system (FPS), special dry box, and the external pressure transducer and temperature probe with a thermal bridge were described. Technical details together with the preliminary calibration measurements of the DMA HP unit at atmospheric pressure were presented.

A classical formula between the density and the period of oscillation based on a spring-mass model along with the polynomial correlations of instrument parameters A and B were considered. For the preliminary measurements, the dry air and ultrapure water were used as two reference fluids, similarly as in case of the atmospheric VTD with the glass U-tube. Temperature dependent parameters A and B were evaluated over a temperature range from 2 to 90 °C. The initial barometric calibration was inspected on a series of testing measurements with four reference fluids; namely dry air, water, ethanol, and ethylene glycol. The collected data helped to carry out the uncertainty analysis of the determined density. For less viscous fluids, the standard uncertainty was found to be of around 0.37 kg/m^3 over the temperature range from –5 to + 140 °C. Good reproducibility of the measurements carried out with selected fluids proved sufficient purity of the U-tube cell and correctness of the measurement procedure.

In the subsequent research, measurements with a closed filling system under various pressures will be performed. A physically-based model developed May et al. [9] will be tested on our instrument by using the vacuum and one reference liquid calibration.

Acknowledgement

The study was supported by the Czech Science Foundation grant no. GA19-05696S and the institutional support RVO: 61388998.

References

1. G. Morrison, D.K. Ward, Fluid Phase Equilib. **62**, 65 (1991)
2. V.G. Niesen, J. Chem. Thermodyn. **21**, 915 (1989)
3. A. T. Sousa, P. S. Fialho, C. A. Nieto de Castro, R. Tufeu and B. Le Neindre, Fluid Phase Equilibr., **80**, 213 (1992)
4. B. Lagourette, C. Boned, H. Saint-Guirons, P. Xans and H. Zhou, Meas. Sci. Technol. **3**, 699 (1992)
5. C. Bouchot, D. Richon, Fluid Phase Equilib. **191**, 189-208 (2001)
6. Y. A. Sanmamed, A. Dopazo-Paz, D. Gonza´lez-Salgado, J. Troncoso and L. Romani, J. Chem. Thermodyn. **41**, 1060 (2009)
7. C.D. Holcomb, S.L. Outcalt, Fluid Phase Equilib. **150**, 815 (1998)
8. S.L. Outcalt, M.O. McLinden, Ind. Eng. Chem. Res. **46**, 8264–8269 (2007)
9. E.F. May, W.J. Tay, M. Nania, A. Aleji, S. Al-Ghafri, J.P.M. Trusler, Rev. Sci. Instrum. **85**, 095111, (2014)
10. O. Prokopová, A. Blahut, M. Čenský, V. Vinš, Proc. Int. Conf. Experimental Fluid Mechanics 2019, Franzensbad (Czech Rep.), November 19-22, 2019.
11. O. Prokopová, A. Blahut, M. Čenský, V. Vinš, Comments on temperature calibration and uncertainty estimate of the vibrating tube densimeter operated at atmospheric pressure, J. Chem. Thermodynamics (2022) (to be submitted).
12. C.C. Sampson, X. Yang, J. Xu, M. Richter, J. Chem. Thermodyn. **131**, 206 (2019)
13. R. Gonfiantini, Nature **271**, 534 (1978)
14. IAPWS G8-10: Guideline on an Equation of State for Humid Air in Contact with Seawater and Ice, Consistent with the IAPWS Formulation 2008 for the Thermodynamic Properties of Seawater, (2010) http://www.iapws.org/
15. National Institute of Standards and Technology, Report of Investigation, Reference Materials 8535, 8536, 8537, NIST Standard Reference Materials Program, Gaithersburg, MD (2005 revision of 1992 original report). Report available at https://www-s.nist.gov/srmors/view_report.cfm?srm=8535

16. IAPWS SR6-08: Revised Supplementary Release on Properties of Liquid Water at 0.1 MPa, (2011), http://www.iapws.org/
17. D. Sommer, R. Kleinrahm, R. Span, and W. Wagner, J. Chem. Thermodynamics, **43**, 117 (2011)
18. M. Kleiber and R. Joh, VDI-Wärmeatlas, 11., bearbeitete und erweiterte Auflage, ch. Stoffwerte von sonstigen reinen Fluiden, pp. 357–488. Springer Verlag Berlin Heidelberg (2013)

Quantitative Schlieren imaging based on fringe projection

Pavel Psota[1*], *Marek* Stašík[1], *Jan* Kredba[1], *Vít* Lédl[1], *Darina* Jašíková[1], *Michal* Kotek[1], and *Václav* Kopecký[1]

[1]Faculty of Mechatronics, Informatics and Interdisciplinary Studies, Technical University of Liberec, Studentská 2, 461 17 Liberec 1, Czech Republic

Abstract. An effective yet simple quantitative synthetic Schlieren imaging technique is presented in this paper. A multiplexed fringe pattern displayed on a common laptop screen is observed through a disturbed medium resulting in a change of the fringe pattern. Fourier analysis of such pattern provides high-resolution and almost real time measurement of rays' deflection due to the medium refractive index variation. The developed technique was applied for measurement of a compressible fluid discharge from a narrow channel in a wind tunnel.

1 Introduction

The term "Schlieren" refers to any optical inhomogeneities in a transparent medium and Schlieren photography relies on the the fact that light rays are bent whenever they encounter changes in density of a fluid. Various modifications of Schlieren techniques [1–3] have been used for flow visualization over a hundred year. With the rise of digital sensing technology, Background Oriented Schlieren (BOS) techniques are emerging [4–6]. BOS is an optical flow diagnostic technique used to measure density gradients in a flow field by tracking the apparent distortion of a target dot pattern. The distortion of the dot pattern is typically estimated by cross-correlating an image of the dot pattern without the density gradients (the reference image) with a distorted image viewed through the density gradients (the gradient image). In this paper we propose BOS technique based on fringe projection with spatial carrier frequencies. Such fringe pattern is similar to the pattern obtained by the interference of two tilted plane waves. A Fourier domain analysis allows for accurate yet fast data processing. This approach was applied for measurement of supersonic compressible fluid discharge from a narrow channel in a wind tunnel. The measured density distribution was compared to interferometric results. In addition to the direct use of this technique in various research topics, it can also be used along with interferometry [7] of high-gradient fields in order to subtract mapping error within interferometric data evaluation.

2 Methods

Fig. 1 shows the interaction between a light ray and a schliere. A schliere can be considered as gradient of the index of refraction in a transparent medium. Many experimental arrangements including wind tunnels are designed in order to keep the refractive index along z-axis constant. In Fig. 1 a two-dimensional schliere with a positive gradient of the index of refraction $\partial n/\partial y > 0$ in y-direction is sketched. Before the light ray interacts with the schliere it moves straight and parallel into the z-direction.

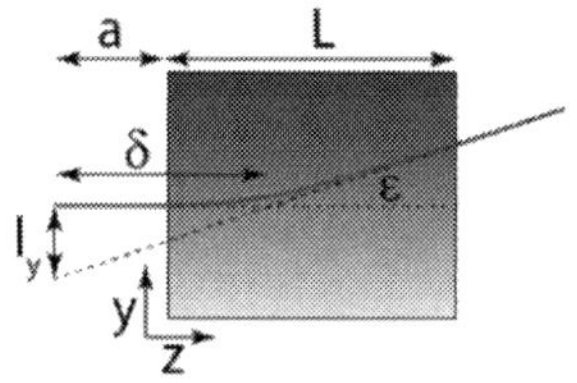

Fig. 1. Geometry of light refraction of a 2D schliere with a gradient in y-direction.

When the ray travels within the schliere it is bent towards the regions of higher index of refraction with a curvature

$$\frac{\partial^2 y}{\partial^2 z} = \frac{1}{n_0}\frac{\partial n}{\partial y} \quad (1)$$

and after travelling a distance of L it is bend by the angular ray deflection ε_y

$$\varepsilon_y = \frac{L}{n_0}\frac{\partial n}{\partial y}. \quad (2)$$

When observed through the schliere, the banded ray seems to be shifted about

$$l_y \approx \delta\varepsilon_y = \frac{L\delta}{n_0}\frac{\partial n}{\partial y}, \quad (3)$$

where $\delta = a + L/2$.

For simplification, let's keep the one-dimensional case (omitting $\partial n/\partial x$). We will expand the description in both dimensions x, y later in the text. Let us assume that a fringe pattern with spatial carrier frequency k_{y0} :

$$I(y) = A + \cos(k_{y0}y) \quad (4)$$

is located on the back side of an experimental section, see Fig. 2a. When observing the fringe pattern through an area

* Corresponding author: pavel.psota@tul.cz

with the presence of flow, the observed fringe pattern is modified due to the schliere effect as:

$$I(y) = \mathrm{A} + \cos\left(k_{y0}y + \varphi_y(y)\right) \quad (5)$$

where $\varphi_y(y) = k_{y0}l_y(y)$ is phase due to the ray deflection l_y in y-direction, see the green ray in Fig. 2a. It is obvious that the phase φ_y is the key to retrieve the ray deflection. Rewriting (5) using complex exponential yields in

$$I(y) = A + 1/2\, C(y) + 1/2\, C^*(y), \quad (6)$$

where $C(x, y) = \exp\left[i\left(\varphi_y(y) + k_{y0}y\right)\right]$. Equation (6) in Fourier domain:

$$\hat{I}\left(k_y\right) = \hat{A} + \hat{C}\left(k_y - k_{y0}\right) + \hat{C}^*\left(k_y + k_{y0}\right) \quad (7)$$

(with the spatial frequency coordinates k_y) is composed of a central DC term - $\hat{A}$ and two conjugated components $\hat{C}$, $\hat{C}^*$ located symmetrically from the center of the spectrum, see Fig. 2b. The roof symbol ^ denotes Fourier spectrum and the superscript * denotes complex conjugation. The introduced spatial carrier frequency k_{y0} plays a key role in the clear separation of all spectral components $\hat{A}$, $\hat{C}$, $\hat{C}^*$in the Fourier Domain and must be sufficiently high to avoid any mutual overlap of the components. On the other hand, k_{y0} cannot be too high in order to meet the Nyquist criterion (resolve the cosine pattern by the camera sensor).

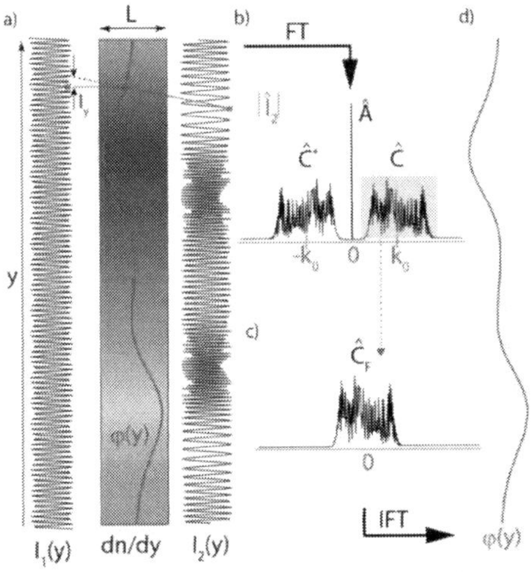

Fig. 2. Principle of the fringe projection technique in 1D (along y-axis): a) fringe pattern before I1 and after I2 propagation through the nonhomogeneous media; b) the Fourier amplitude spectrum of the fringe patter; c) the Fourier amplitude spectrum after a band-pass filtering and shifting; d) phase distribution retrieved from filtered complex field.

Once the spectral components are separated, it is possible to filter out the desired spectral component $\hat{C}$ (containing the phase φ) using bandpass filter around the carrier spatial frequency k_{y0} (see the green interval in Fig. 2b). Considering the shift property of the Fourier Transform, the filtered spectrum $\hat{C}_F$ (subscript F denotes filtered spectrum without $\hat{A}$ and $\hat{C}^*$) can be shifted about the spatial frequency in order to remove the linear phase carrier, see Fig2c. The filtered and shifted spectrum $\hat{C}_F$ isn't any longer Hermitean so its inverse Fourier Transform will contain non-zero both real and imaginary parts and hence the phase (see Fig. 2d) can be retrieved as:

$$\varphi_y(y) = arctan\left(\frac{Im\{C_F(y)\}}{Re\{C_F(y)\}}\right). \quad (8)$$

The spatial frequency k_{y0} as seen by a camera depends on magnification of imaging system and/or may not be accurately know. Moreover, some optical aberrations without the presence of a phenomena due to e.g. optical windows may be introduced. Therefore, a double shot measurement is recommended. Let us assume a measurement at the steady/reference state without presence of a phenomenon resulting in C_{F0} and another measurement performed with the presence of the phenomenon that is processed in the very same way to get C_{F1}. The measurement with the phenomenon can be related to the reference state measurement as

$$\varphi_y(y) = arctan\left(\frac{Im\{C_{F1}(y)C^*{}_{F0}(y)\}}{Re\{C_{F1}(y)C^*{}_{F0}(y)\}}\right). \quad (9)$$

It is important to note that phase φ_y is wrapped within 2π interval and thus, if the phase range exceeded, a spatial unwrapping algorithm [8] must be applied. A displacement l_y of the fringe at each pixel in y-direction can be obtained from the phase φ_y as

$$l_y(y) = \varphi_y(y)/k_{y0}. \quad (10)$$

Assuming geometry in Fig. 1 and (3), the displacement can be used to calculate refractive index slope

$$\frac{dn}{dy} = arctan\left(\frac{l_y(y)}{\delta}\right)\frac{n_0}{L}. \quad (11)$$

As x and y coordinates are orthogonal, the same analysis can be done for x-direction. However, due to dynamic behavior of flow, it is not possible to measure $\frac{dn}{dy}$ and $\frac{dn}{dx}$ one after the other. The solution is a using of the multiplexed fringe pattern resulting in the observed pattern

$$I(x, y) = \mathrm{A} + \cos\left(k_{y0}y + \varphi_y(x, y)\right) + \cos\left(k_{x0}x + \varphi_x(x, y)\right) \quad (12)$$

that is modified by phases φ_x, φ_y due to deflections l_x, l_y in both orthogonal directions. Such fringe pattern is processed independently for each direction and hence the displacements l_x, l_y, eventually refractive index slope maps $\frac{dn}{dx}$, $\frac{dn}{dy}$ can be computed.

3 Measurement of flow discharge

The experimental arrangement is shown in Fig. 3. It is a very simple configuration consisting of an image of the fringe pattern (FP) displayed on a laptop screen, the measured area (MA) and Phototron FASTCAM Mini WX100 high-speed camera with a 70mm focal length lens. The camera resolution is 2048×2048 pixels, each having extension of 20μm.

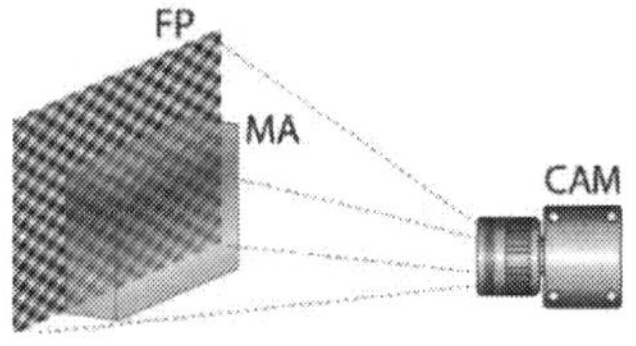

Fig. 3. Experimental arrangement of the fringe projection technique: FP – fringe pattern, MA – measured area (volume), CAM – camera with a lens.

This technique can however be used with any imaging system providing sufficient field of view, spatial and temporal resolution for the particular phenomenon under investigation. In our case, the phenomenon under investigation was a supersonic compressible fluid discharge from a narrow channel in a wind tunnel. The narrow channel was formed by two parallel walls and behind the channel there was a settling chamber, see Fig. 4. The test section had optical windows on both side for the flow and the fringe pattern observation.

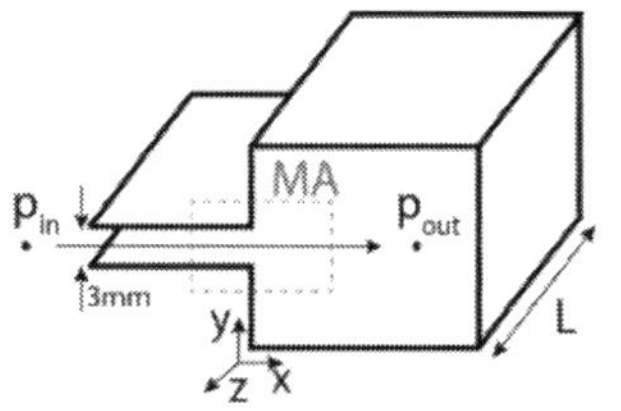

Fig. 4. Basic concept of the narrow channel and the settling chamber measurement. Dashed rectangle represents the investigated area.

During measurement, air flows through the channel due to pressure gradient. Air enters the test section at ambient conditions thus total pressure P_{in}=98.05kPa was equal to the barometric pressure. Static pressure at the outlet of the narrow channel P_{out}=23.87kPa in the settling chamber was also measured for further quantitative evaluation. The intensity fringe pattern was composed of a multiplexed orthogonal cosine fringe patterns with carrier frequencies in *x', y'* directions. The x', y' coordinate system was rotated around 45° to a global reference frame (*x, y*) defined by the narrow channel axis, see Fig 5a.

The fringe pattern was observed through optical glasses on the wind channel walls. The fringe pattern was firstly captured at reference state (P_{in}=P_{out}=98.05kPa) yielding in pattern denoted as I_R. Then the observed intensity fringe pattern was modified due to the presence of the air flow ($P_{in} \neq P_{out}$). The modified fringe pattern I_M was also captured. The both fringe patterns were processed in Fourier domain, see Fig. 5b and phase maps P_1 and P_2 were retrieved using the procedure described above, see Fig. 5c. The phase maps P_1 and P_2 are related to displacements in direction of spatial frequency k-vectors $\boldsymbol{k_1} = (\boldsymbol{k_{x1}}, \boldsymbol{k_{y1}}), \boldsymbol{k_2} = (\boldsymbol{k_{x2}}, \boldsymbol{k_{y2}})$, see Fig. 5b. Magnitudes of the spatial frequencies vectors $|\boldsymbol{k_1}| = |\boldsymbol{k_2}| = 0.45mm^{-1}$ can also be obtained from the Fourier spectrum. Finally, displacements l_1, l_2 and refractive index slopes $\frac{dn}{dx'}, \frac{dn}{dy'}$ can be computed using (10) and (11).

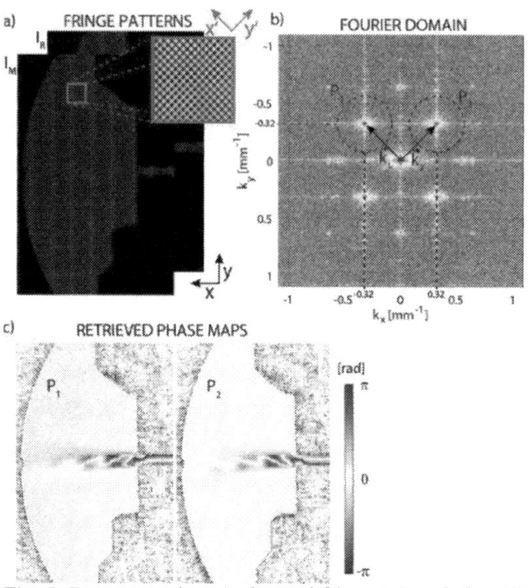

Fig. 5. Data processing: a) observed fringe pattern before IR and after IM the flow was presented; b) the amplitude Fourier spectrum with carrier frequencies; c) retrieved phase maps that proportional to fringe displacement in x', y' direction, respectively.

Usually, refractive index slopes along parallel and perpendicular direction (*x, y* – reference frame) to the channel axis are of the main concern. The reference frame of the fringe pattern (*x', y'*) was therefore rotated about $\theta = 45°$ in order to retrieve slope maps:

$$\begin{pmatrix} \frac{dn}{dx} \\ \frac{dn}{dy} \end{pmatrix} = \begin{pmatrix} cos\theta & -sin\theta \\ sin\theta & cos\theta \end{pmatrix} \begin{pmatrix} \frac{dn}{dx'} \\ \frac{dn}{dy'} \end{pmatrix}. \quad (13)$$

The slope maps are shown in Fig. 6 (left column). In order to verify the approach, an interferometric measurement in the same flow regime was realized.

It is important to note that the interferometric arrangement had very different imaging system from the fringe projection measurement yielding in different spatial resolution. In addition, both measurements (albeit with the same controllable parameters) were performed at different times. Therefore, fringe projection and interferometric results are not directly comparable. However, qualitative comparison of the fringe projection method (left column) and interferometric measurement (right column) evinces a high degree of agreement. Unlike deflection (Schlieren) techniques, interferometers measure the refractive index (not just ray deflection due the refractive index gradient) and thus maps in Fig. 6 (right column) were obtained as slope of interferometrically measured refractive index.

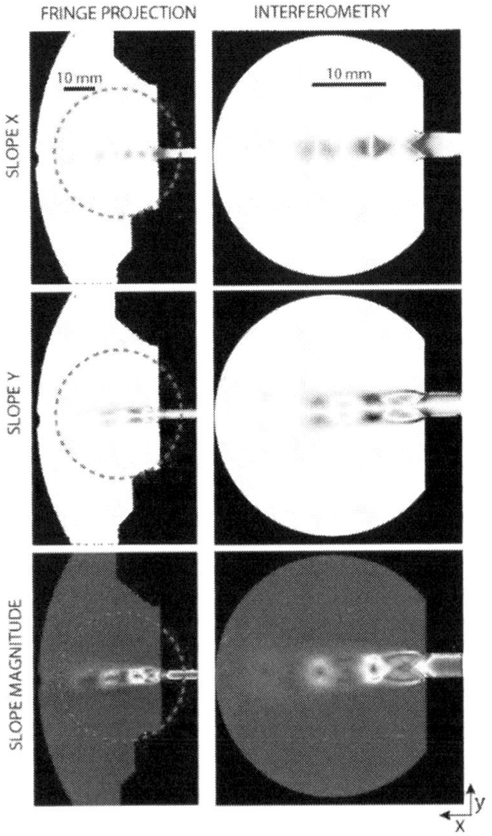

Fig. 6. Slope maps dn/dx, dn/dy and the magnitude of the gradient vector measured by the fringe projection technique and by interferometry. The red dashed circle indicates interferometer's field of view.

In the opposite approach, we obtained the density distribution (see Fig. 7) by numerical integration of slope maps in Fig. 6 measured by the fringe projection technique and using the Gladstone-Dale equation:

$$\rho(x,y) = \frac{p_{out}}{RT_0} + \frac{\Delta n(x,y)}{K}. \quad (14)$$

In (14), K=0.00022m^3/kg represents the Gladstone-Dale constant, T_0=296K denotes the ambient air temperature, R=287.1 J/(kg·K) is the specific gas constant and $\Delta n(x,y)$ is the refracitve index change map resulted from the numerical integration.

The numerical integration is sensitive to local outliers and therefore there are some erroneous areas close to edges/walls in the image.

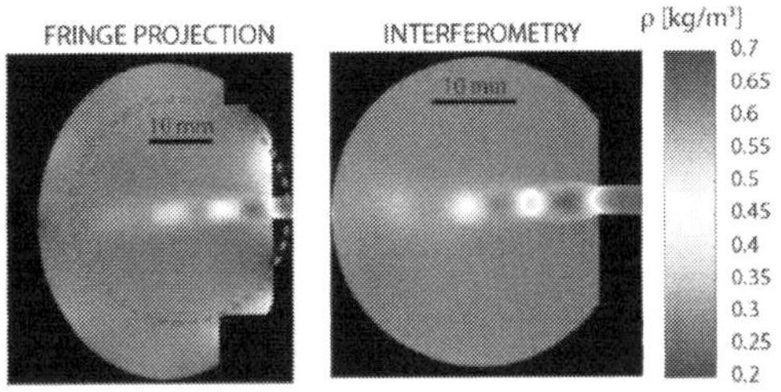

Fig. 7. Density distribution computed from slope maps measured by the fringe projection (left), and density distribution measured directly by interferometry (right).

4 Conclusion

This paper presents a very effective technique for characterization of fluid flow. The technique based on fringe projection is sensitive to ray deflection caused by gradient of the index of refraction in a transparent medium. A well-defined multiplexed cosine fringe pattern is observed through a measured area and modification of the pattern indicates the refractive index variation caused by a phenomenon. The displacement of the fringe pattern in orthogonal directions can be evaluated in each pixel using Fourier analysis. Assuming parameters of the experimental arrangement, the displacement can further be used for quantitative analysis of various phenomena. The developed fringe projection technique was successfully applied for the research of a fluid flow in a narrow channel including a free jet at the channel exit. The experimental section was built in a wind tunnel allowing for a high pressure difference and thus high flow velocity through the channel. The wind tunnel experiment construction assures mostly constant refractive index along the optical axis and thus quantitative analysis was performed. The results were compared to results obtained by interferometer with a good agreement. In the future, this approach will also be used along with interferometry of high-gradient fields in order to subtract mapping error and refine interferometric measurements.

Acknowledgments

We gratefully acknowledge the support of the Grant Program of the Technical University of Liberec (project number PURE-2020-3010). We thank the people from the Aerodynamics Laboratory in NovýKnín (part of The Institute of Thermomechanics of the Academy of Sciences of the Czech Republic) for their help and discussion in setting up experiments in the wind tunnel.

List of symbols

x, y, z *[m]*	Cartesian coordinate system
n *[1]*	Refractive index
L *[m]*	Length of the measured area
l_y *[m]*	Ray shift
ε *[rad]*	Angle of deflection
$I, \hat{I}$ *[1]*	Intensity and its Fourier image
$A, \hat{A}$ *[1]*	Mean intensity (DC term) and its term in Fourier domain
k $[m^{-1}]$	Wavenumber
$\mathbf{k}$ $[m^{-1}]$	wavevector
φ *[rad]*	Optical phase
θ *[rad]*	Angle of rotation
P_{in} *[Pa]*	Total pressure
P_{out} *[Pa]*	Outlet pressure
ρ $[kg{\cdot}m^{-3}]$	Density
T_0 *[K]*	Temperature
K $[m^3/kg]$	Gladstone-Dale constant
R *[kg·K]*	specific gas constant
$\hat{X}$	Fourier spectrum of X
X^*	Complex conjugation of X

References

1. W. Merzkirch., *Handbook of Experimental Fluid Mechanics* (Springer, Berlin, Heidelberg, 2007)
2. G. S. Settles, and M. J. Hargather , Measurement Science and Technology, **28** (2017)
3. A. Eder, and M. Jordan, *Techniques and Applications* (Springer, Berlin, Heidelberg, 2001).
4. S. B. Dalziel, G. O. Hughes and B.R. Sutherland, Experiments in Fluids, **28** (2000)
5. H. Richard, and M. Raffel, Measurement Science and Technology **12**, (2001)
6. J. A. Letelier, P, Herrera, N. Mujica, and J.H. Ortega, Exp Fluids **57**, 18 (2016)
7. P. Psota, P. Dančová, G. Cubreli, V. Lédl, T. Vít, R. Doleček, and O. Matoušek, International Journal of Thermal Sciences **145**, 106029 (2019)
8. R. M. Goldstein, H. A. Zebker, and C. L. Werner, Radio Science **23**, (1988)

Experimental investigation of the wake behind a cooled cylinder

Jaroslav Pulec[1*], *Petra* Dančová[1], and *Václav* Vinš[2]
[1]Technical University of Liberec, Studentská 2, 460 01 Liberec Czech Republic
[2]Institute of Thermomechanics of the Czech Academy of Sciences, Dolejškova 5, 182 00 Prague Czech Republic

Abstract. An experimental channel intended for investigation of a wake behind a cooled cylinder was designed and constructed in this work. The setup is designed for measurements with air at low Reynolds numbers. The measuring techniques based on the standard visualization with a fog generator and the constant temperature anemometry (CTA) were used. The flow performance was investigated behind two cylinders placed inside a channel with square cross section. One cylinder was kept at a constant flow temperature, while the second cylinder could be cooled using a water circuit with a thermostatic bath. The dependency of the Strouhal number on the Reynolds number was evaluated and compared to the literature data and the empirical correlations. The concept of an effective temperature in a non-isothermal flow was also applied. Unfortunately, a considerable difference between the preliminary measurements and the literature data was detected. The air-flow was found to have unsatisfactory high turbulence intensity. Consequently, some modifications of the current channel design are proposed.

1 Introduction

An effect of a boundary layer separation in a case of non-isothermal flow was described by many authors [1]. The purpose of this work was to develop an experimental apparatus in the laboratory of the Technical University of Liberec enabling observation and measurement with good accuracy. The presented results were obtained as part of the master's degree thesis of the corresponding author J. Pulec [2]. The theoretical backgrounds and technical solutions were inspired mostly by the work by Trávníček et al. [3].

Reynolds number is an important characteristic number in the case of flow around cylinder. It is ratio of frictional and resistance forces. Reynolds number is defined as:

$$\mathrm{Re} = \frac{U \cdot D}{\nu_\infty}, \tag{1}$$

where U (m · s^{-1}) is a free-stream velocity, D (m) is a cylinder diameter and ν_∞ (m^2 · s^{-1}) is a kinematic viscosity related to a free-stream temperature T_∞. Due to a cylinder cooling the free-stream temperature is different from a cylinder wall temperature T_w. Heat transfer between the cooled cylinder and the air causes local temperature drop connected with variation of the kinematic viscosity and the Reynolds number.

Kinematic viscosity ν is given by the following relationship:

$$\nu = \frac{\mu}{\rho}, \tag{2}$$

where ρ (kg · m^{-3}) is the fluid density and μ denotes the dynamic viscosity (Pa · s).

In connection with the above-mentioned changes of the Reynolds number a concept of effective temperature was established [4-7]. Based on this, effective Reynolds number was defined:

$$\mathrm{Re_{eff}} = \frac{U \cdot D}{\nu_\mathrm{eff}}, \tag{3}$$

where ν_{eff} (m^2 · s^{-1}) is an effective kinematic viscosity that is related to an effective temperature T_{eff}. The effective temperature can be evaluated from an empirical correlation determined experimentally. It is described by following relationship:

$$T_\mathrm{eff} = T_\infty + c_\mathrm{eff}(T_\mathrm{w} - T_\infty), \tag{4}$$

where $c_{\mathrm{eff}} = 0.28$ is an empirical coefficient. Critical value of effective Reynolds number $\mathrm{Re_{C,eff}}$ was detected as $\mathrm{Re_{C,eff}} = 47.5 \pm 0.7$ [6]. Further, an empirical relationship between the Strouhal number St and the effective Reynolds number $\mathrm{Re_{eff}}$ for heated and unheated cylinders was found [6]:

$$\mathrm{St} = 0.266 - \frac{1.016}{\sqrt{\mathrm{Re_{eff}}}}, \tag{5}$$

The Strouhal number is defined as follows:

$$\mathrm{St} = \frac{f \cdot D}{U}. \tag{6}$$

In equation (6), f (Hz) denotes the vortex shedding frequency.

* Corresponding author: jaroslav.pulec@tul.cz

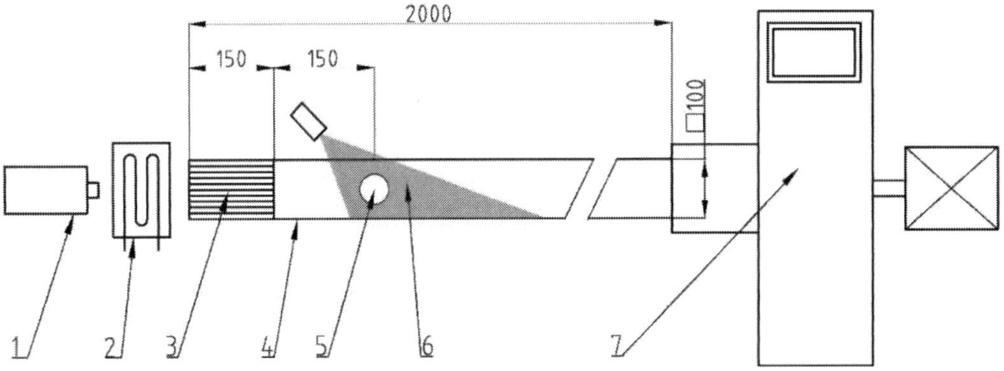

Fig. 1. Experimental apparatus. *1 fog* generator, *2* heater, *3* honeycomb, *4* experimental channel, *5* experimental area, *6* laser sheet, *7* radial fan and electric motor.

A dimensionless temperature (referred also as the temperature ratio) T^* is often used to describe the temperature situation of the non-isothermal flow. It is defined as:

$$T^* = \frac{T_W}{T_\infty}. \quad (7)$$

We note that for $T^* = 1$, the flow becomes isothermal as $T_w = T_\infty$.

Empirical dependence of free-stream kinematic viscosity was found by Goldstein and Cho [8]. It is described by follow equation:

$$\nu_\infty = 0.1556\left(\frac{T_\infty}{298.16}\right)^{1.7774}\left(\frac{1.013\cdot 10^5}{p_{atm}}\right) \cdot 10^{-4}, \quad (8)$$

where p_{atm} (Pa) is atmospheric pressure.

A relationship for theoretical flow velocity can be deduced from equations (3) to (8) in the following manner [3]:

$$U = \left[\frac{1}{4}\left[-\frac{E^*}{D^*}\cdot\sqrt{\frac{\nu_\infty(0.72+0.28T^*)^{1.7774}}{D}}\right] + \sqrt{\left(\frac{E^*}{D^*}\right)^2\frac{\nu_\infty(0.72+0.28T^*)^{1.7774}}{D} + 4\left(\frac{fD}{D^*}\right)}\right]^2, \quad (9)$$

where $E^* = -1.016$ and $D^* = 0.266$ are empirical constants [3].

Since $\nu \sim T^{1.7774}$ (8), the ratio of the Reynolds number and the effective Reynolds number can be written as follow [7]:

$$\frac{\mathrm{Re}}{\mathrm{Re}_{\mathrm{eff}}} = \frac{\nu_{\mathrm{eff}}}{\nu_\infty} = \left(\frac{T_{\mathrm{eff}}}{T_\infty}\right)^{1.7774}. \quad (10)$$

Using equations (4) and (10), equation (5) can be modified as follows [3]:

$$\mathrm{St} = 0.266 - \frac{1.016}{\sqrt{\mathrm{Re}}}(0.72 + 0.28T^*)^{0.8887}, \quad (11)$$

which represents an empirical definition of the Strouhal number for a non-isothermal flow around a cylinder.

2 Experimental apparatus

A simplified scheme of the experimental apparatus is shown in Fig. 1. A radial fan was used for generation and regulation of air flow in our experiment. It was driven by 2.2 kW induction motor Siemens with frequency changer enabling smooth variation of speed up to 2800 rpm.

Measurement and visualization were conducted in a measuring channel in the way of minimization of disturbances. For easier attainment of steady flow, the air channel was conceived as sucking, so a channel outlet is connected to a fan inlet. The crossing was realized by an acrylic sheet with the periodically drilled apertures for throttling the flow. The channel was 2000 mm long and had a square cross-section 100 x 100 mm. The channel was made of adhesive-bonded acrylic sheets. Temperature endurance of the sheets is 343 K for long-term application, and it softens at 373 K. Bottom and back walls were black colored for the visualization purposes. The channel inlet was created by a plastic honeycomb for a flow smoothing.

In order to achieve higher temperature difference between the cylinder and the airflow, an air heater was placed before the channel inlet. It was a 2 kW heating coil with sheet-metal case, which allowed for a constant flow temperature of around $T_\infty = 310$ K.

A measuring area with observed bodies was located 300 mm from the channel inlet as an unadvisable vorticity was found to form at a longer distance. The observed bodies, i.e. the cylinders, were made of a brass pipe with outer diameter of 2 mm. For the effective comparison of the character of the wake in the cases of isothermal and non-isothermal flow, two cylinders were used. Their orientation in the measuring area is depicted in Fig. 2 and

Fig. 4. The upper cylinder was connected to a thermostatic bath Julabo F 34-MA capable to hold a constant temperature of a coolant with a stability of ± 0.02 K. The lower cylinder was left at the free-stream temperature.

3 Experimental methods

3.1. Temperature measurement

Dewetron DEWE-5000 with software Dewesoft 6.6.7 was used for a temperature measurement. The data acquisition system allows for recording and plotting of the temperature variation over time. Six K-type thermocouples were placed inside the channel and measured the free-stream temperature T_∞. Positions of the thermocouples are shown in Fig. 2.

3.2 Visualization

The flow was visualized by fog generator Safex S2010. It was situated before the air heater. Stream of fog was sucked into the channel. The fog particles were illuminated by a laser sheet created by a pulse laser with a cylindrical lens located inside the measuring area. The laser beam had a wavelength of 532 nm. Scheme of the measuring area with the laser sheet in Fig. 2.

Visual data was recorded by a high-frequency camera Dantex Dynamic Nanosense MkIII with Nikon lens. It was able to make a film with a frame-frequency up to 15 kHz.

For setting the record parameters, the software Motion Studio 2.13 by IDT vision was used. Frequency of 300 Hz and resolution of 1280 × 644 px were employed.

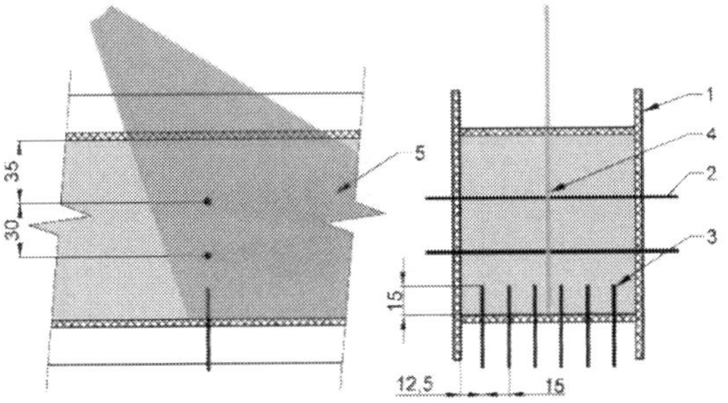

Fig. 2. Measuring area for visualization. *1* channel wall, *2* cylinders placed within the flow, *3* thermocouples, *4* laser sheet, *5* vortex structures illuminated by the laser sheet.

3.3 Velocity measurement

Velocity measurement was performed using a constant temperature anemometry (CTA) by hot-wire sensor (HWA mode). In Fig. 3, location of the velocity sensor is shown. The velocity measurement was carried out without a presence of the cylinders. The airflow was heated to adjust conditions similar to visualization and frequency measurements. According to the coordinate system in Fig. 3, the end of the sensor was traversed in the y-axis direction with 5 mm steps in the interval $y = (15 \div 75)$ mm. x-axis location was fixed at $x = 15$ mm.

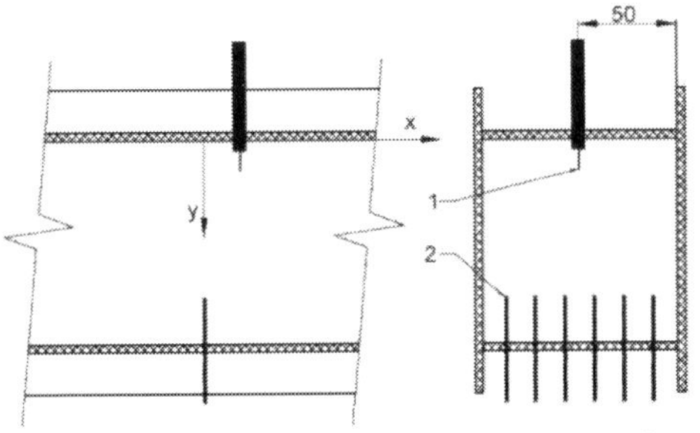

Fig. 3. Measuring area for velocity measurement. *1* sensor HWA, *2* thermocouples.

3.4 Vortex shedding frequency measurement

Two independent methods of frequency measurement were used. The first method was a frequency determination from the video, the other one was fast Fourier transform (FFT) of measured data by CTA. Both methods were used separately in different times and under slightly different conditions. Therefore, the results cannot be directly compared to each other.

HWA sensor location for the frequency determination is shown in Fig. 4. The coordinate system is the same as in Fig. 3. Sensors' coordinate locations were $x = 15$ mm, $y = 35$ mm and $y = 65$ mm for option A and option B, respectively. Frequency measurement was not conducted simultaneously but in the independent measurements, each for a period of 13 seconds.

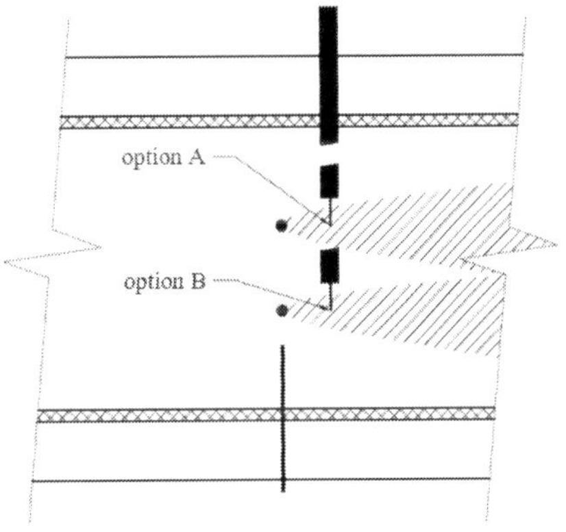

Fig. 4. Measuring area for frequency measurement. *Option A* measurement behind the cooled cylinder, *option B* measurement behind the reference cylinder.

4 Results

4.1 Visualization

All following experiments were performed at a constant rotational speed of the induction motor. At first, the cylinder connected to the cooling loop was left at the flow temperature, i.e. $T_W = T_\infty$, and afterwards the temperature was varied to $T_W = 301.1$ K, $T_W = 291.1$ K and

$T_W = 281.1$ K. Due to low frequency differences, the boundary cases are discussed further.

In Fig. 5 there are two pictures. Both of von Kármán vortex streets were generated in an isothermal flow. Time step between the pictures was 0.09 s approximately. Yellow vertical lines intersect the same couple of vortices. Apparently, the vortex streets have the similar proportions; it means vortex shedding frequencies and width of wakes.

Fig. 6 shows comparison of an isothermal and non-isothermal wakes. Upper vortex street represents a wake behind the cylinder cooled to T_w = 281 K, the lower cylinder is uncooled. In the non-isothermal case, the higher vortex shedding frequency and the greater width of wake can clearly be observed.

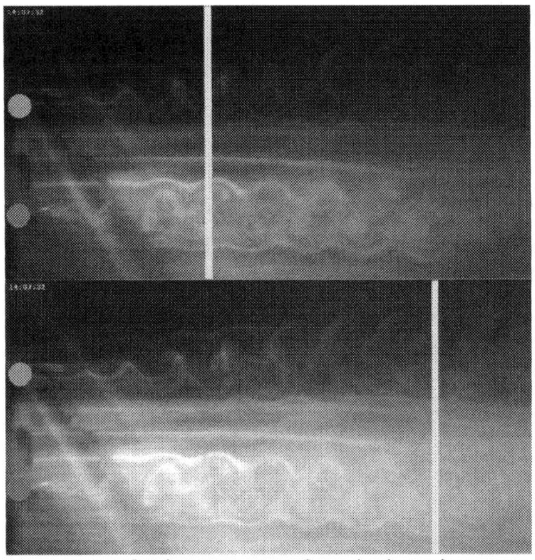

Fig. 5. Von Kármán vortex street for an isothermal case at two-time frames shifted by approximately 0.09 s. *Upper blue point* cooled cylinder (in this case, the cooling was switched off), *lower red point* reference cylinder.

Fig. 6. Von Kármán vortex street for a non-isothermal case. *Upper blue point* cooled cylinder, *lower red point* reference cylinder.

4.2 Measured data analysis

Mean values of the flow velocity inside the channel measured at various vertical positions by the HWA sensor are shown in Fig. 7. As mentioned in chapter 3.3, y = 35 mm and y = 65 mm are the y-axis locations of the cylinders. As can be seen, the velocity profile was not uniform in the vertical direction as it varied from 0.6 to 0.66 m·s^{-1}. The velocity values were measured individually step by step. The velocity fluctuations could be caused either by an unstable fan operation or by an unsteady temperature profile within the channel.

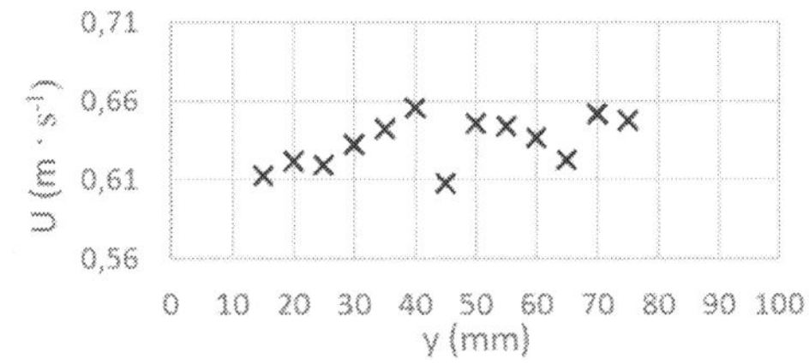

Fig. 7. Velocity profile in vertical direction.

Both measured and calculated values from the visualization experiment are summarized in Table 1. Table 2 contains data obtained from the CTA measurement. Each measurement is marked by a number and a letter. In Table 1, values on rows with the same number are evaluated from the same recording. In these cases, the conditions were identical. The upper cylinder is marked with letter *a*. Letter *b* denotes the lower reference cylinder.

From comparison of lines 1*a* and 1*b*, it is apparent that the vortex shedding frequency was approximately equal as case 1 represented an isothermal flow. Reversely when temperature ratio was decreased the frequency increased. It can also be seen that the effective Reynolds number markedly increased. Consequently, the temperature difference caused a similar effect as the increased flow velocity.

In Table 2, there are data sorted in a similar way, when a different letter marks the different time period of measurement Therefore, the thermal conditions slightly differ.

From a comparison of the free-stream viscosity, the effective viscosity, and the Reynolds numbers the change of effective values due to cooling becomes obvious. Frequencies of cooled and uncooled cylinders can be compared only partly since the measurement of wake behind cylinder *a* and cylinder *b* were made under different conditions of free-stream and the velocity also varied as shown in the CTA results. Ideally, a decrease in effective viscosity should cause a frequency increase. Due to inaccuracies in the flow temperature and the velocity distribution, the frequency change was difficult to evaluate accurately.

Graphs in Fig. 8 and Fig. 9 show overall results obtained both from the visualization and from CTA. The experimental data are compared to the results by Trávníček et al. [3] and Williamson [9]. Results of the Strouhal number correlations (5) and (11) are also shown for comparison. Fig. 8 shows the St–Re relationship, while in Fig. 9 the concept of the effective temperature and the effective Reynolds number is applied, i.e. the St-Re_{eff} relationship is shown. In an ideal case, the measured data points should lie on a curve of the same color in Fig. 8. However as can be seen both in Fig. 8 and Fig. 9, the new experimental results obtained both from the visualization and CTA lie by considerably lower

Tab. 1. Results from the visualization measurement; a – upper cooled cylinder, b – reference isothermal cylinder.

#	U $(m \cdot s^{-1})$	T_w (K)	T_∞ (K)	T^*	T_{eff} (K)	ν_∞ $(m^2 \cdot s^{-1})$	Re	ν_{eff} $(m^2 \cdot s^{-1})$	Re_{eff}	f (Hz)	St
1a	0.622	310.7	310.7	1.00	310.7	$1.762 \cdot 10^{-5}$	70.39	$1.762 \cdot 10^{-5}$	70.39	35.85	0.1156
1b	0.642	310.7	310.7	1.00	310.7	$1.762 \cdot 10^{-5}$	72.66	$1.762 \cdot 10^{-5}$	72.66	35.63	0.1113
2a	0.622	309.8	309.8	1.00	309.8	$1.752 \cdot 10^{-5}$	70.76	$1.753 \cdot 10^{-5}$	70.76	35.40	0.1142
2b	0.642	301.2	309.8	0.97	307.4	$1.752 \cdot 10^{-5}$	73.04	$1.728 \cdot 10^{-5}$	74.07	35.78	0.1118
3a	0.622	311.4	311.4	1.00	311.4	$1.768 \cdot 10^{-5}$	70.12	$1.768 \cdot 10^{-5}$	70.12	35.90	0.1061
3b	0.642	291.15	311.4	0.93	305.7	$1.768 \cdot 10^{-5}$	72.39	$1.712 \cdot 10^{-5}$	74.79	36.77	0.1149
4a	0.622	309.1	309.1	1.00	309.1	$1.745 \cdot 10^{-5}$	71.07	$1.745 \cdot 10^{-5}$	71.07	34.50	0.1113
4b	0.642	281.2	309.1	0.91	301.2	$1.745 \cdot 10^{-5}$	73.37	$1.667 \cdot 10^{-5}$	76.78	36.33	0.1135

Tab. 2. Results from the CTA measurement; a – upper cooled cylinder, b – reference isothermal cylinder

#	U $(m \cdot s^{-1})$	T_w (K)	T_∞ (K)	T^*	T_{eff} (K)	ν_∞ $(m^2 \cdot s^{-1})$	Re	ν_{eff} $(m^2 \cdot s^{-1})$	Re_{eff}	f (Hz)	St
1a	0.642	281.2	311.7	0.90	303.1	$1.771 \cdot 10^{-5}$	72.49	$1.686 \cdot 10^{-5}$	76.16	36.62	0.1141
2a	0.642	281.2	311.8	0.90	303.2	$1.773 \cdot 10^{-5}$	72.42	$1.687 \cdot 10^{-5}$	76.11	37.84	0.1179
3a	0.642	281.2	311.8	0.90	303.2	$1.772 \cdot 10^{-5}$	72.43	$1.687 \cdot 10^{-5}$	76.11	37.84	0.1179
1b	0.622	311.1	311.1	1.00	311.1	$1.766 \cdot 10^{-5}$	70.50	$1.766 \cdot 10^{-5}$	70.50	37.84	0.1216
2b	0.622	311.2	311.2	1.00	311.2	$1.766 \cdot 10^{-5}$	70.49	$1.766 \cdot 10^{-5}$	70.49	37.84	0.1216
3b	0.622	311.2	311.2	1.00	311.2	$1.766 \cdot 10^{-5}$	70.50	$1.766 \cdot 10^{-5}$	70.50	39.06	0.1255

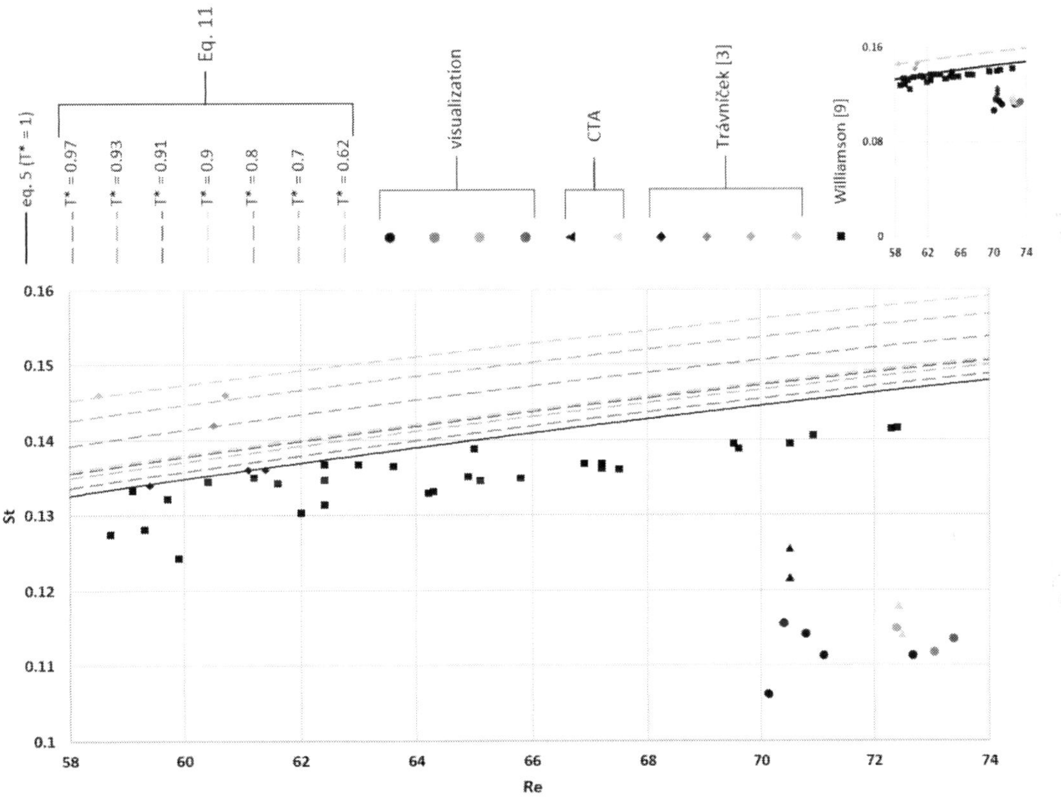

Fig. 8 Dependence of the Strouhal number St on the Reynolds number Re.

Strouhal numbers that the theoretical values (curves) and the literature data by Trávníček et al. [3] and Williamson [9]. The deviations of Strouhal number is in the range from 20.2 to 26.6 % for the visualization method and from 13.4 to 23.7 % for CTA. Literature data by Williamson show a maximum deviation of 7.7 % (mean deviation is around 2.7 %) and the data by Trávníček et al. differ by less than 1 %. The measured flow velocity deviated from equation (9) by 14.6 ÷ 17.3 % in case of CTA. These considerable differences are suspected to be caused by an unsteady character of the flow with relatively high turbulence intensity. A potential rising flow coming from the heater at the channel inlet could occur close to measurement area. The turbulence intensity was estimated to lie in the range from 8.1 to 14.6 %. In the region close to the cylinders the turbulence intensity could be of around 8 to 9 %.

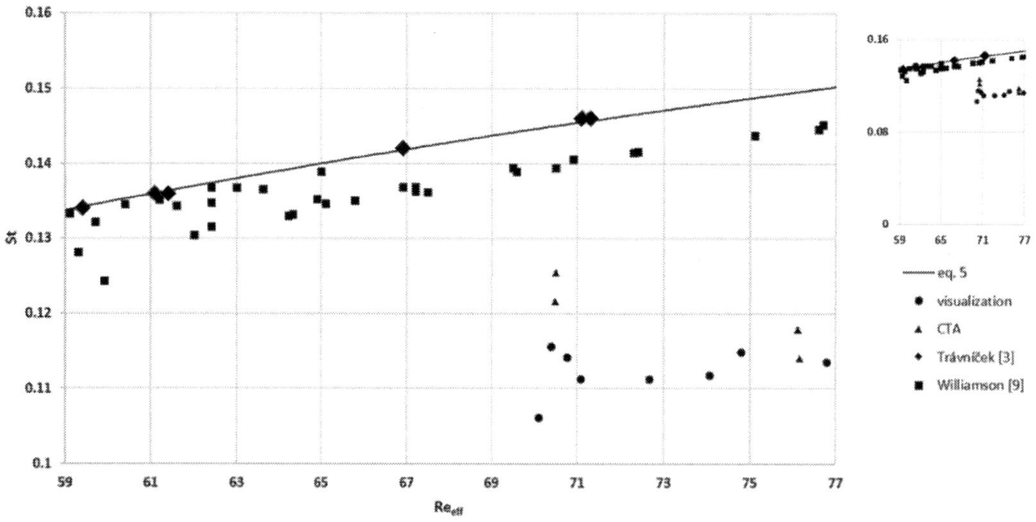

Fig. 9 Dependence of the Strouhal number St on the effective Reynolds number Re_{eff}.

5 Conclusion

An experimental channel intended for investigation of a wake behind a cooled cylinder was designed and constructed. The channel was used at temperatures up to 313 K and flow velocities from 0.62 to 0.64 m · s^{-1}. The channel was equipped with two cylinders. Temperature of one cylinder was varied by employing water cooling. The wall temperature was set to $T_W = T_\infty$ (isothermal case), $T_W = 301.1$ K, $T_W = 291.1$ K a $T_W = 281.1$ K. The second cylinder was used as a reference as its temperature was kept constant equal to the air flow temperature $T_W = T_\infty$. The flow characteristics were studied using the visualization technique with a fog generator. The flow velocity and the vortex shedding frequency were determined using the constant temperature anemometry.

The experimental setup was found to be suitable for the air-flow visualization. The preliminary measurements confirmed influence of the cylinder cooling on the wake character. However, the collected data showed a considerable deviation from the theoretical values for the Strouhal number of up to 26.6 % at the given Reynolds number. Current design of the channel was found to be inconvenient for an accurate measurement of characteristic magnitudes of a body flow. Modifications of the channel design, such as improvement of the front heater, employment of thermal insulation on outer walls of the channel, or installation of a flow stabilization element at the inlet are planned in order to improve the flow performance, especially at the inlet section.

Acknowledgment

This research was carried out at the Technical University of Liberec as part of the project "Experimental, theoretical and numerical research in fluid mechanics and thermomechanics, no. 21291" with the support of the Specific University Research Grant, as provided by the Ministry of Education, Youth and Sports of the Czech Republic in the year 2019 and with the institutional support RVO: 61388998.

References

1. R. Munson, T. H. Okiishi, W. W. Huebsch, A. P. Rothmayer, *Fundamental of Fluid Mechanics* (Wiley, 2012)
2. J. Pulec, *master's degree thesis* (TU Liberec, 2018)
3. Z. Trávníček, A.-B. Wang, W.-Y. Tu, Exp Fluids **55**, 1679 (2014)
4. J. C. Lecordier, L. Hamma, P. Paranthoën, Experiments in Fluids **10**, 224 (1991)
5. J. C. Lecordier, F. Dumouchel, P. Paranthoën, International Journal of Heat and Mass Transfer **42**, 3131 (1999)
6. J. C. Lecordier, L. W. B. Browne, Le S. Masson, F. Dumouchel, P. Paranthoën, Experimental Thermal and Fluid Science **21**, 227 (2000)
7. A.-B. Wang, Z. Trávníček, K. C. Chia, Physics of Fluids **12**, 1401 (2000)
8. R. J. Goldstein, H. H. Cho, Experimental Thermal and Fluid Science **10**, 416 (1995)
9. C. H. K. Williamson, Journal of Fluid Mechanics **206**, 579 (1989)

Preliminary measurements of the thermal field in a rectangular channel by the Planar Laser-Induced Fluorescence method

Jaroslav Pulec[1*], *Pavel* Mony[1], *Václav* Vinš[2], and *Petra* Dančová[1]

[1]Technical University of Liberec, Studentská 2, 460 01 Liberec Czech Republic
[2]Institute of Thermomechanics of the Czech Academy of Sciences, Dolejškova 5, 182 00 Prague Czech Republic

Abstract. This paper describes the preliminary measurements of the temperature field in water channel by the Planar Laser-Induced Fluorescence (pLIF) method in the one-colour mode using commercial LaVision system. The aim of the work is to find suitable parameters of the experiment (concentration of fluorescent dye, laser power, arrangement of the experiment, etc.) for measuring the temperature field in a closed glass channel of rectangular cross-section with a heated bottom-wall and to verify an applicability of this method in general. The intention is to carry out measurements in the transverse plane of the duct such that the specific thermos-convective structures of the Rayleigh-Bénard-Poiseuille convection, i.e. mixed convection, are detected. For simplicity, the initial measurements were performed in the natural convection regime, i.e. the water in the channel was not driven by the circulation pump. In this case, a larger temperature gradient was created near the wall compared to the force convection case when a cold liquid flows continuously into the experimental area. The presented work serves as a methodological guide for future measurements and provides valuable experience for the design of a new channel for investigation of the thermal flow instabilities.

1 Introduction

Thermoconvective structures in ducts with heated or cooled walls are experimentally investigated for several decades. Rayleigh-Bénard-Poiseuille (RBP) flow is a specific case of convection in a parallel-plate horizontal channel with a heated bottom-wall. It is a mixed convection consisting of a primer Poiseuille forced flow [1] and a secondary flow caused by heat transfer near the wall, also known as Bénard cells [1, 2].

Many experimental studies on this phenomenon have been published since 1980s. Osborne and Incropera [3] performed an investigation of top and bottom-wall heating effect in a wide rectangular channel. Shortly afterwards, Incropera et al. [4] published a paper dealing with a specific longitudinal rolls. The work was specified in paper of Maughan and Incropera [5]. Similar works can be mentioned from the recent years, namely by Benderradji et al. [6], Mey et al. [7] and Taher and Abid [8]. Authors of the last study used the pLIF method to measure temperature fields.

The Planar Laser-Induced Fluorescence (pLIF) is a radiometric method for the measurement of temperature and concentration fields. The principle consists of measurements of the energy emission intensity $I\,(\mathrm{W \cdot m^{-3}})$ of a fluorescent dye, which is excited by a coherent light source according to the relationship $I = I_0 C \phi \varepsilon$, where $I_0\,(\mathrm{W \cdot m^{-2}})$ is the incident light flux, $C\,(\mathrm{kg \cdot m^{-3}})$ is the concentration of dye in a solution, $\varepsilon\,(\mathrm{m^2 \cdot kg^{-1}})$ is an absorption coefficient and ϕ (1) is the quantum efficiency, which is temperature dependent for some dyes (e.g., Rhodamine B has a sensitivity of 2 % per 1 Kelvin) [9]. The pLIF method, whether in one-color or two-color mode, has been more widely used in recent years to measure a variety of fluid dynamic and thermodynamic phenomena. Some of them are briefly summarized in Table 1.

Tab. 1. An overview of selected pLIF experiments.

author	content	method	RhB concentration ($\mu g \cdot l^{-1}$)
Xu et al. [10]	emission spectra for different concentration of RhB	1-col	2000 ÷ 10 000
Amiri et al. [11]	exchanger measurement	1-col	10
Zhang et al. [12]	effect of viscosity on fluorescence intensity	1-col	10 ÷ 50
Sakakibara and Adrian [9]	convection over a heated horizontal surface	2-col	50
Sakakibara et al. [13]	thermally stratified pipe flow	1-col	100 ÷ 1000

A relatively large variance in dye concentrations can be observed in the selected experimental studies. It should

* Corresponding author: jaroslav.pulec@tul.cz

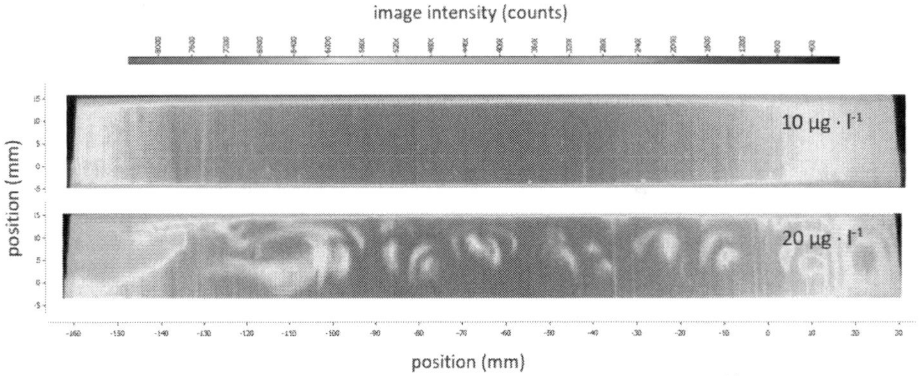

Fig. 2. Intensity response to temperature change at 10 and 20 $\mu g \cdot l^{-1}$.

be mentioned that the fluorescence intensity increases with concentration [10], but the choice of Rhodamine B concentration depends on the power of the laser and the thickness of the measured liquid layer. The LaVision manufacturer states the recommended concentration of Rhodamine B in the range of 10 to 1000 $\mu g \cdot l^{-1}$. This study was performed as a test measurement to find suitable parameters and arrangements and to verify its applicability for our experiment when the thermal flow instabilities will be measured at various Reynolds numbers and heat fluxes in a rectangular water channel.

2 Experimental equipment and setup

The experiment was performed using the PIV system LaVision which contains a hardware and a software extension for pLIF measurement. The pLIF system included:

- camera sCMOS 5,5 MPx, resolution 2560 × 2160 pixels, 16 bit, exposure time 15 μs to 100 ms, spectral range 370 nm to 1100 nm
- 50 mm lens f/1,4D,
- Nd:YAG double-cavity laser, 340 mJ per pulse, wavelength 532 nm, maximum repetition frequency 15 Hz, pulse duration up to 10 nm, 1,8 m optical arm,
- long-pass filter 540 nm
- software DaVis 10.

The data was recorded by single-frame mode with a frequency of 15 Hz.

In Fig. 1 there is a scheme of a current experimental setup. The measured space is a glass rectangular duct with a cross-section of 200 × 20 mm^2. On the outer surface of a bottom-wall, a resistive heating foil is placed. A constant heat flux of 13.4 W passes through 10 mm thick glass wall. A laser is positioned above the channel such that a vertical laser sheet is created in the cross-section of the channel in the space above the heated area. An sCMOS camera is placed to capture the cross-section with as little distortion as possible. A simple case study of the natural convection, i.e. with the circulation pump turned off, was investigated in this work.

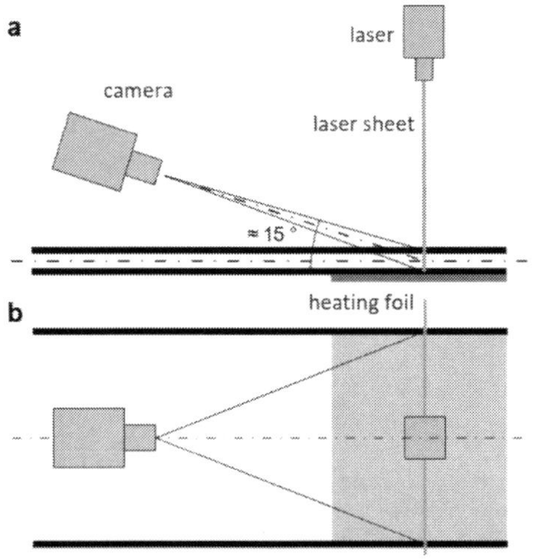

Fig. 1. Scheme of the experimental setup; ***a*** side view, ***b*** top view.

3 Analysis of suitable parameters

One of the important results of this work was to find a suitable concentration and to adjust the corresponding aperture of the camera lens. The Rhodamine B was gradually mixed into the heated experimental channel until a sufficiently strong signal response to the temperature change was found. The laser power and the lens aperture settings were also regulated for each concentration. The concentration ranging from 2.5 to 20 $\mu g \cdot l^{-1}$ was verified. Fig. 2 shows the response of the recorded image intensity to the concentrations of 10 and 20 $\mu g \cdot l^{-1}$. The settings parameters used for the final measurement are described in Table 2.

Table 2. Final parameters for the natural convection case.

parameter	value
lens aperture	f/1.4
laser power	160 mJ per pulse
Rhodamine B concentration	20 $\mu g \cdot l^{-1}$

4 Temperature calibration and image correction

After finding the appropriate dye concentration with the best temperature response, the temperature calibration of pLIF was performed. By gradual heating of water in the channel, five calibration data points for five different water temperatures were captured. Temperature was measured by four thermocouples located in several places over the channel cross section with a standard uncertainty of ± 0.2 K. The temperature field was continuously monitored and after its stabilization, the intensity distribution was recorded. A linear calibration curve was obtained from the temperature measurements. It should be noted that precise calibration requires many calibration points over wide temperature range in order to evaluate a second order polynomial. The linear function was selected for preliminary tests as it assures a continuously decreasing trend over the entire range.

In addition, it is necessary to record correction images namely background image (for digital noise subtraction) and sheet image (for correction of laser sheet geometry).

5 Results and discussion

The preliminary measurements were carried out in natural convection regime. The heat power of 13.4 W corresponded to the supercritical Rayleigh number for natural turbulent flow (Ra = 1,100,000 with a critical value of $Ra_{crit,turb}$ = 300,000 [1]). The data were taken with frequency 15 fps and subsequently evaluated and corrected. Fig. 3 shows 11 images each corresponding to the 10th frame, i.e. a measurement sequence of 7.3 seconds.

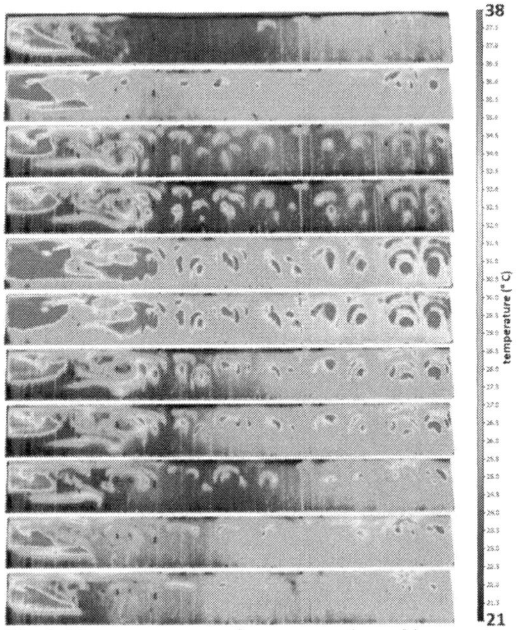

Fig. 3. 11 temperature map images over time with a constant time step of 0.66 s.

Temperature is symbolized by colour scale ranging from dark blue (21 °C) to red (38 °C). The images show thermoconvective structures caused by natural convection. Formations similar to Bénard's cells appear on the temperature map as a mushroom-shaped structures formed by a warmer liquid rising due to a density variation. The images show fluctuations in the calculated temperature in the space between the structures. This is suspected to happen due to the fluctuations of laser pulse power over time.

Fig. 4 shows the course of intensity along the horizontal axis of the channel cross-section. The upper curve represents an average value of intensity across the width of the channel. The average intensity was calculated from a section of data without visible thermoconvective structures. The lower curve shows the variance of the minimal and maximal intensity at each point. The graph indicates that the local intensity deviation is up to 25 %. The average values are related to the geometry of the laser sheet. A maximum of light intensity was in the centre of laser sheet, a minimum on the channel sides. The detected feature can be improved using the software sheet image correction. The time fluctuation requires either an energy monitor in the laser path or a two-color pLIF method. Stationary processes can also be solved by time averaging.

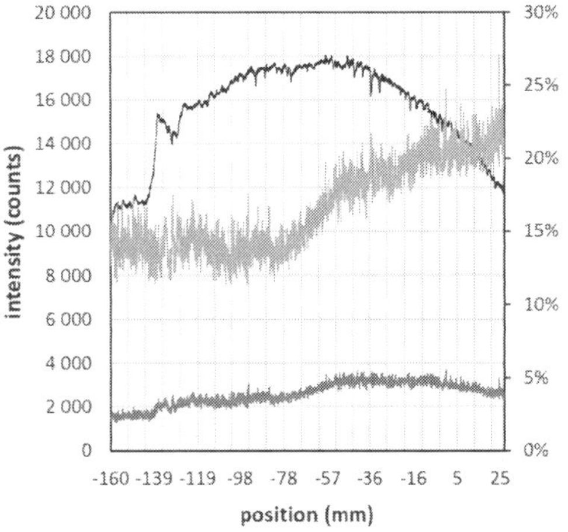

Fig. 4. Intensity along horizontal channel centreline. *The upper black curve*: average image intensity; *the lower red curve:* difference between the maximum and minimum of intensity fluctuation; *central green curve:* percentage deviation on secondary axis.

It has to be mentioned that the presented natural convention had a higher temperature gradient near the wall than it is expected for the mixed convection case. When the circulation pump is switched on, the wall cooling will be more intensive. For mixed convection regime, the pLIF system has to be set much more sensitively. The temperature calibration needs to be based on at least ten calibration points over a sufficiently wide range, which exceeds the expected temperature interval during the experiment. Due to the distortion of the image by the refraction of light at the interfaces of the channel, it was decided to modify the optical setup arrangement.

The further measurement will be performed with the two-colour pLIF and the flow will be monitored from the side-wall along central vertical plane in order to observe the growth of the temperature boundary layer. In this case, the camera views will be approximately perpendicular to all interfaces and there will be minimal flattening of the monitored space.

6 Conclusion

The natural convection pLIF test measurements in the channel of a basic geometry were performed. A wide rectangular glass channel with a bottom-wall section heated over a resistive heating foil was used. The fundamental parameters of the pLIF setup were found including the appropriate concentration of the fluorescent dye. The applicability of the method for measurement of the temperature field of the mixed convection was verified and the limits of the current arrangement and the selected one-color pLIF mode were described. In further work, pLIF will be tested in a two-color mode together with a modified side-wall arrangement of the optical setup.

Acknowledgement

This work was supported by the student grant of the Technical University of Liberec under the project No. SGS-2019-5073 and the institutional support RVO: 61388998.

References

1. Çengel, Turner a Cimbala, Fundamentals of the Thermal Fluid Sciences, Boston: McGraw-Hill, 2017.
2. Incropera, DeWitt, Bergman a Lavine, Fundamentals of Heat and Mass Transfer, Hoboken: John Wiley & Sons, 2006.
3. Osborne a Incropera, „Laminar, mixed convection heat transfer for flow between horizontal parrallel plates with asymmetric heating,“ *J. Heat and Mass Transfer,* sv. 28, č. 1, pp. 207-219, 1985.
4. Incropera, Knox a Maughan, „Mixed-Convection Flow and Heat Transfer in the Entry Region of a Horizontal Rectangular Duct,“ *J. Heat Transfer,* sv. 109, č. 2, pp. 434-439, 1987.
5. Maughan a Incropera, „Secondary flow in horizontal channels heated from below. J. Experiments in Fluids,“ *J. Experiments in Fluids,* sv. 5, č. 5, pp. 334-343, 1987.
6. Benderradji, Haddan, Taher, Médale, Acid a Papini, „Characterization of fluid flow patterns and heat transfer in horizontal channel mixed convection,“ *J. Heat and Mass Transfer,* sv. 44, č. 12, pp. 1465-1476, 2008.
7. Mey, Westhoff a Wagner, „Experimental Investigation of Flow Structure Formation in a Heated Duct Flow,“ v *Dillmann A., Heller G., Kreplin HP., Nitsche W., Peltzer I.: New Results in Numerical and Experimental Fluid Mechanics VIII. Notes on Numerical Fluid Mechanics and Multidisciplinary Design*, Berlin, Springer, 2013.
8. Taher a Abid, „Experimental determination of heat transfer in a Poiseuille-Rayleigh-Bénard flow,“ *J. Heat and Mass Transfer,* č. 54, pp. 1453-1466, 2018.
9. Sakakibara a Adrian, „Whole field measurement of temperature in water using two-color laser,“ *Experiments in Fluids,* č. 26, pp. 7-15, 1999.
10. Xu, Cai, Liu a Yan, „Measurement of Water Temperature using Planer Laser Induced Fluorescence,“ *AIP Conference Proceedings,* č. 1207, pp. 189-192, 2010.
11. Amiri, Taher a Mongeau, „Quantitative visualization of temperature field and measurement of local heat transfer coefficient over heat exchanger elements in sinusoidal oscillating flow,“ *Experimental Thermal and Fluid Science,* č. 85, pp. 22-36, 2017.
12. Zhang a F. Simoens, „Viscosity and Temperature Influences on PLIF,“ *Procedia Engineering,* č. 28, pp. 683-687, 2012.
13. Sakakibara, Hishida a Maeda, „Measurements of thermally stratified pipe flow using image-processing techniques,“ *Experiments in Fluids,* č. 16, pp. 89-96, 1993.
14. Koffi, Abid, Médale a Papini, „Mixed convection flow in a horizontal rectangular channel subjected to a horizontal thermal gradient,“ *J. Heat and Mass Transfer,* sv. 47, č. 10, pp. 1251-1260, 2011.

Survey of experimental channels employed on investigation of a non-isothermal laminar flow and design requirements for a new experiment

Jaroslav Pulec[1*], *Petra* Dančová[1], and *Václav* Vinš[2]
[1]Technical University of Liberec, Studentská 2, 460 01 Liberec Czech Republic
[2]Institute of Thermomechanics of the Czech Academy of Sciences, Dolejškova 5, 182 00 Prague Czech Republic

Abstract. In this article, a literature survey of the experimental channels used for investigation of a non-isothermal flow is presented. Particular interest is focused on the flow stability with a heated wall by a uniform heat flux. The aim of this work is to prepare a comprehensive overview of the technical design and experimental settings used in previous studies. Collected information shall help to design and manufacture new horizontal channel with a rectangular cross-section with a width of around 200 mm and a height of 20 mm. The channel will be equipped with a constant heat flux section on a bottom wall for investigation of the vortex structures and the flow instabilities using modern optical methods. The channel will operate with liquid (most probably water) in the range of low Reynolds numbers. Parallel laminar flow with Poiseuille velocity profile at the inlet of a heated section and a uniform heat flux predefine the essential requirements on the channel design.

1 Introduction

Flow stability with a presence of a heat flux near the wall plays on one hand an important role in many technical applications, such as design of effective cooling and heating, and on the other hand represents an interesting fluid mechanics phenomena that have not been sufficiently described so far. In order to describe and predict the thermal stability under various conditions, the accurate experimental measurements still represent the most promising approach. The numerical simulations, i.e. the computational fluid dynamics (CFD), still suffer of rather significant inaccuracies when it comes to prediction of the heat fluxes and the thermal flow stability. Hence the simulations need to be verified on sample experimental examples, e.g., with a simple geometry and with well-defined both flow and thermal characteristics.

A summary of previous studies presented in this work will serve as an important inspiration in the design and construction of a new rectangular channel intended for the experimental investigation of a non-isothermal flow at low Reynolds numbers. The motivation for our long-term research is to obtain new accurate experimental data for a non-isothermal flow by detecting a transition from a laminar pure forced convection, to a laminar mixed convection, and finally to a fully turbulent unsteady mode. The new data would help to better describe the character of the non-isothermal flow with a constant heat flux and could possibly be employed on the verification of numerical simulations.

2 Review of non-isothermal rectangular channels

2.1 Water channels

Osborne and Incropera [1] studied water flow in a rectangular duct heated from the bottom and the top. Sidewalls of the test section were made of 12.7 mm plexiglass plates. The top and bottom walls consisted of 9 mm thick aluminium sheet to which a heater and a thermal insulation were attached. The channel consisted of an inlet developing section, a heated test section, and a tail section avoiding possible negative effects of flow structures at the channel output. After the tail section, the water flew through a system of pipes into the tank with a coil maintaining the temperature of the circulating water. The tank was connected to a return loop with a pump, a valve, and a flowmeter system, which returned the flowing water back to the duct inlet section equipped with a flow straightener. A simplified scheme of the water circuit together with construction layers of the heated wall are shown in Fig. 1.

Maughan and Incropera [2] and Incropera et al. [3] studied the vortex structures of a mixed convection in a water flow. For this purpose, a rectangular duct heated from the bottom adapted for the flow visualization with a dye stream was used. The top and side walls of the channel were made of transparent acrylic sheets. The bottom heated wall was manufactured with a requirement for a uniform heat flux. Its layers are schematically described in Fig. 2.

* Corresponding author: jaroslav.pulec@tul.cz

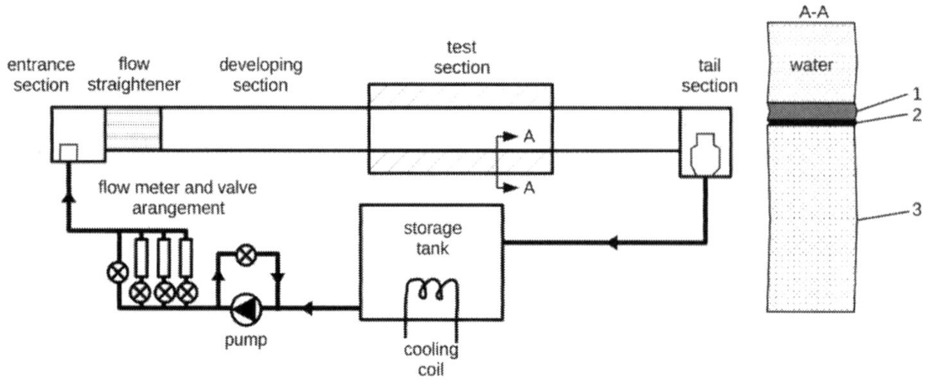

Fig. 1. Scheme of Osborne's water circuit. *1* aluminium sheet, *2* resistance heater, *3* thermal insulation.

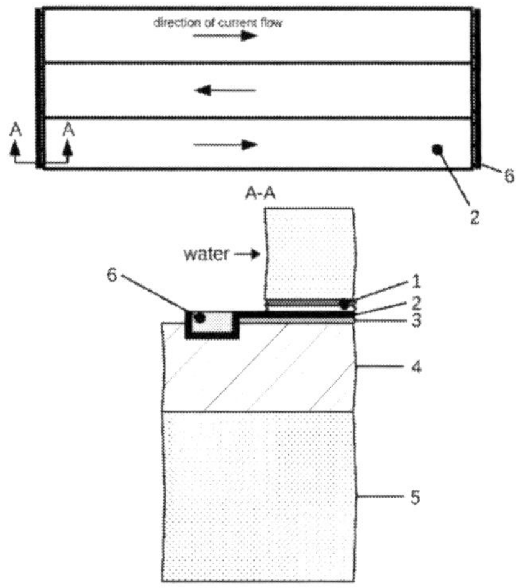

Fig. 2. Scheme of layers of heating segment of Maughan and Incropera experiment. *1* Mylar/adhesive composite, *2* steel foil, *3* transfer tape, *4* plexiglass substrate, *5* Styrofoam insulation, *6* copper bus.

The heating element was realized by three strips of 0.05 mm thin stainless-steel strips oriented longitudinally side by side with 1 mm wide gaps. At the end, the stainless-steel strips were connected in series to the copper buses. On the inner side, the layer with steel strips was covered with a 0.13 mm thick mylar/adhesive composite. It was applied to layers of adhesive transfer tape (0.05 mm), plexiglass substrate (12.7 mm) and polystyrene thermal insulation (50.8 mm). The temperature conditions at the bottom part of the channel were examined using liquid crystal sheets. A negligible overheating in the bus area and a minimum effect of 1 mm gaps between the stainless-steel strips were indicated. The thermocouples for measuring the wall temperature were placed on the outer side of the steel foil. They were attached using a thermal conductive epoxy. The original water circuit by Osborne and Incropera [1] shown in Fig. 1 was used. However, the test section was placed right after the flow straightener.

Another experiment for an investigation of the thermal stability in a water flow was carried out by Benderradji et al. [4]. The rectangular channel consisted of a developing, test and an end section as shown in Fig. 3. The test section was also made up of several layers. The heated bottom consisted of transparent polycarbonate (10 mm), three electric heaters and copper sheet (2 mm). Upper and lateral walls were made of 10 mm plexiglass. The channel was adapted for dye visualization again. Except for the water tanks, the power supply, the flowmeter and the valves for setting the desired flow rate, the experiment was placed on a table with a vibration control system.

A rectangular water channel by Koffi et al. [5] created to study the flow influenced by side-wall heating is depicted in Fig. 4. The experimental section was constructed of 3 mm plexiglass and the heated section was realized by strata made of a copper sheet, strip electrical resistor and thermal insulation. The experiment was adjusted for measurement with PIV (Particle Image Velocimetry) and the temperature was recorded by infrared camera.

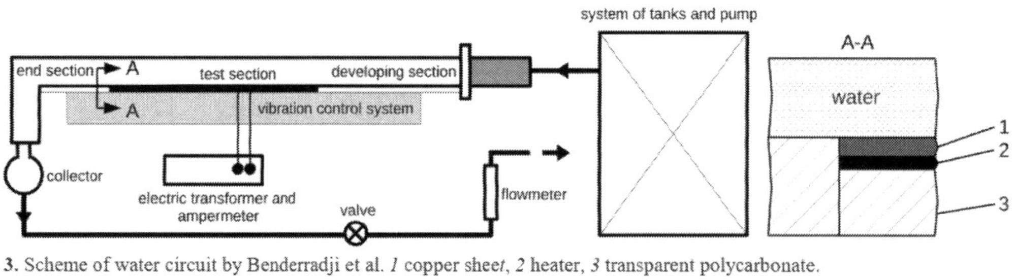

Fig. 3. Scheme of water circuit by Benderradji et al. *1* copper sheet, *2* heater, *3* transparent polycarbonate.

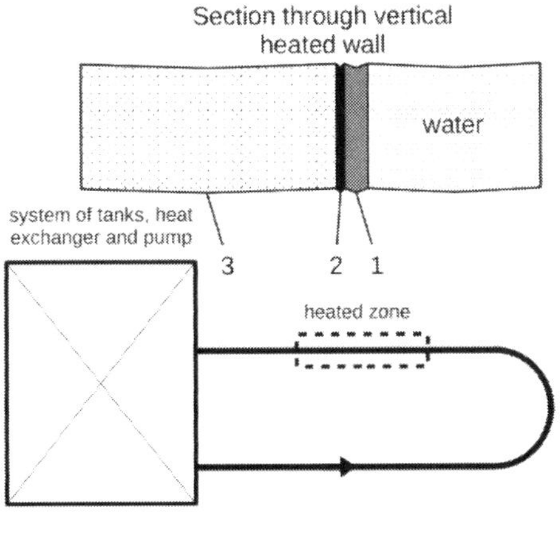

Fig. 4. Scheme of the water circuit by Koffi et al. *1* copper sheet, *2* strip electrical resistor, *3* thermal insulation..

2.2 Air channels

The experimental channels described in this section were designed to study non-isothermal air flow. The first is the channel by Hwang et al. [6], which allowed to vary a channel height and to investigate influence on the flowing structures. The test section is shown in Fig. 5. The bottom wall was made of 0.5 mm thick polished and black lacquered aluminium sheet. The heater was made of 0.65 mm Ni-Cr resistance wire wrapped around a 5 mm thick asbestos plate. A 20 mm asbestos plate was placed from the outside as the thermal insulation. The heating section was divided into three segments in the main flow direction and they were controlled individually by a voltage regulator. The top and side walls were made of 6 mm plexiglass.

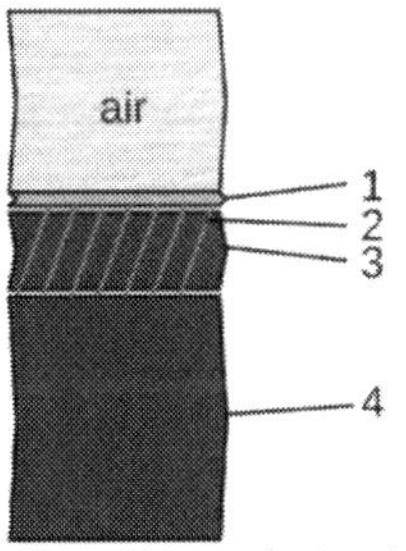

Fig. 5. Scheme of heating wall layers in the channel by of Hwang et al. *1* polished and black lacquered aluminium sheet, *2* Ni-Cr resistance wire, *3* asbestos plate, *4* asbestos thermal insulation.

In 2006, Chen et al. published [7] a study about the effect of a solid prism insertion on various channel position to explore the possible stabilization of the vortex flow. The 15 mm copper bottom wall was electrically heated by a DC power supply and the 3 mm glass upper wall was cooled by water which was running through a 3 mm gap between the glass wall and a 2 mm plexiglass plate, as shown in Fig. 6. The channel was braced up with a copper alloy frame.

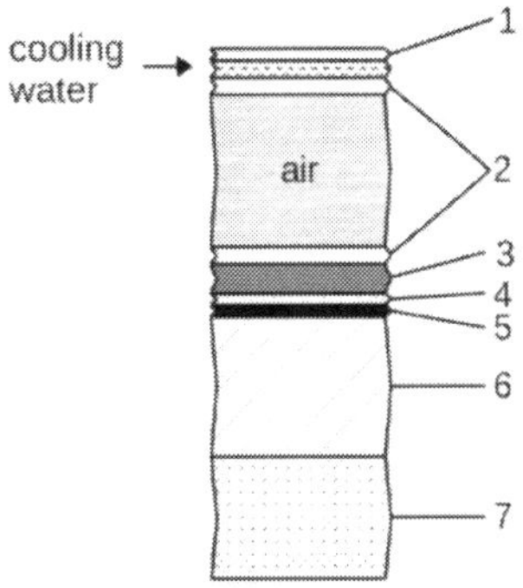

Fig. 6. Scheme of layers of Chen et al. heating bottom wall and cooling top wall. *1* plexiglass plate, *2* glass plate, *3* copper sheet, *4* mica sheet, *5* heater, *6* bakelite, *7* thermal insulation

Mergui et al. [8] studied the influence of the channel width, therefore the channel allowing lateral movement of the sidewalls was used. The heated bottom wall was realized with 10 mm copper sheet and independent heaters. Other walls were made of 5 mm polycarbonate. The upper wall was cooled by a water loop because of an extension of experiment conditions (upper cooling causes the same effect on the longitudinal roll structures as lower heating). Scheme of the heated wall is depicted in Fig. 7.

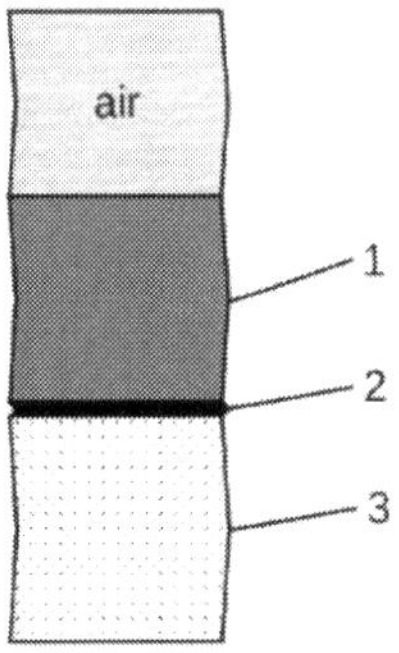

Fig. 7. Scheme of layers of Mergui's heating wall. *1* copper sheet, *2* heater, *3* thermal insulation.

3 Water channel developed for preliminary experiments

Within the frame of the master's degree thesis of A. Žemličková [9] defended at the Department of Power Engineering Equipment of the Technical University of Liberec, a water channel schematically shown in Fig. 8 was recently designed and manufactured. The channel was made of 10 mm thick glass sheets and was intended for preliminary measurements of a non-isothermal water flow. Unfortunately, the channel was found to suffer of several construction problems which precluded measurement of accurate data that could be used, for example, for reliable verification of the numerical simulations. The problematic design issues can be summarized as follows:

- The heating foil was attached externally on an outer side of the channel. Due to the low thermal conductivity of the glass, a poorly defined heat transfer to the flowing water could be achieved.
- Channel walls were joined in a way that approximately 1 mm thick layer of glue appeared in the inner corners of the channel, which did not allow to observe the boundary layer in the immediate vicinity of the heated bottom wall.
- The construction allowed only limited access to internal surfaces of the channel necessary for the channel maintenance and cleaning.
- The channel construction was not sufficiently stiff for the safe manipulation.

We note that it is not a scope of this work to discuss results obtained with this channel. For more details, please see ref. [9]. Only the construction aspects are discussed here.

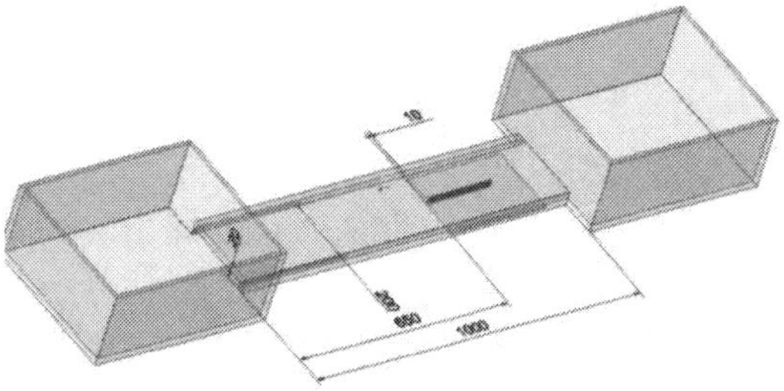

Fig. 8. Model of channel for initial experiments [9]

4 Requirements on channel construction

Based on the literature survey and own experience with the water channel described in section 3, the design of a new water channel shall take into account following requirements:

a) sufficiently long developing and exit sections for a fully-developed laminar Poiseuille velocity profile at the beginning of the heated test section,
b) inner surfaces without disturbing edges or steps between individual sections and sufficiently smooth polished inner surfaces,
c) uniform and well-defined heat flux at the bottom of the heated segment,
d) elimination of hardly measurable heat losses into the channel construction and ambient, i.e. installation of a duct insulation, especially at the channel lower part,
e) enough thermal conductivity of the bottom wall in order to maximize the heat transfer into the flowing liquid,
f) avoid heat conduction before and after the heated section. A possible solution could be an installation of a secondary water circuit for cooling the lower part of the channel before and after the heated test section. A secondary circuit could be probably designed as an additional loop of the main circuit in order to maintain the same temperature as inside the main channel.
g) simple maintenance and cleaning of the test section, ideally the possibility of opening the top wall of the channel,
h) perpendicular joining of walls without a corner glued joint that would obstruct the view of the boundary layer,
i) continuous regulation of the flow rate and the heat flux,
j) transparent side and top walls to allow for precise visualization,
k) sufficiently large water tanks before and after the channel to achieve approximately constant temperature of the circulating water and to limit possible flow fluctuations caused by the pump,
l) support platform for transversal and longitudinal manipulation of the channel for scanning by cameras across the experimental section,
m) compact dimensions due to space limitations in the laboratory and an optical table length of around 1500 mm,
n) channel support frame probably made of aluminium profiles ensuring stiffness of the whole apparatus and allowing its safe mobility.

5 Conclusion

A review of the technical design of water and air channels intended for investigation of a non-isothermal flow with a constant heat flux on a boundary wall was prepared. The technical solution of the necessary elements such as the wall heating or cooling were described. Based on the previous works of other researcher and on own experience acquired with a preliminary water channel made of glass, a set of important requirements on the design of a new water channel was defined.

The new channel shall allow to collect data for the non-isothermal flow over a heated surface. At present, preparatory calculations and construction works necessary for the channel design are undertaken.

Acknowledgment

This research was carried out at the Technical University of Liberec as part of the project "Experimental, theoretical and numerical research in fluid mechanics and thermomechanics, no. 21291" with the support of the Specific University Research Grant, as provided by the Ministry of Education, Youth and Sports of the Czech Republic in the year 2019 and with the institutional support RVO: 61388998.

References

1. D. G. Osborne, F. P. Incropera, *Laminar, mixed convection heat transfer for flow between horizontal parallel plates with asymmetric heating*, (1985)
2. J. R. Maughan, F. P. Incropera, *Secondary flow in horizontal channels heated from below*, (1987)

3. F. P. Incropera, A. L. Knox, J. R. Maughan, *Mixed-Convection Flow and Heat Transfer in the Entry Region of a Horizontal Rectangular Duct*, (1987)
4. A. Benderradji, A. Haddah, R. Taher, M. Médale, C. Abid, F. Papini, *Characterization of fluid flow patterns and heat transfer*, (2008)
5. F. Koffi, C. Abid, M. Médale, F. Fapini, *Mixed convection flow in a horizontal rectangular channel*, (2011)
6. G. J. Hwang, C.-L. Liu, *An Experimental Study of Convective Instability*, (1976)
7. S. W. Chen, D. S. Shu, J. T. Lir, T. F. Lin, *Buoyancy driven vortex flow and its stability in mixed convection*, (2006)
8. S. Mergui, X. Nicolas, S. Hirata, *Sidewall and thermal boundary condition effects on the evolution* (2011)
9. A. Žemličková, *master's degree thesis*, (2019)

Automatic field measurement of groundwater discharges in a tunnel

Petr Rálek[1,*], *Milan* Hokr[1], *Lubomír* Slavík[1], and *Aleš* Balvín[1]

[1]Faculty of Mechatronics, Informatics and Interdisciplinary Studies, Technical University of Liberec, Studentská 1402/2, 461 17 Liberec 1, Czech Republic

Abstract.
The paper describes a complex project of flow rate measurement applied for groundwater discharge into a tunnel, for both individual seepage points and the collection channel totalizing the flow. Various standard methods are used according to a large range of flow rate magnitudes (tipping bucket, danaid, V-shaped weir) with partial improvements and technical adaptations. Data collection is provided by individual measurement units with ethernet interface sending data to a database server through the internet. Experience of 10 years with several upgrades of either individual measurement or the electronic collection and transmission system is summarized.

1 Introduction

Flow rate measurement in hydrology and hydrogeology is one of the basic field research procedures [1]. Open-channel flow in streams is well established, including automation and data transmission, typically via the GSM network [2]. On the other hand, for the automated measurement of water discharge from the rock through the tunnel wall it is not routinely applied. For groundwater flow studies, the tunnel inflow spatial and temporal variations can be useful as extended information [3] compared to common field methods of borehole/well pressure tests [4], borehole water levels, and surface springs flow rates. Collected data from this study are e.g. used as an input to numerical models [5], [6].

The presented study is a part of projects related to geological disposal of the spent nuclear fuel and to the geothermal energy. In both cases, the detailed understanding of groundwater flow and the related physical processes (thermal, mechanical) in heterogeneous rock is important.

First part of the paper describes technical features of the measurement methods, automation, and data transmission. Then we present our experience from over 10 years long continuous measurement in order to review the applicability of various methods or how the data are influenced by the selected measurement method or by the complex groundwater phenomenon itself [7].

1.1 Bedřichov site

The measurement has run in the water supply tunnel in Jizera mountains, Czech Republic. The tunnel connects Josefův Důl reservoir with the water treatment plant in Bedřichov. It is about 2600 m long, with the height difference of 40 m. Pipe for the supply of raw water from the reservoir to the treatment plant leads through the tunnel. A detailed description of the site geology is provided in the report of Klomínský et al. [8].

Various phenomena are measured within the whole project: water inflow rates, pH, electric conductivity, natural tracer concentration, temperature, rock fracture displacements, seismicity etc., some in a cooperation with other research teams and institutions.

Particularly, the long time monitoring of groundwater discharging into the tunnel has been held in numerous observation points (see Fig. 1). It is based on the choice of basic observation points[1] Vn from the work [8], which were further supplemented with newly labelled observation points[2] Wx.

1.2 Character of the inflow

The in-situ measurement deals with flow rates at the numerous seepage sites with different character (from dropping through flowing individual seepages to the collecting channel flow representing the total tunnel inflow) and also with the large range of flow rate magnitude (from microlitres to litres per second):

1. Discharge points from the walls or ceiling of the tunnel. Depending on the flow rate magnitudes, they can be divided into:

- dripping (V1–V3; see Fig. 5),
- flowing (V4–V7, W142–W2313; see Fig. 5 or Fig. 7).

Depending on their rate variability, they can be divided into:

*e-mail: petr.ralek@tul.cz

[1] $n = 1 - 8, 10$

[2] x [m] is the position in the tunnel

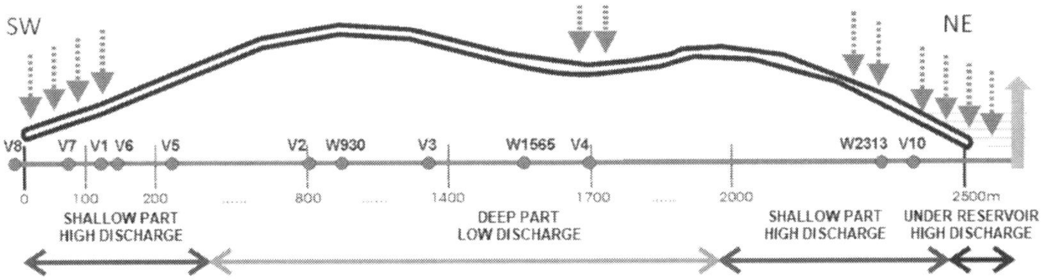

Fig. 1. Basic diagram of Bedřichov tunnel with the positions of selected discharge points and representation of the overburden.

Fig. 2. Water level measurement in the channel. On the left: ultrasonic probe (in the observation point V8). In the middle: a cylinder with the load string (H104). On the right: the pressure probe (H248).

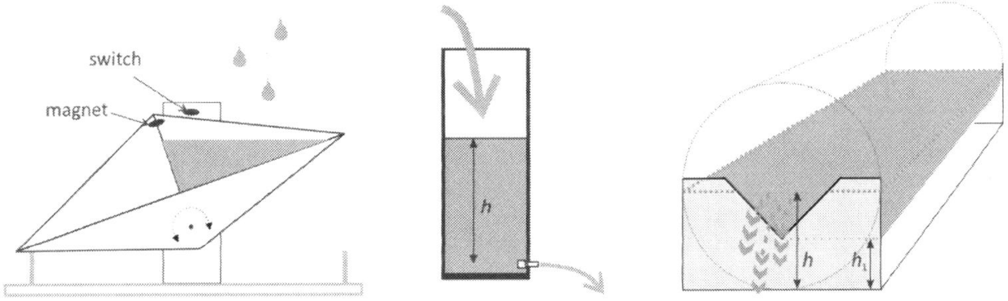

Fig. 3. From left to right: tipping bucket, danaide with one outlet, V-shaped weir; h is the actually measured quantity.

- discharge points in the deep part of the massif; with low variability (V2–V4, W930, W1565),
- discharge points in the shallow part; with large variability (V1, V6, V7),
- discharge points on the borderland between deep and shallow parts; with medium variability (V5, W2210, W2313).

2. Total discharge into the tunnel

Collecting channel at the bottom of the tunnel captures all discharges into the tunnel. By subtracting the flow values from two points in the channel, the total discharge in a given section of the tunnel can be obtained. The observation points[3] in the channel are labelled as Hx (Fig. 2).

[3] x [m] is the position in the tunnel

2 Groundwater discharge measurement

2.1 Methods of flow rate measurement

Various methods, reflecting the character and magnitude of inflow, are used (see Fig. 3):

- Tipping buckets: by filling the buckets with a known volume V and tipping them by its weight, a signal is induced by a reed switch and a magnet. They can record either (a) each tipping time t_i of the bucket (this regime has been actually used by us); the flow rate is $Q(t_i) = \frac{V}{t_i - t_{i-1}}$ or (b) the number of times n the bucket is tipped over a specified time Δt; the flow rate is then $Q = \frac{nV}{\Delta t}$. Differences are discussed in paragraph 4.1.
- Danaids: measuring vessels (cylinders) with one or more outflow orifices (with known sectional area S). They record the water level h in the vessel. The flow rate is then computed as $Q = \mu S \sqrt{2gh}$, where μ is the leaving coefficient of the orifice (0.9 for our case).
- Measuring cylinders[4]: they record increasing water level h in the cylinder; after the cylinder is filled, it is emptied and the whole process periodically repeats. With known sectional area S, the flow rate is $Q = \frac{S\Delta h}{\Delta t}$, where Δh is the level difference over the time inverval Δt.
- V-shaped weir: the flow rate is calculated from the height level $\hat{h} = h - h_1$ in the channel above the measured profile. For the profiles with 90° cut and $\hat{h} > 20$ cm, the flow rate can be computed as $Q = 1.343\, \hat{h}^{2.47}$.

Some of the methods have been modified by authors [10]. We present our experience of the use of each particular method. The examples of installations are shown in Fig. 2 and Fig. 5.

2.2 Water level measurement

The water level in the channel or in the vessel can be measured by various methods (for schemes, see Fig. 4; installations in Fig. 2):

- pressure probes compensated by atmospheric pressure or relative pressure; water level h can be then obtained from $h = \frac{p_{meas} - p_{atm}}{\varrho g}$, or $h = \frac{p_{meas}}{\varrho g}$, respectively,
- ultrasonic level sensor; $h = d - h_s$, where h_s is the vertical position of the sensor and d is the measured distance from the sensor to the water surface evaluated internally by the device from the reflected pulse travel time,
- load strings by a partially submerged weight; the load is measured by the resonance frequency of the loaded string (the excitation and electrical signal processing is a built-in operation of the device); then $\Delta h = (R_0 - R_1)G$, where R_0 is the initial value, R_1 is the actual value and G is the unique parameter for every device.

Methods differ both in their accuracy and in their installation (e.g. spatial) demands. This is in more details discussed in paragraph 4.3.

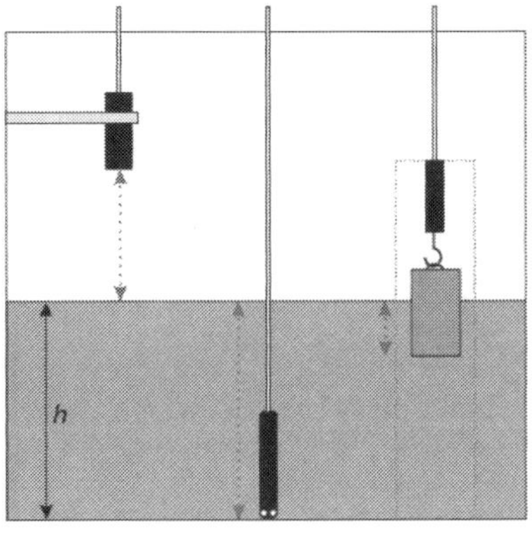

Fig. 4. Water level measurement. From left: ultrasonic probe, pressure probe, load string.

Fig. 5. Methods of the water discharge measurement. On the left: tipping bucket (observation point V2). In the middle: danaide (V7). On the right: measuring cylinder (W930).

[4]original construction [9] patented by the Czech Geological Survey

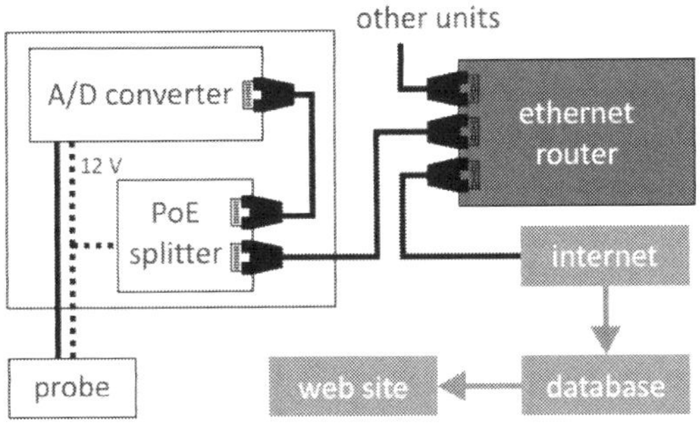

Fig. 6. Configuration of electric and data collection network, with an example of one measuring point.

3 Automatic data collection

During time, the measurement in a large part of the observation points was automatized, in particular in the 0-930 m chainage positions with electric and data network, while in the remaining part, loggers with internal memories and regular manual data download are used.

The first generation of the measurement and collection, established in 2010, is presented in [11]. It used the RS485 bus, own development of electronic modules, and GSM connection to the internet at the tunnel entrance.

The second generation is based on the ethernet network and commercial measuring units. At a particular discharge location, the system consists of (Fig. 6): sensor (depending on the method) + electrical connection (A/D converter or pulse counter) + ethernet connection. The data are stored in a database (kept since the first generation system) with public presentation at web page http://bedrichov2.tul.cz/ (results of [12], in trial mode; an example of visualised data is shown in Fig. 14).

The measurement methods are based on water level measurement (cylinders, weirs) or discrete pulse generation by tipping buckets. The data collection of an analog measurement is realized with the AD4ETH unit [13] with A/D converter from 4-20mA (= output of pressure LMP307i [14] or ultrasonic S18UAI [15] level sensors) sending HTTP GET requests regularly to web server with running PHP script which enters data to a PostgreSQL database. Longer interval of 10 minutes is used for cylinders with steady level and higher range and shorter interval of 1 minute is used for weirs to cut-off noise by averaging (see paragraph 4.1). The unit provides programmable linear conversion from the current range to the level value (mm).

The pulse counting is realized with Papago TH2DIDO [16], also with integrated ethernet interface and HTTP GET communication, sending data asynchronously, i.e. the counter value and time stamp from the device real-time clock, after each pulse. This avoids any loss of accuracy by sampling errors. Conversion from raw data (level in mm, counter) to flow rate units is implemented in the database through a trigger function.

Fig. 7. Complete equipment of the automatic measurement. Up: V6 discharge with probes. In the middle: cables leading from probes directly or trough Greisinger [17] transmitters into the water resistant box. Bottom: the box containing the PoE splitter, the A/D converter for general quantities and the Papouch A/D converter (here with its own PoE and the temperature sensor).

https://doi.org/10.1051/epjconf/202226401038

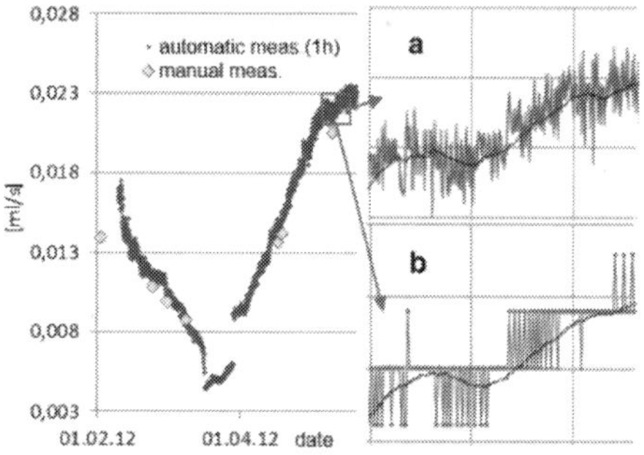

Fig. 8. Dripping seepage V2 – discharge level progression with detailed part of the course for two different record regimes of the tipping bucket (a – every tipping time; b – number of tips per 1 h). Daily floating averages (black lines) are rather similar.

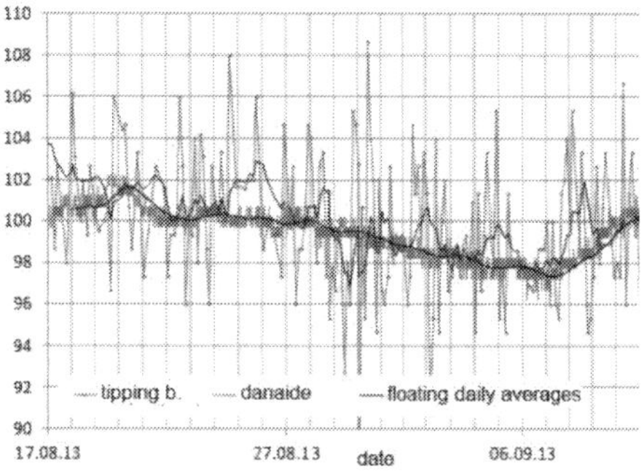

Fig. 9. V6 discharge variance [%] (rel. to time period average). Comparison between tipping bucket and danaide.

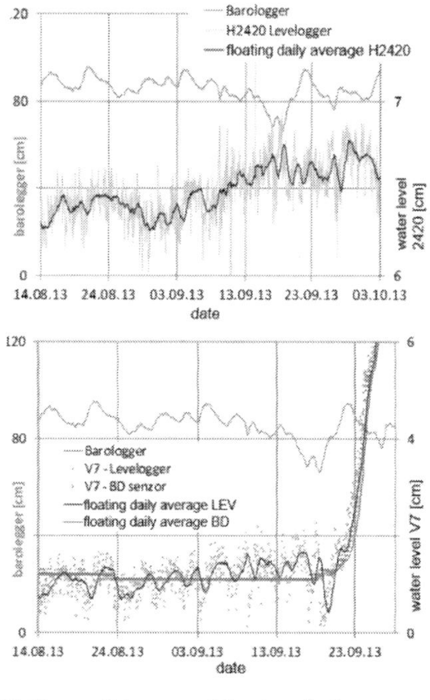

Fig. 10. Up: parallel courses of the atmospheric pressure and the automatically compensated water level in H248. The "overcompensation" is distinctive. Bottom: the same period recorded for V7 in comparison with relative pressure probe record (red line).

4 Our experience from the measurement

In the following paragraphs, we present some comments on the data post-processing, based on our experience from a long time in-situ measurement.

4.1 Data post-processing

The raw data, obtained by the measurement, include some noise or fluctuations, which have to be distinguished and post-processed. These fluctuation can be from various origins:

- Data influenced by the measurement method

  Typical example is the data from the tipping bucket (see Fig. 8). Whether we record time of each tip (case **a** on the Fig. 8) or their sum over a particular time period (case **b** Fig. 8), the flowing averaging is needed to obtain the well arranged graph. The case B provide more smooth curve, while the case A can catch the short time discharge fluctuations.

  Measuring vessels or danaids provides more smooth data than tipping buckets, given by their character. The smoothness relies purely on measurement frequency (we don't reflect the qualities of probes here). An illustrative example for the V6 discharge is shown in the Fig. 9.

- Data influenced by the measurement equipment

  This case is typical for the various methods of the water level measurement. The values from probes recording the relative pressure are almost ready to use without averaging (see red line in Fig. 10 bottom). In comparison, data from Levelogger probes, which need compensation to the atmospheric pressure, are relatively strongly influenced directly by this compensation. Natural atmospheric pressure fluctuation are qualitatively transposed (with opposite sign) to the water level data - something, which we can call overcompensation (see Fig. 10; the projected fluctuation are in a bold contrast with almost constant water level recorded by the LMP307i probe). The compensated data (after averaging) are suitable rather for long-term trends.

  There are usually different claims on the measurement sensitivity in time ranges of different lengths, which can be illustrated with the data from the measuring vessel: see Fig. 11 (the upper part for the period of several month and the bottom part for the period of years). For short time fluctuation, we have to record the actual water level; for long time trends, it is sufficient to record the filling-times.

  The main disadvantage of ultrasonic probes is that they record the noise caused by the water level surging (see Fig. 13). By contrast, in the string load measurement, the weight momentum compensates instant oscillations of the water level.

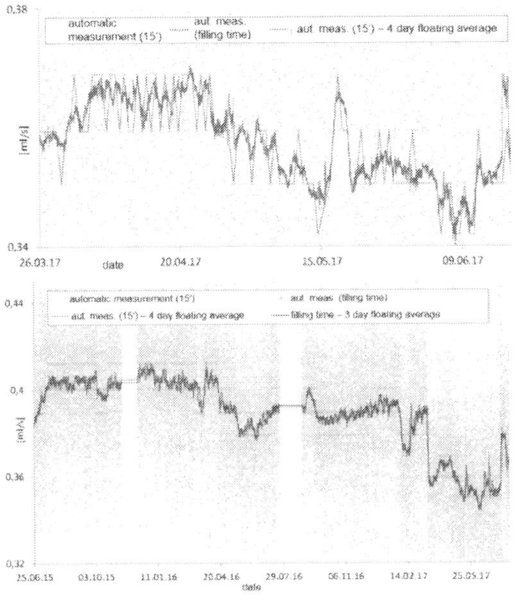

Fig. 11. Discharge level at W930. A comparison between continuous water level measuring (blue dots) and filling time recording (yellow dots with red line). Up: 3-month period. Bottom: 2-year period.

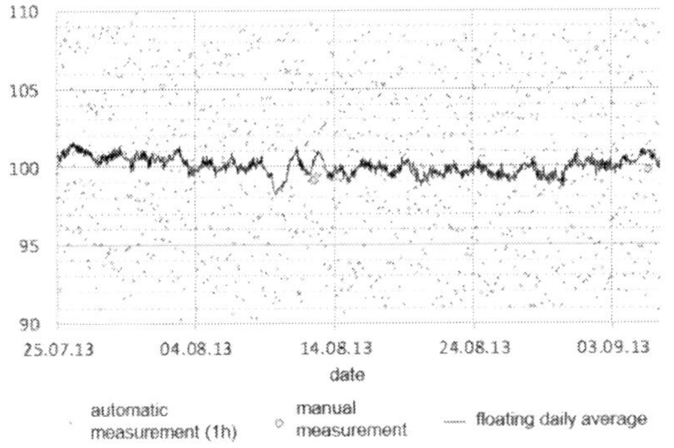

Fig. 12. V4 discharge variance (rel. to time period average %). Daily floating average embodies distinct fluctuation, caused by the atmospheric pressure compensation.

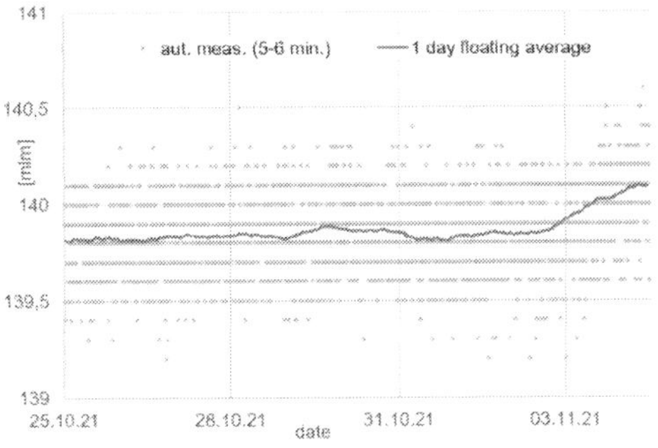

Fig. 13. H248 water level variance (ultrasound probe).

- Data influenced by the discharge itself

  Next to types of fluctuations caused by the measurement, mentioned above, the variability of the data can be caused by the momentary fluctuations of the discharge itself (which can even easily comprehend those measurement-caused fluctuations). This is characteristic for the discharge V4: momentary fluctuations, caused by the periodic aeration of the pipe, are fairly observable (with the variable period in low tens of seconds). On the other hand, the long-time flow rate is very stable.

  The Fig. 12 represents a 2-week record. While the data (recorded with 15 min. period) shows the variation about ±10% from the 2-week average, the real discharge level variation is about ±1% (documented by the manual measurement).

4.2 Measuring equipment efficiency

- Tipping buckets:

  They have been proven and suitable for dripping discharge points. In the case of flowing discharge points, their possibilities are limited: the shuttle mechanism becomes worn when using a tipping bucket with a small volume (in relation to the rate of the measured flow); tipping buckets with a larger volume are not sensitive to changes in the flow rate.

- Danaids:

  They have been proven for a wide range of rates of flowing discharge points, with a suitably chosen hole size and working height of the water level they are sensitive to changes in flow rates.

- Measuring cylinders:

  Their application field covers the borderland between tipping buckets and danaids. They are suitable for flowing or dripping discharge points with a stable or slowly changing flow rates. Sensitivity to changes in the flow rate depends on the frequency of data recording. For long-term trends in the yield of stable discharge points, records of cylinder fill times are sufficient.

- V-shaped weir:

  The used profile is not very sensitive to changes in flow rate, and even a slight change in the water level (even in the case of a ripple) is recorded in the fluctuation data. There is a need of the water level measurement accuracy better then millimeters, which is hard to achieve for some measuring principles (see next paragraph 4.3).

4.3 Measurement accuracy

The accuracy comparison of particular methods of water level measurement is presented in Table 1.

Ultrasonic probes have high relative accuracy. Moreover, they are not in the contact with water, which can avoid an origin of sediments or coagulation and prolong their working life.

Pressure probe has less accuracy in comparison with remaining measuring principles. Coagulation origin can

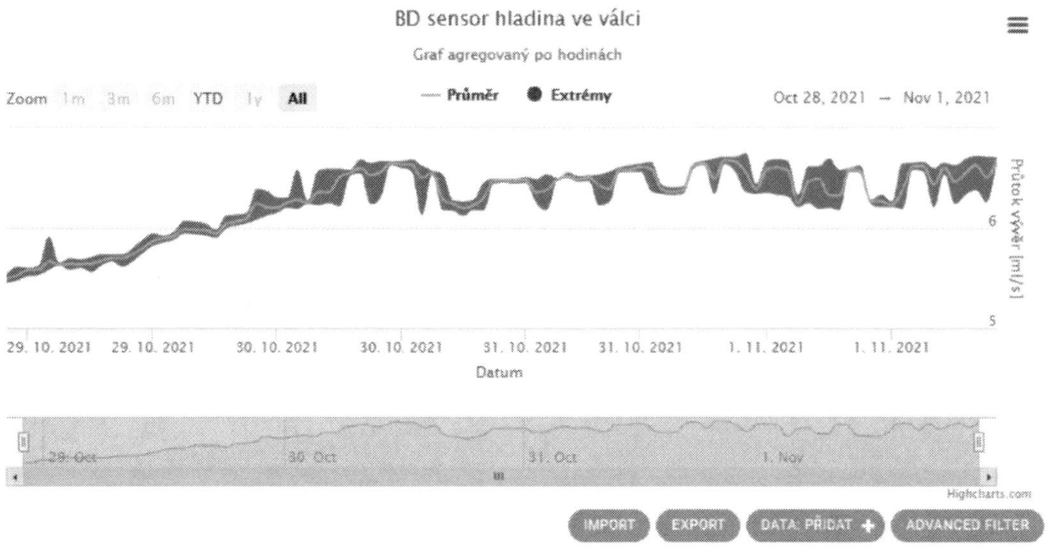

Fig. 14. An example of data presentation from the database web site – flow rate at V6, 1-hour average and min/max range in the same sampling.

Table 1. An accuracy overview of various probes (adopted from technical documentation).

Method	working range [cm]	resolution [mm]	relative accuracy [%]	absolute accuracy [mm]
US probe Turck S18UAI [15]	30	–	0.08	0.25
US probe US1200, Fiedler-Mágr [18]	120	1	0.2	2.4
pressure probe LMP307i, BD Sensors [14]	100	1	0.1	1
pressure probe Levelogger Gold/Edge, Solinst [19]	500	1	0.05	2.5
load string device Geokon [20]	15	0.04	0.1	0.15

significantly shorten the working life of the probe, since the membrane of the probe is in the open contact with water. On the other hand, they have very simple installation and higher-order measuring range. Levelogger probes have a longer service life than BD-Sensor probes (some of malfunctioning probes have been replaced after about 3 years of working). Presented accuracies for Levelogger probes suppose the ideal connection with the atmosphere or the perfect time coordination with the barometric logger, which is mostly not practicable.

Load string principle has an advantage in the very high absolute accuracy (with rather small working range). Its disadvantages lie in a rather difficult installation with considerable space demands and a need of the proprietary evaluation unit (it has not been connected to the presented online data collection system).

5 Conclusion

The presented measurement project demonstrated a heterogeneous telemetry system with various flow conditions and measurement principles. Generic units with configurable ethernet interface have been used, in contrast with a common use of proprietary protocols and data storage services provided by the measuring unit producers or supplier. The water level measurement was limited by the sensors' accuracy and durability, but the experience proved its value for observation and evaluation the temporal variations of groundwater discharge.

Acknowledgements. This work was co-funded by the Radioactive Waste Repository Authority (as a partial result of project "Research support for Safety Evaluation of Deep Geological Repository") and by the Ministry of Education, Youth and Sports, and EU within the project RINGEN+, CZ.02.1.01/0.0/0.0/16_013/0001792. Authors also thank to SČVK a.s., the water treatment plant company, for providing access to the tunnel.

References

[1] I. Strangeways, Measuring the Natural Environment, Cambridge University Press, 374 p., ISBN 0-521-57310-6 (2000)

[2] M. Acosta-Coll, F. Ballester-Merelo, M. Martinez-Peiro, E. De la Hoz-Franco, Real-Time Early Warning System Design for Pluvial Flash Floods-A Review SENSORS **18 (7)**, Art. 2255 (2018)

[3] D. Mas Ivars, Water inflow into excavations in fractured rock – a three-dimensional hydro-mechanical numerical study. Int. J. Rock Mech. Min. Sci **43**, 705–725 (2006)

[4] L. Rukavickova, J. Holecek, P. Holeckova, P. Najser, L. Gvozdik, T. Paces, Comparison of hydraulic conductivity of rock matrix and fractured blocks of granitic rocks, International Journal of Rock Mechanics and Mining Sciences **144**, Art. 104743 (2021)

[5] M. Hokr, A. Balvín, I. Škarydová, Tunnel inflow in granite – fitting the field observations with hybrid model of discrete fractures and continuum, Fractured Rock Hydrogeology. IAH - Selected Papers on Hydrogeology, **20**, CRC Press, pp 240–256 (2014)

[6] A. Balvín, M. Hokr, K. Šteklová, P. Rálek, Inverse modeling of natural tracer transport in a granite massif with lumped-parameter and physically based models: case study of a tunnel in Czechia, Hydrogeology Journal, Springer-Verlag GmbH Germany, **1**, 22 p, ISSN: 1431-2174 (2021)

[7] M. Hokr, L. Rukavičková, V. Havlová et al., Development and testing of methods for rock characterisation – Results of monitoring in Bedřichov tunnel, RAWRA Technical report, **289/2018**, 155 pp. (2018)

[8] J. Klomínský, F. Woller (eds) Geological studies in the Bedřichov water supply tunnel. RAWRA Technical report 02/2010. Czech Geological Survey, Prague, 103 p., ISBN 978-80-7075-760-4 (2010)

[9] J. Holeček et al., Způsob měření průsaků vody, zejména do podzemních děl a zařízení k provádění tohoto způsobu. No. 305696. Česká geologická služba, ARCADIS CZ a.s., Ústav struktury a mechaniky hornin AV ČR, v. v. i., Technická univerzita v Liberci, ISATech, s.r.o (2015) – in Czech

[10] P. Rálek and M. Hokr, Methods of Water Inflow Measurement in the Bedřichov Tunnel, EGRSE - Explor. Geophys., Remote Sens. Environment, **2**, pp 30–39, Czech Association of Geophysicists (2013)

[11] R. Špánek, M. Hernych, M. Hokr, P. Svoboda, M. Tyl, J. Řimnáč, J. Štuller, Bedřichov Tunnel – Continual Automated Measurement of Physical Quantities, Exploration Geophysics, Remote Sensing and Environment, 2011/2, pp. 73-82, Czech Association of Geophysicists (ČAAG), ISSN 1803-1447 (2011)

[12] L. Pelc, Zpracování a vizualizace dat z vodárenského přivaděče Bedřichov (Processing and Visualization of Data from Bedřichov Water Tunnel), Diploma Thesis, Technical University of Liberec, 43 p. (2021) – in Czech

[13] Papouch s.r.o., AD4ETH: Měřicí převodník s Ethernetem, online: https://papouch.com/ad4eth-merici-prevodnik-s-ethernetem-p4607/

[14] LMP 307, online http://www.bdsensors.cz/vyska-hladiny/ponorne-sondy/detail/produkt/lmp-307/ (cit. 2018)

[15] Product S18UIA, http://pdb2.turck.de/cz/DE/products/000000060 00004090002003a (cit. 2018)

[16] Papouch s.r.o., Papago TH 2DI DO ETH: Environment monitor, online: https://papouch.com/papago-th-2di-do-eth-environment-monitor-p3159/

[17] https://www.greisinger.de/p/transmitter/6/ (cit. 2021)

[18] Ultrazvukové hladinoměry US1200, US3200 a US4200, https://www.fiedler-magr.cz/cs/produkty/hladinomery/snimace-vysky-hladiny/us1200-3000 (cit. 2018)

[19] Levelogger® Edge, https://www.solinst.com/products/dataloggers-and-telemetry/3001-levelogger-series/levelogger-edge/ (cit. 2018)

[20] Weir Monitor (VW) Model 4675LV, https://www.geokon.com/4675LV (cit. 2018)

CFD analysis of an anemometric sensor design for air flow measurement in a heat exchanger

Jan Sedlacek[1,*], *Michal* Kus[1], *Miroslav* Strob[2], and *Richard* Matas[1]

[1]University of West Bohemia, NTC, Univerzitni 8, 301 00 Plzen, Czech Republic
[2]4JTech s.r.o., Ringhofferova 115/1, 155 21 Praha, Czech Republic

Abstract. This paper focuses on the issue of flow measurement in automotive heat exchangers. The measurement is complicated due to the limited space in the heat exchanger area and the non-uniform velocity profile. Therefore, only some measuring techniques can be used, which must be adapted to the specific conditions. The paper presents the initial stages of the development of a sensor for measuring the flow in a heat exchanger. These are CFD simulations used to determine the nature of the flow around the sensor and verify its effect on the flow field in its surrounding and on the entire heat exchanger. The results of these initial simulations are presented and discussed.

1 Introduction

Due to the limited efficiency of passenger car drives and other systems, residual heat is generated during their operation. It must be dissipated to the surrounding environment. This is solved using heat exchangers, usually a set of exchangers according to the characteristics of a particular car. The heat exchangers are normally located in the front of the car, where it is easier to ensure their proper function. This is a sufficient flow of air around their heat exchange surfaces. It is influenced not only by the heat exchangers themselves, but also by the flow around the car and inside the engine compartment, the shapes of the grilles, the fans and other elements. Experimental measurements directly on the exchangers are necessary to verify their correct function [1]. The measurement usually takes place directly at or in the relevant heat exchanger. However, these measurements are quite complicated because there is very limited space for the installation of measuring probes in the area of exchangers [2].

Only a few methods can be used for direct measurement of the air flow integral value [3]. One of them is hot-wire (-film) anemometry [4, 5]. Some previous work focused on the use of hot wire anemometers with traversing probes along the exchanger surface pointed out, that this measurement is in fact impossible in the engine compartment and it has significant measurement error due to considerable inhomogeneity of the flow behind the exchanger. Both of these problems should be eliminated by installing hot-wire (-film) sensors in the exchanger. The development presented in this paper was therefore focused on installing a sufficient number of sensors into the heat exchanger to ensure the accuracy of direct measurement of the total air flow. It is thus an intrusive method with the necessary removal of fins with louvres between the channels with the cooled medium.

The presented work focuses on the initial stage of the development of hot film sensors for this type of measurement. This stage involved determining the concept of the sensor from the point of view of the sensing element and the electronics. Based on this concept, it was possible to determine the expected dimensions of the sensor and design it. The sensor model was then implemented in a selected type of heat exchanger and CFD simulations [6] were performed to optimize its design and verify its properties and impact on the modified exchanger.

2 Sensor design, models

As mentioned, the sensor model was based on the concept of the hot film element and electronics. A hot film was chosen because the company has thin layer sensors of its own development. The dimensions of the hot film plate were designed 7x9 mm. A suitable structure of the hot film was designed on the surface so that the film reached resistances suitable for the current regulator, which ensures its heating. The hot film plate is placed directly on the printed circuit board with current regulator and electronics for measurement and communication.

The dimensions of the printed circuit board were set at 20x20 mm with a thickness of 3 mm. These dimensions already include the final surface treatment of the sensor, i.e. coating the board by a protective and insulating housing. The plate with the hot film element protrudes 0.3 mm above the surface of the encapsulated printed circuit board. Thin side plates have been designed to secure the position of the sensor between the channels. These can be modified according to the

* Corresponding author: sedlacek@ntc.zcu.cz

specific type of exchanger without changing the sensor itself. The effect of the length of these side plates was assessed during the initial CFD simulations. Based on the results, their length was determined to be the same as the depth of the sensor, i.e. 20 mm. The model of the resulting sensor design is shown in Figure 1.

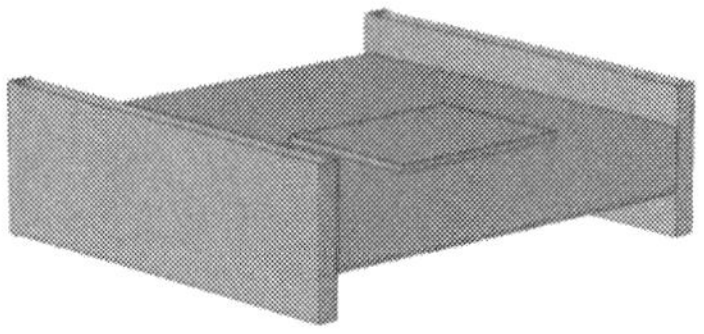

Fig. 1. Model of the hot film sensor.

A detailed model of a part of the selected heat exchanger was prepared to assess the influence of the designed sensor on the change of flow through the exchanger at the place of its installation and the change of the pressure loss of the exchanger. A segment of size 32.2 x 23.25 mm (depth of the exchanger 26 mm) was selected from the exchanger, for which detailed models without and with implemented hot film sensor were prepared. Both segment models are shown in Figure 2.

Fig. 2. Heat exchanger segment models: without sensor and with hot film sensor implemented (part of the segment with a view of the sensor location).

The segments were placed in the overall model, which includes the input and especially the output part to affect the flow behind the segments. Symmetries and periodicities were applied to the walls to reduce the influence of boundary conditions. Normal velocity was defined at the inlet and atmospheric pressure at the outlet. The computational mesh consists of 9.3 million polyhedral cells. The overall model is shown in Figure 3.

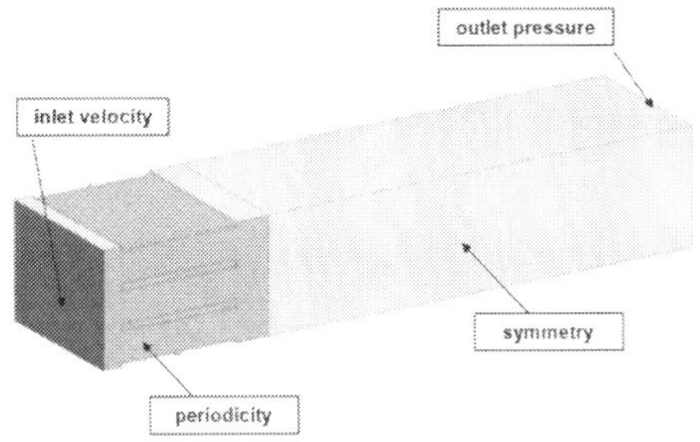

Fig. 3. The overall model with boundary conditions.

3 CFD simulations - segments

CFD simulations on models with heat exchanger segments were performed in the ANSYS Fluent 2021 R1 program [7]. The flow was considered incompressible (air) and turbulent – k-ω SST turbulence model [8]. Inlet velocities were considered in the range of 2.5 to 10 m/s. The outlet pressure was defined as atmospheric.

Selected variants of the model were also solved with heat transfer, when a constant temperature of 80 degrees Celsius was set on the surface of the water channels.

The simulations were not solved only for the mentioned model geometries, the analyses also included variants for assessing the location of the hot film sensor between channels (front, rear), sensor inclination (mounting error) and hot film element plate position (longitudinal and transverse orientation).

The following results are for the indicated model geometries and without heat transfer only.

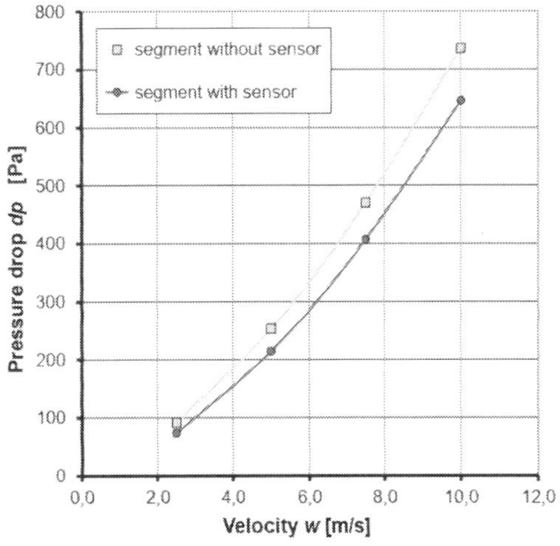

Fig. 4. Pressure drop of the segment models with and without the sensors.

An important result of CFD simulations was the finding of the difference in segment pressure loss. The graph in Figure 4 shows that the segment with the sensor has a lower pressure drop, the difference compared to the segment without the sensor is about 20 % depending on the conditions.

The following figure 5 shows flow lines in the model's segments coloured by the velocity.

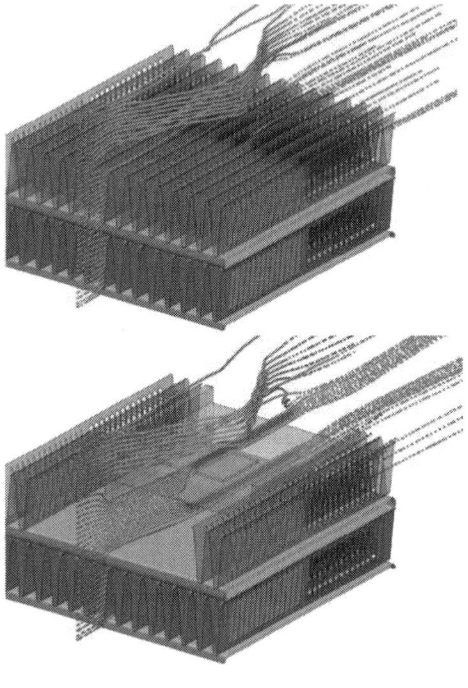

Fig. 5. Flow lines coloured by velocity: segment without sensor at the top, with sensor at the bottom.

The results found on both segment models are important, but do not comprise the overall impact of sensor installation on the heat exchanger. However, the application of the used methodology to the whole exchanger is not possible in principle. The size of the computational task i.e. the number of computational elements would reach extreme values.

4 CFD simulations – porous zones

To determine the effect of sensors installation on the whole exchanger, a simplified model with use of porous zones for different parts of the exchanger was prepared. The model was again in version without and with sensors. Based on the results of both model segments listed in the previous chapter, the parameters of two types of porous zones were determined - the heat exchanger part without the sensor and the heat exchanger part with the installed sensor.

To maintain the highest similarity of this model to the real exchanger, water channels were preserved and the porous zones replace only the fins and louvres volume.

Figure 6 shows the segment with porous zones with the sensor. The segment without the sensor was prepared and solved in a similar way. With the help of these sub models, the whole model of the exchanger was assembled. A total of 70 sensors were considered for optimal coverage of the area of the heat exchanger. In this case, the individual sensors cover the same area and the integral determination of the air flow is thus simple and correct.

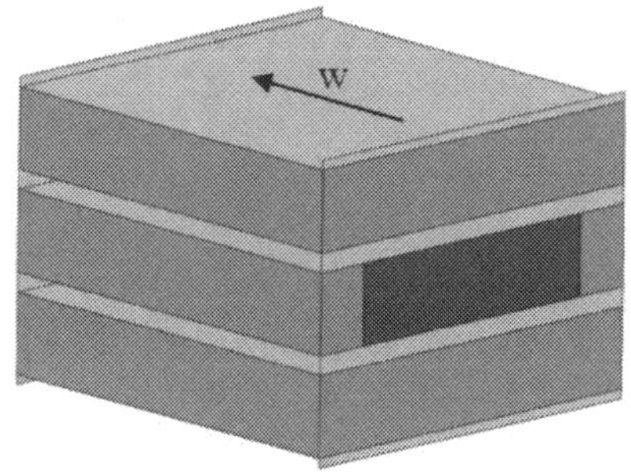

Fig. 6. Segment model with fin and louvres and sensor area simplified by porous zone.

A schematic view of the distribution of the sensors over the heat exchanger surface is shown in Figure 7.

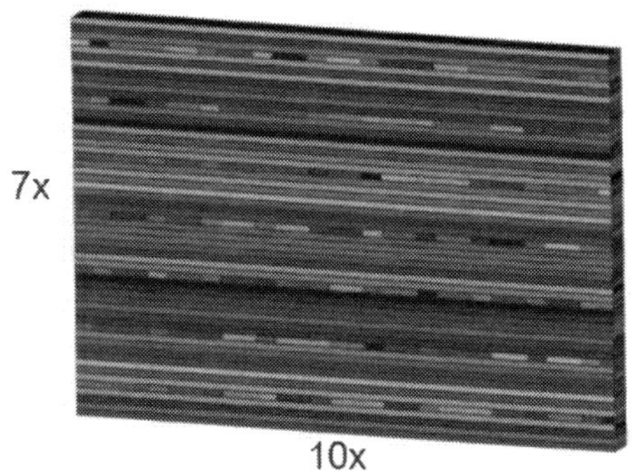

Fig. 7. Geometry of the overall porous heat exchanger model with sensors in a 10x7 matrix.

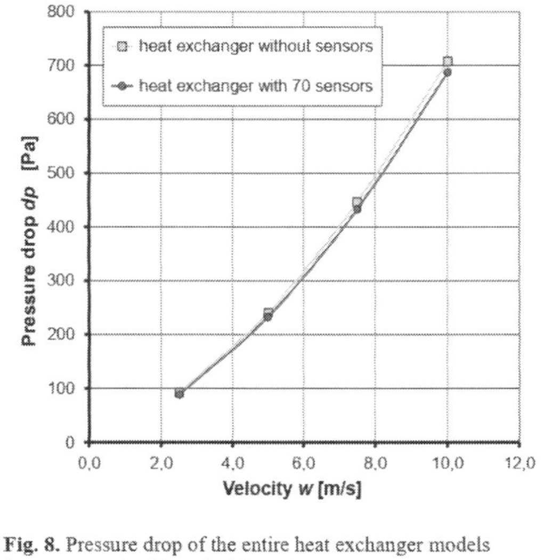

Fig. 8. Pressure drop of the entire heat exchanger models (porous zones) with and without the sensors.

CFD numerical simulations for the whole exchanger model were performed under the same conditions as the simulations for the segments. Heat transfer was not considered at all for these model variants. In addition to a detailed analysis of the velocity field in the heat exchanger area, the basic result was again the pressure drop of the heat exchanger with and without sensors.

The comparison is shown in the graph in Figure 8. The result shows that when the entire area of the exchanger is considered, the influence of the sensors is very small. The pressure loss difference is 3.3 %. Given the idealized mounting of sensors in computational models, it can be expected that the difference will be probably even lower in real design. The velocity field in the cross-section of the exchanger is shown in Figure 9. The figure shows higher velocities behind the exchanger in the area of the sensors, which is still visible far beyond the exchanger.

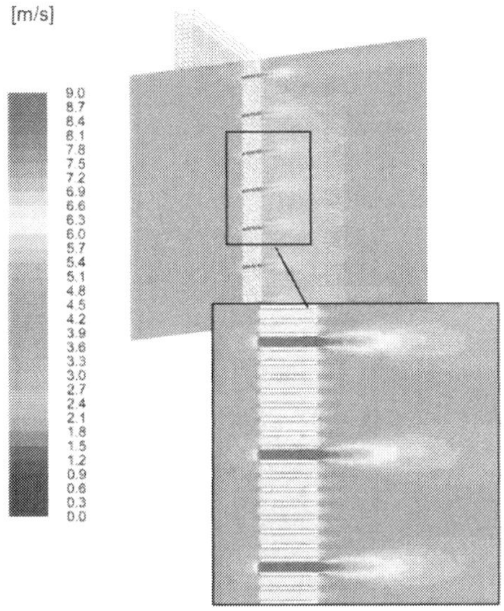

Fig. 9. Velocity contours in vertical section through the exchanger in the axis of the sensor column.

5 Conclusions

The previous chapters gave a brief description of the initial stages of the development of a hot film-based sensor for measuring the total air flow through automotive heat exchangers. Based on the concept of the hot film element itself and the necessary electronic part, a sensor model was developed. It was implemented in the exchanger segment model and subjected to computational analyses. The aim was to find a suitable shape and location of the sensor in the exchanger and to verify the sensitivity of future measurements to the design of the sensor and the errors of its mounting. After verifying the design on a small segment, a model of the entire exchanger was prepared. Due to the multiple increments of the model, it was necessary to use simplification, fins and louvres are modelled as porous zones. The results show a small deviation (3.3 %) of the pressure loss of the exchanger with installed sensors from the exchanger itself. At the same time the results show an acceleration of the flow in the slots with the sensor, which is also significant in the area behind the exchanger. The next step in the analysis is to supplement the model with fan holder and fans (Sahara). The calculations should lead to an evaluation of the determined higher velocities behind the sensors. At the same time, various operating conditions will be analysed, including the influence of fans.

Acknowledgement

This project is funded with state support from the Technology Agency of the Czech Republic and the Ministry of Industry and Trade within the TREND Program (project FW03010199).

References

1. Sortor, M., *SAE International Journal of Materials and Manufacturing*, vol. 4, no. 1, SAE International, 2011, pp. 1221–30, http://www.jstor.org/stable/26273854.
2. Eton Y., et al., *Proceedings of the 14th Australasian Fluid Mechanics Conference*. 2001.
3. Kuthada T. , *SAE Int. J. Passeng. Cars - Mech. Syst.* 6(1):88-96, 2013, https://doi.org/10.4271/2013-01-0598.
4. Tselentis A., *Application of Hot–Wire (–Film) Flowmeters to Water Velocity Measurement in Wells*. Journal of Hydrology. 58 (1982).
5. Que R., Zhu R., *A Two-Dimensional Flow Sensor with Integrated Micro Thermal Sensing Elements and a Back Propagation Neural Network. Sensors.* Sensors 2014, 14(1), 564-574; https://doi.org/10.3390/s140100564
6. Perrotin, T., Clodic, D. , *International Journal of Refrigeration*, Volume 27, Issue 4, 2004, Pages 422-432, ISSN 0140-7007, https://doi.org/10.1016/j.ijrefrig.2003.11.005.
7. ANSYS Fluent User's Guide 2021 R1, ANSYS, Inc., 2021.
8. Menter F. R., *Zonal Two Equation k-ω Turbulence Models for Aerodynamic Flows,* AIAA Paper 93-2906.

Feasibilities of foam-forming bacteria disintegration by ultrasonic atomization

Petr Schovanec[1,*], *Darina* Jasikova[1], *Michal* Kotek[1], *Karel* Havlicek[2], and *Jakub* Eichler[3]

[1] Faculty of Mechatronics, Informatics and Interdisciplinary Studies, Technical University of Liberec, Institute of New Technologies and Applied Informatics, Studentska 1402/2, Liberec 1, 461 17, Czech Republic
[2] Department of Environmental Technology, Technical University of Liberec, Institute for Nanomaterials, Advanced Technology and Innovation, Studentska 1402/2, Liberec 1, 461 17, Czech Republic
[3] Faculty of Mechatronics, Informatics and Interdisciplinary Studies, Technical University of Liberec, Institute of Mechatronics and Computer Engineering, Studentska 1402/2, Liberec 1, 461 17, Czech Republic

Abstract. This article presents the feasibilities of bacterial disintegration using ultrasonic atomization. Ultrasound creates acoustic cavitation, which by its effects represents a high gradient of pressure in the form of a shock wave and a local force effect. In our case, the ultrasonic excitation produces a fine micron mist of 3 micrometers. The generated microcavities create a cavitation cloud that is able to disintegrate the foam-forming bacteria or even inactivate bacterial cells. The main goal of the research is the use of ultrasound to disintegrate the foam-forming bacteria during the impact of acoustic cavitation on living cells.

1 Introduction

The cavitation phenomena can be defined as a collection of effects connected to the origin, activities, and collapse of macroscopic bubbles in a liquid [7-10]. There are many ways to generate bubbles. For example, a rotating propeller, a vibrating surface, local deposition of energy, electrical discharge, or ultrasound. In our case, the cavitation bubbles are generated by ultrasound. Ultrasound usually leads to forming a cavitation cloud where the bubbles are not separated. These bubbles create structures that react collectively; however, individual bubbles are fundamental elements of these structures [3,4].

The research in the field of cavitation previously focused mainly on the investigation of shock waves, bubble behavior in the surrounding of solid surface, and the bubble behavior in variously viscous liquids [1-3]. However, the current investigations also focused on the use in the field of foam-forming bacteria disintegration to be included.

The advantage of using ultrasound is the absence of other chemicals that otherwise need to be added to the system. Foam-forming bacteria is disintegrated in a physicomechanical way, and with suitable settings of operating frequencies and ultrasound intensities, it is a very gentle method.

2 Experimental

The experimental part is based on previous research [5, 6], which contains bubble interaction surrounding of solid surface.

In this experiment, we used the fine micron mist and its microcavities phenomena to disintegrate foam-forming bacteria. The main potential is in use in wastewater treatment plants. Previously, bacteria sterilization research focused on using cavitation to defoam and sterilize the environment [14].

The experiment contains a laboratory power supply, oscillator, and piezoelectric transducers. The laboratory power supply powers the oscillator, which generates a signal with a frequency of 1.65 MHz. This signal is conducted to piezoelectric transducers, which generate ultrasonic waves. Piezoelectric transducers are placed at the bottom in a vessel containing 1.4 liters of water and foam-forming bacteria from a wastewater treatment plant. The setup of the whole experiment is seen in Fig. 1.

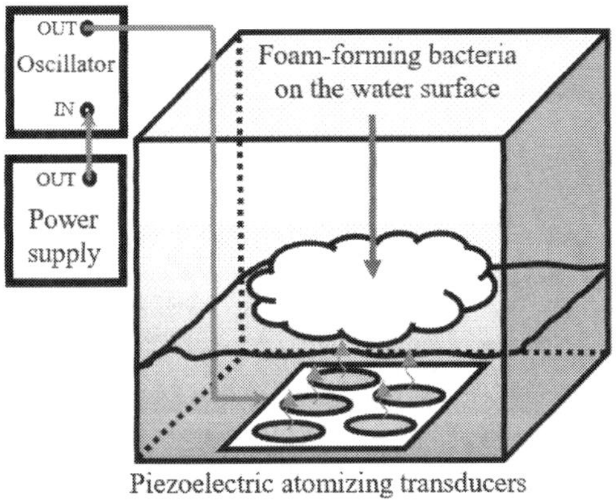

Fig. 1. The experimental setup.

* Corresponding author: petr.schovanec@tul.cz

2.1 Effects of ultrasonic

Thermal, mechanical, or physicochemical effects occur during the ultrasound. The thermal effects occur when ultrasound passes through living tissue due to energy absorption, leading to the tissue's heating. It happens mainly at the tissue interface but also when passing through homogeneous tissue. The level of the energy absorption depends on the frequency of the ultrasound. Absorption and dispersion in tissues increase with increasing frequency [10].

The mechanical effects of ultrasound can include cavitation, in which rapid pressure changes occur due to the compaction and dilution of the environment, i.e., the oscillation of molecules. During this physical phenomenon, vacuum cavities are formed in flowing fluids or places of fluids with rapid pressure changes, which by their extinction damage cell structures.

While during the physicochemical effects of ultrasound are accelerated chemical reactions due to the excitation of molecules. Physicochemical effects in combination with thermal effects cause local heating of the tissues and thus lead to increased blood flow to the tissues and accelerated metabolism.

2.2 Underwater ultrasonic transducer

The behavior of the ultrasonic transducer located underwater at the bottom of the vessel is seen in Fig. 2. The behavior of the transducer can be easily influenced by two parameters, the power input of the transducer or the height of the water level. Here we have chosen several different water level heights. The ultrasonic wave directed at the water surface causes a water column to disintegrate into water particles and mist.

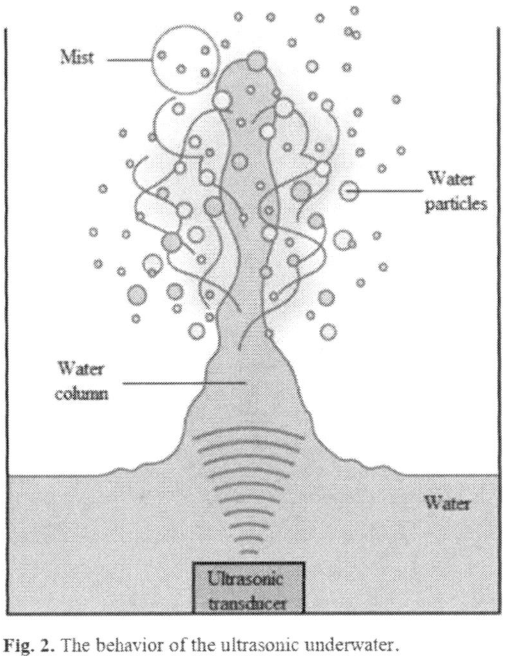

Fig. 2. The behavior of the ultrasonic underwater.

2.3 Parameters of ultrasonic transducers

In addition to their size and diameter, the main parameters of ultrasonic transducers include resonant frequency, resonant impedance, capacitance, average water droplet size, and maximum input power can be seen in Table 1.

Table 1. Basic parameters of ultrasonic transducers.

Producer and Type	Midas Com., Ltd. M165D25	APC Int., Ltd. #50-1014
Resonance Frequency	1.65 MHz ± 3%	1.65 MHz ± 10%
Resonance Impedance	2 Ω	< 5 Ω
Capacitance @ 1kHz	2 nF ± 20%	1.9 nF ± 30%
Dissipation Factor @ 1kHz	0.5% max.	< 0.5%
Input Power (maximum)	30 Watt	29 ± 1 Watt
Average Water Droplet Size	3 μm	2 μm

2.4 Foam-forming bacteria

Foam at wastewater treatment plants containing a mixture of fibrous bacteria (Microthrix, Gordonia, Skermania) was used to test the effect of ultrasonic micro-cavitation on the vitality of these bacteria. Fibrous bacteria in the foam are highly resistant microorganisms, where their removal at wastewater treatment plants is necessary. Monitoring the influence of various parameters (settings) of generated a fine micron mist on these bacteria gives an idea of their viability under given conditions in a particular environment [11-13].

The foam composed of bacteria and resistant interconnected structures easily creates a clearly defined transition with the aqueous environment. This complex structure behaves similarly to a partially solid surface. In view of this fact, the behavior of ultrasonic micro cavitation bubbles can be expected to impact a solid surface described in more detail by [5].

We focused on the behavior of the ultrasonic micro cavitation concerning the disintegration of foam-forming bacteria on the water surface, both in terms of visualization and the study of the behavior of ultrasonic transducer submerged in the liquid, which causes locally accelerated by vibration.

3 Results

Several different water level heights were tested. Ultrasound was generated towards the foam-forming bacteria placed on the water surface. Ultrasound creates acoustic shock waves followed by an increasing cavitation bubble. The action of ultrasound leads to the disintegration of the foam-forming bacteria. The magnitude of the impact of ultrasound on the foam-forming bacteria depends on the distance from the water surface.

Ultrasound has minimal effect on foam-forming bacteria over long distances because the foam has only been moved away from the action of ultrasound. When ultrasound acts on foam-forming bacteria over shorter distances, it was disintegrated to a certain extent. The most significant disintegration effects were recorded at a length of two centimeters. We had performed PIV measurements for this distance, as shown in Fig. 3 and Fig. 4.

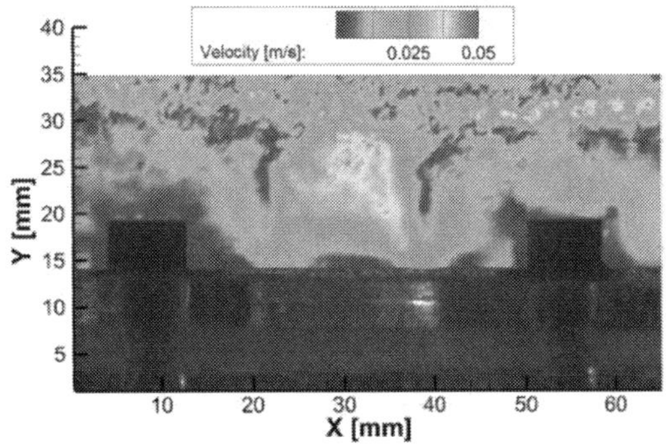

Fig. 3. Scalar map of velocities obtained using the PIV method.

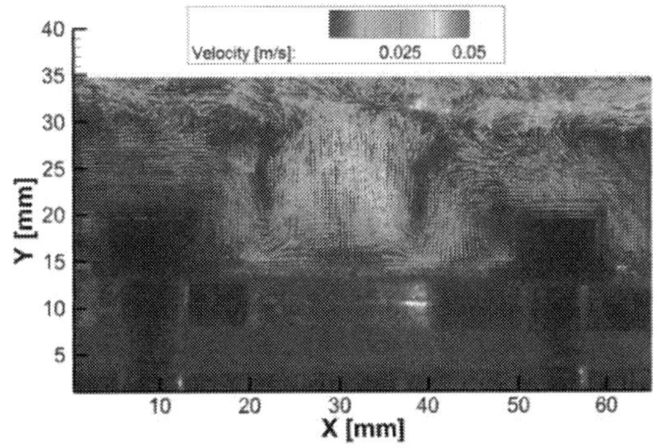

Fig. 4. Vector map of velocities obtained using the PIV method.

Fig. 3 and Fig. 4 shows the liquid's flow velocities from the one ultrasonic transducer towards the water surface on which the bacterial foam is spread. We can infer that the flow velocities between the ultrasonic transducer and the water surface reach values of about 0.04 $m \cdot s^{-1}$. It is important to emphasize that the device contains five piezoelectric atomizing transducers located at the vessel's bottom.

Fig. 5 shows the foam-forming bacteria on the water surface before the ultrasonic cavitation. The ultrasound transducer was set at a frequency of 1.65 MHz and was applied to the foam-forming bacteria for one minute. Fig. 6 shows foam-forming bacteria after ultrasound application, which can be observed as affecting fibrous bacteria's viability.

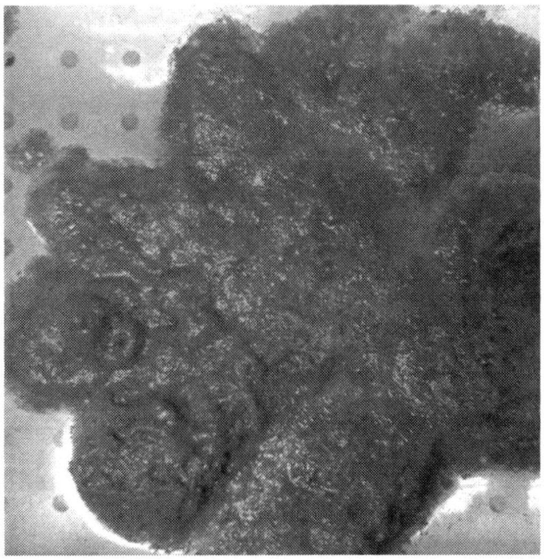

Fig. 5. The photograph was taken before ultrasonic cavitation.

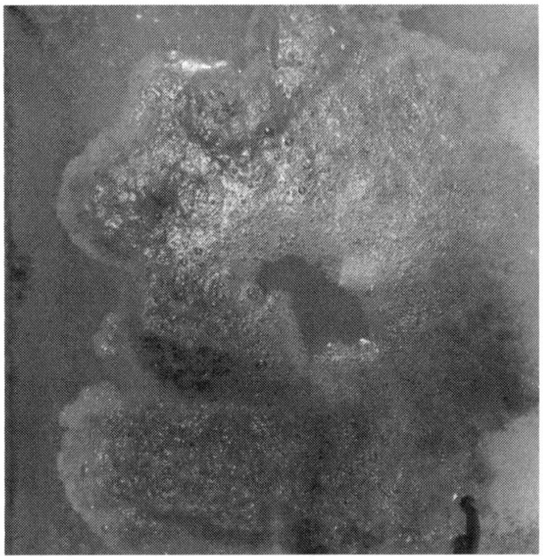

Fig. 6. The photograph was taken after ultrasonic cavitation.

Despite these results in Fig. 6, whether the bacteria were wholly eliminated was uncleared. We evaluated live and dead bacteria for this reason. The evaluation was done using confocal microscopy of foam-forming bacteria. The green color indicates live and red color dead bacteria (Microthrix, Gordonia, Skermania) contained in the foam-forming bacteria in Fig. 7 and Fig. 8.

Fig. 7 shows the record of live/dead confocal microscopy. It's a reference sample of a foam-forming bacteria that was not affected by ultrasound. While Fig. 8 shows the effects of five ultrasonic transducers after one minute of operation, and as can be seen, many bacteria have been eliminated.

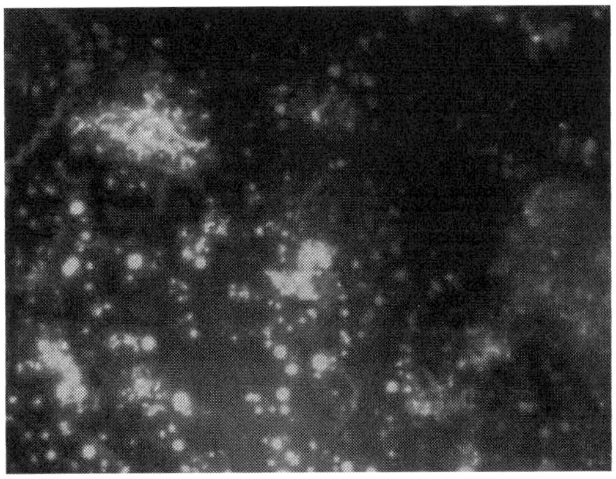

Fig. 7. The evaluation of live/dead bacteria, reference image before the ultrasound.

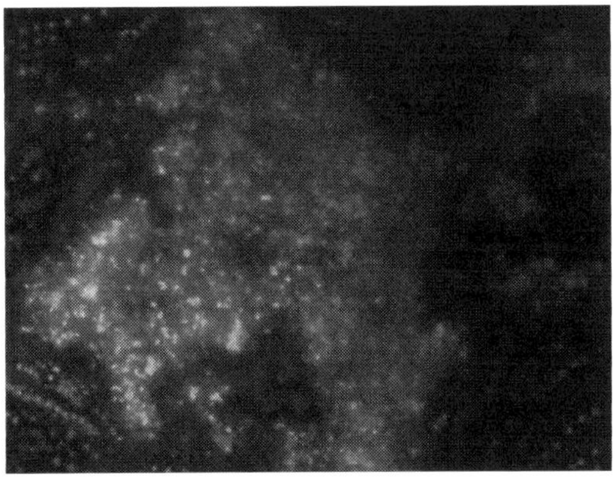

Fig. 8. The evaluation of live/dead bacteria after one minute of ultrasound.

4 Conclusion

Visualization is an essential and helpful tool in the study of fluid flow. Here we used the PIV method to calculate the flow velocity, which is shown using a vector field. Furthermore, the interaction of flow with the water surface was determined.

Ultrasonic cavitation was used to disintegrate foam-forming bacteria and determine the system's most suitable and effective input parameters. The viability of the foam-forming bacteria (Microthrix, Gordonia, Skermania) depending on the properties of the setting ultrasound parameters were studied.

During our research, it was shown that the ultrasonic effect followed by cavitation disintegrates bacteria. The cavitation effect plays an essential role in the disintegration of bacteria because we were used very low power of acoustic intensity; the result was totally evident in Fig. 6 and Fig. 8.

The effect on the viability of the fibrous bacteria was most pronounced at a distance of 2 cm. Fiber viability was not wholly suppressed, but we were observed bacterial deformation and total loss of viability.

Acknowledgements.

This work was supported by the Ministry of Education, Youth and Sports of the Czech Republic and the European Union - European Structural and Investment Funds in the frames of Operational Programme Research, Development and Education - project Hybrid Materials for Hierarchical Structures (HyHi, Reg. No. CZ.02.1.01/0.0/0.0/16_019/0000843), and this work was partly supported by the Student Grant Competition of the Technical University of Liberec under the project No. SGS-2020-3059.

References

1. D. Jasikova, *Proceedings of SPIE*, **10151**, (2016)
2. D. Jasikova, *EPJ Web of Conferences*, **143**, (2017)
3. D. Jasikova, *EPJ Web of Conferences*, **180**, (2018)
4. P. Schovanec, W. Garen, *ACC Jrnl* **25**, 58 (2019)
5. P. Schovanec, D. Jasikova, M. Kotek, V. Kopecky, *Proceedings of conference EFM 2019*
6. P. Schovanec, D. Jasikova, *MATEC Web of Conf.*, **328**, 05003 (2020)
7. Ch. E. Brennen, *Cavitation and bubble dynamics*, (Cambridge University Press, 2015)
8. P. K. Kennedy, *Prog.QuantElectr.* **21**, 3, (1997)
9. Vogel, A., *App. Physics B: Lasers and Optics* **68**, 271-280 (1999)
10. W. Lauterborn, *Archives of Acoustics* **33**, (2008)
11. K. Havlicek, L. Svobodova, *Materials & Design*, **194**, (2020)
12. J. Nemecek, M. Nechanicka, et al. *Chemosphere*, **237** (2019)
13. K. Havlicek, M. Nechanicka, *VTEI*, **62**, (2020)
14. J. B. Winterburn, P. J. Martin, *Asia-Pacific Jrnl of Chem. Eng*, **4**, 184-190 (2009)

Aerodynamic and dynamic loading in a blade cascade designed for flutter research

Petr Šidlof[1*], *David* Šimurda[2], *Jan* Lepicovsky[2], and *Martin* Štěpán[1]

[1]Technical University of Liberec, NTI FM, Studentská 2, 461 17 Liberec 1, Czech Republic
[2]Institute of Thermomechanics, Czech Academy of Sciences, Dolejškova 5, 182 00 Prague 8, Czech Republic

Abstract. Flow-induced vibration of turbine and compressor blades, so called blade flutter, represents a serious problem for designers and operators of large turbomachines. The research of mechanisms leading to this dangerous aeroelastic instability, which can occur especially in modern long and slender blades, is hindered by lack of experimental data. A new experimental setup for controlled flutter testing has been designed in cooperation of the Institute of Thermomechanics of the Czech Academy of Sciences and Faculty of Mechatronics of the Technical University of Liberec. The test section consists of five planar blades placed in a transonic wind tunnel, with high-frequency torsional oscillation of the middle blade driven by an electric motor. The contribution presents the results of first measurements, namely the static pressure distribution for various inlet Mach numbers, aerodynamic moments and deformation of the middle blade due to inertial loads during high-frequency oscillation.

1 Introduction

Flow-induced vibration of turbine and compressor blades, so called blade flutter, represents a serious problem for designers and operators of large turbomachines [1]. The research of mechanisms leading to this dangerous aeroelastic instability, which can occur especially in modern long and slender blades, is hindered by lack of experimental data.

Building on previous experience with the NASA Transonic Flutter Cascade [2], a new experimental setup for controlled flutter testing has been designed in cooperation of the Institute of Thermomechanics of the Czech Academy of Sciences and Faculty of Mechatronics of the Technical University of Liberec during last two years [3]. Together with numerical simulations [4], the setup will serve for investigation of the aerodynamic and structural conditions leading to this dangerous aeroelastic phenomenon. The current contribution reports on first measurements of the pressure distribution, aerodynamic moments and deformation of the middle blade due to inertial loads during high-frequency oscillation.

2 Experimental setup

2.1 Mechanical setup

The experimental setup consists of a test section, which has a width of 160 mm (along the blade span) and height 195 mm mounted in the suction-type high-speed modular wind tunnel of the Institute of Thermomechanics (see Fig. 1). The blade cascade consists of five flat blades with rounded leading and trailing edges. The chord length of the blades is $c = 120$ mm and thickness $h = 5$ mm. The geometry of the blade cascade and location of the static pressure probes monitoring the inlet and outlet isentropic Mach number are shown in Fig. 2.

Figure 1. Test section mounted in the wind tunnel.

The incidence angles of individual blades can be varied, while the stagger angle of the entire cascade stays the same. Therefore, the torsional deformation of real blades can be readily simulated. The middle blade is driven by a high-speed electric motor and can be either fixed with a specific incidence angle, or undergo forced torsional oscillation with amplitudes $1 - 3$ deg and frequencies up to 200 Hz. More details can be found in [3,4].

* Corresponding author: petr.sidlof@tul.cz

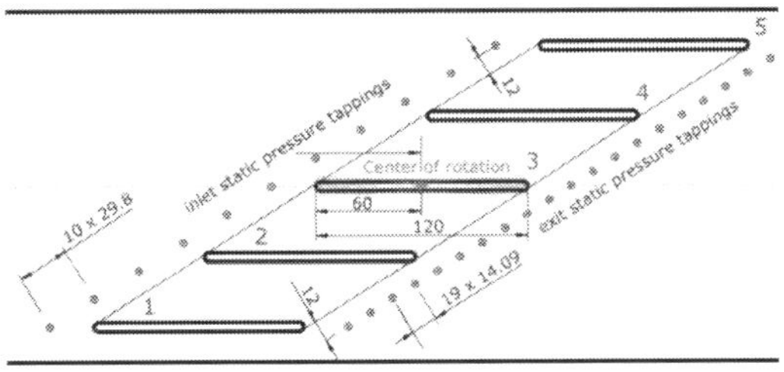

Figure 2. Geometry of the blade cascade.

2.2 Pressure, torque and motion sensors

In addition to monitoring the static pressures upstream and downstream of the blade cascade, the blades are equipped with static pressure probes, whose distribution on the blade can be seen in Fig. 3. The blade can be rotated by 180 degrees to measure the static pressure distribution both on the upper and lower surface.

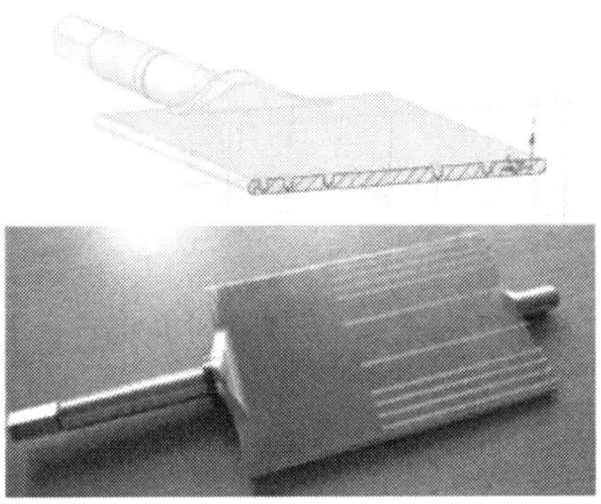

Figure 3. Static pressure probes mounted in the blades.

To monitor the torque, a full strain-gauge bridge is installed on the blade drive shaft. In the case of static blades, the strain gauges measure the total aerodynamic moment. When the middle blade undergoes forced torsional oscillation, the strain gauge signal contains both the aerodynamic and inertial torque, caused by the mass (moment of inertia) of the blade.

The position and deformation of the blade during forced oscillation is measured by a laser triangulation sensor (MicroEpsilon ILD 2310-40), monitoring the vertical displacement of the leading edge in the middle of the channel.

Figure 4. Stain gauge bridges mounted on blades No. 3 and 4.

3 Results

3.1 Static pressure measurement

The results of static pressure distribution over the blade surface are reported for three different flow regimes with outlet isentropic Mach numbers ranging between $M_{2is} = 0.35 - 0.87$. In all these cases, the static incidence angle of the middle blade was $\alpha = -3^o$ (leading edge directed downwards). The results in Fig. 5 show a zone of low pressure on the lower surface close to the leading edge, whose value increases with the increasing flow velocity. Integration of the pressure values over the whole surface yields the total aerodynamic moment, listed in Tab. 1.

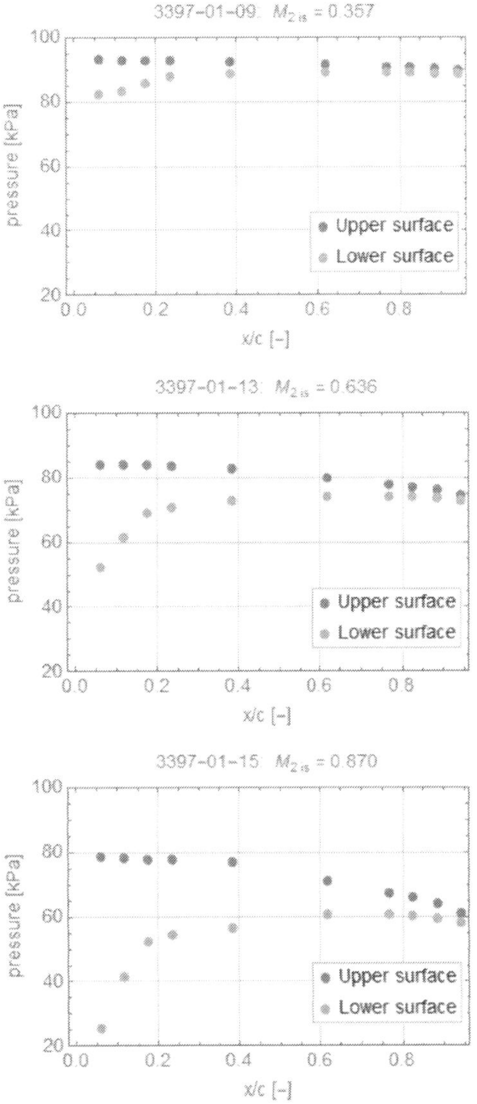

Figure 5. Distribution of the static pressure along the upper and lower surface of the middle blade for three flow regimes (values of the outlet isentropic Mach number M_{2is}).

Table 1. Aerodynamic torque on the middle blade evaluated from static pressure measurements.

ID	M1is [-]	M2is [-]	Blade static angle [deg]	Aerodynamic moment [N.m]
3997-01-09	0.3	0.357	-3	1.76
3397-01-13	0.499	0.636	-3	4.96
3397-01-15	0.596	0.870	-3	8.54

3.2 Measurement of aerodynamic torque by strain gauges

In a different set of measurements, the aerodynamic torque on the middle blade fixed with a static incidence angle $\alpha = -3^{\circ}$ was measured for the same three flow regimes using the full strain gauge bridge installed on the blade trunnion, as described in section 2.2. The results are summarized in Tab. 2. The data agree well both in the trend and in the order of magnitude. However, compared to the reconstructed values of aerodynamic moment in Tab. 1, the values measured by strain gauges are 30-40% lower.

Table 2. Aerodynamic torque on the middle blade measured by strain gauges for three flow regimes.

ID	M1is [-]	M2is [-]	Blade static angle [deg]	Aerodynamic moment [N.m]
3397-03	0.3	0.355	-3	1.01
3397-04	0.5	0.644	-3	3.36
3397-05	0.6	0.880	-3	5.18

3.3 Measurement of blade oscillation and inertial torque in still-air conditions

When the middle blade undergoes forced torsional oscillation, the strain gauges register not only the aerodynamic moments, but also the inertial torque. In order to assess the order of magnitude of these two components, a series of measurements was realized in still-air (zero flow velocity) conditions with a prescribed amplitude at the root of the blade $\alpha_0 = 1^{\circ}$ and frequency range $f = 10 - 210$ Hz. Despite the high stiffness of the structure, as the frequency increases and approaches the first natural frequency of the blade, the blade starts to twist. The waveforms and amplitudes of the leading edge deflection for increasing frequency are reported in Fig. 6.

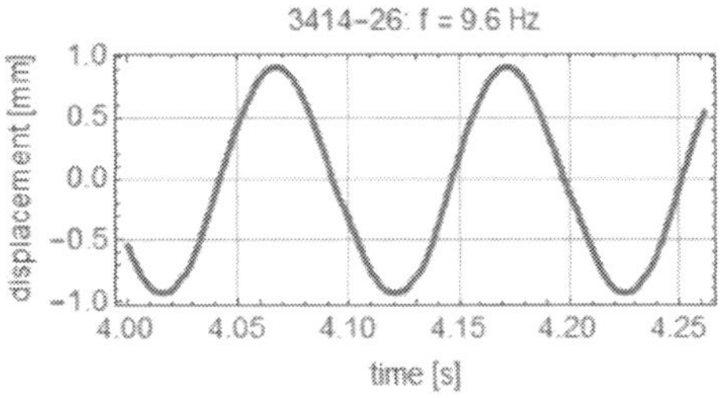

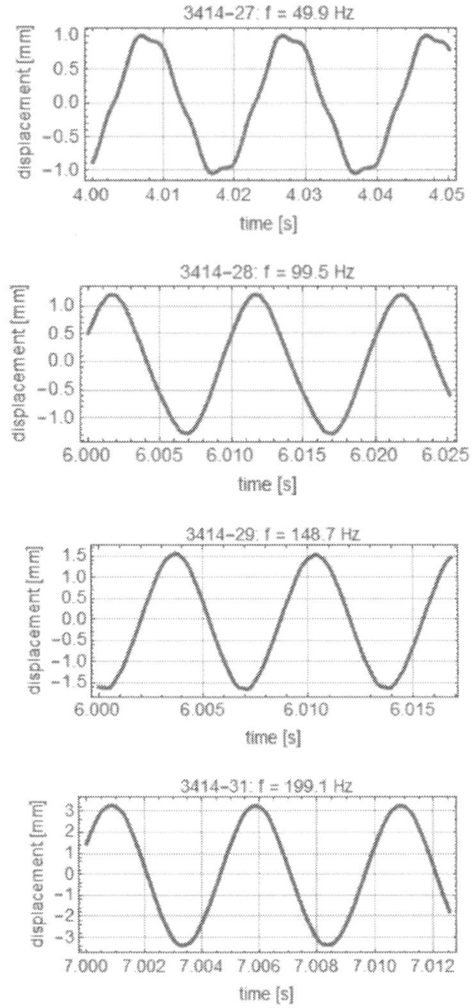

Figure 6. Deflection of the blade leading edge during oscillations in the frequency range $f = 10 - 200$ Hz.

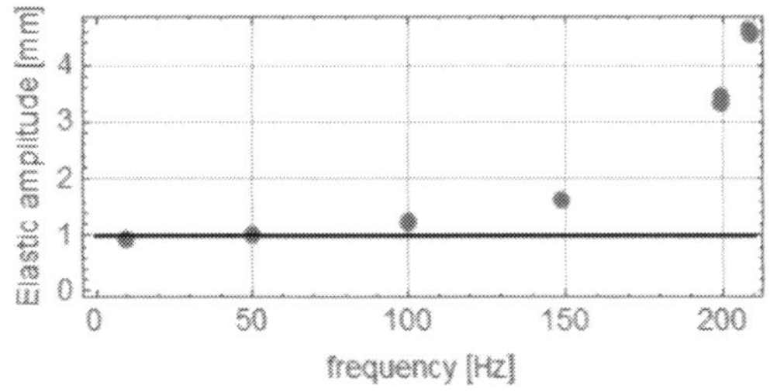

Figure 7. Amplitude of the blade oscillation measured as a function of frequency of vibration. Black line corresponds to the amplitude of a perfectly rigid blade.

For low frequency ($f = 10$ Hz), the blade oscillates as an almost perfect rigid body with an amplitude of 1 mm. At higher frequencies, however, the amplitudes sharply increase (see Fig. 7).

The signals from the strain gauge are shown in Fig. 8. For low frequencies, the torque waveform is not harmonic, which is caused mainly by the radial clearance in the driving mechanism. For higher oscillation frequencies, the inertial torque quickly increases (see also Fig. 9) and approaches the expected harmonic waveform.

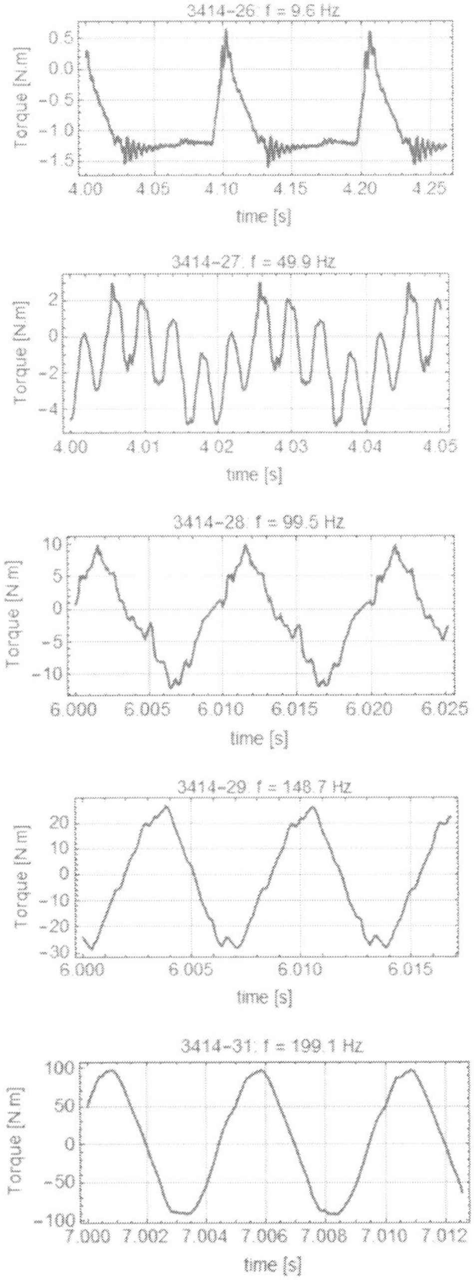

Figure 8. Torque measured by the strain gauges during oscillations in the frequency range $f = 10 - 200$ Hz in still-air conditions.

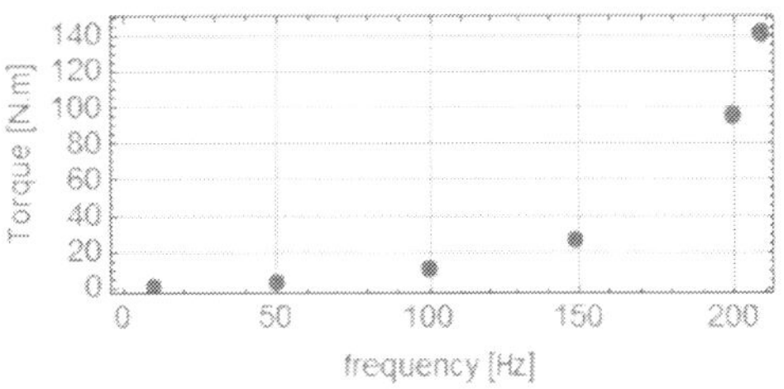

Figure 9. Amplitude of the torque as a function of frequency of vibration.

4 Discussion and conclusions

Using a new setup for blade flutter investigation, the aerodynamic moments were measured using two methods and compared to the inertial torque. In this first test, the torque measured by the strain gauges was 30-40% lower than the aerodynamic moment evaluated from static pressure measurements on the blade surface at identical flow conditions. Presumably, this was caused by inaccurate calibration of the strain gauges, and also by the fact that the number of pressure probes is limited and integration over the blade surface introduces errors.

For static incidence angles ranging between $-3^o\,..+3^o$ and outlet Mach numbers $M_{2is} = 0.3 - 0.6$, the aerodynamic moment is $1 - 6\,\mathrm{N\cdot m}$. For lower frequencies of oscillation up to $f = 100$ Hz, this is comparable to the torque due to inertia. For higher frequencies, however, the inertial torque quickly increases and exceeds the aerodynamic moments by one or nearly two orders of magnitude. The rate of increase is higher than quadratic – this is because the frequency of the forced oscillation already approaches the lowest natural frequency of the blade. For upcoming measurements of aerodynamic moments during blade oscillation, it will be challenging to separate the inertial and aerodynamic components of the torque signal.

Acknowledgements

The research has been supported by the Czech Science Foundation, project 20-11537S "*Experimental research on flutter excitation function in turbomachinery*".

References

1. I. McBean, K. Hourigan, M. Thompson, F. Liu, Journal of Fluids Engineering **127** (2005)
2. J. Lepicovsky, E.R. McFarland, V.R. Capece, T.A. Jett, R.G. Senyitko, NASA TM-211894 (2002)
3. Lepicovsky, P. Šidlof, D. Šimurda, M. Štěpán, M. Luxa, AIP Conf. Proc. 2323 (2021)
4. J. Fürst, M. Lasota, J. Lepicovsky, J. Musil, J. Pech, P. Šidlof, D. Šimurda, Processes **9**, 1974 (**2021**)

Pressure losses reduction in a steam turbine compact valve

Vaclav Slama[1, 2*], *David* Simurda[3], *Ladislav* Tajc[2], *Bartolomej* Rudas[1], *Jindrich* Hala[3], *Martin* Luxa[3] and *Tomas* Radnic[3]

[1]Doosan Skoda Power s.r.o., Tylova 1/57, 301 28 Pilsen, Czech Republic
[2]University of West Bohemia, Power System Engineering Department, Univerzitni 8, 306 14 Pilsen, Czech Republic
[3]Institute of Thermomechanics of the Czech Academy of Sciences, Dolejskova 1402/5, 182 00 Praha 8, Czech Republic

Abstract. Valves are integral parts of modern steam turbines. They provide flexible operations and fast load changes which have become more often due to the increasing share of renewable energy. Along with this, the low pressure loss is required for guaranteed operation conditions. This paper is concerned with investigation of pressure losses in the compact valve, which is used as a valve for the intermediate-pressure turbine part, and its geometry modification to lower overall pressure losses. The investigation is based on results of numerical simulations which were carried out in the Doosan Skoda Power Company using a package of ANSYS software tools. Source of increased pressure losses in the compact valve is identified and a way of its reduction is proposed based on the comparison of the current results with data from a typical control valve model. Data analysis is provided in the paper.

1 Introduction

The energy industry cannot reasonably survive without modern steam turbines with high flexibility and efficiency [1]. Therefore, steam turbines have to be enhanced in all their parts including valves. As can be found in different sources [2-6], valves for steam turbines have become objects for a wide research and development. Doosan Skoda Power along with other external cooperative organizations, in this case with the Institute of Thermomechanics of the Czech Academy of Sciences, has also a strong tradition of valves research [7-10].

Fig. 1. Compact valve assembly [11].

The current research activities include namely the investigation of flow field in complex modern valve assemblies with focus on pressure losses as well as pressure fluctuations. Once the performance of an existing valve assembly is assessed, the attempt to reduce the pressure loss can be done.

Wide range of measurements on the current valve assembly with the compact control valve (Fig. 1) has been already carried out and published in [11, 12]. In these publications, details of comparisons between numerical simulations and experimental data have been discussed as well. It was concluded that the differences are reasonable and numerical simulations are suitable for further studies. As a consequence, the attempt to change the valve geometry to reduce pressure loss was done and it is presented in this paper.

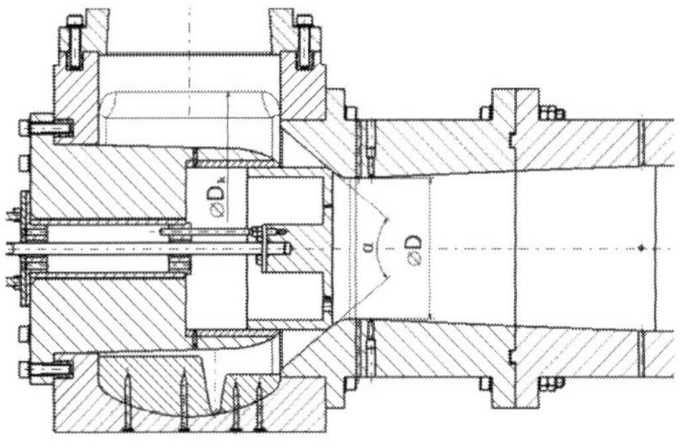

Fig. 2. Cross-section through the compact valve [12].

Since the volumetric flow rate of steam is significantly greater in the intermediate-pressure turbine part in compare with the high-pressure turbine part, the valves have to be also larger. Due to this, the valves are more expensive. In order to keep production costs low, the compact valve assemblies have been designed. The compact means that the ratio between D_k/D, see Fig. 2, is smaller than in typical control valves. For the same

* Corresponding author: vaclav.slama@doosan.com

diameter D, it is shown in Fig. 3. Since D_k is smaller, a free space for steam in the control valve chamber is also smaller. It was identified as a possible reason why pressure losses are greater than in typical control valves [12].

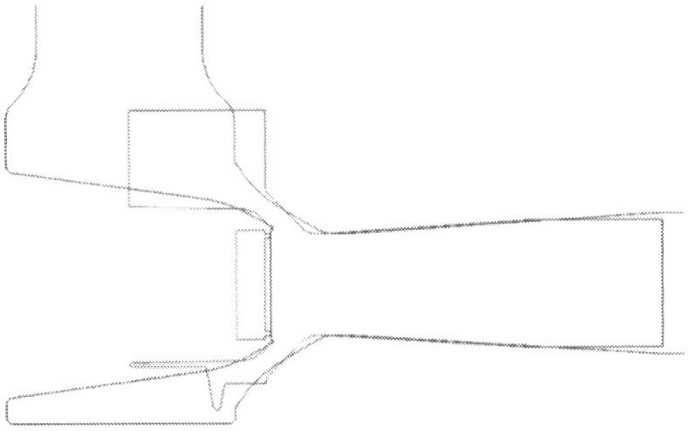

Fig. 3. Relative comparison of cross-section through the valve chamber, a typical control valve (blue) and a compact control valve (red) [12].

Dimensions D_k, α and D (see Fig. 2) cannot be enlarged if the valve has to be compact. However, the valve cone suspension can be changed, thus the space in the valve chamber can be enlarged. It is shown in numerical models (Figs. 5 and 6) and results with pressure flow fields (Figs. 8 and 9) where the space for the flow field is greater in the new chamber. There is assumption that this may reduce pressure losses. Thanks to the numerical simulations, it can be validated.

2 Numerical model

The numerical models were created according to the experimental one. The original compact valve is shown in Fig. 4. It consists of inlet and outlet pipelines which are shown in the blue colour, one stop valve which is shown in the red colour and one control valve which is shown in the green colour. The model with the changed control valve is shown in Fig. 5. It can be seen that the valve cone suspension is smaller therefore the free space in the valve chamber is greater.

A hexahedral as well as tetrahedral meshes, see Fig. 6, were used to create the numerical model. The typical mesh quality criteria and also the recommendation of the turbulence model to keep y+<5 was satisfied [12].

The flowing medium was defined as dry air with thermodynamic parameters according to the experiment. A high resolution scheme was used to compute the steady state RANS solution with a compressible turbulent model k-ω SST. Inlet and outlet boundary conditions were specified according to measured data [11, 12]. For this study, ten different boundary conditions for two different control valve models were calculated. In all cases, the control valve is fully open.

Simulation convergence criteria were quantified by 10^{-4} residual target value. Furthermore, the global mass flow rate imbalances and imbalances of required pressure losses were also monitored to be less than 0.1%.

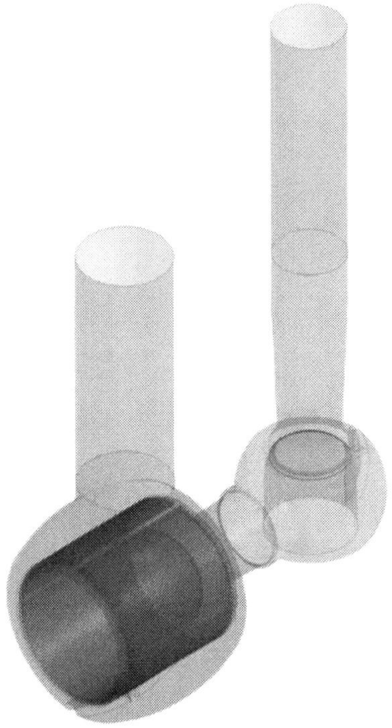

Fig. 4. Compact valve model assembly with the original control valve chamber.

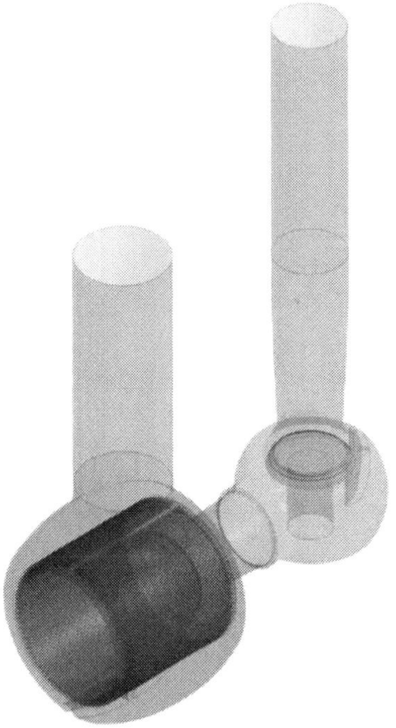

Fig. 5. Compact valve model assembly with the control valve chamber with greater space.

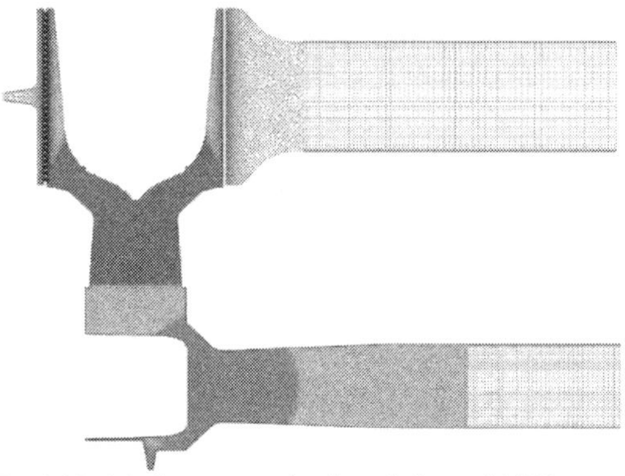

Fig. 6. Mesh in the cross-section through the model [12].

3 Results and discussion

The evaluation of the calculated results can be depicted in the form of valve characteristics. It is widely used to generalize results for pressure loss models for control valves. A valve characteristic is a graph of the dependency between pressure ratio $\varepsilon = p_2/p_1$ and mass flow ratio $q = Q/Q_{cr}$ for different valve cone lifts. In this study, only the case with maximal valve cone lift is presented. More valve cone lifts are presented in [12].

p_2 is the pressure downstream of the control valve (at the interface between the control valve domain outlet and the outlet pipeline inlet, see Fig. 4), p_1 is the pressure upstream of the control valve (at the interface between the stop valve domain outlet and the control valve domain inlet, see Fig. 4), Q is the mass flow rate through the valve and Q_{cr} is the critical mass flow rate which is defined as:

$$Q_{cr} = A_{dif} \cdot \sqrt{\frac{p_{1t}}{v_{1t}}} \cdot \sqrt{\gamma \cdot \left(\frac{2}{\gamma+1}\right)^{\frac{\gamma+1}{\gamma-1}}} \tag{1}$$

A_{dif} is the cross-sectional area of the control valve diffuser throat with diameter D (see Fig. 2), p_{1t} is the inlet total pressure, v_{1t} is the inlet total specific volume and γ is the heat capacity ratio.

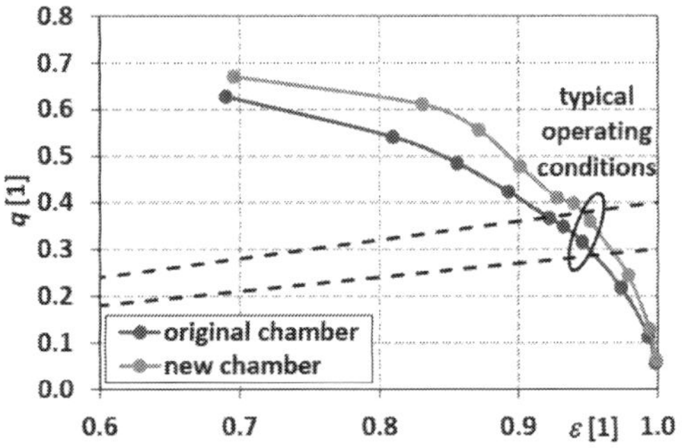

Fig. 7. Comparison of valve characteristics for fully open valve.

The comparison of calculated valve characteristics is shown in Fig. 7. It can be seen that for the same pressure ratio, the mass flow ratio is greater in the valve with the new chamber. It means that the pressure loss is smaller. In practice, for the required mass flow rate, the pressure ratio is greater which presents a lower pressure loss.

For the typical operating conditions (see Fig. 7), the pressure flow field in the cross-section of the original valve chamber and the new chamber is shown in Fig. 8 and Fig. 9, respectively. It can be seen that there is greater asymmetry and significant vortex under the cone in the valve with the original chamber (see Fig. 8). On the contrary, the flow field is more symmetric and without significant vortex in the chamber with the greater space. Due to the greater space, the flow can be more stabilized and equalized before going under the cone to the diffuser. As a result, the pressure losses are lower. It was detected in all cases with different mass flow ratios (see Fig. 7).

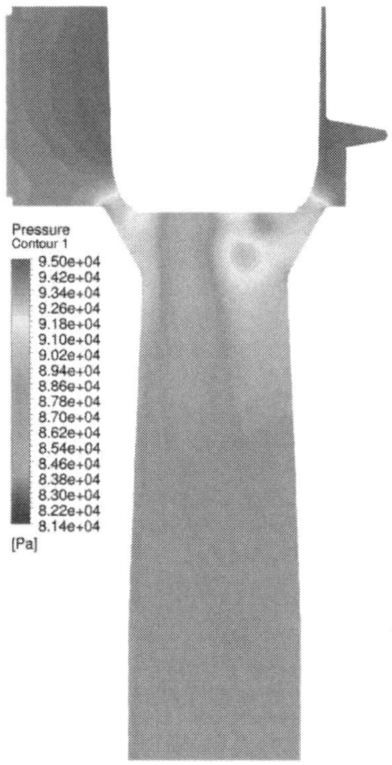

Fig. 8. Pressure flow field in the compact valve with the original chamber.

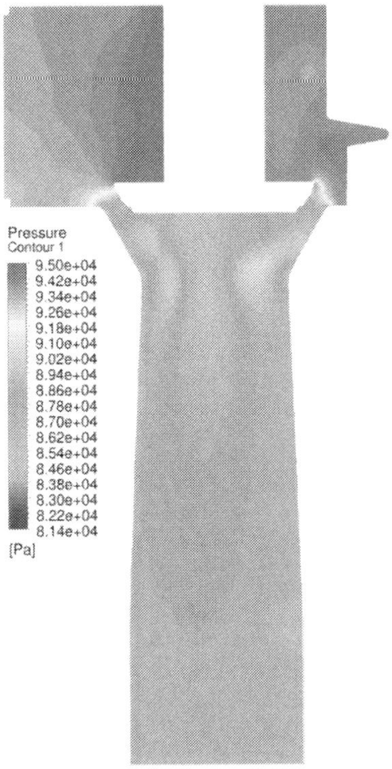

Fig. 9. Pressure flow field in the compact valve with the new chamber.

4 Conclusions

There is a potential in reducing of pressure losses in the modern compact valve. Thanks to the numerical simulations it was found out that the pressure loss is reduced when the space around the valve cone in the valve chamber is enlarged. It can be achieved by only changing the valve suspension, therefore keeping the valve to be compact.

This research could be done thanks to the Technology Agency of the Czech Republic, National Center for Energetics TN01000007/05. The authors also thank the Doosan Skoda Power Company and the Czech Academy of Sciences, Institute of Thermomechanics for giving a permission to publish this paper.

References

1. D. G. Wilson, T. Korakianitis, *The design of high-efficiency turbomachinery and gas turbines*, The MIT Press (2014)
2. M. Clari, T. Polklas, F. Joos, Three dimensional flow in the diffuser of a steam turbine control valve, *Turbo Expo: Power for Land, Sea, and Air*, GT2011-45617, pp. 2327-2334 (2012)
3. Ch. Windemuth, M. Lange, R. Mailach, Investigation of unsteady pressure fluctuations in a simplified steam turbine control valve, *J. Eng. Gas Turbines Power*, GTP-20-1603, 143(8): 081017 (2021)
4. F. Suzuki, S. Nishida, S. Fukao, M. Tsuruta, Verification of combined main steam valve pressure distribution and vibration characteristics with downscale model test on air condition, *Turbo Expo: Power for Land, Sea, and Air*, GT2019-90406, V008T29A009 (2019)
5. S. Wallat, B. C. Domnick, C. Munch, D. Brillert, A test rig concept to study fluid structure interactions in a steam turbine valve, *Turbo Expo: Power for Land, Sea, and Air*, GT2018-75094, V008T29A003 (2018)
6. Peng Wang, Hongyu Ma, Yingzheng Liu, POD and extended-POD analysis of pressure fluctuations and vortex structures inside a steam turbine control valve, *Turbo Expo: Power for Land, Sea, and Air*, GT2018-75312, V008T29A011 (2018)
7. P. Kollross, L. Tajc, An Experimental research of the DSPW steam turbine control valve, *AIP Conference Proceedings* **2189**, 020012 (2019)
8. V. Slama, L. Mrozek, B. Rudas, D. Simurda, J. Hala, M. Luxa, Experimental and numerical study on pressure losses and flow fluctuations in a high-pressure valve assembly of steam turbine governing system, *Turbo Expo: Power for Land, Sea, and Air*, GT2020-14474, V009T23A007 (2021)
9. V. Slama, L. Mrozek, L. Tajc, M. Klimko, P. Zitek, Flow analysis in a steam turbine control valve with through-flow valve chamber, *ASME J of Nuclear Rad Sci*, 7(2): 021401 (2021)
10. L. Mrozek, V. Slama, K. Sedlak, L. Tajc, Analysis of losses in steam turbine control valves, *AIP Conference Proceedings* **2323**, 060003 (2021)
11. V. Slama, L. Mrozek, B. Rudas, D. Simurda, J. Hala, T. Radnic, Investigation of pressure losses and flow fluctuations in an intercept valve assembly of an intermediate-pressure turbine part, *ASME IMECE*, IMECE2020-23359, V008T08A026 (2021)
12. V. Slama, D. Simurda, L. Mrozek, L. Tajc, J. Hala, T. Radnic, Pressure losses and oscillations in a compact valve of a steam turbine, *MATEC Web of Conferences* **345**, 00027 (2021)

Counter flow atomizer: Effect of the area ratio of the outlet orifice and the inlet air canal

Karolína Smutková[1*], *Milan* Malý[1], *Ondřej* Cejpek[1], and *Jan* Jedelský[1]

[1]Brno University of Technology, Faculty of Mechanical Engineering, Czech Republic

Abstract. The need to reduce the energy demand of processes also creates pressure to make atomizers more efficient. A promising development path could be counterflow atomizers (CFA). An atomizer, or synonymously a nozzle, is a device that turns a liquid into a spray. The CFA is a twin-fluid type of atomizer because both the liquid and the atomizing gas are fed into it. Initial results suggest that CFA is capable of the same quality of atomization but using half the air mass flow rate compared to conventional twin-fluids atomizers when operated at identical inlet pressure. This means half the energy requirements with the same efficiency. This atomizer also shows a great promise in the atomization of highly viscous substances such as waste-based fuels and biomass oils. In CFA, the air expands twice; first, at the discharge from the air inlet canal into the mixing tube, and second, at the discharge from the atomizer to the surrounding atmosphere. Therefore, one of the main control parameters is the area ratio of the exit orifice and the air inlet canal. This study experimentally investigates the effect of this ratio on the spray quality for two different CFA atomizers using Phase Doppler Anemometry (PDA), which provides the velocity and size of individual droplet in the spray. The atomizers were operated at the air inlet pressure of 100 and 200 kPa and gas-to-liquid mass (GLR) ratios of 5, 10 and 20%. The effect of the double expansion can be well observed in the pressure differences between the air inlet and the pressure inside the mixing tube. The length of the air counterflow insert had a significant effect on the atomizer performance. For the shorter counterflow channel, a minimal effect on flow was observed; this atomizer behaved like a conventional twin-fluid atomizer and all expansions occurred downstream of the exit orifice. The longer counterflow canal changed the internal expansion ratios, larger droplets, wider spray and higher discharge coefficients (Cd) were obtained.

1 Introduction

At the present time, when the pressure for low energy consumption and renewability, which is often associated with higher viscosity in liquids, is the most intense in the history of mankind, the counterflow atomizers seem to be one of the promising paths for development. For CFA, it is interesting to note that there are only brief references to them from the 20th century and only few recent articles.

Hoxie et al. [1] compared the characteristics of CFA against commercially available internal mixing air assist atomizers (COM). They used water and glycol, and exit diameters close to ours. They designed the CFA in Fig. 1 where the geometry is without our more significant mixing area. Drops were imaged using a standard shadowgraph technique utilizing a pulsed laser as a light source to provide a diffuse background.

The CFA with a larger exit orifice showed large differences in performance at the lowest pressure tested compared to the COM in terms of sensitivity to the change in GLR, the D32 dropped by 35% for a change of 10% in GLR. For the more viscous medium, the CFA produced a narrower droplet distribution. Compared to the water measurements, the drop in GLR was accompanied by a drop in ID32.

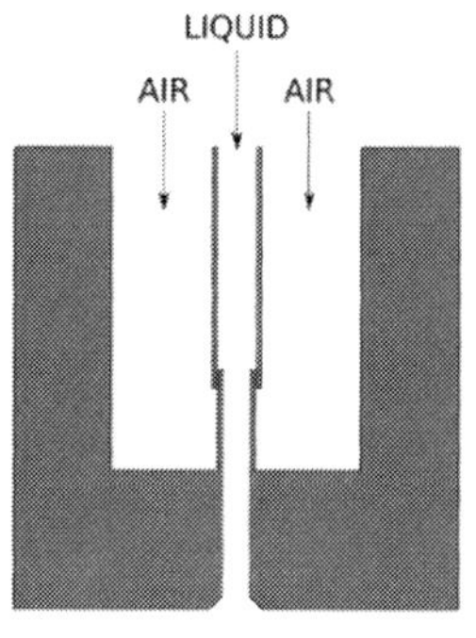

Fig. 1. A. Hoxie et al - CF atomizer.

The most significant outcome of their study is that, even for large flow rates, d32 values remain at or less than 20 microns, despite using GLR values less than 20%. The high sensitivity of the d32 to GLR suggests that much smaller values of d32 may be attained without significant increases in GLR. Efficiency over a COM atomizer is increased by nearly 100%. The authors describe this performance trend as increasing with increasing viscosity, but they do not know the reason for this phenomenon.

* Corresponding author: 200171@vutbr.cz

Rangarajan et al. [2] also studied CFA that they compared to the 'flow-blurring' (FB) atomizer studied byother investigators for four test liquids of viscosity ranging from 1 to 133.5 mPa.s. Both atomizers are shown in Figure 2. Detailed measurements of droplet size distributions were carried out using laser shadowgraphy, along with high-speed flow visualization. In their study, the two-dimensional (2D) simulations are conducted by the "compressible Multiphase Inter Foam" solver in the OpenFOAM-6 toolbox. They used water, propylene glycol and glycerol/water (85% - 15% v/v). For a fixed AQ_{liquid} 60 ml/min, the *GLR* was between 25 and 100%.

Regarding the dependence of D_{32} on *GLR*, they obtained similar results to [1]. The CFA produces a spray whose characteristics are relatively insensitive to fluid viscosity over the range studied, for gas-liquid mass flow ratios between 0.25 and 1. They found that the CFA generates a spray with D_{32} that are very weakly sensitive to the changes of viscosity over two orders of magnitude. This is consistent with the hypothesis mentioned in their paper that the counterflowing velocity and density profiles establish a flow with high levels of turbulent stresses. The mixing process appears essentially incompressible, with a scaling that depends primarily on the volumetric flow rates of atomizing air.

High-fidelity simulations suggest that the primary mechanism responsible for spray formation is a Kelvin-Helmoltz-type instability of the liquid jet inside the nozzle, rendering it insensitive to liquid viscosity.

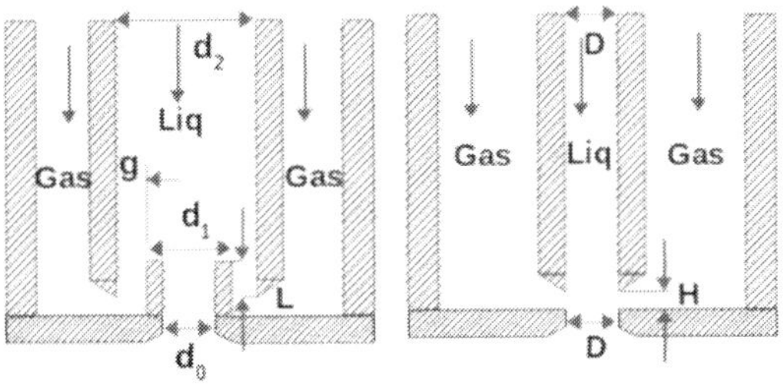

Figure 2. Left - CF atomizer as tested in [2], right – FB atomizer.

In this study, it was focused on a part of the internal geometry of these atomizers, namely the ratio of the areas of the outlet orifice and the air inlet canal, as this is a relatively unexplored issue. As can be seen in Fig. 3, the liquid and gas supply to these atomizers is parallel, the counterflow effect is caused by the internal geometry, more precisely by the air supply channel to the mixing chamber. It was been investigated the dependence of Integral Sauter Mean Diameter (*ISMD; ID*$_{32}$), Spray Cone Angle (*SCA*) and C_d on *GLR* at pressures (p_1) 100 and (p_2) 200 kPa for 2 nozzles differing in air inlet channel length and outlet diameters.

2 Experimental setup

The experimental data included in this paper were acquired using a specially designed test bench in the Spray laboratory at Brno University of Technology. All experiments were performed at room temperature (22 °C). The following paragraphs describe the equipment and instrumentation used and include a short discussion over the atomizers and PDA system.

2.1 Atomizers

For this paper, two CFAs with slightly different geometries were used; their design was based on the expertise of articles dealing with similar issues [1, 2].

For simplicity, markings reflecting the atomizer type and outlet diameter have been introduced: **Dx.y**_CF**z**, where **x.y** is a diameter of exit orifice [mm] and **z** = 1 → shorter counterflow canal; = 2 → longer counterflow canal.

It was focused on the ratio of the areas of the exit orifice and the air inlet canal. These affect the double expansion ratio, which implies the atomization quality and energy requirements. It was hypothesized that, with the lengthening of the counterflow canal, more of the expansion would take place in the mixing chamber, thus achieving better spray parameters compared to conventional atomizers. It was also predicted that the droplet size would increase with the exit orifice.

As seen in Fig. 3. below, our CFA has one axial liquid inlet in the central axis and six axial symmetrically placed air inlets leading into the annulus. The turnover of the jet is caused by the central canal.

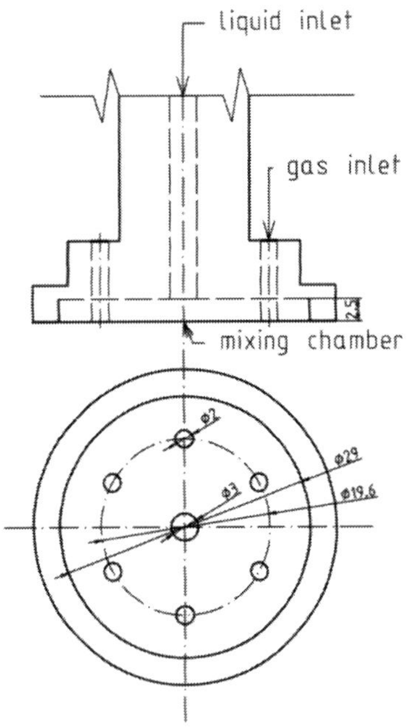

Fig. 3. Body of the counterflow atomizer.

The atomizers differed in the diameter of the outlet opening and the length of the counterflow channel (Fig. 4.). The nozzle with the shorter counterflow canal was manufactured with 1.0, 1.5, and 2 mm exit orifice diameter, while the nozzle with the longer counterflow canal was manufactured with 1.4, 1.7, and 2.1 mm. All the variants were measured at *GLR* of 5, 10 and 20% and air pressures of 100 and 200 kPa.

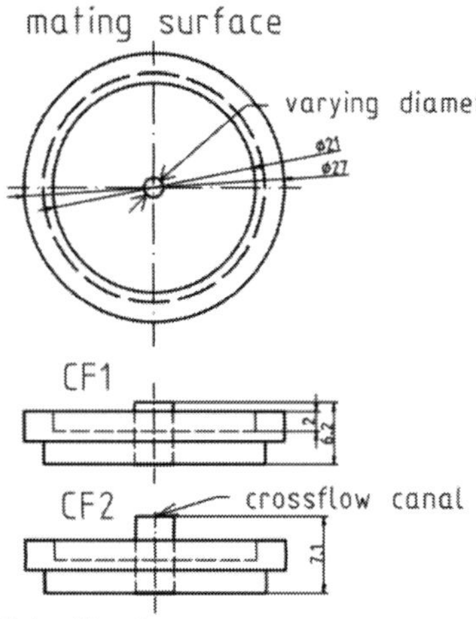

Fig. 4. Exit orifice plate.

2.2 PDA system

The size and velocity of the spray droplets were probed using a two-component fibre-based commercial PDA (Dantec Dynamics A/S Skovlunde, DK), Fig. 5 The PDA is able to acquire the axial, radial or tangential velocity components in the coincidence mode along with the simultaneous drop sizes; in this case, only 1D measurement was used as the radial component was negligible.

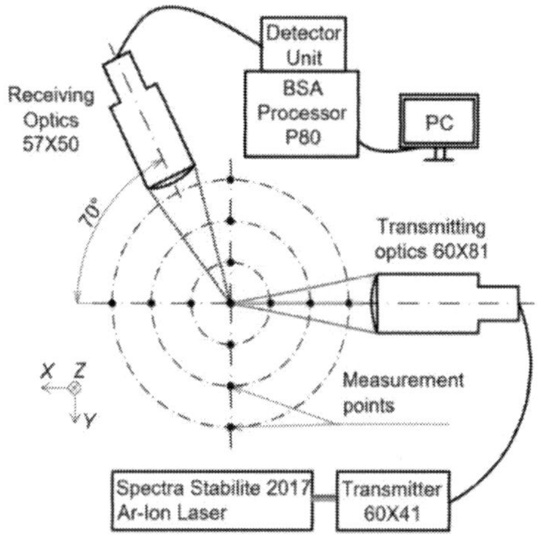

Fig. 5. Setup of the PDA measurement with a coordinate system [4].

Parameter	Value	
Laser power output	0.35 W	
Scattering angle	72°	
Receiver mask	C	
Receiver spatial filter	0.2 mm	
The focal length of transmitting/receiving optics	500/500 mm	
Wavelength	488 nm	514.5 nm
Velocity component	**Axial**	**Radial**
Velocity centre	25.7 m/s	0 m/s
Velocity span	51.4 m/s	24.4 m/s
Sensitivity	950 V	1080 V
SNR	0 dB	0 dB
Signal gain	8 dB	10 dB
Level validation ratio	8	8

Fig. 6. The main system parameters.

The spray was probed at axial distances of Z = 100 mm from the exit orifice along two radially orthogonal axes. Measurements were taken between -27 and 27 at a 3 mm step between two adjacent points. In each measurement point, either 20 000 samples were acquired, or a 12 second acquisition duration was achieved, whichever required the shorter time interval.

The PDA measurement was controlled by the Dantec BSA software 5.2.

For ID_{32} the standard equation was used [1]:

$$ISMD = ID_{32} = \frac{\sum_{i=1}^{n} r_i f_i d_{30,i}^3}{\sum_{i=1}^{n} r_i f_i d_{20,i}^2} \quad (1)$$

where r_i is a radial distance from the spray centreline, f_i is the data-rate, d_{20} is a surface mean diameter and d_{30} is a volume mean diameter.

For *SCA* was used a simplified equation [4]:

$$SCA = 2 \cdot arctg\left(\frac{\frac{90 - Q_{cumulative_{n+1}}}{(Q_{cumulative_n} - Q_{cumulative_{n+1}}) \cdot (p_{n+1} - p_n)} + p_{n+1}}{h}\right) \cdot 180 \cdot \pi \quad (2)$$

Where h is the height coordinate.

The equation for C_d:

$$C_d = \frac{Q_{liquid}}{\frac{\pi \cdot d_{exit\_orifice}^2}{4} \cdot \sqrt{2 \cdot p_{liquid} \cdot p_{liquid}}} \quad (3)$$

https://doi.org/10.1051/epjconf/202226401043

3 Results and discussion

All selected parameters were considered in relation to *GLR*, as it corresponds to the energy operational requirements.

The first discussed parameter is ID_{32}, see Fig. 7 and Fig. 8, as it provides the global representation of the Sauter mean diameter by its mass-weighted averaging over the entire radial profile [3].

The CF1 atomizer produced a spray with lower ID32 compared to CF2 at both operating pressures, see Fig. 7. and Fig. 8. This can be attributed to the flow characteristic in the mixing chamber, which is related to the pressure difference and velocity profile. For all investigated CF1 modes, it can be assumed that the expansion occurs downstream of the exit orifice, since the pressure inside the mixing chamber was identical to the air pressure. For CF2, the expansion ratios varied, reflecting pressure drops of 0.167, 0.29 and 0.592 on average for all outlet diameters, in order. In all cases, the pressure difference grew in proportion to the growth of GLR. It can be concluded that, as the exit orifice increases, more expansion occurs in the mixing chamber.

An approximate relationship reflecting ID_{32} versus *GLR* and the exit orifice was obtained in the following form.

$$ID_{32} = 143 \cdot GLR^{-0.35} \cdot d_{exit\_orifice}^{0.3} \qquad (4)$$

The magnitude of ID_{32} has a decreasing trend with increasing *GLR*. The highest ID_{32} of about 140 µm was obtained at 100 kPa pressure, *GLR* 5 for the CF2 atomizer. The lowest values below ID_{32} of 60 µm were obtained for *GLR*, around 200 kPa pressure and CF1 nozzle with 1.0- and 1.5-mm exit orifice diameter.

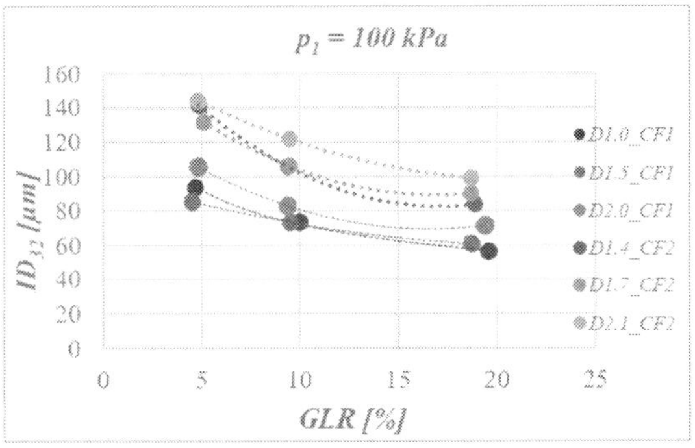

Fig. 7. The influence of *GLR* on ID_{32} for p_1 = 100 kPa.

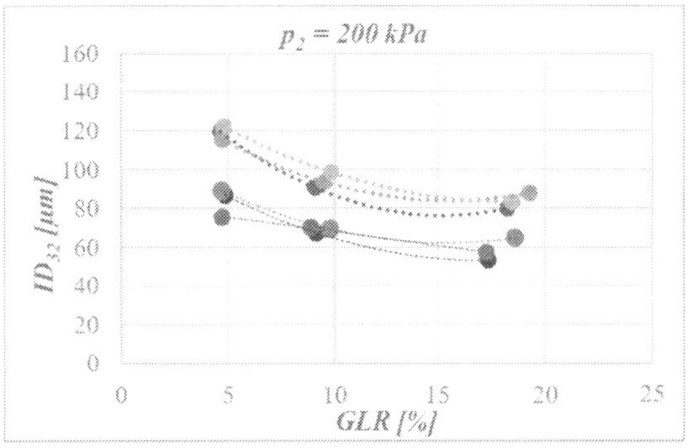

Fig. 8. The influence of *GLR* on ID_{32} for p_2 = 200 kPa.

In general, it was expected an increase in *SCA* with *GLR* for two-fluid atomizers [5]. Looking at Fig. 9 for p_1, this only occurs for the D1.5_CF1 and D1.0 CF1 atomizers, where a slight downward trend for D2.0_CF1 can be seen. For all the CF2 atomizers, a convex trend can be observed. For p_2 (Fig. 10), a larger SCA appears but the trends of the *SCA* are similar to those of p_1 = 100 kPa.

In order to find a definite cause for this trend, it is necessary to test several modes and investigate the potential changes in the internal flow.

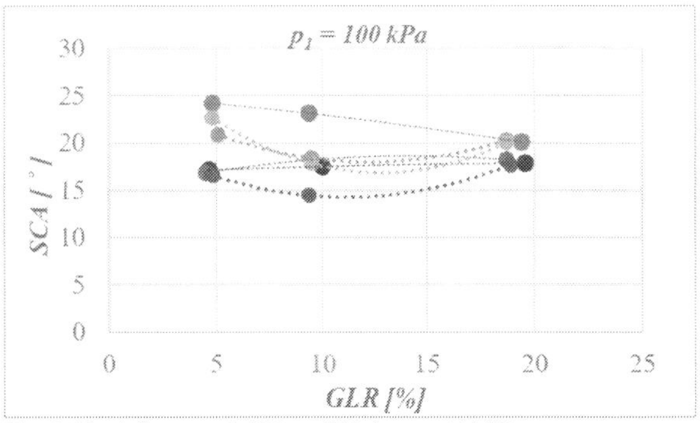

Fig. 9. The influence of *GLR* on *SCA* for p_1 = 100 kPa.

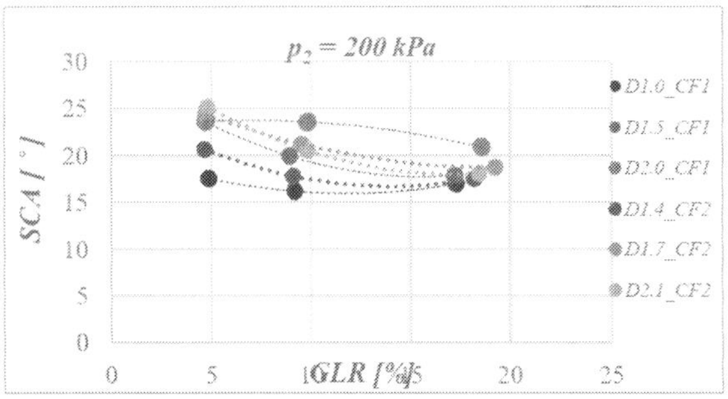

Fig. 10. The influence of *GLR* on *SCA* for p_2 = 100 kPa.

C_d is decreasing with *GLR* increase for all tested variants, see Fig. 11. and Fig. 12.

This is typical for twin-fluid atomizers. The lowest C_d values are obtained for atomizer D1.5_CF1, the highest for D1.4_CF2. The CF2 exhibits a systematically higher C_d compared to CF1. The increasing exit orifice diameter tends to reduce the C_d values, which is comparable with effervescent atomizers where C_d increases with decreasing GLR and also with decreasing exit orifice [6].

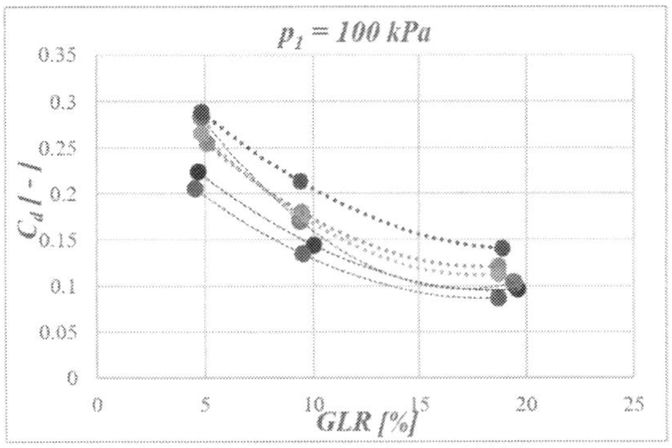

Fig. 11. The influence of *GLR* on C_d for p_1 = 100 kPa.

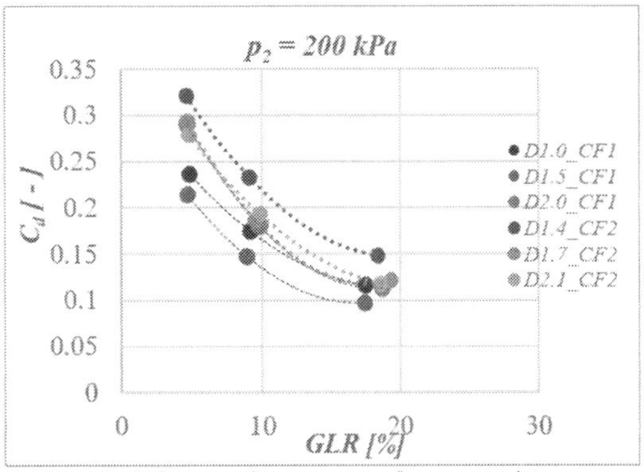

Fig. 12. The influence of *GLR* on *SCA* for p_2 = 100 kPa.

4 Conclusion

In this study, the hypothesis of a significant effect of internal expansions on the performance of the atomizer was confirmed. The CF1 atomizer was observed to have a lower internal expansion rate and thus it behaved more as expected for a conventional twin-fluid atomizer.

The ID_{32} for both types of atomizers increased as the orifice dimeter increased and the *GLR* decreased. Larger droplets were obtained for CF2. This was observed for both operating pressures.

For *SCA*, we observed the strongest effect of the ratio of external to internal expansion. The *SCA* shown an increasing trend only partially for CF1 which has the expansion mostly occurring downstream of the exit orifice. The CF2 atomizers performed a rather convex trend. C_d increased with decreasing *GLR* and increasing pressure.

Highspeed imagining would be useful for further investigation, as it appeared that, due to the long counterflow channel in CF2, the expanding air localized the liquid to the centre, forming a narrow spray cone and affecting the Cd progression.

In future research, there is a necessity to further investigate the effect the counterflow canal. Also, the internal flow should be studied to explain the mixing process inside the atomizer for different expansion rates.

Acknowledgement

This work was supported by the Ministry of Education, Youth and Sports of the Czech Republic under the project LTAIN19044 funded from the program INTER EXCELLENCE (INTER–ACTION).

References

1. A. Hoxie, E. Johnson, V. Srinivasan, P. Strykowski, *ICLASS, 14, (*2018)
2. R. Rangarajan, H. Zhang, P. L. Strykowski, A. Hoxie, S. Yang, V. Srinivasan, *Com., Fu. And Em.*, ***4B**, (2020)*
3. J. Jedelsky, M. Jicha, *Applied Energy*, **132(C)**, (2014)
4. M. Maly, M. Sapik, O. Cejpek, G. Wigley, J. Katolicky, J. Jedelsky, *Experimental Thermal and Fluid Science*, **106**, (2019)
5. A. H. Lefebvre, V. G. Mcdonell. Atomization and Sprays, **2**, (2017)
6. M. Ochowiak, L. Broniarz-press, J. Rozanski, *Experimental Thermal and Fluid Science*, **34(8)**, (2010)

List of abbreviations

CFA	counterflow atomizer
COM	commercial internal mixing air assist atomizer
FB	flow-blurring atomizer
GLR	gas-to-liquid mass ratio
PDA	Phase Doppler Anemometry

List of symbols

C_d		discharge coefficient
$d_{exit\_orifice}$	m	diameter of the exit orifice
d_{20}	m	volume mean diameter
d_{30}	m	surface mean diameter
d_{32}	m	Sauter Mean Diameter
f	Hz	datarate
ISMD, ID_{32}	m	Integral Sauter Mean Diameter
p	Pa	pressure
$Q_{cumulative}$	kg/s	cumulative liquid flow rate
Q_{liquid}	kg/s	flow rate of the liquid
r	m	radial distance
SCA	°	spray cone angle
ρ	kg/m³	density

Development of a 10 kW class axial impulse single stage turboexpander for a micro-CHP ORC unit

Jan Spale[1*], *Guk Chol* Jun[2], *Vaclav* Novotny[2], *Philipp* Streit[3], *Andreas P.* Weiß[3], and *Michal* Kolovratnik[2]

[1] Czech Technical University in Prague, University Centre for Energy Efficient Buildings, Trinecka 1024, Bustehrad 27343, Czech Republic
[2] Czech Technical University in Prague, Faculty of Mechanical Engineering, Technicka 4, Praha 6, 16607, Czech Republic
[3] University of Applied Sciences Amberg-Weiden, Center of Excellence for Cogeneration Technologies, Kaiser-Wilhelm-Ring 23, 92224 Amberg, Germany

Abstract. Development of micro ORC systems with 1-15 kW power output for micro-cogeneration and waste heat recovery at the Czech Technical University in Prague, University Centre for Energy Efficient Buildings (CTU UCEEB) has over ten years of history with many successes. These include 6 different ORC units, all with in-house designed rotary vane expanders (RVE) of many versions throughout this development. Among main advantages of the RVE belong relatively simple and robust design at low cost even at very small series of single-unit production and all that with acceptable efficiency. The ORC units operate with hexamethyldisiloxane (MM) working fluid at high pressure ratios and expansion ratios and the isentropic efficiency of RVE has a limit at these conditions around 60%, often however only at values around 50%. While this might be enough on a cost side for commercialization of this technology, in pursuit of higher efficiency solutions, different expander technology needs to be selected. A turbo-expander is a logical choice with prospect of higher efficiency. At the same time, a literature review has found a lack of reported detailed experimental data for micro (5-50 kW) turbo-expanders, possibly hindering global development towards economically feasible solutions. A project named Dexpand, "Optimised expanders for small-scale distributed energy systems" aims at these issues by objectives in designing, optimizing, manufacturing and testing several ORC expanders with MM and isobutane and their subsequent performance mapping and comparison. One major task is a design of a turboexpander for a 120 kWth biomass fired microcogeneration ORC unit currently operated at the CTU UCEEB. An axial impulse single stage turboexpander was selected as a suitable choice, providing a prospect of a decent efficiency at technically manageable rotational speed and size. This paper provides a detail of currently performed design activities, starting from boundary conditions specification, over development and optimization of a 1D model, preliminary 2D CFD calculations and finishing in a state of a robust and detailed 3D CFD model with a real gas model. Note that the working fluid, high molar mass organic vapour, is highly non-ideal in its behaviour and the flow conditions with pressure design ratio around 13 is highly supersonic (nozzle outlet isentropic Mach number exceeds 2). The current results based on 3D CFD indicate a prospect of an isentropic efficiency 71% at mechanical power output of 11 kW. Lastly, ongoing and future work is outlined, which includes aerodynamic optimization based on the developed 3D CFD model and construction design of the entire turbine assembly.

1 Introduction and overview

Organic Rankine cycle (ORC) power systems became an unrivalled technical solution and an industrial standard in several applications, such as low temperature heat utilization in geothermal systems, combined heat and power (CHP) systems in the scale of several MWs down to hundreds of kW or waste heat recovery (WHR) power systems down to dozens of kilowatts. ORC systems in small to micro scale domestic CHPs face mainly economical barriers with economy-of-scale and large cost per installed kilowatt in micro scale units. [1,2]

When focusing on micro scale ORC CHP with electrical output in the order of less than 10 kW, many laboratory units and prototypes have been built and tested. Regardless of these R&D efforts, these micro scale ORC power systems mostly have not seen commercialization, the rest have not yet been proven to be economically feasible or are until today very scarce on the market. Mainly because their installations face economical barriers with economy-of-scale, since downscaling the ORC power systems to micro scale results in high specific costs associated with low initial production quantities and large cost per installed kilowatt. [3] Another great reason which disfavours small scale ORC power systems is the lack of reliable, efficient and cost-effective expansion machines.

For medium and large-scale ORC power systems, a turbine is the state-of-the-art vapour expansion device. For the small to micro scale systems though, volumetric expanders dominate the market. Main reason for that is the possibility to derive the design of a small volumetric

* Corresponding author: jan.spale@cvut.cz

expander from an off-the-shelf volumetric compressor, common machine in refrigeration and compressed air technology, connected with low investment costs and reliability. [4]

Turboexpanders for small scale ORC plants are nowadays rather scarce and seldom used. Other reason for that may be that volumetric machines are able to operate in a wide range of conditions with decent efficiency compared to narrow design points of turboexpanders. Apart from that, volumetric machines are bringing mostly drawbacks in an increased complexity of the system, additional lubrication system is usually needed, wear occurs on the contact surfaces and the sealing is rather complicated.

This contribution discusses a preliminary thermodynamic and aerodynamic design of a turbine for an already existing small-scale ORC CHP plant, currently in operation with a rotary vane expander (RVE) and hexamethyldisiloxane (MM) as a working fluid as described in Figure 1 in a process flow diagram. The replacement with a turbine solution promises not only an increase in efficiency of the expansion process but maybe even more importantly an increase in reliability of the operation of the expander by omitting any contact surfaces between the stator and rotor. As no wear on these surfaces occurs, an increase in the lifespan of the machine is expected as well. All of these results are the key turning points for a turboexpander development. [5]

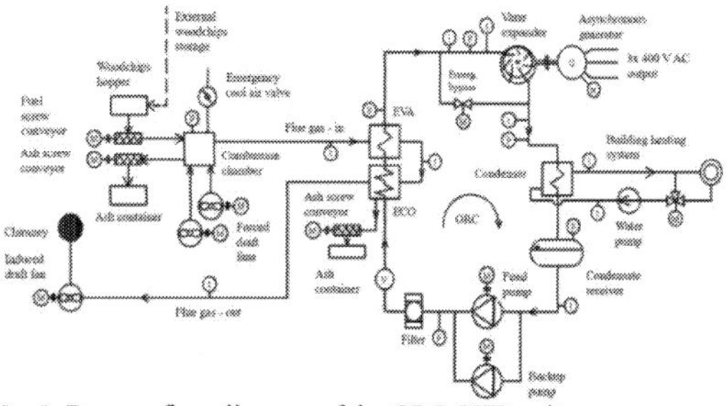

Fig. 1. Process flow diagram of the ORC CHP unit.

2 Turboexpander design

This chapter presents the design methodology and procedure for an ORC micro scale turboexpander. The design procedure of the ORC micro turboexpander follows several generations from the proof of concept unit for preliminary cold air testing towards the prototype of the turbine working with organic vapour.

2.1 Boundary conditions and turboexpander parameters

The boundary conditions to the turboexpander given from the cycle conditions are displayed in Table 1. Main relevant parameters of the cycle that determine the turbine design are the following: expander inlet and outlet pressure, inlet temperature, nominal mass flow rate and the nature of the working fluid; these cycle boundary conditions are supposed to be conserved with a new expander as well.

Table 1. Turboexpander design boundary conditions defined by the cycle performance.

Parameter	Value	Units
Expander inlet pressure p_{in}	650	kPa
Expander inlet temperature T_{in}	190	°C
Expander inlet superheating T_{SH}	10	K
Expander outlet pressure P_{out}	50	kPa
Working fluid	MM	-
Working fluid mass flow rate m_{wf}	0.3	$kg.s^{-1}$

According to Table 1 one may observe that the design pressure ratio for the turboexpander is around 12, and the inlet temperature of the superheated organic vapour is 190°C. For these inlet conditions, the speed of sound is 123 m/s, and it is quite obvious from this preliminary estimate that the fluid flow at the nozzle outlet for an isentropic enthalpy drop of around 50 kJ/kg will be highly supersonic. For ORC turbines with such high Mach numbers (above 1.4), it is necessary to design the nozzles as uniquely-shaped convergent-divergent to respect the nature of the supersonic fluid flow. [6]

The working fluid selection – hexamethyldisiloxane (MM) impacts the design of the turbomachine heavily. Due to its nontoxicity, excellent thermal and chemical stability, and limited flammability, it can be possibly considered as the most suitable fluid for such high-temperature ORC from an engineering point of view. [7–9] Other reasons for its application were also a reasonable price, that it remains liquid under ambient conditions, solubility of oil and previous unpleasant experiences with hydrocarbons as working fluid.

The working fluid can be considered as a dry fluid – meaning it works with a positive slope of the saturated vapor curve (ds/dT). This brings several peculiarities – such as an ensured expansion into the superheated steam region without any danger of condensation nuclei or droplets formation.

Another factor is the high molar mass and the molecular complexity of the organic fluid which results in low volumetric flow rates (an order of magnitude lower compared to steam at the same p, T) and high volumetric ratios (though not as high as for steam), low enthalpy drops along the expansion line and very low speed of sound. These aspects highly influence the turbomachinery design.

The complex organic fluids show a very non-ideal behaviour especially in the region which is of the upmost interest in turbomachinery — the vapour single phase region. As the consequence, rather complicated real gas Equations of State (EoS) have to be utilized during the design phase (nowadays possible thanks to fluid properties libraries such as REFPROP or CoolProp). Siloxanes as members of so-called Bethe-Zel'dovich-

Thompson (BZT) fluids family exhibit in the single vapour phase region a negative fundamental derivative in gas dynamics Γ - therefore all the thermodynamic properties show high sensitivity to its values and thus for precise enthalpy calculation, many experimental data and specific EoS (e.g. Peng-Robinson-Stryjek-Vera EoS) are needed. [7,10,11]

The implications of that are severe and of various character. On one hand, it helps to design very compact and little loaded machines resulting in rather cheap turbomachinery. On the other hand, low speed of sound means that the presence of a supersonic flow is inevitable and the blade design is nonconventional with convergent-divergent nozzles. Therefore, the occurrence of shock waves and their interactions with boundary layer can be observed in the fluid flow.

High volumetric ratios for very complex molecules lead to difficulties in processing the large difference in the volumetric flow rate in a single-stage machine which often has to be compensated by a significant change of the blade height along the streamline, partial admission and very high flow deflection angles. The variation of blade height comes with a penalty in the form of lower stage efficiency as it leads to an increased vorticity of the flow caused by the perpendicular velocity component and therefore increased secondary losses. The losses are furthermore increased by the high deflection blade shapes.

For micro to small scale ORC applications - be it automotive internal combustion engine bottoming ORC, domestic CHP units or small scale concentrated solar power plants - the expander cost makes the major share of the cost of the entire unit. Therefore, the aim of the authors' design is not to maximize the isentropic efficiency of the machine at any costs but rather to optimize it for reasonable price while reaching sufficient efficiencies and reliability. A single stage turbomachine is thereby preferred in such systems over multistage optimised turboexpanders.

High rotational speeds are often avoided during the design of a steam turbine. On the contrary for ORC turbines this is not the case, as the blade loading is usually very low due to much lower enthalpy drop and the temperature ranges in which the turboexpanders operate are also much lower compared to conventional steam turbines. Hence, the blades are much less mechanically stressed and high rotational speeds are not such a large concern from the mechanical standpoint. Also, the turboexpander as a whole is generally smaller and it is possible to apply permanent magnet high speed generators.

For all of the above stated reasons, a design approach of an ORC turbine cannot be simply derived from the design methodology of a steam turbine and has to follow a completely different method taking into the account all the aspects, specifics and peculiarities of the organic fluid flow. Traditional zero-dimensional statistical diagrams such as Smith charts or Baljé map can provide only a very rough estimate for the isentropic efficiency, because they are calibrated for large scale steam turbines. [9] The effect of highly supersonic flows at the stator outlet and strong shocks, large blade height variation and high deflection buckets should not be neglected even in the rough estimate.

2.2 1-D meanline design of the ORC turboexpander

This chapter discusses the preliminary thermodynamic and aerodynamic design of an axial impulse stage turbine for the given ORC CHP unit and boundary conditions. The 1-D meanline design is based on energy and mass balance in each state of the turbomachine and with several assumptions:

- Steady states at the inlet, stator outlet and rotor outlet
- The thermodynamic properties are calculated for midspan diameter
- Conservation of rothalpy principle is applied
- Conservation of moment of momentum
- Cycle boundary conditions are to be conserved and not affected by the replacement of the expander
- The kinetic energy of the working fluid at the inlet is neglected
- Any heat losses from the turboexpander to the environment are neglected
- All geometrical properties have to be kept within the limit of manufacturability
- The kinetic energy at the outlet is considered not to be any recovered
- Rotor buckets designed with constant channel width and equal but opposite relative flow angles
- The incidence and deviation angles are assumed to be zero
- Loss correlations adopted from turbine drives of rocket turbopump

The model was created using MS Excel with VBA and REFPROP add-in for fluid properties.

Since the designed turboexpander is meant to be affordable and simple to manufacture, assembly and operate, the desired combination of high power output and a compact design makes the impulse turbine a favourable choice compared to the reaction turbine. These may offer higher efficiency, at the cost of a larger machine. An axial configuration is chosen for the simplicity of manufacturing and easier design of the turboexpander. Also, partial admission can be easily adopted for axial impulse stage turbines. Finally, there were previous experiences from the experimental trials with cold aerodynamic tests of axial impulse stage turbines during the works on the master's thesis of the author.

High pressure ratios, transonic flow velocities and compressible flow effects along with the proximity of the saturation curve lead to the complicated combination of a non-ideal gas and compressible flow. In the commonly and commercially adopted 1-D meanline models, one of the two usually assumes the absence of the other. In the designed model, real fluid behaviour is therefore assumed, with a non-ideal gas model, the heat capacity ratio thus cannot be considered constant and is calculated for each state of the turbine.

For the initialization of the turboexpander design, some of the design variables had to be chosen with respect to the rotational speed, diameter of the machine and some other geometrical components. Another very important parameter to be estimated in the first step of the design is the isentropic efficiency of the turboexpander. These initial values and guesses are reported in Table 2. Some of these parameters were kept constant throughout the design phase and some were optimized, such as the rotational speed of the turbine and the blade height.

Table 2. Chosen input parameters and guesses; parameters denoted with * were later optimized – optimized values in brackets.

Parameter	Initial value	Units
Rotational speed n*	24000 (28000)	rpm
Midspan diameter Dmid	100	mm
Nozzle outlet flow angle α2	13	°
Isentropic efficiency guess	70	%
Partial admission guess e*	58.5 (97.5)	%
Blade height ratio*	0.1 (0.06)	-
Minimum blade height h_min	5	mm
Rotor blades aspect ratio AR	2	-

To better illustrate the design of the axial impulse stage, Figure 2 below shows the resulting velocity triangles of the turbine.

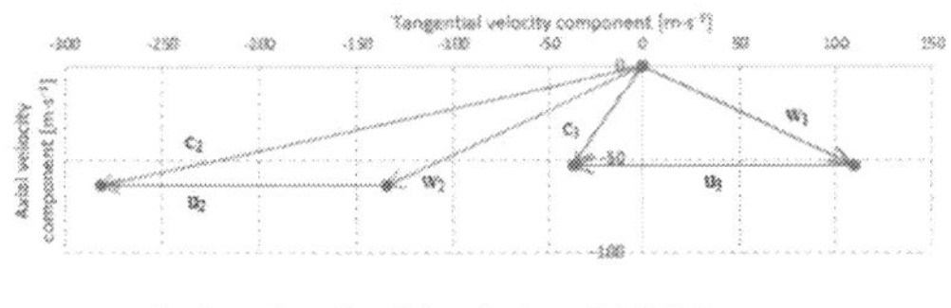

Fig. 2. Velocity triangle of the designed ORC impulse stage.

2.3 Design of the flow components

The stator is designed as a series of convergent-divergent de Laval nozzles. As the flow suffers sever real gas effects and is highly supersonic, the resulting nozzle geometry is very unconventional, cambered and unlike any stator blade geometries that could be found in an airfoil atlas. The geometry of the nozzle is thus designed as a combination of the divergent part, which for such highly supersonic flow impacts the efficiency of the nozzle the most and therefore more effort is put into designing it, and the convergent part.

The rotor blades were designed as constant channel width buckets with geometry derived from [12]. The shape of the buckets is again unconventional and highly cambered since it experiences large flow deflection. Thus, conventional airfoil for turbine rotor blades cannot be used, and the geometry is designed specifically for this type of application. The blades are simple cylindrical as they are easy to manufacture with standard technology. To reflect the increase in pressure within the rotor wheel due to the friction losses of the supersonic fluid flow, an increase in crossflow area is designed by means of linearly increasing blade length along the chord line of the blade. The aspect ratio was chosen as 2 as recommended by Agromayor and Nord in [13] for this type of axial impulse supersonic turbines in ORC. The blades were designed as unshrouded due to the manufacturability constraints.

3 Numerical simulation

Figure 3 shows the geometry of the computational model domain. The model consists of two computational domains S and R. Domain S contains 2 stator blade channels in the shape of a Laval nozzle. The rotating domain R contains 7 intermediate blade channels typical of a straight turbine stage. Both domains have the same tangential pitch of $2\pi/5$. This means that the computational model includes 1/5 of the flow section of the turbine.

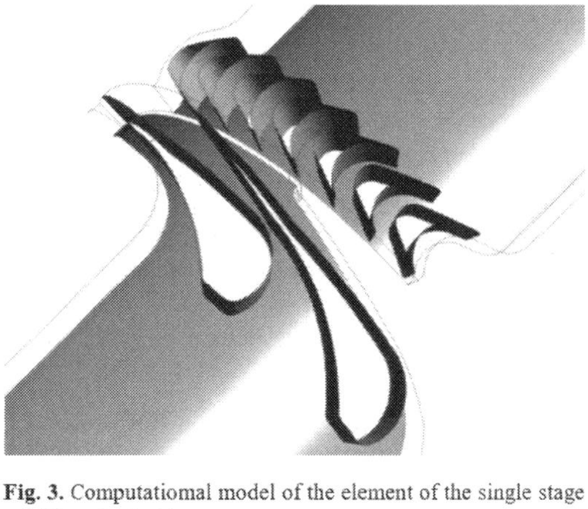

Fig. 3. Computatiomal model of the element of the single stage axial impulse turbine.

3.1. Numerical model settings and boundary conditions

The computational domains are connected using the Frozen rotor interface model. The rotating domain R is set to a very high speed of 28000rpm. The SST k-ω turbulent model is used for the simulation. Since a very high Mach number is expected in the flow section of the turbine, a High-Speed Wall Heat Transfer Model is also used. A pressure-inlet type boundary condition with total pressure and total temperature is set at the input of the computational model. At the output of the calculation model, a pressure-outlet boundary condition with static pressure is used. The set input and output parameters of the working medium are shown in Table 1.

3.2. Fluid properties of the organic vapour

The real gas properties of hexamethyldisiloxane (MM) must be included in the CFD analysis, as it plays an important role in the quality and accuracy of the analysis. An automatic generator of RGPgen v2.0 (Real Gas Properties) for ANSYS CFX was used in this simulation. This RGPgen, developed by Numsolution, can generate a set of material properties of any working medium in a form that is directly accepted by ANSYS CFX.

3.3. Mesh generation

The networks of both computational domains are prepared in ANSYS ICEM. Both domains are meshed with hexahedral cells and the computational model contains a total of 24.4 million hexahedral cells. The surface network detail of both computational domains is shown in Fig. 4.

Fig. 4. Close-up onto the generated mesh of the stator-rotor interface.

4 Results & discussion

Results of the numerical simulation are reported within this chapter. These results are to be validated by an experimental campaign as the next goal of the project to develop a turbine for the aforementioned ORC CHP unit.

To describe the nature of the fluid flow through the turbine stage, contours of the Mach number in the stationary frame on midspan position are presented in Figure 5. Siloxane vapour at the admission pressure enters the stator nozzles, where in the convergent part of the nozzle increases its speed up to the speed of sound in the nozzle throat. Divergent part of the nozzle is designed as a linear segment increasing the nozzle area up to the designed outlet nozzle area, achieving Mach 2 at the outlet.

Significant shockwave structures at the stator vanes trailing edge are apparent. These shockwaves greatly determinate the fluid flow at the rotor inlet. There are again strong oblique shockwaves at the leading edge of the rotor buckets. These oblique shockwaves affect the fluid flow inside the channel – interaction of the shockwave with the boundary layer at the suction side of the bucket results in a flow separation, therefore secondary losses occur in the rotor wheel.

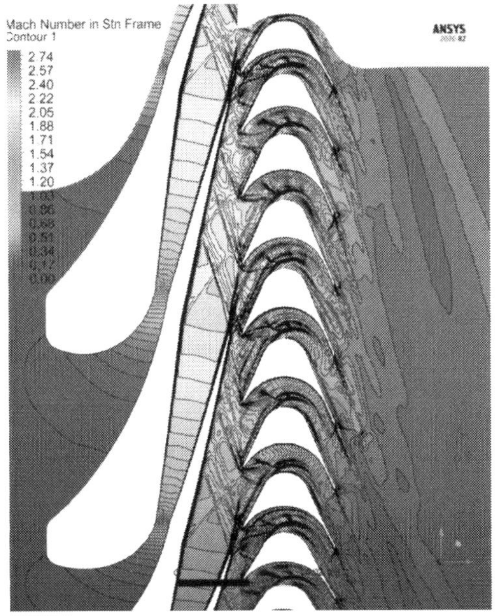

Fig. 5. Midspan contours of Mach number in stationary frame.

To assess the fluid flow in the spanwise direction, components of the velocity vector along the blade span are plotted, as well as the velocity angles of the fluid flow in Figure 6 below. Thanks to the optimization of the rotor blade outlet flow angle, the effect of the flow separation at the suction side of the rotor blade was decreased, as can be observed on the rather monotonic and flat behaviour the courses of the velocity compounds exhibit along the span at the leading and especially the trailing edge.

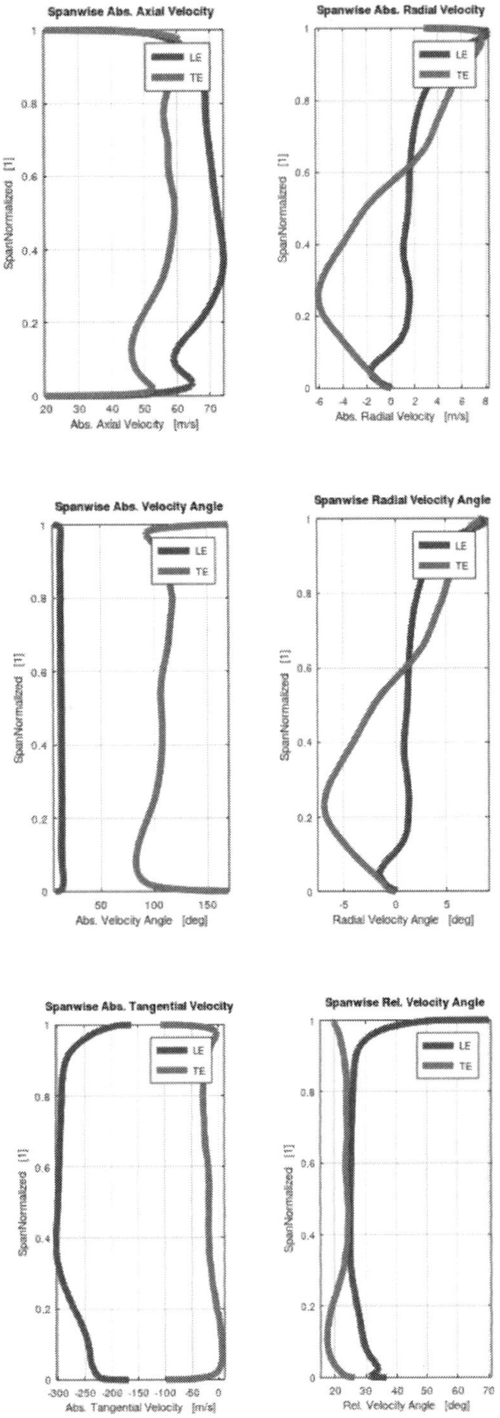

Fig. 6. Variation of the averaged components of the velocity vector along the span of the rotor blade at the leading edge and the trailing edge.

Results of the 3D CFD model were finally compared with the result of the optimized 1D model with loss correlations. Some of the parameters are listed in Table 3 below.

Table 3. Comparison of the achieved parameters inbetween the 1D meanline model and the CFD simulation.

Parameter	1D model	CFD model	Units
Nozzle outlet/Rotor inlet			
Mach number	1.95	2.01	-
Static pressure	55	64.3	kPa
Static enthalpy	298.9	296.2	kJ/kg
Rotor outlet			
Mach number	0.46	0.48	-
Static pressure	55	54.9	kPa
Static enthalpy	302.7	298.2	kJ/kg
Isentropic efficiency ηis	69.2	71.1	%
Mechanical power output	10.8	11.2	kW

4.1. Conclusion and future works

This paper presented an overview of the current progress in the DEXPAND project focused on the development of optimised ORC expanders for various solutions, in this case an axial impulse stage turbine for a biomass CHP ORC plant, and especially the numerical CFD simulation carried out in ANSYS CFX with real gas fluid properties. The rated mechanical power output of the turbine according to the CFD simulation is assumed to exceed 11 kW and total-to-static isentropic efficiency of over 71%.

Future works in this regard consist of further optimization using a Concepts NREC AxCent tool, designing the whole turbine assembly, manufacturing and assembling it and finally delivering an experimental campaign at the biomass CHP ORC unit at CTU UCEEB with a prospect of replacing the current volumetric expander.

Acknowledgement

The DEXPAND project benefits from a € 1 500 000 grant from Iceland, Liechtenstein and Norway through the EEA Grants and the Technology Agency of the Czech Republic within the KAPPA Programme.

References

1. Macchi E, Astolfi M. Organic Rankine Cycle (ORC) Power Systems: Technologies and Applications. Woodhead Publishing; (2016).

2. Colonna P, Casati E, Trapp C, Mathijssen T, Larjola J, Turunen-Saaresti T, et al. Organic Rankine Cycle Power Systems: From the Concept to Current Technology, Applications, and an Outlook to the Future. J Eng Gas Turbines Power;**137**:100801, (2015).

3. Padinger R, Aigenbauer S, Schmidl C. Best practise report on decentralized biomass fired CHP plants and status of biomass fired small-and micro scale CHP technologies. IEA Bioenergy;**32**, (2021).

4. Weiß AP. Volumetric expander versus turbine – which is the better choice for small ORC plants. 3rd ASME ORC Conf. Brussels, p. 1–10, (2015).

5. Mascuch J, Spale J, Vodicka V, Zeleny Z, Novotny V. Paper ID : 19 , Page 1 SCALING-UP A BIOMASS FIRED MICRO-CHP ORC FOR BETTER PERFORMANCE TOWARDS COMMERCIALIZATION, **19**:1–10, (2021).

6. Astolfi M, Macchi E. Efficiency Correlations for Axial Flow Turbines Working With Non-Conventional Fluids. Asme Orc 2015.1–12. (2015).

7. Colonna P, Guardone A, Nannan NR. Siloxanes: A new class of candidate Bethe-Zel'dovich-Thompson fluids. Phys Fluids;**19**, (2007).

8. Keulen L, Gallarini S, Landolina C, Spinelli A, Iora P, Invernizzi C, et al. Thermal stability of hexamethyldisiloxane and octamethyltrisiloxane. Energy;**165**:868–76, (2018).

9. Erhart TG, Gölz J, Eicker U, Van Den Broek M. Working fluid stability in large-scale organic rankine cycle-units using siloxanes - Long-term experiences and fluid recycling. Energies;**9**:1–16, (2016).

10. Colonna P, Nannan NR, Guardone A, Lemmon EW. Multiparameter equations of state for selected siloxanes. Fluid Phase Equilibria;**244**:193–211, (2006).

11. Cammi G, Conti CC, Spinelli A, Guardone A. Experimental characterization of nozzle flow expansions of siloxane MM for ORC turbines applications. Energy;**218**:119249, (2021).

12. Hauer J. Analytische Studie zur optimierten Turbinenauslegung. Hochschule Amberg – Weiden, (2011).

13. Agromayor R, Nord LO. Preliminary Design and Optimization of Axial Turbines Accounting for Diffuser Performance. Int J Turbomachinery, Propuls Power;**4**:32, (2019).

Conception of minichannel cooling for a PVT heat exchanger

Dariusz Strąk[1]*, and *Magdalena* Piasecka[1]

[1]Kielce University of Technology, Faculty of Mechatronics and Mechanical Engineering, Al. 1000-lecia Panstwa Polskiego 7, 25-314 Kielce, Poland

Abstract. In the paper, a new construction of a compact plate heat exchanger dedicated to PVT cooling was numerically tested. The efficiency of photovoltaic panels decreases as their temperature increases, but cooling improves efficiency. A model counter-current heat exchanger with rectangular minichannels was presented. A heated wall of the heat exchanger was a photovoltaic cell. There were laminar flows of Fluorinert FC-72 and distilled water in the minichannels. FC-72 was heated by the photovoltaic panel. The CAD model of a minichannel heat exchanger was proposed. CFD calculations were performed using Simcenter STAR-CCM+ software. The fluid temperature and velocity profile in the minichannels were determined to finally obtain the values of the heat transfer coefficient. The main parameters taken into consideration in the computations concerned: meshes, physics, and geometry. Finally, local heat transfer coefficients were determined.

1 Introduction

The increased demand for energy, the limited amount of fossil resources and the excessive pollution of the environment have caused great interest in renewable energy sources in recent years. One type of renewable energy source is solar energy, which can be converted into electricity through the use of photovoltaic cells. During sunny days, photovoltaic panels can heat up to a temperature of about 50 - 60 °C. When solar panels are heated, their efficiency and service life decrease. It is estimated that as the temperature of the panels increases every 1°C, their efficiency decreases by approximately 0.5% [1]. To increase the efficiency of photovoltaic cells, the surface of the photovoltaic cell should be cooled. It is also assumed that as the temperature of the PV panels increases every 10 °C, the degradation rate doubles. A side effect of the sunlight that falls on PV panels is the thermal energy produced in this process, in addition to electricity. The design of photovoltaic thermal (PVT) devices is usually based on classical photovoltaic panels. Moreover, a special cooling module can be installed below the photovoltaic cells to improve their efficiency.

Skoplaki and *Palyvos* in [1,2] studied the external conditions that affect the operation of photovoltaic panels. By cooling the solar cells with a stream of fluid, such as air or water, the electricity yield can be improved. However, a conceptually better design is to reuse the heat energy that the coolant emits [3,4]. In liquid PVT modules, the heat sink is characterized by a lower heat transfer coefficient, lower density, lower heat capacity, and lower thermal conductivity. Water is used most often as a working medium in liquid modules, which is characterized by higher density, increased heat capacity, and thermal conductivity compared to air PVT modules [5]. *Bergene* and *Løvvik* in [6] developed a detailed physical model of the PVT/w flat collector by which the overall efficiency was estimated. *Garg* et al. conducted mathematical and experimental studies of PVT systems [7,8]. *De Vries* investigated the operation of several PVT collector designs. The single-glazed structure has been shown to be superior to the structure without glazing (which has an unfavorable thermal efficiency) or double glazing (which has an unfavorable electrical efficiency). By transient analysis, *Prakash* in [9] pointed out that the air collector design has a lower thermal efficiency than the water collector due to the inferior heat transfer between the thermal absorber and the air stream. *Rockendorf* et al. in [10] compared the performance of a thermoelectric collector (producing heat first, then electricity) and a PVT/w collector (in a sheet-and-tube design). The electric power of the PVT/w collector was found to be significantly higher than that of the thermoelectric collector.

Generally, the reported thermal efficiency of practical PVT/liquid systems is generally in the range of 45 - 70% for unglazed and glazed collector designs. For flat PVT/air systems, the optimum thermal efficiency can reach 55%.

Akbarzadeh and *Wadowski* investigated the design of a new panel cooling based on a heat pipe [11]. The thermosiphon or gravity-assisted heat pipe is a completely passive system that transfers heat from one place to another. The proposed cooling system for the rear part of the solar cells included two interconnected heat exchangers, preempted and filled with R-11, R-22, and water refrigerants. *Bai et al.* in [12] presented a simulation study of the use of PVT/in collectors as water heating devices in a solar-assisted heat pump

* Corresponding author: dstrak@tu.kielce.pl

system. *Hollick* evaluated the improvement of the energy efficiency of a system in which solar cells were attached to a metal trapezoidal sheet on the vertical facades of buildings [13].

In [14], Eulerian two-fluid model coupled with the extended wall boiling model was used to simulate the departure from nucleate boiling (DNB) in vertical heated tubes under high pressure using Simcenter STAR-CCM+. CFD calculations were performed with different wall heat fluxes. A two-fluid model coupled with an extended wall boiling model was used to simulate DNB in vertical heated tubes with both uniform and nonuniform heat fluxes. In [15] in order to obtain data to correlate cross-flow resistance coefficient, a computational fluid dynamic simulation using Star-CCM+ was performed for the flow of water through a 5 × 5 rod bundle geometry supported by a spacer grid with split mixing. The two layer k-epsilon turbulence model was used with an all y+ automatic wall treatment function in Simcenter STAR-CCM+. Correlations were presents to determine the resistance coefficient to cross-flow.

This paper aims to present a new construction of a compact plate heat exchanger dedicated to PVT cooling. The photovoltaic cell constitutes the heated wall of a counter-current heat exchanger with rectangular minichannels. The CAD model of a heat exchanger with minichannels was proposed. CFD calculations were performed using Simcenter STAR-CCM+ software. The fluid temperature and velocity profile in the minichannels were determined. Furthermore, temperature distributions on the heat exchanger's plates – fluids contact surfaces were achieved. Finally, local heat transfer coefficients were obtained.

At Kielce University of Technology, research on flow boiling heat transfer in minichannels has been provided for many years. For the analysis of the experimental data, analytical - numerical methods such as: Trefftz and Piccard methods [16], Trefftz and Beck methods [16,17], in which the calculations were based on the use of Trefftz functions. Furthermore, the commercial ADINA program was applied in numerical calculations to perform advanced mathematical procedures using experimental data [18,19]. Similarly, the results of the CFD and heat transfer calculations for the test section with a group of minichannels performed in Simcenter STAR-CCM + software were presented in [20].

2 Experimental stand and preliminary tests

The schematic diagram of the experimental stand is shown in Fig. 1. The most important circuits realized in the experimental setup are two closed loops of working fluids, including:

i) 'hot' fluid loop, in which the FC-72 fluid circulates,
ii) 'cold' fluid loop, in which distilled water circulation occurs.

Other systems implemented at the setup include: the data and image acquisition system and the power supply and control system.

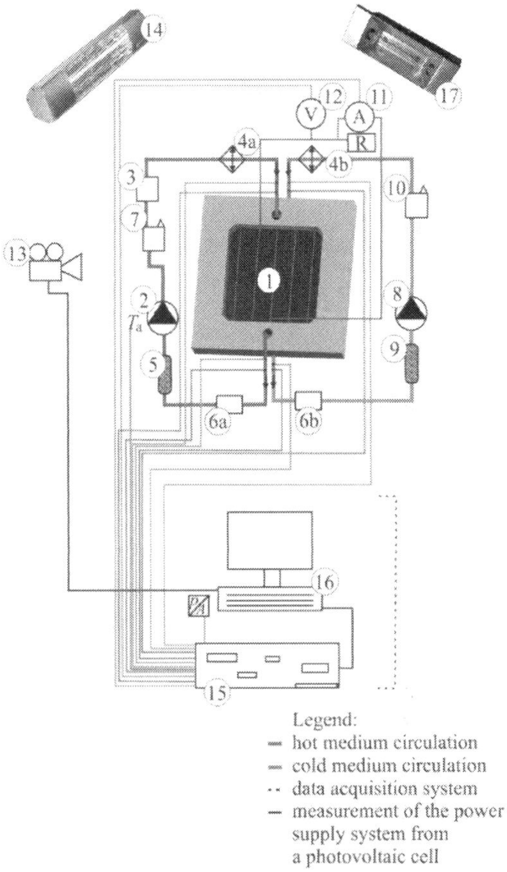

Fig. 1. Schematic diagram of the experimental stand: 1 – test module with minichannels, 2,8 – circulating pumps, 3 – flow regulator, 4 – heat exchanger, 5,9 – filters, 6a, 6b – flow meter, 7,10 – air separators, 11 – ammeter, 12 – voltmeter, 13 – thermal imaging camera, 14 – infrared heater, 15 – data acquisition station, 16 – computer, 17 – halogen lamp, T_a – ambient temperature, p_a – ambient pressure, R – receiver load (a light bulb).

The view of the test module with minichannels for cooling photovoltaic panels is shown in Fig. 2.

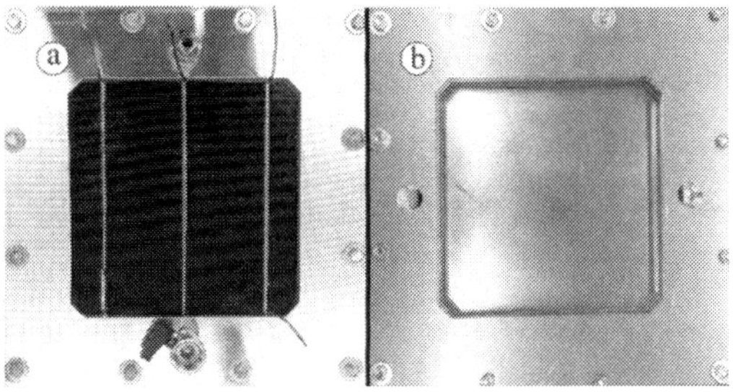

Fig. 2. View of the test module: a) side of the solar panel, b) side of a copper plate.

https://doi.org/10.1051/epjconf/202226401045

Figure 2a presents a view of the test module from the side of the PV panel and Fig. 2b shows the back part of the module with a visible copper plate. The PV panel has dimensions of 156 mm in width and 156 mm in length, as well as 4 mm in thickness. A counter-current flow of the working fluids in minichannels is required. The flow of both fluids is laminar.

The individual components of the test module are shown in Fig. 3. Its main elements are two parallel copper plates (2, 4) and the PV cell plate (6). A group of 12 parallel minichannels (six 'hot' minichannels and six 'cold' ones) with a rectangular cross section (each 1.5 mm deep, 20 mm width and 70 mm length) is produced between them.

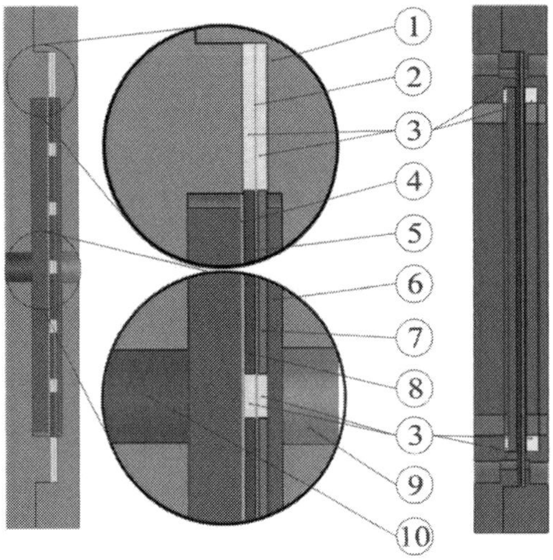

Fig. 3. View of the layers of the PV-T panel cooled with minichannels: 1 – cover (aluminium), 2 – middle plate (copper), 3 – silicone gasket, 4 – outer plate (copper), 6 – PV cell panel, 7 – 'hot' minichanel, 8 – 'cold' minichanel, 9 – inlet/outlet of the 'hot' minichanel, 10 – inlet/outlet of the 'cold' minichanel.

Two selected minichannels of each type are illustrated in Fig. 3 (7, 8). In the minichannels, there is a flow of the FC-72 following fluids: the cooling fluid FC-72 (in the 'hot' minichannels) and distilled water (in the 'cold' minichannels). Additional elements of the test module are four gaskets made of silicon (3).

During experiments, the part of the test module with FC-72 fluid), the outer wall of the minichannel (glass with PV cell) (6), Fig. 3) is heated with an infrared heater (power heating in the range 700 W - 1200 W) and illuminated by a 1000 W halogen bulb.

Heat transfer between the flowing media proceeds through a copper plate (2) with thickness of 0.3 mm. The outer plate of the cold part of the exchanger (4) with thickness of 0.5 mm is also made of copper. The PV panel (6) consists of several thin layers, the first of which is tempered glass (3.2 mm in thickness); the next layer consists of 0.12 mm thick silicon cells, and the lower part contacts the fluid in the minichannel. Type K thermocouples and pressure gauges have been installed at the inlet and outlet of the collectors that supply the minichannels (inlet and outlet collectors). The data from selected experimental series were applied in numerical computations.

3 Simcenter STAR CCM+ numerical calculations

Simcenter STAR-CCM+ software is used for numerical simulation of fluid dynamics based on a wide range of simulation models, including CFD, computational constant mechanics (CSM), electromagnetics, heat transfer, multiphase flow, particle dynamics, reaction flow, electrochemistry, aeroacoustics. It is a simulation tool that provides the most comprehensive set of physical models for computer engineering (CAE) [21].

The numerical computations were provided in the Simcenter Star-CCM+ software. The elements of the 3D model were meshed in version 2020.2.1 Build 15.04.010 (win64/intel18.3vc14-r8 Double Precision).

To perform numerical calculations, a model of the test module with minichannels, intended to cool photovoltaic panels, was assumed in Simcenter STAR-CCM+. The fluid regions consist of the volumes of two fluids: FC-72 and water, for which the flow rates are set, as well as the pressure at the outlet and the fluid temperature at the inlet. In addition, a uniform static temperature on the outer surface of the photovoltaic panel was assumed. A condition of the 'mass flow inlet-type' has been established at the inlets of regions with two liquids, while the 'pressure outlet' has been used as the outlet condition.

Table 1 presents the characteristics of the plates as essential elements of the test module and the experimental data that concern them.

Table 1. Characteristics of the plates and experimental data.

Properties/ Experimental data	**Heated plate PV**	**Copper plate**	**Closing plate Cu**
Thickness, main physical properties			
Thickness, σ [m]	0.004	0.0003	0.0005
Thermal conductivity, λ [W/(m^{2}K)]	1.4	380	380
Density [kg/m^3]	2500	8940	8940
Main experimental data			
Temperature, T_{in} [°C]	15	15	15
Temperature, T_{out} [°C]	60	23	18
Static temperature [°C]	50	-	-

Table 2 shows main fluid properties and experimental data. The flow of fluids in the calculations has been described by means of the K-Epsilon model.

Table 2. Main fluid properties and experimental data.

Physical properties/ Experimental data	Fluid FC-72	Distilled water
Main physical properties		
Thermal conductivity, λ [W/(m²K)]	0.06	0.59
Density, [kg/m³]	1713 -1720	999.1 - 992.2
Main experimental data		
Mass flow rate, Q_m [kg/s]	0.0092	0.00556
Temperature, T_{in} [°C]	15	15
Temperature, T_{out} [°C]	29	23
Gauge pressure, p_{in} [bar]	0.1	0.1
Gauge pressure, p_{out} [bar]	0.16	0.61

The geometry of the model heat exchanger with minichannels is presented in Fig. 4.

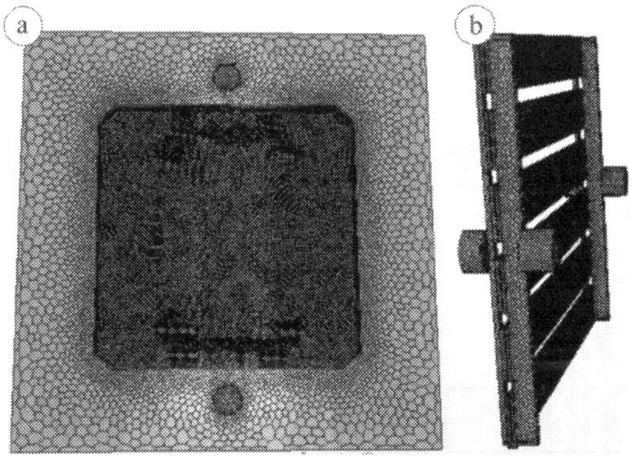

Fig. 4. Model of the test module for numerical calculation in Simcenter STAR-CCM+: a) polyhedral mesh view, b) fluid region flowing in 'hot' minichannel and 'cold' minichannel.

The shape of the exchanger consists of several elements with varying thicknesses. The discretisation of their volume used a polyhedral grid, created based on the tetrahedral grid, using all its advantages in the ability to quickly map the shape of complex geometry. However, by constituting a higher order grid, polyhedral causes the possibility to generate a more accurate result with a smaller number of elements. Additionally, in order to increase the accuracy of the impact of the near-wall layer on the flow profile, the grid was densified around the surfaces of the heated plate, as well as the copper centre plate and the one closing the exchanger. The resulting discretisation effect is shown in Fig. 5.

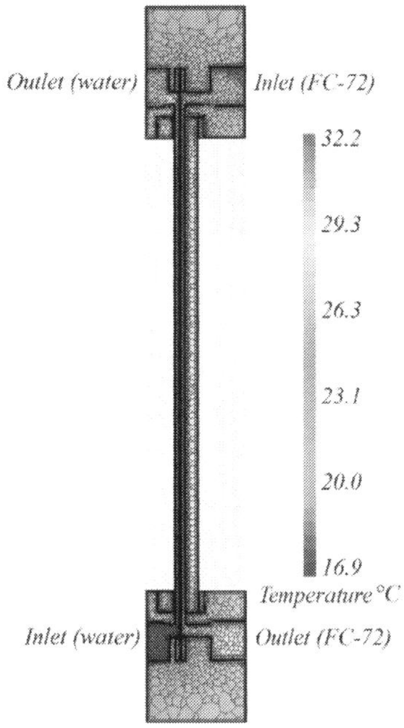

Fig. 5. View of the mesh of the layers of the test module, Simcenter STAR-CCM+.

3.1 CFD calculation results

In Simcenter Star CCM+, computations can be performed in a steady or unsteady condition, taking into account the impact of time on the flow process. In the calculations, a steady state was assumed. For fixed mass flow and inlet pressure values, it was possible to calculate the outlet pressure and the temperature of the fluids. The main objective was to determine the heat transfer coefficient for the individual layers of contact between the fluids and the plates. Multiple iterations of numerical calculations for the same model, with the input mass flow values, enable obtaining the pressure characteristics as a function of flow. Figure 6 presents the residual dependence as a function of iteration. Up to 6500 iterations have been achieved.

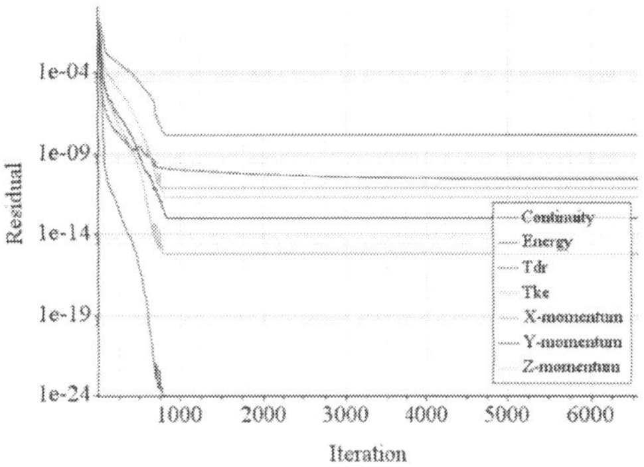

Fig. 6. The residuals as a function of iteration.

The energy was stabilised in the 800th iteration and stayed at 10^{-11}. Continuity stayed at 10^{-8}, for the K-Epsilon model: T_{dr} - the turbulent dissipation rate is at a level of 10^{-24} and T_{ke} - turbulent kinetic energy is at 10^{-16}.

Simcenter Star CCM + software enables visualization of heat and flow parameters. The pressure and flow distributions as the well as temperature distributions in selected cross sections were illustrated as the results of numerical computations, of the basis on the experimental data, shown in Tables 1 and 2. The temperature distribution on the surface of the PV panel/FC-72 flowing in the 'hot' minichannel is given in Fig. 7 and on the contact surface the copper plate/water in the 'cold' minichannel - in Fig. 8.

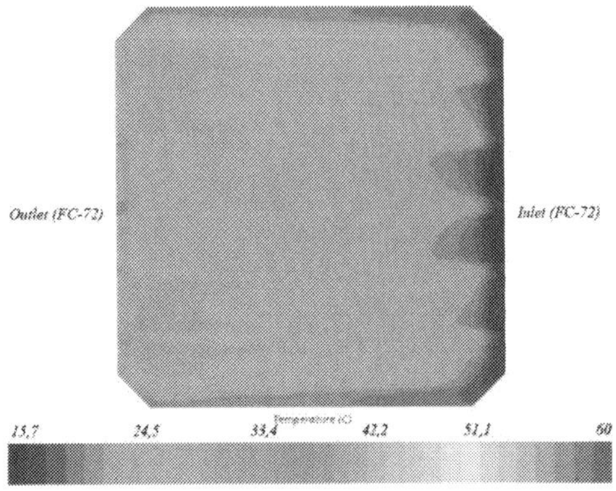

Fig. 7. Temperature on the contact surface: PV panel/FC-72 ('hot' minichannel).

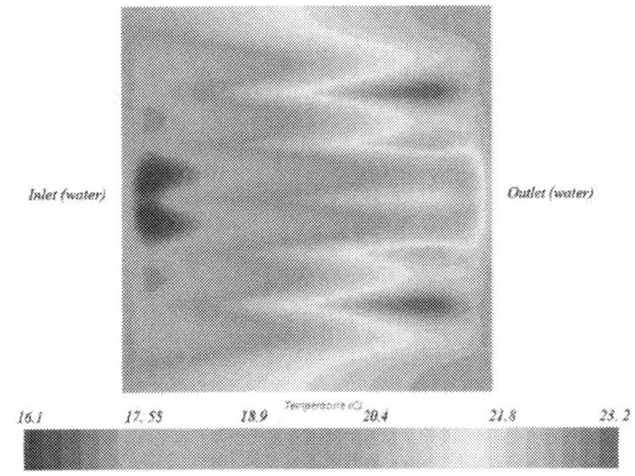

Fig. 8. Temperature on the contact surface: the copper plate/ distilled water ('cold' minichannel).

Analyzing the dependences shown in Figs. 7 and 8, it can be noticed that the temperature increases with the distance from the inlet. The temperature difference on the PV panel/FC-72 contact surface between the inlet and the outlet is approximately 18°C, Fig. 7. It can be underlined that the FC-72 fluid is heated along the flow in minichannels mainly on the contact surface with the PV panel. Compared to the previous difference in temperature, the temperature difference on the contact surface between the copper plate and the distilled water, from the inlet to the outlet, is lower reaching 8°C, Fig. 8.

Calculations of the heat transfer coefficients were performed. The dependence of the heat transfer coefficient as a function of the distance from the minichannel inlet is shown in Fig. 9, for selected fluid layer (cross-section of the minichannel) as follows:

- at the contact surface: the PV panel - FC-72, Fig. 9a, local heat transfer coefficients $\alpha_1(x)$;
- at the contact surface: FC-72 - the copper plate, Fig. 9b, local heat transfer coefficients $\alpha_2(x)$;
- at the contact surface: the copper plate - water, Fig. 9c - local heat transfer coefficients $\alpha_3(x)$.

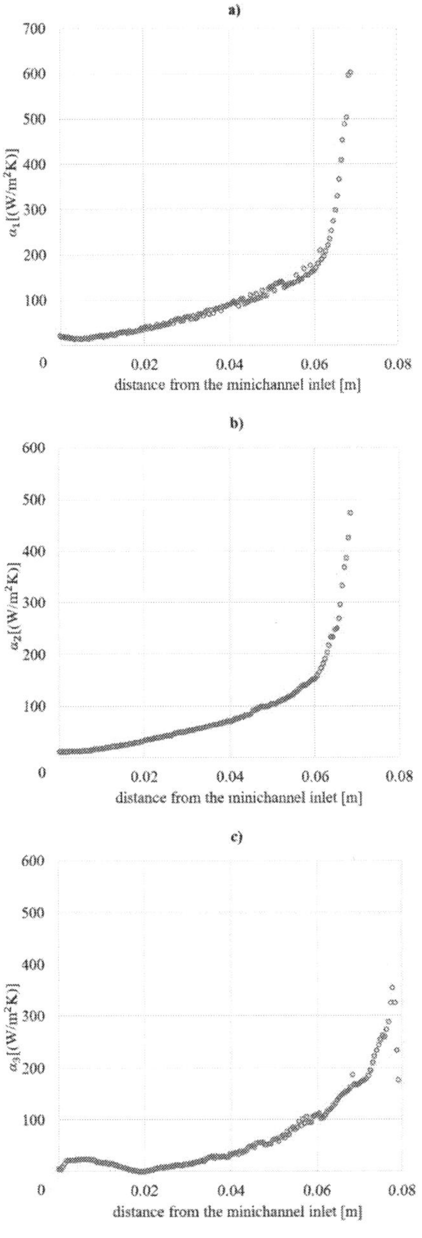

Fig. 9. Heat transfer coefficient versus the distance from the minichannel inlet: a) α_1 - PV panel/FC-72, b) α_2 - FC-72/copper plate, c) α_3 - copper plate/water.

When analyzing dependences showing the heat transfer coefficient versus the distance from the minichannel outlet to inlet, Fig. 9, it is noticed that the heat transfer coefficient decreases with the distance from the minichannel inlet, reaching maximum values near the inlet. Heat transfer coefficient values are in the ranges as follows:

- referring to the contact surface: the photovoltaic panel and FC-72: $\alpha_{1,min}$ = 11 W/(m^{2}K), $\alpha_{1,max}$ = 602 W/(m^2 K),
- referring to the contact surface: FC-72 and the copper plate: $\alpha_{2,min}$ = 8.3 W/(m^{2}K), $\alpha_{2,max}$ = 492.5 W/(m^2 K),
- referring to the contact surface: the copper plate and water: $\alpha_{3,min}$ = 3.7 W/(m^{2}K), $\alpha_{3,max}$ = 346.4 W/(m^2 K).

By comparing both experimental data on the heat transfer coefficient shown in Figs. 9a and 9b, the values of α_2 are lower than those of α_1. As in the case of the heat transfer coefficient α_1, the values of α_2 increase with increasing distance from the channel inlet and the heat flux supplied to the heated plate. When analyzing the data of the heat transfer coefficient presented in Fig. 9c, it can be seen that the coefficient α_3 are lower than α_1 and α_2 values.

The selected dimensionless numbers, mainly Reynolds, Prandtl and Graetz numbers, were determined separately for the minichannel with FC-72 and distilled water. In Table 2 shows the average values of these dimensionless numbers. Reynolds number values indicate laminar flow in both minichannels. Furthermore, the Fanning friction factors for both channels and the overall heat transfer coefficient were calculated and listed in Table 3.

Table 3. Values of selected dimensionless numbers, Fanning friction factors and the overall heat transfer coefficient.

Dimensionless Number	Average Value
Reynolds number (FC-72) Re_{FC}	1299.3
Reynolds number (water) Re_w	577
Prandtl number (FC-72) Pr_{FC}	11.7
Prandtl number (water) Pr_{FC}	8.1
Graetz number (FC-72) Gz_{FC}	185
Graetz number (water) Gz_w	55
Fanning friction factor (FC-72) f_{FC}	0.018
Fanning friction factor (water) f_w	0.047
Overall heat transfer coefficient	**Unit [W/(m^2·K)]**
Overall heat transfer coefficient k	523.6

The Reynolds number was calculated as follows:

$$Re = G \cdot d_h / \mu_l \quad (1)$$

where: G - mass flux, d_h - the hydraulic diameter of the minichannel, μ_l – liquid dynamic viscosity.

The following formula was used to calculate the Prandtl number:

$$Pr = \mu_l \cdot c_{p,l} / \lambda_l \quad (2)$$

where: λ_l - liquid thermal conductivity, cp_l - liquid specific heat.

Graetz number was defined according to the formula:

$$Gz = d_h \cdot Re \cdot Pr / L \quad (3)$$

where L - the minichannel length.

When designing heat exchangers, it is important to determine the overall heat transfer coefficient and the Fanning friction factor. For the counter-current heat exchanger, the overall heat transfer coefficient related to the heat transfer area A was calculated as follows:

$$k = Q / A \cdot \Delta T \quad (4)$$

where ΔT means logarithmic mean temperature difference calculated as in [22] and Q is the average of the heat fluxes from the hot and cold minichannels.

The Fanning friction factor f was obtained using the following formula [23,24]:

$$fRe = 24(1\text{-}1.3553K + 1.9467K^2 - 1.7012K^3 + 0.9564K^4 - 0.2537K^5) \quad (5)$$

where K is the aspect ratio equal to the ratio of the channel width to the depth.

When performing numerical calculations, validation of the results should be taken into account. It should be underlined that the results of the calculations presented in this article were based on experimental data. One of the validation procedures of CFD calculations is to provide a mesh dependence study. This step ensures that the discretization error is known and that it can be quantifiable. In particular, the methodology used is GCI, developed by Roache [25]. Using this methodology, the results of the heat transfer calculation related to the mini-channel heat sink performed in the Simcenter Star-CCM+ software were described in [20]. Furthermore, when comparing the results with those of other authors who applied rectangular minichannels in heat exchanger construction, it can be noticed that the results are similar in the "hot" minichannel as in works [14, 15, 26-32].

In summary, the presented research results should be treated as preliminary. In further research, the analysis of heat transfer in wider thermal and flow parameters, for other working fluids, with a variable spatial orientation of the test module and other dimensions of minichannels.

4 Conclusions

This paper aims to present a new construction of a compact plate heat exchanger dedicated to PVT cooling. The photovoltaic cell constitutes the heated wall of a counter-current heat exchanger with 12 rectangular minichannels. To perform numerical calculations, the CAD model of the test module was created in Simcenter STAR-CCM+ software. Due to the CFD calculations on the data from experiments, the temperature distributions on the heat exchanger plates and fluid contact surfaces were achieved. The temperature distributions indicate that the temperature increases with the distance from the inlet. Finally, local heat transfer coefficients were determined, presented on graphs, and discussed. It was noticed that the heat transfer coefficient increased with distance from the inlet. Further investigations with a wide range of experimental thermal flow and geometrical parameters, including spatial orientation, are planned.

Acknowledgment

The research reported herein was supported by a grant from the National Science Centre, Poland, No. UMO-2018/31/B/ST8/01199. The calculations were made using the Simcenter Star-CCM+ software developed by Siemens PLM Software Inc., provided by GMSystem which is a supplier of Siemens Digital Industries Software in Poland.

References

1. E. Skoplaki, J.A. Palyvos, Renew. Energy. **34**, 23–29 (2009)
2 E. Skoplaki, J.A. Palyvos, Sol. Energy. **83**, 614–624 (2009)
3 T.T. Chow, Appl. Energy. **87**, 365–379 (2010).
4 T.T. Chow, G.N. Tiwari, C. Menezo, Int. J. Photoenergy. 307287, 1-17 (2012)
5 M. Souliotis, Y. Tripanagnostopoulos, S.A. Kalogirou, 1-4/2009 **1**, 28–31 (2010).
6 T. Bergene, O.M. Løvvik, Sol. Energy. **55**, 453–462 (1995)
7 H.P. Garg, R.S. Adhikari, Renew. Energy. **11**, 363–385 (1997)
8 H.P. Garg, R.S. Adhikari, Int. J. Energy Res. **23**, 1295–1304 (1999).
9 J. Prakash, Energy Convers. Manag. **35**, 967–972 (1994)
10 G. Rockendorf, R. Sillmann, L. Podlowski, B. Litzenburger, Sol. Energy. **67**, 227–237 (1999).
11 A. Akbarzadeh, T. Wadowski, Appl. Therm. Eng. **16**, 81–87 (1996)
12 Y. Bai, T.T. Chow, C. Ménézo, P. Dupeyrat, Int. J. Photoenergy, 265838 (2012)
13 J.C. Hollick, Renew. energy. **15**, 195–200 (1998).
14 Q. Li, M. Avramova, Y. Jiao, P. Chen, J. Yu, Z. Pu, J. Chen, Nucl. Eng. Des. **326**, 403–412 (2018)
15 V.Y. Agbodemegbe, X. Cheng, E.H.K. Akaho, F.K.A. Allotey, Nucl. Eng. Des. **285**, 134–149 (2015)
16. M. Piasecka, S. Hożejowska, B. Maciejewska, A. Pawińska, Energies, **14**, 1832 (2021)
17. B. Maciejewska, K. Strak, M. Piasecka, Int. J. Numer. Methods Heat Fluid Flow. **28** (2018)
18. M. Piasecka, B. Maciejewska, P. Łabędzki, Energies. **13**, 6647 (2020)
19. B. Maciejewska, P. Łabędzki, A. Piasecki, M. Piasecka, EPJ Web Conf. **143** (2017)
20. M. Piasecka, A. Piasecki, N. Dadas, Energies,**15**, 536 (2022)
21. https://www.plm.automation.siemens.com/global/en/products/simcenter/STAR-CCM.html.
22. J. H. IV Lienhard, J. H. V. Lienhard, 5rd ed. Phlogiston Press Cambridge, MA, USA (2019)
23. W. Rybiński, J. Mikielewicz, Arch. Thermodyn. **35**, 29–42 (2014)
24. R.K. Shah, A.L. J. Fluids Eng. **102** (2) 431–455 (1978)
25. P. J. Roache, J. Fluids Eng. Trans. ASME, 116, **3**, 405–413, (1994)
26. M. Piasecka, S. Hożejowska, A. Pawińska, D. Strąk, Energies, **15**, 1340 (2022)
27. K.R. Anderson,; M. Devost, W. Pakdee, N. Krishnamoorthy, J. Electron. Cool. Therm. Control. **03**, 144–154 (2013)
28. P. Dąbrowski, R. Kumar, Int. Commun. Heat Mass Transf. **129**, 105685 (2021)
29. J. Duan, J. Zhao, X. Li, S. Panchal, J. Yuan, R. Fraser, M. Fowler, Energies, **14**, 4187 (2021)
30. H.A. Zondag, Renew. Sustain. Energy Rev. **12**, 891–959 (2008)
31. M. Fattahi, K. Vaferi, M. Vajdi, F. Sadegh Moghanlou, A. Sabahi Namini, M. Shahedi Asl, Ceram. Int. **46**, 11647–11657 (2020)
32. H. Alizadeh, M. Alhuyi Nazari, R. Ghasempour, M.B Shafii, A. Akbarzadeh, Sol. Energy, **206**, 455–463 (2020)

On dynamics of the flow in gap between two buildings in tandem configuration

Václav Uruba[1, 2*], *Pavel* Procházka[1], *Václav* Svoboda[3], and *Renata* Gnatowska[4]

[1]Department of Fluid Dynamics, Institute of Thermomechanics of the Czech Academy of Sciences, Dolejškova 5, 182 00 Prague 8, Czech Republic

[2]Department of Power System Engineering, Faculty of Mechanical Engineering, University of West Bohemia, Universitní 8, 306 14 Plzeň, Czech Republic

[3]Faculty of Nuclear Sciences and Physical Engineering, CTU in Prague, Břehová 78/7, 115 19 Prague 1, Czech Republic

[4]Czestochowa University of Technology, Faculty of Mechanical Engineering and Computer Science, 21 Armii Krajowej Av, 42-200 Czestochowa, Poland

Abstract. Two buildings differing in height in tandem configuration are subjected to a turbulent boundary layer flow. The physical modelling in a wind tunnel has been performed. Particle Image Velocimetry in horizontal planes has been applied. The time-mean flow-field is dominated by the contrarotating vortex pair in the near-wake of the upstream building forming strong back-flow in this wake. Dynamical activity is located close to the bigger downstream building below the level of the smaller building roof. The dynamical structures in the space between them are studied using advanced analytical methods, Oscillation Pattern Decomposition (OPD). The 3 important OPD modes were detected differing by topology and frequency. The 2 OPD modes are characterized by vortical structures formed within the wake of the smaller upstream building and moving downstream towards the bigger building. The last OPD mode shows strong pulsation transversal flow in front of the bigger building. Typical frequency of the cyclo-stationary behaviour could be characterized by the Strouhal number in the range from 0.09 up to 0.15.

1 Introduction

With increasing urbanization and density of population, the understanding of the problem of flow fields in urban environments, so called micrometeorological situation, becomes more important. Increased interest in has been observed over the past few decades. The results of such studies could be applied in the design process of new buildings as well as reconstruction and optimization of the existing urban areas and city centres. This issue plays an important role in environmental protection, defining so-called wind comfort around and between buildings, the life quality in urban areas as well as specific economic aspects of utilization of the defined zones. The studies of the wind environment around buildings are generally performed as wind tunnel experiments since it is the most well-established way to simulate a natural wind. The Reynolds number of wind tunnel modelling is several orders of magnitudes lower than this in the real situation, however its value is supercritical, turbulence in wakes is well developed and thus the results on turbulent diffusion are relevant, however the relevance is limited, see the discussion e.g. in [1,2].

Wind-tunnel measurements are performed using classical point techniques such as hot-wire or hot-film anemometry very often, see e.g. [1-5]. Recently, the Particle Image Velocimetry (PIV) technique is used, [6,7], as it provides much more information about the flow topology than the point methods.

The objective of this paper is to provide a detailed experimental analysis of the flow interactions of buildings models situated one after another in a so-called tandem arrangement. This configuration is rather common in practice. The main aim is to clarify the flow field between the buildings with help of advanced experimental technique, as the time-resolved PIV technique in several measuring planes to capture both spatial and dynamical features.

The preceding study [8] was performed on the same geometry, however vertical flow around the buildings has been studied within the vertical plane of symmetry. Dominant vortical structures have been detected both in time-mean and velocity fluctuation fields. The dynamical analysis has been performed using Proper Orthogonal Decomposition (POD) method based on turbulence kinetic energy assessment.

The presented paper is concentrated on flow in horizontal planes in gap between the buildings.

2 Experimental setup

The existing blown-down facility located in Shear Flows and Turbulence laboratory of the IT AS Prague was used for the experiments.

* Corresponding author: uruba@it.cas.cz

2.1 Physical model

The model of the buildings in tandem configuration is placed to the flow in the test-section inlet of the blown-down facility. The cross-section of the closed test-section is 250 x 250 mm^2 and 4.5 m long.

The model of buildings is placed in the test-section axis, 4 m from the inlet and it is shown in Figure 1, where positions of the planes of measurement are shown in green.

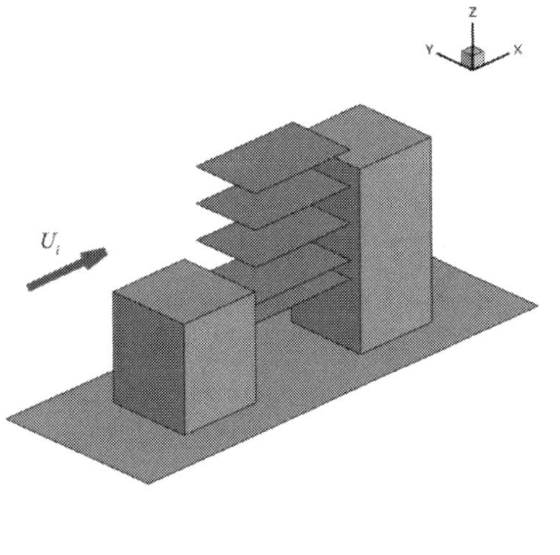

Fig. 1. The three-dimensional image of the experiment setup, planes of measurement are in green.

The two buildings with the same square ground plane with side $d = 25$ mm and heights $1.2d$ and $2d$ are shown, respectively. The smaller building is placed upstream, the flow is along the building's axis. The gap between the buildings is $1.5d$ is the streamwise direction.

The coordinate system is shown in Figure 2, where the view from above towards the model in $-z$ direction is shown. The coordinates are dimensionless, indicating multiples of the building ground plane side d.

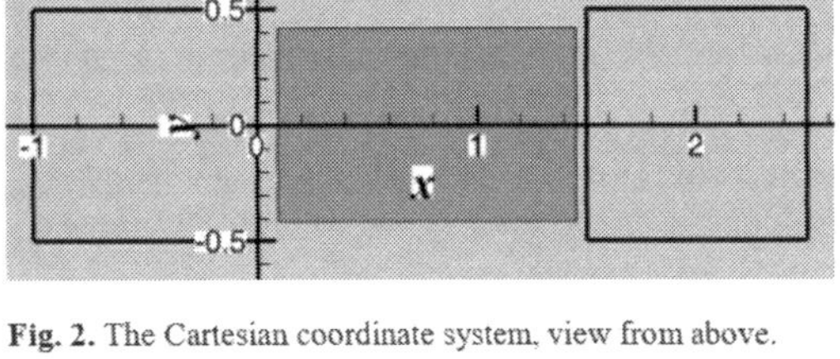

Fig. 2. The Cartesian coordinate system, view from above.

The mean velocity profile in the edge of the upstream, smaller building (position $x = -1$) is shown in Figure 3. The heights of the smaller ($1.2d$) and bigger ($2d$) buildings are shown only for comparison. The velocity profile is well-developed turbulent boundary layer velocity profile, conventional thickness is about $2d$. In z direction the profile is constant in the central region $z = +-3d$. The velocity U is dimensionless, indicating multiples of the incoming velocity $U_i = 10$ m/s outside the boundary layer.

The mean velocity profile was measured using PIV and HW measuring techniques, see [9].

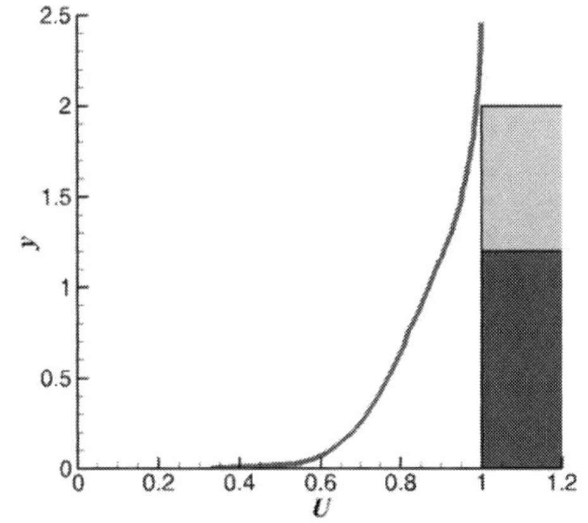

Fig. 3. Mean velocity profile on the model inlet.

The Reynolds number based on the velocity U_i and buildings side d was about 16 thousands.

2.2 Experimental technique

For evaluation of velocity distributions within the measuring planes (see Figure 1) the Particle Image Velocimetry technique was applied, hereinafter PIV.

The velocity vector field was measured using time-resolved PIV method. The measurement apparatus consists of laser and CMOS camera by Dantec. The laser is New Wave Pegasus, Nd:YLF double head with wavelength of 527 nm, maximal frequency 10 kHz, shot energy is 10 mJ (for 1 kHz) and corresponding power is 10 W per one head. The camera Phantom V611 with resolution of 1 280 x 800 pixels is able to acquire double-snaps with frequency up to 3 kHz (full resolution) and it uses internal memory 8 GB. The data were acquired and post-processed in Dynamic Studio software. The evaluated velocity fields consisting of 159 x 99 vectors were acquired in frequency 2 kHz, one record contained 4 000 double-snapshots representing 2 s in real time. The SAFEX particles, oil droplets 1 μm in diameter, have been used.

2.3 Methods of analysis

Besides the classical statistical data analysis based on averaging, the advanced methods were applied.

The Oscillation Pattern Decomposition method (hereinafter OPD) is used to study the dynamical properties of the flow-field. The OPD method yields series of OPD modes. Each OPD mode is characterized by its topology in complex form (consisting of real and imaginary parts), frequency f and attenuation of the pseudo-periodic (oscillating) behaviour. Attenuation or amplitude decay is described by so called e-folding time τ_e representing the mean time period of the mode amplitude decay by factor "*e*". The other decay characteristic is dimensionless "periodicity" p which expresses the e-folding time in multiples of periods of the OPD mode defined by its frequency.

$$p = \tau_e \cdot f \tag{1}$$

The frequencies are expressed in dimensionless form as Strouhal numbers *Sh*, defined as follows:

$$Sh = \frac{f \cdot d}{U_i} \tag{2}$$

The details on OPD method could be found in [9,10].

3 Results

All results are to be presented in dimensionless form. For this purpose, the model dimension d and inlet velocity U_i were used.

First, the statistical data, mean values and variances, are to be shown. Then, the results of advanced dynamical analysis OPD will be presented.

3.1. Time-mean patterns

The method of averaging was applied on velocity vector field series. The results are to be shown as vector fields with vector-lines added arbitrarily (blue lines).

Two typical results in the lanes $z = 0.8$ and 1 are shown in Figures 4 and 5, respectively.

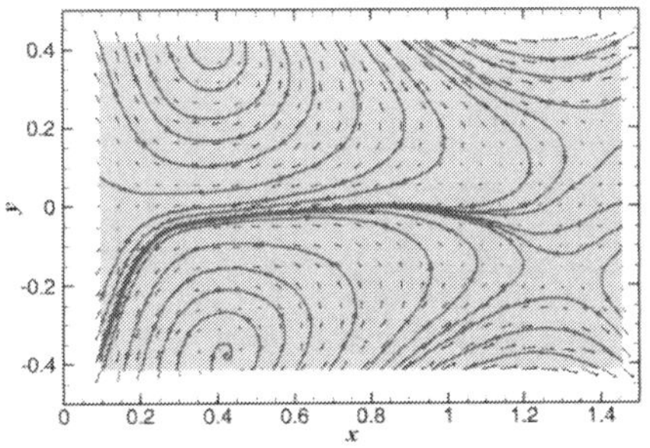

Fig. 4. The mean velocity field in the plane $z = 0.8$.

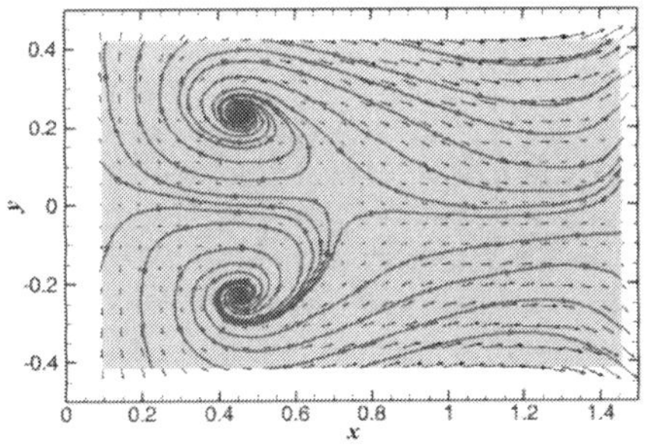

Fig. 5. The mean velocity field in the plane $z = 1$.

The measuring planes bellow the smaller building roof are strongly affected by the buildings presence. The plane $z = 0.8$ shows the contra-rotating vortices located at $(x;y)$ positions (0.4;+-0.4) generating strong back-flow on the axis – see Figure 4. Close to the downstream building, the saddle point is detected at $x \approx 1.3$.

The vortices are well pronounced in the plane $z = 1$, they are closer to each other, again generating back-flow just behind the upstream building. The vortices centres are located at positions (0.4;+-0.25), see Figure 5.

The higher measuring planes ($z > 1.2$) show nearly unperturbed flow above the upstream building. In Figure 6, there is the flow-field in the plane $z = 1.4$, the flow declination could be detected close to the higher building, for $x > 1.1$

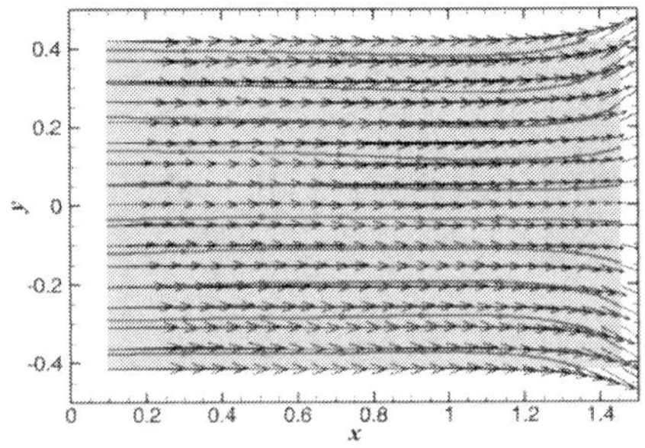

Fig. 6. The mean velocity field in the plane $z = 1.4$.

The measuring planes for higher z exhibit even smaller perturbations of the mean flow.

The Turbulent Kinetic Energy (*TKE*) is a measure of local dynamical activity of the flow-field, which plays an important role in diffusion processes. The *TKE* is defined using all 3 velocity components fluctuations as half of sum of velocity components variances in a given point in space. As we have at our disposal only 2 velocity components in the plane of measurement (the out-of-plane component is not detected) we have used the following definition of the *TKE*:

$$TKE = 3/4\left(\overline{u^2} + \overline{v^2}\right), \tag{3}$$

here u and v are fluctuations of streamwise and spanwise velocity components, respectively. Then, $\overline{u^2}$ is the streamwise velocity component variance. Dimensionless velocity is considered. This definition of *TKE* presumes the variance of the third, unresolved velocity component to be average of the two resolved.

The distribution of the *TKE* in the plane $z = 1$, is shown in Figure 7. The near wake behind the upstream building is relatively calm, maximal fluctuation activity is detected just in front of the higher building in the position $x = 1.2$.

https://doi.org/10.1051/epjconf/202226401046

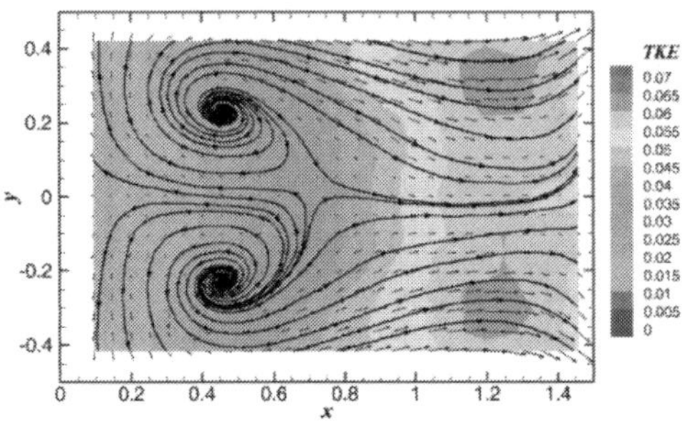

Fig. 7. Turbulence Kinetic Energy in the plane $z = 1$.

The distribution of TKE in the whole measured domain is shown in Figure 8, where the iso-surfaces of TKE are drawn.

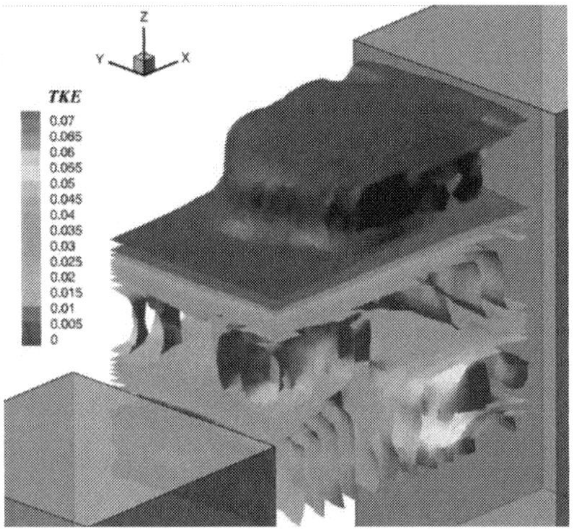

Fig. 8. Isosurfaces of Turbulent Kinetic Energy in the whole measurement domain.

Indeed, maximal fluctuation activity could be detected just in front of the higher building and bellow of the smaller building roof level. Above the smaller building roof, the *TKE* is decreasing considerably, slightly below the higher building level, the *TKE* is approaching zero.

The dynamical analysis will be performed for the level $z = 1$.

3.2. Dynamical patterns

The flow-field in the position $z = 1$ has been subjected to detailed dynamical analysis using the OPD method.

The 16 OPD modes have been evaluated, in Figure 9 the modes are represented in the plane *Sh-p*, definitions see (1) and (2).

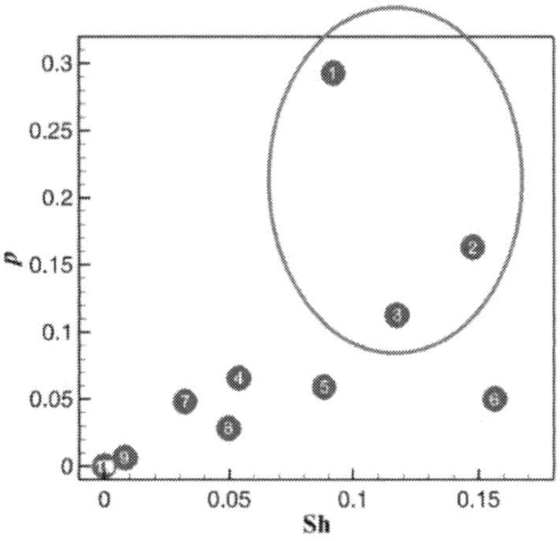

Fig. 9. OPD spectrum.

The periodicity bellow 0.1 indicates rapid decay of a an OPD mode in question, such modes are dominated by random behaviour, and as such they play minor role in the flow dynamics. So, we decided to analyse only 3 OPD modes with the p above 0.1, within the red zone in Figure 9.

The first OPD mode topology, the most stable one, is depicted in Figure 10, the real and imaginary parts, respectively. It consists of the fluctuation velocity vector-field, the vector-lines are added arbitrarily for better clarity.

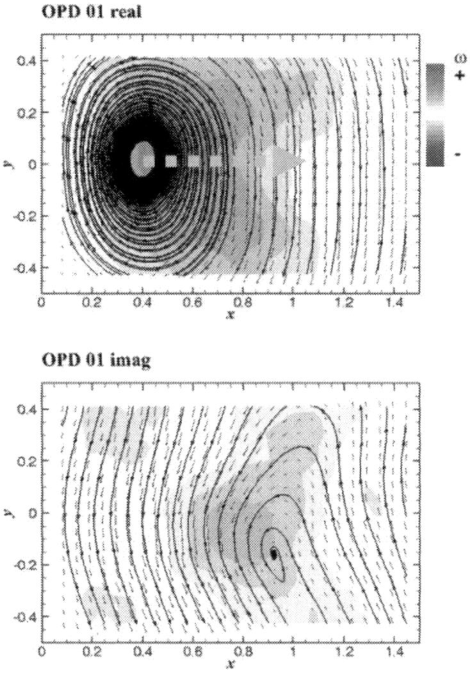

Fig. 10. Topology of the 1st OPD mode, real and imaginary parts.

In representations of the OPD modes topologies, the vorticity component perpendicular to the plane of measurement is added for better clarity, indicating positive (in red) and negative (in blue) orientations, however in arbitrary units. The real and imaginary parts of an OPD mode represent phase 0 and $\pi/2$ of the cyclostationary process, respectively. More information of OPD modes interpretation could be find e.g. in [9,10].

The mode OPD 1 is dominated by a vortex moving along the x axis. It originates within the near wake of the smaller building as a strong vortex and moving towards the bigger building, weaken and changing orientation of rotation in the same time. The yellow broken arrow indicates the vortex trajectory. The frequency corresponds to the Strouhal number about 0.092.

The OPD mode 2, in Figure 11, is characterized by a pair of contrarotating vortices. Their dynamics is similar to the mode OPD 1, they raise just behind the smaller building and moving in the streamwise direction towards the bigger one. Unlike the mode OPD 1, the vortices strengthen approaching the bigger building and tend to its sides. The Strouhal number is about 0.15.

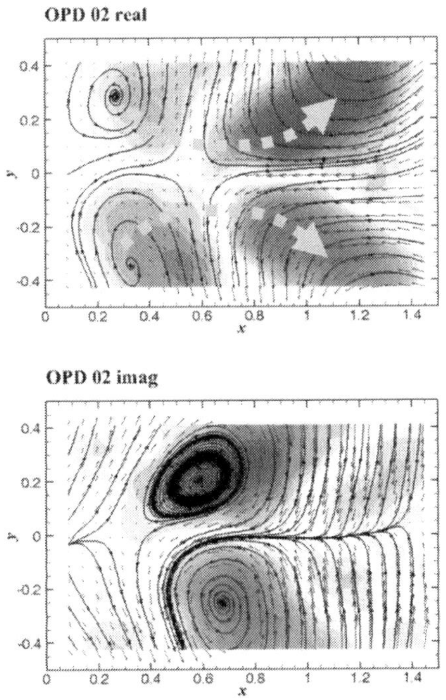

Fig. 11. Topology of the 2nd OPD mode, real and imaginary parts.

The OPD mode 3 topology is depicted in Figure 12. This mode is dominated by strong spanwise flow in front of bigger building. The direction is changing from positive to negative y axis orientation. The Strouhal number is about 0.12.

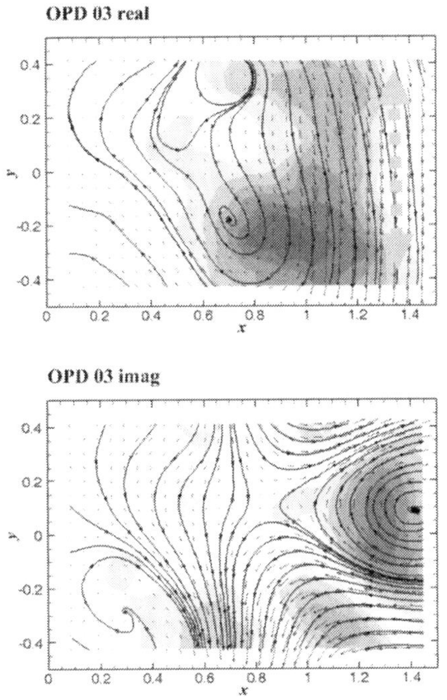

Fig. 12. Topology of the 3rd OPD mode, real and imaginary parts.

The higher order OPD modes are characterized by periodicity lower than 0.1 or even 0. This means that the damping of the mode is very quick in terms of period of cyclo-stationary behaviour. The real part of topology simply appears on random basis in time and then disappear gradually.

4 Conclusions

The presented paper shows some aspect of flow in the gap between two buildings in tandem configuration. Both mean velocity field and flow dynamics are addressed with emphasis to the region just below the smaller building roof level.

The time-mean flow-field is dominated by the contrarotating vortex pair sitting in the near-wake of the upstream building forming strong back-flow in this wake. Dynamical activity is located close to the bigger downstream building bellow the level of the smaller building roof. Above this level, the turbulence activity decay very quickly and it is very weak on the higher building roof level.

The pseudo-periodical or cyclo-stationary flow dynamics contains 3 important modes differing by topology and frequency. The vortical structures are formed within the near wake of the smaller upstream building and they are moving downstream towards the bigger building. In the 1st OPD mode the single vortex travels from the upstream wake to the downstream building. The 2nd OPD mode shows the vortex contrarotating pair moving from the wake around the sides of the bigger building. The 3rd mode exhibits strong transversal pulsations just in front of the bigger building.

The typical frequency of cyclo-stationary behaviour could be characterized by the Strouhal number in the range $\mathrm{Sh} \in (0.09; 0.15)$

The presented work was financially supported by student project SGS-2019-021 (Improving the efficiency, reliability and service life of power machines and equipment 5), UWB in Pilsen. The investigation was supported by National Agency for Academic Exchange under Grant no. PPI/APM/2019/1/00062.

References

1. Y. Uematsu, M. Yamada, H. Higashiyama, T. Orimo, *J. Wind Eng. Ind. Aerodyn.*, **44** (1-3) 2289-2300 (1992)
2. B. Blocken, T. Stathopoulos, J. Carmeliet, *J. Aerosp. Eng. – ASCE,* **21** (4) 280-287 (2008)
3. A. Jarża, R. Gnatowska, *Chem Process Eng-Inz.* **27** (3/1), 721-735 (2006)
4. R. Gnatowska, *Int. J Turbo Jet. Eng.* **28** (1) 21-29 (2011)
5. R. Gnatowska, *Pol J Environ Stud*, **24** (4), 1571-1577 (2015)
6. J.P.A.J. Van Beeck, G. Dezso, P. Planquart, *PHYSMOD 2009*, Von Karman Institute (VKI) (2009)
7. R. Gnatowska, M. Sosnowski, V. Uruba, *E3S Web of Conferences*, **14** 01034 (2017)
8. R. Gnatowska, R. Kellnerova, V. Uruba, *EPJ Web of Conferences*, **180** 02029, (2018)
9. V. Uruba, *EPJ Web of Conferences*, **25** 01095 (2012)
10. V. Uruba, *J. of Fluids and Structures*, **55**, pp. 372-383 (2015)

Contribution on the kinematics of the polishing process on a polishing machine with horizontal overarm

Lukáš Veselý[1], *Ondřej* Matoušek[1, 2], *Tomáš* Vít[1, 3*], and *Miloš* Müller[3]

[1]asphericon s.r.o., 463 12 Jeřmanice, Czechia
[2]Technical university of Liberec, Institute of New Technologies and Applied Informatics, 461 17 Liberec, Czechia
[3]Technical university of Liberec, Dept. of Power Engineering Equipment, 461 17 Liberec, Czechia

Abstract. Understanding and controlling of the polishing process on conventional NC controlled machines is an important step to optimize production, reduce machine time and increase quality of production. The paper deals with the basic analysis of the problem and its comparison with the experiment performed during the real process on the machine and with the numerical MBS simulation of the process.

1 Polishing process

The technology of polishing of various materials is one of the oldest production processes. It therefore seems surprising that these processes are still not described by the corresponding analytical models, or a credible procedure for numerical simulations is not created. It is possible to find a number of publications that relate to individual partial problems of polishing. However, a comprehensive overview of procedures, supported by proper validation, is lacking. A closer look at the problem shows that the whole process is affected by a number of parameters, some of which are random in nature.

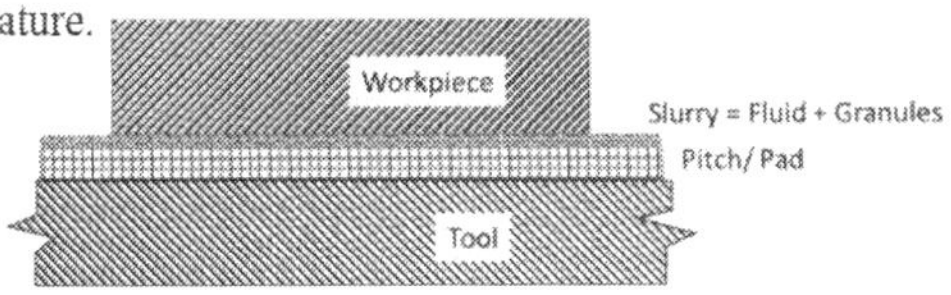

Fig. 1. Basic components of the polishing process.

In [1] it is possible to find a clear description of the components that are involved in the vast majority of polishing processes (see Fig. 1). The authors state that four most important components of the polishing chain include a workpiece, in our made from optical glass, fluid which represents the chemical composition of demineralized water and additives like pH controllers, inhabitants to control chemical interactions and surfactants.

The third main component of the polishing process are solid granules. Solid granules act together with the fluid component of polishing slurry. The task of the granules in the slurry is mainly the mechanical removal of material from the polished surface. The most important parameters of the granules are the chemical composition, size and shape of the individual particles and the concentration of the particles.

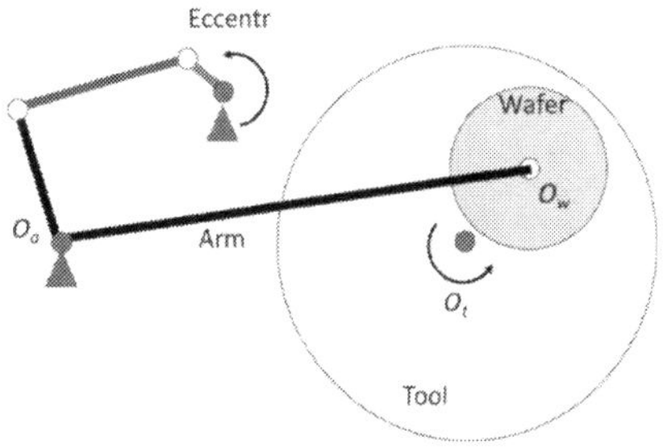

Fig. 2. Layout of NC polishing machine with horizontal overarm.

Fourth important part of the polishing chain is the polishing tool. The polishing tool provides relative motion between the granules and the workpiece. In most cases, the polishing tool consists of a rigid wall made of steel, alumina or silicon carbide, which serves as a carrier for a layer of pitch or various pads (polyurethane foam). Pads usually contain a variety of groove patterns to ensure even distribution of the polish slurry.

2 Kinematics of polishing

The basic arrangement of the polishing process is presented on Fig. 2. The tool rotation remains constant during the process, as does the shaft rotation, which transmits the speed to the eccentric arm of the lever mechanism. The oscillation of the arm and the associated speed of movement of the arm is thus determined by the geometric proportions of the individual arms.

The polished optical surface does not gain momentum per se. The only driving force of rotation is the rotational movement of the machining tool and movement of the arm. The kinematics of the polishing process without arm movement is very often described and analysed in the

* Corresponding author: tomas.vit@tul.cz

literature. For example, more detailed analysis of the problem is published in [2].

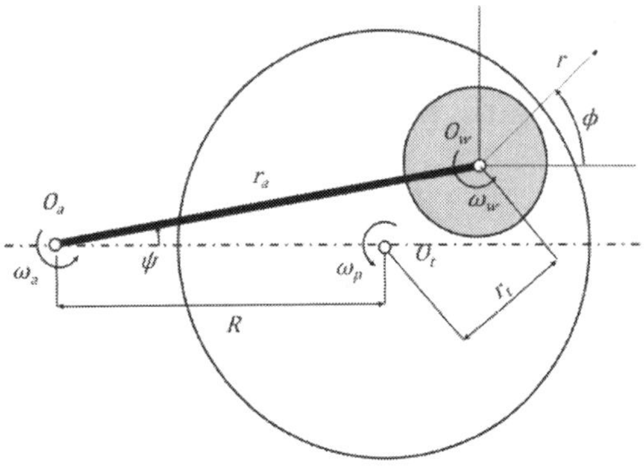

Fig. 3. Kinematics of the polishing mechanism.

Based on the kinematics of the horizontal overarm mechanism (Fig. 3) it is possible to separate the complex motion into three parts. Rotation of the workpiece, rotation of the polishing tool and motion of the arm. The resulting velocity correspond to the combination of the components mentioned above.

The velocity components of the workpiece (lens) in cylindrical coordinate system (r, ϕ) originated at the centre of the workpiece O_w are:

$$v_{w,r} = 0;\ v_{w,\varphi} = \omega_w r \quad (1)$$

Velocity components of the arm in the same coordinate system (r, ϕ) should be expressed as:

$$v_{a,r} = \omega_a r_a \sin\varphi\ ;\ v_{a,\varphi} = \omega_a(r + r_a \cos\varphi) \quad (2)$$

and velocity components of the polishing tool as:

$$v_{t,r} = \omega_t r_t \sin\varphi\ ;\ v_{t,\varphi} = \omega_t(r + r_t \cos\varphi). \quad (3)$$

Distance between centre of rotation of the workpiece and the tool should be expressed as $r_t = (R^2 + r_a{}^2 - 2Rr_a \cos\psi)^{1/2}$, where $R = |O_a O_t|$ denotes distance between rotation centre of the arm and the tool and $\omega_a = \frac{\delta\psi}{\delta t}$.

Components of the relative velocity should be expressed as:

$$v_{R,r} = \omega_a r_a \sin\varphi - \omega_t r_t \sin\varphi\ ; \quad (4)$$

$$v_{R,\varphi} = \omega_w r + \omega_a(r + r_a \cos\varphi) - \omega_t(r + r_t \cos\varphi) \quad (5)$$

The magnitude of the relative velocity is:

$$v_R = \{[r(\omega_w + \omega_a - \omega_t) + \omega_a r_a \cos\varphi - \omega_t r_t \cos\varphi]^2 + [\omega_a r_a \sin\varphi - \omega_t r_t \sin\varphi]^2\}^{1/2} \quad (6)$$

The tangential component of viscous friction could be then written as:

$$dF_\varphi = \frac{\mu p(r\delta r\delta\varphi)[(\omega_a r_a - \omega_t r_t)\cos\varphi + r(\omega_w + \omega_a - \omega_t)]}{v_R} \quad (7)$$

And corresponding torque on the wafer:

$$M_w = \int_0^{r_w}\int_0^{2\pi} \mu p r^2 \left[\frac{(\omega_a r_a - \omega_t r_t)\cos\varphi}{v_R} + \frac{r(\omega_w + \omega_a - \omega_t)}{v_R}\right] d\varphi dr \quad (8)$$

It is obvious that for $\omega_w = \omega_t - \omega_a$ the torque of the wafer is $M_w = 0$.

In real operation, the connection between the arm and the workpiece body is realized via a ball joint. Additional torque is generated in the real process due to the friction in the ball joint. The moment M_w will therefore be nonzero and $\omega_w < \omega_t - \omega_a$.

3. Numerical simulation

Several approaches can be chosen for numerical simulation of the problem. Numerical simulation using tailored code requires detailed knowledge of the problem. On the other hand, it provides complete control over the progress of the solution, the resulting model is simpler and less demanding on computing power and time.

The use of commercial MBS systems enables very fast and user-friendly preparation of models. However, this is redeemed by the high computation time and the loss of control over the simulation process. Due to the complexity of the problem, this can lead to meaningless results.

We have chosen both approaches when solving this problem. However, custom code simulation is currently only applicable for polishing of flat surfaces

Fig. 4 shows the vector fields of the peripheral speeds of the tool (blue) and the workpiece (red).

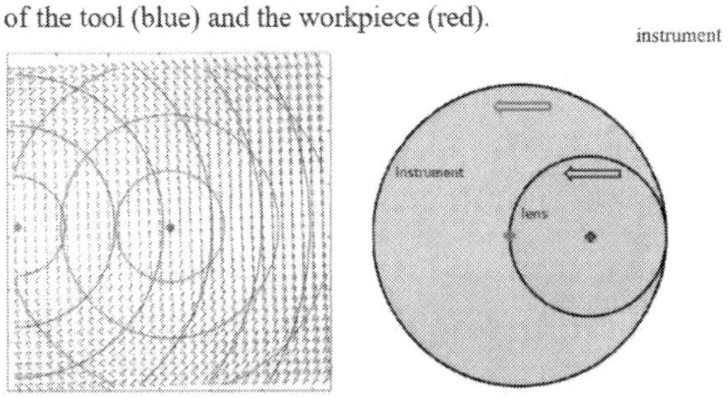

Fig. 4. Left, velocity vectors, blue color corresponds to tool, red to workpiece, crosses represent element centers, right, instrument and lens arrangement in Synchrospeed mode. The arrows indicate the direction of rotation.

Only the contact area of the tool and the polished surface is displayed. The red and blue contours, respectively, show the increasing circumferential speed of the workpiece and the tool, and it is clear from them how they overlap. The crosses of the respective colours show the centres of the individual elements. It is possible to get a good idea of the principle of the whole method from the direction of the individual arrows. If we focus on the development of velocities on the central axis, there is visible a clear development of circumferential velocities. At the centre of the tool, the circumferential speed is zero, but because the polished surface has its edge here, its circumferential speed is, on the contrary, maximum. As you move further along the axis, the circumferential speed of the workpiece decreases linearly, but the circumferential speed of the tool also increases linearly and the direction of movement of the surfaces is opposite in these positions.

Results from MBS simulation using Ansys Rigid Dynamics module are presented at Fig. 5 and Fig. 6. In contrast to the simulation in the MATLAB code, the beginning of the rotation of the tool in the time of 7 s and the associated transients is evident.

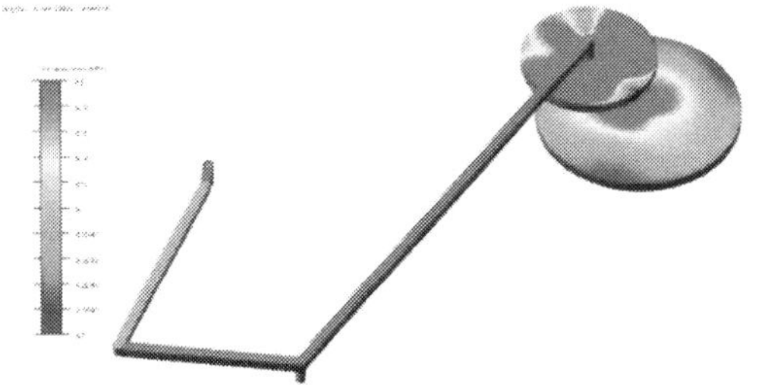

Fig. 5. MBS model of the polishing mechanism.

Presented results show the evolution of angular velocity as a function of time during polishing of planar surfaces. Due to the autorotational oscillating kinematics, the position of the polished surface is also time dependent.

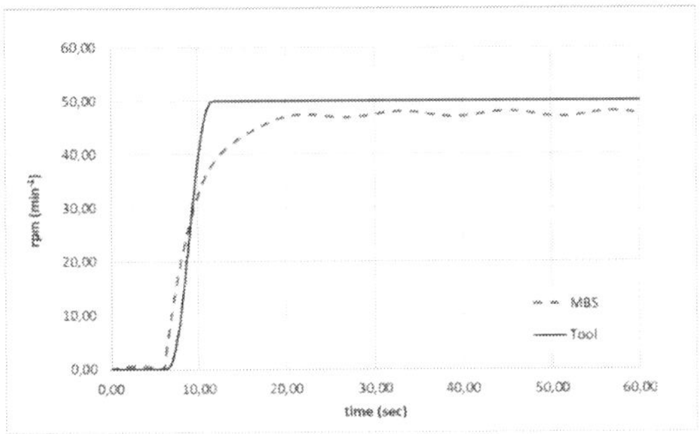

Fig. 6. results of MBS simulation.

4. Experiment

The whole experiment took place in a laboratory with a constant temperature of 24 °C. A 200 mm diameter tool with glued 0.58 g / cm^3 polyurethane was used on the NLP 400 HS-1. The density of the Auerpol PZ 110 polishing slurry used was 1.04 g/cm^3. To reduce the number of variables during polishing, the tool speed was set to a constant value of 50 rpm. Likewise, the rotation speed of the eccentric arm was set to a constant value of 5 rpm. The magnitude of the lens oscillation also remained unchanged throughout. The measurements were performed in eight positions of the tool relative to the workpiece $|r_a - R|$ = (-40; -30; -20; -10; 0; 10; 20; 30 mm). The positions are marked in Fig. 7.

However, only the results for $|r_a - R|$ = 30 mm are presented in this paper.

4.1. Measurement of the angular velocity of the workpiece

The position of the overarm was derived from acceleration measured by the accelerometer attached to the overarm. The signal was quite noisy due to vibrations generated during the sliding movement of the lens across the instrument (Fig. 8). It was necessary to perform a FFT to determine the frequency of the oscillating motion from the resulting frequency spectrum and to apply bandpass IIR filter to remove frequencieses generated by vibrations.

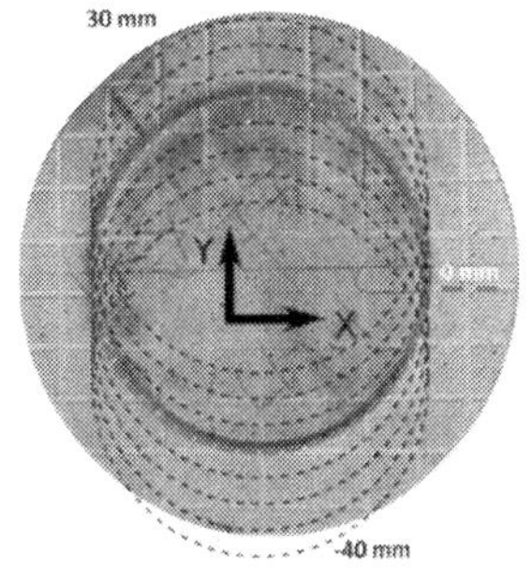

Fig. 7. Different Tool-Workpiece positions.

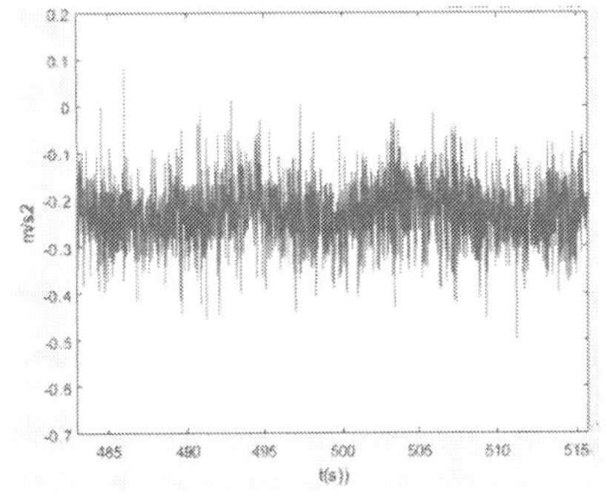

Fig. 8. Non processed results from angular velocity measurement.

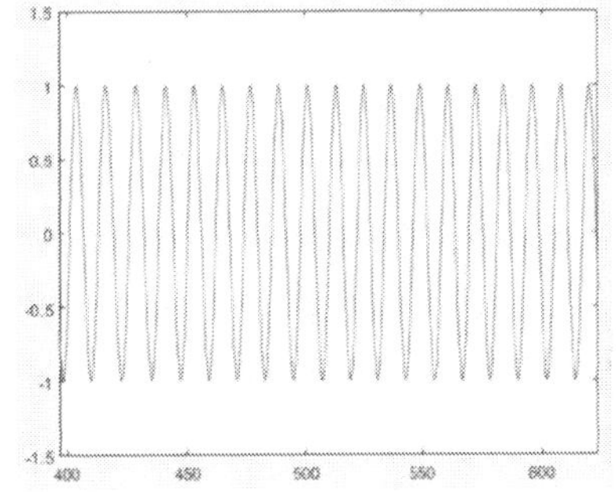

Fig. 9. Final filtered signal.

The Hilbert transform of the signal determined the waveform envelope function and removed the modulation. From an experimental point of view, only the relative position of the lens is important, so the loss of absolute position information due to modulation removal is irrelevant. The result of processed signal which shows relative position of the overarm as a function of time is presented in Fig. 9.

4.2. Measurement of the angular velocity of the workpiece

Angular velocity of the workpiece was measured by a non-contact optical sensor located in close proximity to the rotating workpiece. The workpiece was marked with tags and the time between the individual tags was evaluated during the experiment. From this, the angular velocity was subsequently evaluated.

From the theoretical assumptions and from the results of simulations, it is evident that the course of speed changes is sinusoidal. The frequency of the sine signal is given by the speed of rotation of the eccentric arm. The time period for the calculation was therefore chosen on the basis of the Nyquist criterion. After calculating the average speeds over the entire twenty-minute record, the appropriate data interpolation using the sine functions was performed.

Subsequently, a FFT analysis of the interpolated data (Figure 3) of the average angular velocity as a function of time was performed to confirm the results. This analysis showed the maximal amplitude at 0.0828 Hz, which corresponds to an eccentric arm rotation speed of 4.968 rpm.

5. Results

Fig. 10 shows comparison of the angular velocity achieved by different methods.

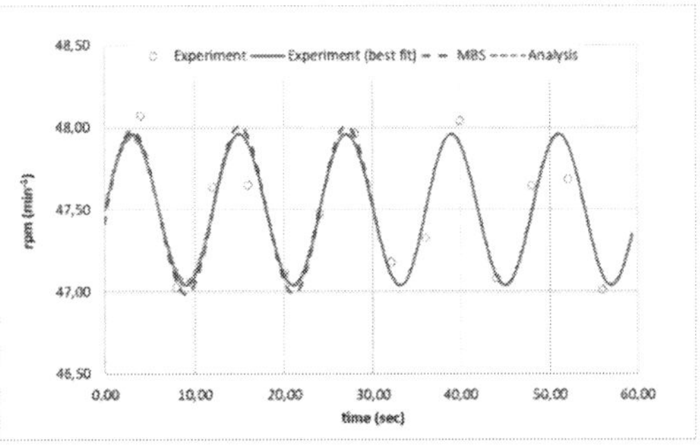

Fig. 10. Comparison of the achieved results.

It shows that different approaches could lead to comparable results. As it is difficult to determine the friction in the ball joint connecting the overarm and the workpiece the results from experiment were used to find this value. Presented comparison shows that results of analytical solution should be used for fast determination of the basic parameters of polishing process. However, the MBS is shown as powerful but expensive and time-consuming tool.

6. Conclusion

The nature of the rotational speed of the workpiece on conventional NC controlled machines can be determined by different methods, which give similar results. Results of analysis and numerical simulations are in agreement with results of precise experiments.

The rotational speed is given only by the transfer of force from the rotation of the tool to the lens. If the workpiece moves against the direction of tool rotation, the speed increases. On the contrary, when moving in opposite direction, the speed decrease.

Acknowledgements

The research has been supported by the Ministry of Industry and Trade, project The Country for the Future, FX01030046.

References

1. C. J. Evans, E. Paul, D. Dornfeld, D. A. Lucca, G. Byrne, M. Tricard, F. Klocke, O. Dambon, B.A. Mullany, CIRP Annals, Volume 52, Issue 2 (2003)
2. L. Jiun-Yu, Mechanics, mechanism and modeling of the chemical mechanical polishing process, Massachusetts institute of technology (2001)

Optimization of the polishing process on a polishing machine with horizontal overarm

Lukáš Veselý[1], *Ondřej* Matoušek[1,2], *Tomáš* Vít[1,3*] and *Jan* Novosád[3]

[1]asphericon s.r.o., 463 12 Jeřmanice, Czechia
[2]Technical university of Liberec, Institute of New Technologies and Applied Informatics, 461 17 Liberec, Czechia
[3]Technical University of Liberec, Dept. of Power Engineering Equipment, 461 17 Liberec, Czechia

Abstract. Understanding the basic principles of polishing on machines with autorotating swing kinematics is important for the description and understanding of more advanced polishing technologies. The article deals with the theoretical and numerical analysis of the polishing process and its comparison with the experiment performed during the real process on the machine. The dependence of the mutual position of the tool and workpiece axes on the removal distribution within the polished surface was described and demonstrated

1 Polishing process overview

Polishing is a chemical-mechanical machining of the workpiece surface. The goal is to get a smooth, flawless surface that also meets the shape requirements. Usually during polishing, one of the two parts (tool or product) rotates, while the other part oscillates. Polishing is the final and most precise machining operation, during which the surface of the lens can still be handled. [1, 2].

The basic equation which describes the dependence of the removal rate on process parameters is the Preston equation [3]:

$$\frac{d\xi}{dt} = k_p p v_R, \qquad (1)$$

where ξ is the thickness of the material layer removed, t is the polishing time, p is the pressure acting on the polished element at a given location, v_r is the relative velocity between the polishing tool and the workpiece and k_p is Preston constant.

The validity of the relationship mentioned above was confirmed in many works. Currently, theoretical research focuses on evaluating the influence of chemical and mechanical factors on the polishing process. Details and analysis of individual approaches can be found in [4].

This article aims to show the influence of individual members of Preston equation on the final removal rate of material in the case of polishing with overarm.

1.1 Preston constant

The magnitude of the Preston constant depends on various parameters. According to [4, 5], the Preston constant is affected by the friction coefficient between the workpiece / slurry / and the pad. It is also affected by the friction mod. The magnitude of the Preston constant also depends significantly on the hardness of the machined surface, chemical composition of polishing slurry, character and size of polish grains.

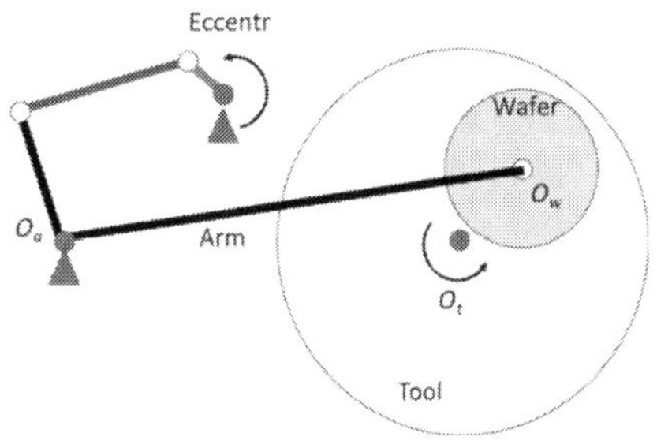

Fig. 1. Layout of NC polishing machine with horizontal overarm.

1.2 Modes of polishing

It is possible to distinguish between three modes of polishing based on interaction between the polishing pad/slurry and workpiece. In the case of so-called contact mode, there is direct contact between the polishing pad and the workpiece. No stable fluid film is formed between the workpiece and the tool, so no pressure gradient is created. This mode occurs at low relative velocity and high pressure. The coefficient of friction is in the range of 0.1.

In the hydroplaning mode (Fig. 2), a constant film of fluid is formed between the workpiece and the tool. Unlike the previous one, this mode is characterized by a high value of relative velocity and a relatively low value of pressure. The coefficient of friction typically ranges from 0.001 to 0.01. The pressure distribution in hydroplaning mode is highly dependent on the attack angle, i.e., the wedge between the workpiece and the tool. The third possible mode is mixing or transition mode,

* Corresponding author: tomas.vit@tul.cz

which arises in unstable states between the two modes mentioned above.

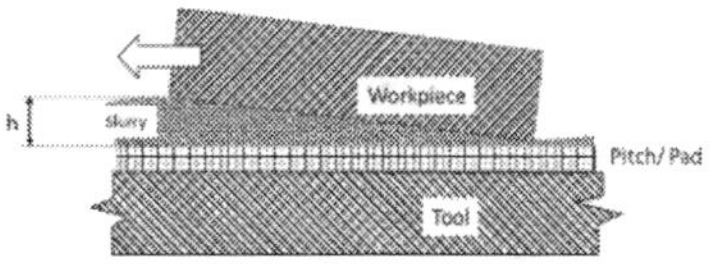

Fig. 2. Hydroplaning mode of polishing.

Results of experiments presented in [2] and [4] shows significant drop of the friction coefficient in the relative velocity range called Coulomb interval. In agreement with the previous mentioned modes the drop of friction coefficient is described as the aqua planning effect, characterized by the formation of a liquid film between the glass surface and the polishing pad because of the high rotational speed.

A technique for the measurement of the dynamic fluid film pressure during polishing was presented in [3]. It was found that the pressure distribution is significantly different between static and dynamic conditions.

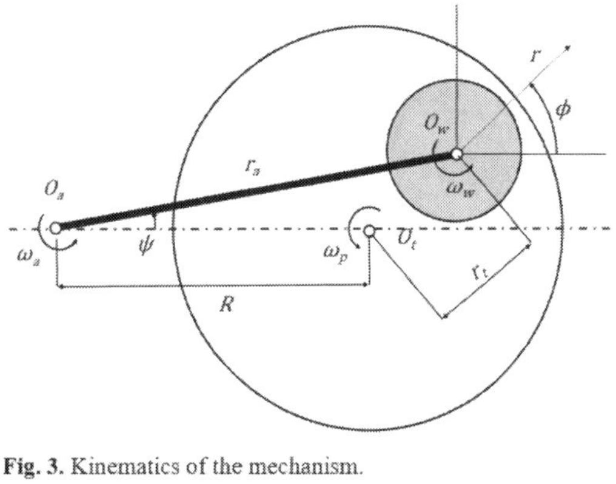

Fig. 3. Kinematics of the mechanism.

Stationary results showed an asymmetrical pressure distribution and some subatmospheric pressures. Dynamic experiments showed pressure increasing monotonically from edge to centre and no sub-atmospheric pressures.

1.3 Kinematics of polishing

The basic arrangement of the polishing process is described in [4]. Regarding the definition of individual components, as shown in Fig. 3 the following relationships can be specified.

The magnitude of the relative velocity is:

$$v_R = \{[r(\omega_w + \omega_a - \omega_t) + \omega_a r_a \cos\varphi - \omega_t r_t \cos\varphi]^2 + [\omega_a r_a \sin\varphi - \omega_t r_t \sin\varphi]^2\}^{1/2} \quad (2)$$

And corresponding torque on the wafer:

$$M_w = \int_0^{r_w}\int_0^{2\pi} \mu p r^2 \left[\frac{(\omega_a r_a - \omega_t r_t)\cos\varphi}{v_R} + \frac{r(\omega_w + \omega_a - \omega_t)}{v_R}\right] d\varphi dr \quad (3)$$

It is necessary to realize that both ω_a and r_t are time dependent harmonic functions generated by the motion of eccentric part of the mechanism.

It is obvious that for $\omega_w = \omega_t - \omega_a$ the torque of the workpiece is $M_w = 0$. In this case the relative velocity is independent on position r on the workpiece and the average relative velocity is uniform over whole surface of the workpiece.

In real operation, the connection between the arm and the workpiece body is realized via a ball joint. Additional torque is generated in the real process due to the friction in the ball joint. The torque M_w will therefore be nonzero and $\omega_w < \omega_t - \omega_a$.

1.4 Pressure distribution.

The pressure distribution between the workpiece and the tool can be described by the so-called Reynolds equation, which is derived for the flow of fluid in very narrow channels. Detailed analysis can be found in [4]

$$\frac{\partial}{\partial x}\left(\frac{h^3}{12\mu}\frac{\partial p}{\partial x}\right) + \frac{\partial}{\partial y}\left(\frac{h^3}{12\mu}\frac{\partial p}{\partial y}\right) = \frac{\partial}{\partial x}\left[\frac{h(u_{x,w}+u_{x,t})}{2}\right] + \frac{\partial}{\partial y}\left[\frac{h(u_{y,w}+u_{y,t})}{2}\right] - u_{x,w}\frac{\partial h}{\partial x} - u_{y,w}\frac{\partial h}{\partial y} \quad (4)$$

The analysis shows that the pressure distribution strongly depends on the value of wedge which is expressed by the change in gap thickness h over the workpiece. It can be easily shown that the increase of the angle between the workpiece and the tool leads to the resultant of the pressure shifts to the trailing edge.

Analytical solution of equation (4) is impossible even for a simple configuration between workpiece and tool. Therefore, it is appropriate to use numerical simulations.

Numerical simulation was performed in an Ansys Fluent environment. The entry conditions were chosen to match the experiment. The angle between the workpiece and the tool, which could not be measured during the experiments, was changed in four steps to describe the trend.

2. Experiment

2.1 Conditions

The whole experiment took place in a laboratory with a constant temperature of 24 °C. A 200 mm diameter tool with glued 0.58 g / cm^3 polyurethane was used on the NLP 400 HS-1. The density of the Auerpol PZ 110 polishing slurry used was 1.04 g/cm^3. To reduce the number of variables during polishing, the tool speed was set to a constant value of 50 rpm. Likewise, the rotation speed of the eccentric arm was set to a constant value of 5 rpm. The magnitude of the lens oscillation also remained unchanged throughout. The measurements were performed in eight positions of the tool relative to the workpiece $|r_a - R|$ = (-40; -30; -20; -10; 0; 10; 20; 30 mm). The positions are marked in Fig. 4.

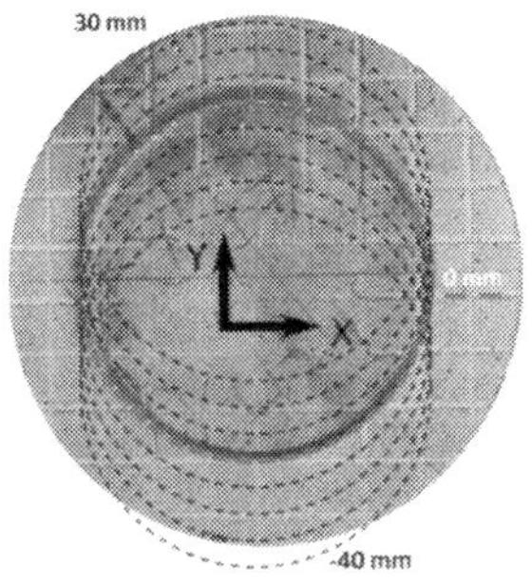

Fig. 4. Different Tool-Workpiece positions.

2.2 Preparation of the sample

The shape of the surfaces was polished based on previous experience with polishing to the best possible level in the preparatory phase of the experiment (Fig. 5. The surface form accuracy was checked with a Xonox VT 1200 PS Fizeau interferometer. The reference surface used was a Xonox PRO Line 4" transmission sphere with a focal length at infinity. The use of a 4 "sphere made it possible to evaluate the entire surface of the measured lens.

Subsequently, the lens was polished for 20 minutes in each position. After each measurement, the shape of the optical surface was measured, and the size of the removal was evaluated.

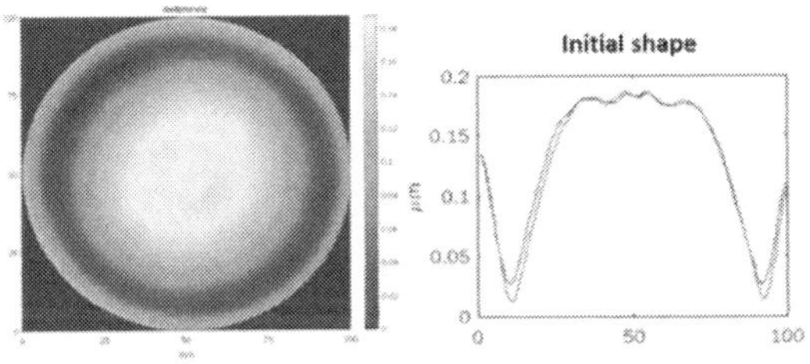

Fig. 5. Initial (before the experiment) shape of the surface.

3. Results

3.1. Experiment

Interferograms and profiles of the surface measured in perpendicular directions are presented at Fig. 6 and Fig. 7. All experiments were performed on the same surface. Change of the surface form after the polishing at specific position $(r_a - R)$are presented.

Results shows high removal rates at the edge of the surface for high values, positive or negative, of $(r_a - R)$. Position of $(r_a - R)$ close to 0 leads to increase of the removal rate at the centre of the workpiece.

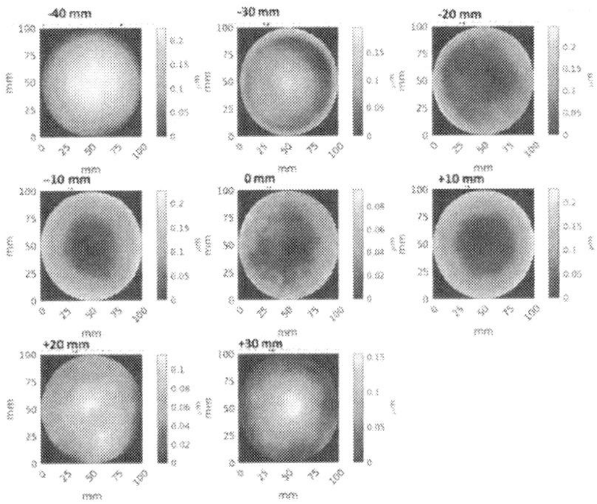

Fig. 6. Interferograms of the polished surface after individual steps. Removal after 20 minutes of polishing is presented for different values $(\boldsymbol{r_a} - \boldsymbol{R})$.

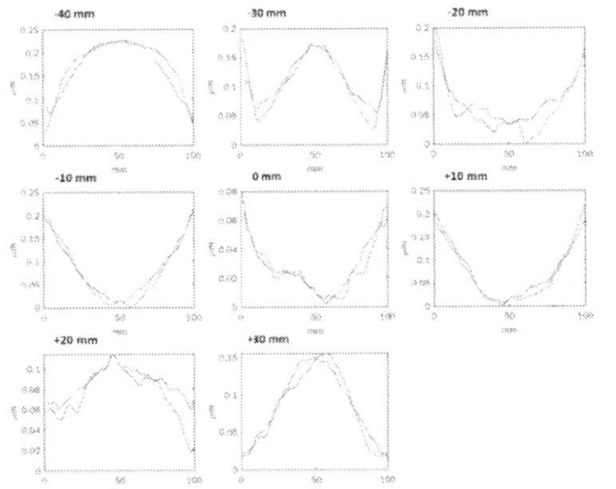

Fig. 7. Removal measured in perpendicular directions for different values of $(\boldsymbol{r_a} - \boldsymbol{R})$.

3.2 Relative velocity

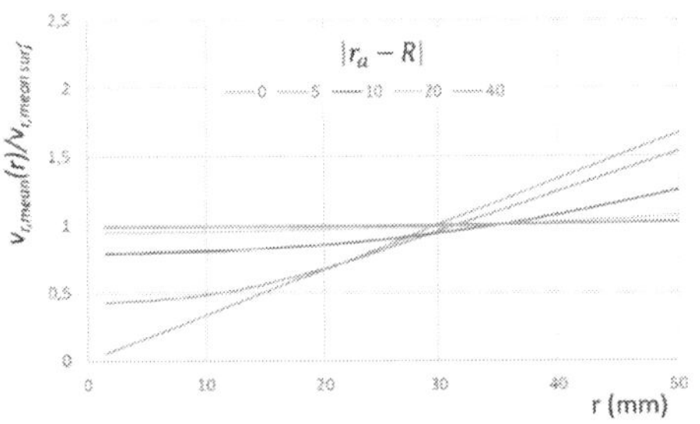

Fig. 8. Mean relative velocity during 50 cycles as a function of $|\mathbf{r_a} - \mathbf{R}|$ and distance from the workpiece centre.

By analysing the equations (2), it was possible to evaluate the mean relative velocity at each workpiece point during a sufficiently long polishing time, which corresponds to 50 cycles of the overarm. The evaluation was performed by numerical integration in the MATLAB environment. The results (Fig. 8) show that, except for states where $|r_a - R|$ is close to zero, the mean values of relative velocity are constant, independent on the position on the workpiece.

The unevenness increases whenω_w starts to me smaller than ω_t, which corresponds to the increasing friction in the ball joint.

3.3 Pressure distribution

The results of numerical simulations presented in Fig. 9 to Fig. 12 show the pressure distribution for different wedges and different values of $|r_a - R|$. The change in pressure as a function of angle is clear. The results also show a shift in the center of gravity of the pressure to the trailing edge with increasing angle.

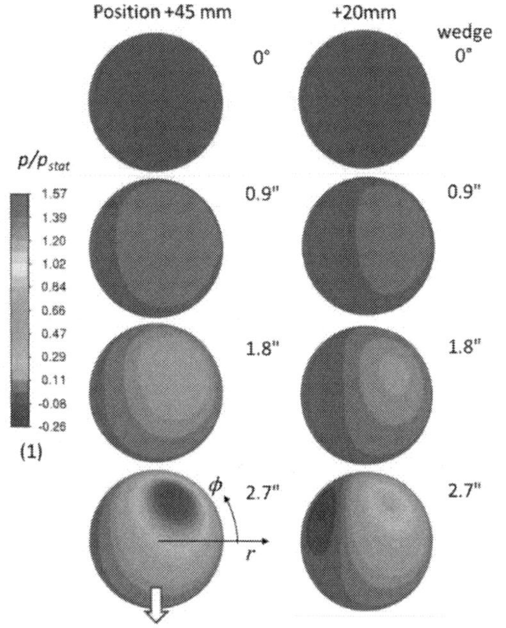

Fig. 9. Contour maps of the static pressure distribution on the workpiece. Position denotes $|\boldsymbol{r}_a - \boldsymbol{R}|$.

4. Conclusion

The influence of the relative velocity and pressure distribution on the value of removal rate of material from the polished surface was analysed in the article. The analysis was compared with the experiment.

Based on the simple kinematics of the overarm mechanism it is possible to calculate the magnitude of the relative velocity and its mean value during the polishing process. It was shown that mean relative velocity increases significantly from the centre to the edge of the workpiece for small arm extension values. For higher values of $|r_a - R|$, the value of the mean relative velocity is independent of position and is almost uniform throughout the entire polished workpiece.

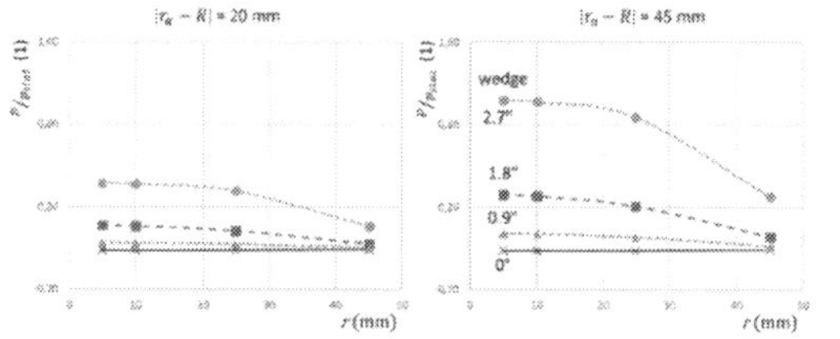

Fig. 10. Mean value of the static pressure as a function of the distance from the workpiece centre and the wedge.

A similar system can be used to analyse pressure. If there is no wedge between the workpiece and the tool, the static pressure on the workpiece surface will show only small differences. The pressure field can thus be considered homogeneous. However, the situation changes if the wedge starts to increase.

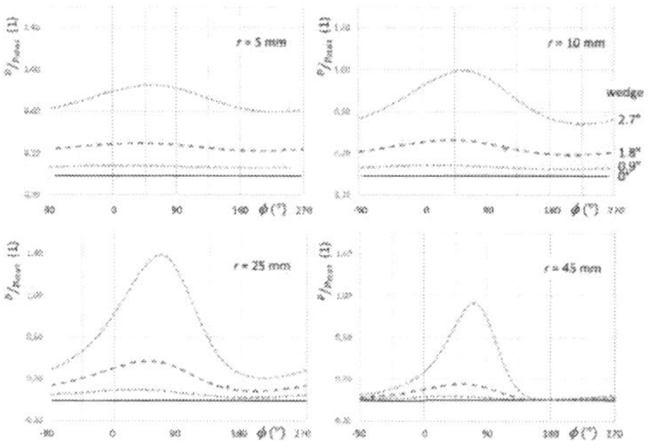

Fig. 11. Static pressure distribution at different positions on the workpiece for $|\boldsymbol{r}_a - \boldsymbol{R}| = 45$ mm.

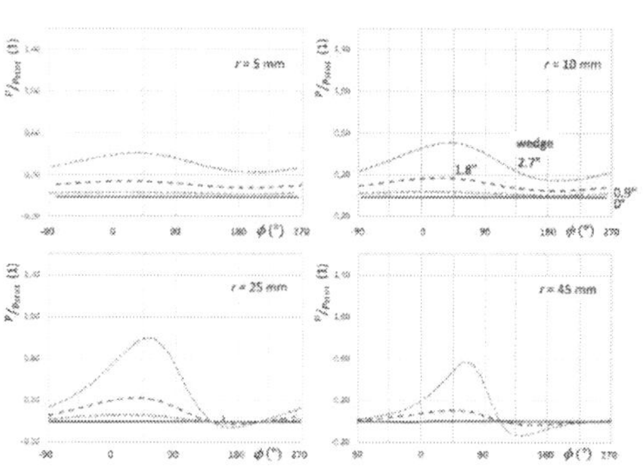

Fig. 12. Static pressure distribution at different positions on the workpiece for $|\boldsymbol{r}_a - \boldsymbol{R}| = 20$ mm.

Even a small value of the wedge leads to a significant change in the pressure distribution. As the wedge increases, the centre of the pressure shifts to the trailing edge. At the same time, the pressure increases.

The value of the wedge in the real process depends on the magnitude of the relative tool movement velocity, i.e., rotation speed of the tool and velocity of the overarm. So, it is possible to assume a higher value of dh when $|r_a - R|$ increases.

The results of the numerical simulations show a significant increase in pressure in the centre of the polished surface with increasing wedge (i.e., with increasing $|r_a - R|$). If we consider both the effect of relative velocity and the effect of pressure, we conclude that it is possible to explain the results of the experiments as follows:

- At $|r_a - R|$ close to 0, an increase in the relative velocity from the center of the wafer is noticeable. On the contrary, there is no wedge between the workpiece and the tool, which leads to an even pressure distribution. Removal rate is therefore higher at the edge of the workpiece than at the centre (states -20; -10; 0; 10 at Fig. 7)

- At large $|r_a - R|$, the relative velocity is evenly distributed. On the contrary, the pressure shows a significant decrease from the centre to the edge. According to (), therefore, there is a significantly greater removal in the centre of the polished area (states -40; -30; +20; +30 at Fig. 7).

The presented results are only of a qualitative nature. In the next work, a detailed analysis will be performed using all the possibilities of CFD. At the same time, experiments will be performed in real processes with the most precisely defined conditions. From these first results, however, it is clear that CFD in combination with the experiment can provide a tangible explanation for such a complex phenomenon as polishing.

Acknowledgements

The research has been supported by the Ministry of Industry and Trade, project The Country for the Future, FX01030046.

References

1. N. Belkhir, D. Bouzid, V. Herold, Determination of the Friction Coefficient During Glass Polishing. Tribology Letters, 33, 55-61. (2008).
2. A. Kelm, R. Boerret, S. Sinzinger, Improving the polishing accuracy by determining the variance of the friction coefficient. Journal of The European Optical Society - Rapid Publications, 7. (2012)
3. F. W. Preston, The Theory and Design of Plate Glass Polishing Machines, J. Soc Glass Technology, Vol. 11, pp. 214-256. (1927)
4. C. J. Evans, E. Paul, D. Dornfeld, D. A. Lucca, G. Byrne, M. Tricard, F. Klocke, O. Dambon, B. A. Mullany, CIRP Annals, Volume 52, Issue 2 (2003)
5. N. Kyungyoon, L. Jiun-Yu, N. Saka, J.-H. Chun, Mechanics, Mechanisms and Modeling of the Chemical Mechanical Polishing Process. (2003).

Experimental set up for the investigation of partial phase changes of phase change materials

Martin Zálešák[1*], *Patrik* Bouchal[1], *Milan* Ostrý[2], and *Jiří* Hejčík[1]

[1]Brno University of Technology, Faculty of Mechanical Engeneering, Energy Institute, Technická 2896/2, 61669 Brno, Czech Republic
[2]Brno University of Technology, Faculty of Civil Engeneering, Institute of Building Structures, Veveří 331/95, 60200 Brno, Czech Republic

Abstract. Thermal energy storage with phase change materials (PCMs) has attracted a lot of attention in the last several decades. Most PCMs do not change phase at a constant temperature but rather in a certain temperature range. It means that the PCM need to transit through its phase change temperature range to fully change phase from the solid state to the liquid state and vice-versa. The situation, in which the phase transition begins and/or ends within the phase change temperature range (in the mushy zone), is usually called a partial phase transition (or a partial phase change). The partial phase transitions occur quite often in real-life thermal energy storage systems with PCMs; especially when a PCM has a wide phase change temperature range. The behavior of PCMs during the partial phase transitions is poorly understood at the moment, because the experimental techniques used for the characterization of PCMs (such as the differential scanning calorimetry – DSC) are difficult to apply for the study of partial phase transitions. The lack of knowledge in this area influences the accuracy of phase change simulation models. The main goal of the experimental investigations, described in the paper, was to obtain data for the development of a simulation model for partial phase changes. The experimental set up for the investigation of partial phase changes of PCMs has been proposed, assembled, and the pilot measurements have been conducted. The experimental set up consists of two water storage tanks (that can be maintained at different water temperatures), a water-PCM concentric tube type heat exchanger and a data acquisition system. The water flows through the central tube of the heat exchanger while the PCM is located in the annular space of the exchanger. The water storage tanks, maintained at the temperatures within the phase change temperature range of a PCM, allow for the investigations of the heat storage cycles consisting of partial phase changes.

1 Introduction

In last several decades, there has been an increasing interest in the thermal energy storage systems integrating phase change materials (PCMs). Thermal energy storage plays an essential role in mitigating the delay between energy supply and demand in many different applications [1]. Since the PCMs are the central part of such systems, the knowledge of their thermophysical properties and thermal behaviour is necessary for the successful numerical analysis. As has been shown by many authors in recent years, some PCMs exhibit an asymmetrical behaviour during the phase transition, often referred to as the thermal hysteresis [2, 3].

Since the most of PCMs undergo the phase transition in a temperature range rather than at a constant temperature, we can differentiate between 2 scenarios [4] which are illustrated in Figure 1. The direct phase transition between solid and liquid states, which begins and ends outside the "mushy" zone is called the complete phase change. Similarly, the situation in which the phase transition begins and/or ends within the "mushy" zone is usually called the partial phase change. The thermal behaviour during the partial phase change is not very well understood at the moment, which lowers the overall accuracy of numerical model of heat transfer in PCMs [5]. The differential scanning calorimetry (DSC) and temperature history (T-history) methods are used quite commonly as experimental methods for the determination of the enthalpy of fusion and its temperature dependency, however the results are usually very difficult to apply considering the partial phase change [2].

Many authors were investigating the partial phase change lately. Thonon et al. [6] conducted a study investigating the phase change hysteresis and supercooling numerically and experimentally. The authors presented the analytical model of the thermal behaviour of a PCM experiencing the partial phase change cycles. Various heating/cooling rates were considered in the study. The experimental validation was carried out on a PCM brick sample, measuring the heat flux and the PCM temperature. The studied material was PEG6000 which is a polymer PCM suitable for domestic hot water storage. The authors reported a good agreement between experimental and simulated results considering the complete phase change. The

*Corresponding author: 162098@vutbr.cz

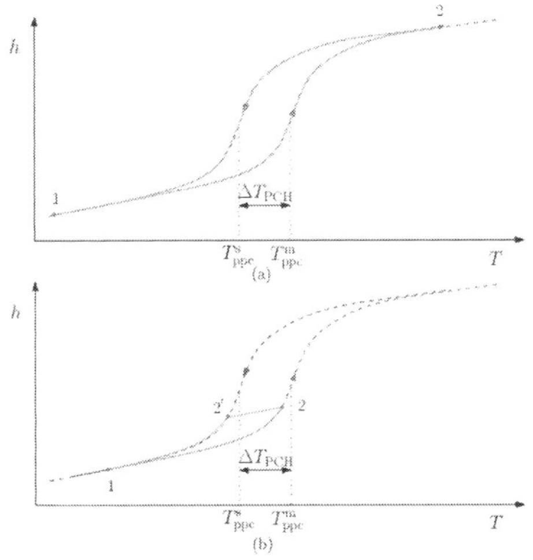

Fig. 1. The h-T dependency during a (a) complete phase change and (b) partial phase change.

numerical modelling of the partial phase change was also shown to be satisfactory, however there were some discrepancies in the thermal behaviour of the cooling process (after partial melting of PCM). Barz [7] experimentally and numerically studied the phase transition behaviour of three commercial paraffin-based PCMs in a plate-fin heat exchanger. The heat exchanger was subjected to series of cyclic heating/cooling cycles within the phase temperature range. The results have shown that the performance of the "curve scale" phase change hysteresis model developed by the authors was superior in comparison with a model based on the phase fraction-temperature curve of the heating curve only. The three phenomenological models for prediction of temperature-induced phase transitions were proposed by Barz et al. [8] - the "curve track", the "curve switch" and the "curve scale" models. The proposed models are directly identified from the data from complete melting and solidification, which reduces the effort for model calibration (eg. manufacturer material properties can be used). The predictive performance of the novel "curve scale" model during 26 partial load cycles has shown to be superior compared with the "curve switch" and the "curve track".

As has been shown in the literature, the behaviour of PCMs undergoing the partial phase change is still not well understood with a very little experimental validation conducted. The experimental methods for the determination of the thermal properties (such as DSC or T-history) have proven to be very difficult to implement for this purpose (considering partial phase change) since they are limited by the sample size (usually only up to 10 mg) and the results significantly depend on the heating/cooling rate. As an alternative, many authors combine the numerical model of the thermal system containing a PCM with experimental investigations of such a system, creating the inverse identification problem. When coupled with a suitable optimization method and well defined initial and boundary conditions, the inverse problem could be solved providing the information about the thermal behaviour of a selected PCM undergoing a partial phase transition. The authors of this study have already investigated a similar inverse identification procedure focusing on a complete phase change, showing promising results [9].

2 Experimental setup

The experimental setup has been constructed, consisting of two separate water storage tanks (which are maintained at different temperatures) connected to a water-PCM concentric tube type heat exchanger with the data acquisition system as is shown in Figure 2. The inlet water temperature of the water-PCM heat exchanger can be controlled by 3-way valve (combining water from the hot and cold water storage tanks). The Pt100 temperature probes were used for the measurement of inlet and outlet water temperatures. The mass flow of water was measured after the water-PCM heat exchanger and in the end the water flow did lead into drain. The water flows though the central tube of the heat exchanger. The studied material was paraffin-based PCM Rubitherm RT 35 HC which was located inside the annular space of the concentric tube heat exchanger with total mass of the PCM equal to 0.98 kg. The material properties of the RT 35 HC are summarised in the Table 1.

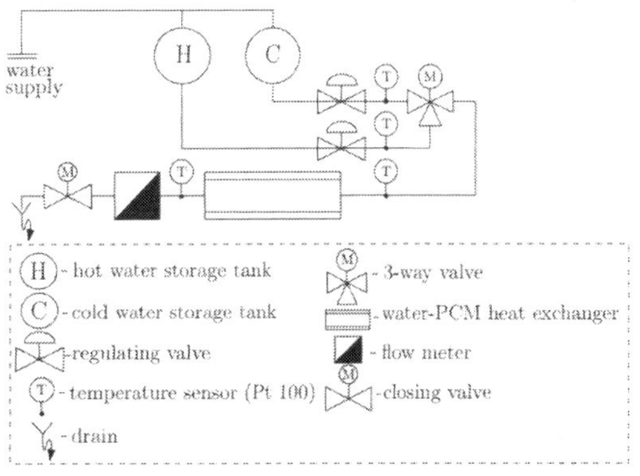

Fig. 2. Schematics of the experimental setup.

3 Methods

The proposed procedure for the inverse identification of the thermophysical properties of a PCM consists of four main parts:

1. Numerical model of the studied thermal system
2. Parametrization of the material properties which describe the thermal behaviour
3. Experimentally measured temperatures in specific locations
4. Suitable optimization method.

Table 1. Rubitherm RT 35 HC manufacturer properties

Material property	Value	Unit
Phase change temperature	35	[°C]
Heat storage capacity*	240	[kJ kg^{-1}]
Specific heat capacity	2000	[J kg^{-1} K^{-1}]
Density	800	[kg m^{-3}]
Heat conductivity	0.2	[W m^{-1} K^{-1}]

*(latent and sensible heat between 27 - 42°C)

The numerical model is based on 2-D heat conduction inside the PCM with governing equation:

$$\rho c_{\text{eff}}(T)\frac{\partial T}{\partial \tau} = \frac{\partial}{\partial x}\left(k\frac{\partial T}{\partial x}\right) + \frac{\partial}{\partial y}\left(k\frac{\partial T}{\partial y}\right), \tag{1}$$

where T is temperature, τ is time, k describes thermal conductivity, ρ is density and x, y are spatial coordinates. The phase change is modelled using the effective heat capacity method, which uses the specific heat capacity and transforms it into the function of temperature $c_{\text{eff}}(T)$. The shape of such function was defined in form of a Gaussian function as:

$$c_{\text{eff}}(T) = c + c_{\text{L}_f}\exp\left\{-\frac{(T - T_{\text{ppc}})^2}{\sigma}\right\}, \tag{2}$$

with c being specific heat capacity, c_{L_f} is height coefficient, T_{ppc} is peak phase change temperature and σ defines the sharpness of the Gaussian function.

The proposed study is dealing mainly with the overall inverse identification procedure, the experimental set up construction and experimental data acquisition and preparation. As can be seen in the Figure 3, the interrupted melting process could be modelled either by using the effective heat capacity method or enthalpy method. The melting process is interrupted at specific temperature T_{sw} and reversed to solidification with chosen transition method being proposed by Bony and Citherlet in [10] also very often referred to as the "curve switch" model [7, 8].

Parameters of curves following the behaviour in Figure 3 are all combined into the parameter vector **p**. The direct evaluation of the numerical model (considering specific set of model parameters) leads to a specific outlet water temperature $T^{\text{water}}_{\text{out,sim}}$. The objective function was defined as a sum of squared differences between the experimentally measured and simulated outlet water temperatures:

$$\text{OF}(\mathbf{p}) = \sum_{p=0}^{\frac{t_{\max}}{\Delta t}} \left(T^{\text{water}}_{\text{out,sim}} - T^{\text{water}}_{\text{out,exp}}\right)^2 . \tag{3}$$

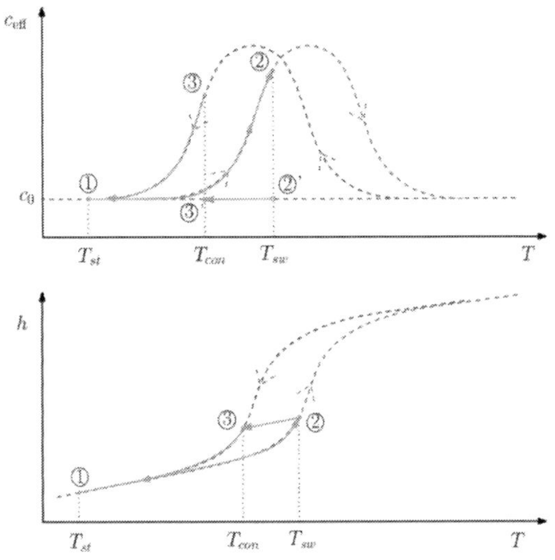

Fig. 3. Parametrization of the interrupted melting process using c_{eff} –T and h–T dependency, according to Bony and Citherlet [10].

Combined with the experimentally measured outlet water temperature $T^{\text{water}}_{\text{out,exp}}$, the objective function is then iteratively evaluated following a suitable heuristic optimization method such as: particle swarm optimization (PSO) method, differential evolution (DE) method or genetic algorithm (GA) until the termination criterion is fulfilled and the vector of optimal parameters $\mathbf{p}_{\text{OPT}}$ is determined.

4 Results

When the experimental setup for the investigation of partial phase changes was assembled the pilot experiments were conducted and the obtained data were evaluated. As can be seen in the Figure 4 and 5, the inlet and outlet water temperatures were measured for several different heating and cooling cycles. During the heating phases, the temperature of 55 °C was set in the water-PCM heat exchanger inlet, utilising only water from the hot water storage tank. Main goal of the first scenario was to fully undergo the phase transition of the PCM, which could be used for the description of thermal behaviour during complete phase change. After about 40 minutes, the PCM in the annular cavity was considered fully molten and the process was reversed, discharging the heat from the heat exchanger with the cold water with temperature of 21 °C until the whole PCM was completely solid again.

In order to investigate the thermal behaviour of PCM during the partial phase change, in total 5 different melting/solidification cycles were measured. The individual scenarios did differ in the charging/discharging time being 1, 2, 3, 5 and 10 minutes.

5 Conclusion

The main outcomes of the presented study can be summarised as:

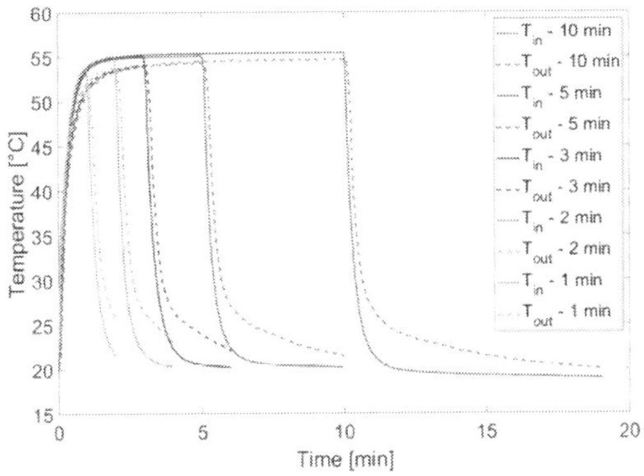

Fig. 4. Inlet and outlet water temperature for different charg-ing/discharging time intervals.

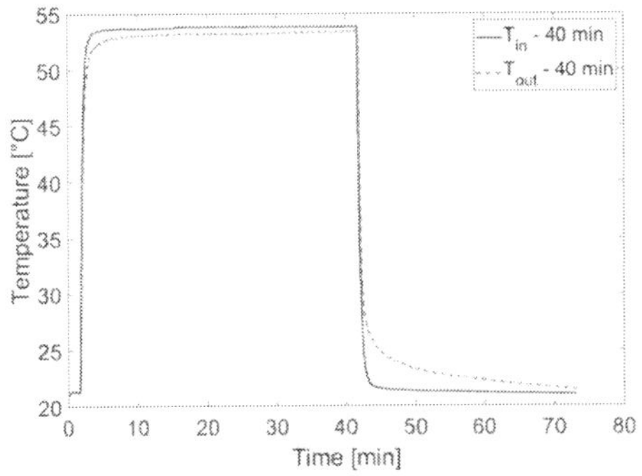

Fig. 5. Inlet and outlet water temperature for complete phase transition

- The experimental set up with the water-PCM heat exchanger was designed and constructed.
- The initial measurements were conducted, starting with the complete phase transition heating/cooling cycle. However, since the objective of the inverse identification procedure is to determine the thermal behaviour of PCM during the partial phase changes, inlet/outlet water temperatures of water-PCM heat exchanger were measured for a series of interrupted (incomplete) heating cycles.
- The inverse identification procedure was proposed and possible parametrization of the effective heat capacity curve was suggested. In the future studies, the inverse problem will be defined (with use of the experimental data measured in this study) and solved using a suitable optimization method (eg. PSO, DE, GA).

Acknowledgement

This work was supported by project FSI-S-20-6295 and by the Czech Science Foundation, grant number 19-20943S "Compatibility of plastics and metals with latent heat stor-age media for integration in buildings.

References

[1] E. Franquet, S. Gibout, J.P. Bédécarrats, D. Haillot, J.P. Dumas, Thermochimica Acta **546**, 61 (2012)
[2] L. Klimeš, P. Charvát, M.M. Joybari, M. Zálešák, F. Haghighat, K. Panchabikesan, M.E. Mankibi, Y. Yuan, Applied Energy **263**, 114572 (2020)
[3] X. Jin, Huoyan, H.X. Shi, X. Zhang, Energy Conversion and Management **106**, 608–614 (2015)
[4] B. Delcroix, M. Kummert, A. Daoud, Journal of Building Performance Simulation **10**, 422 (2017)
[5] F. Goia, G. Chaudhary, S. Fantucci, Energy and Buildings **174**, 54 (2018)
[6] M. Thonon, G. Fraisse, L. Zalewski, M. Pailha, Applied Thermal Engineering **190**, 116751 (2021)
[7] T. Barz, Journal of Energy Storage **34**, 102164 (2021)
[8] T. Barz, J. Emhofer, K. Marx, G. Zsembinszki, L.F. Cabeza, Journal of Building Performance Simulation **12**, 770 (2019)
[9] M. Zálešák, P. Charvát, L. Klimeš, Applied Thermal Engineering **197**, 117392 (2021)
[10] J. Bony, S. Citherlet, Energy and Buildings **39**, 1065 (2007)

AUTHOR INDEX

EDP Sciences – Web of Conferences
17, Avenue du Hoggar
Parc d'Activité de Courtabouf
BP 112
F-91944 Les Ulis Cedex A
France

ISBN 978-1-7138-5738-9

9 781713 857389

CURRAN ASSOCIATES INC.
proceedings.com